Micro Reaction Technology in Organic Synthesis

Micro Reaction Technology in Organic Synthesis

Charlotte Wiles and Paul Watts

CRC Press
Taylor & Francis Group
Boca Raton London New York

CRC Press is an imprint of the
Taylor & Francis Group, an **informa** business

CRC Press
Taylor & Francis Group
6000 Broken Sound Parkway NW, Suite 300
Boca Raton, FL 33487-2742

© 2011 by Taylor and Francis Group, LLC
CRC Press is an imprint of Taylor & Francis Group, an Informa business

No claim to original U.S. Government works

Printed in the United States of America on acid-free paper
10 9 8 7 6 5 4 3 2 1

International Standard Book Number: 978-1-4398-2471-9 (Hardback)

This book contains information obtained from authentic and highly regarded sources. Reasonable efforts have been made to publish reliable data and information, but the author and publisher cannot assume responsibility for the validity of all materials or the consequences of their use. The authors and publishers have attempted to trace the copyright holders of all material reproduced in this publication and apologize to copyright holders if permission to publish in this form has not been obtained. If any copyright material has not been acknowledged please write and let us know so we may rectify in any future reprint.

Except as permitted under U.S. Copyright Law, no part of this book may be reprinted, reproduced, transmitted, or utilized in any form by any electronic, mechanical, or other means, now known or hereafter invented, including photocopying, microfilming, and recording, or in any information storage or retrieval system, without written permission from the publishers.

For permission to photocopy or use material electronically from this work, please access www.copyright.com (http://www.copyright.com/) or contact the Copyright Clearance Center, Inc. (CCC), 222 Rosewood Drive, Danvers, MA 01923, 978-750-8400. CCC is a not-for-profit organization that provides licenses and registration for a variety of users. For organizations that have been granted a photocopy license by the CCC, a separate system of payment has been arranged.

Trademark Notice: Product or corporate names may be trademarks or registered trademarks, and are used only for identification and explanation without intent to infringe.

Visit the Taylor & Francis Web site at
http://www.taylorandfrancis.com

and the CRC Press Web site at
http://www.crcpress.com

Contents

Preface xiii
Acknowledgments xv
Authors xvii
Abbreviations xix

Chapter 1 Introduction to Micro Reaction Technology 1
    1.1   What Is Micro Reaction Technology? 1
    1.2   Fabrication/Construction of Micro Reactors 2
        1.2.1   Glass Devices 2
        1.2.2   Silicon Devices 3
        1.2.3   Polymeric Devices 3
        1.2.4   Metallic Devices 3
        1.2.5   Ceramic Devices 4
        1.2.6   Reactor to World Interface 4
    1.3   Manipulation of Reactants and Products within Flow Reactors 4
        1.3.1   Mixing within Micro Reactors 5
        1.3.2   Flow Types within Biphasic Systems 6
    1.4   Advantages of Micro Reaction Technology 7
        1.4.1   Process Safety 7
        1.4.2   Novel Reaction Conditions 8
        1.4.3   Reduced Waste Generation 9
        1.4.4   On-Site Chemical Processing 11
    1.5   Disadvantages of Micro Reactors 11
        1.5.1   Handling of Insoluble Materials 12
    1.6   Process Intensification 13
    1.7   *In Situ* Reaction Monitoring 14
        1.7.1   Raman Spectroscopy 14
        1.7.2   *In Situ* Analysis by FTIR Spectroscopy 16
        1.7.3   Nuclear Magnetic Resonance Detection 21
        1.7.4   Chromatographic Techniques 22
        1.7.5   Development of Sensors for Process Monitoring 23
    1.8   Commercial Availability of Continuous Flow Reactor Technology 24
    1.9   Outlook 31
    References 31

Chapter 2 Micro Reactions Employing a Gaseous Component 37
    2.1   Gas-Phase Micro Reactions 37
        2.1.1   Gas-Phase Oxidations 38

		2.1.2	Hydrogenation Reactions within Microstructured Reactors 41
		2.1.3	Dehydration and Dehydrogenation Reactions 42
		2.1.4	Fischer–Tropsch Synthesis 43
		2.1.5	Synthesis of Methylisocyanate in a Micro Reactor 43
	2.2	Gas–Liquid-Phase Micro Reactions 44	
		2.2.1	Continuous Flow Chlorination Reactions 44
		2.2.2	Continuous Flow Fluorination Reactions 45
		2.2.3	Ozonolysis within Micro Reactors 45
		2.2.4	Biphasic Carbonylations 48
		2.2.5	Transfer Hydrogenations under Continuous Flow Conditions 55
		2.2.6	Miscellaneous Biphasic Micro Reactions 57
	2.3	Gas–Liquid–Solid Micro Reactions 60	
		2.3.1	Triphasic Oxidations under Flow Conditions 61
		2.3.2	Carbonylations Using Solid-Supported Catalysts 62
		2.3.3	Hydrogenations within Continuous Flow Reactors 63
		2.3.4	Slurry-Based Micro Reactions 66
		2.3.5	Miscellaneous Triphasic Micro Reactions 69
	References 70		

Chapter 3 Liquid-Phase Micro Reactions 77

	3.1	Nucleophilic Substitution 77	
		3.1.1	C–C Bond Formation: Acylation Reactions 77
		3.1.2	C–C Bond Formation: Alkylation Reactions 78
		3.1.3	Enantioselective C–C Bond-Forming Reactions 83
		3.1.4	C–O Bond Formation: Esterification Reactions 85
		3.1.5	C–O Bond Formation: Etherification Reactions 89
		3.1.6	C–O Bond Formation: Epoxide Hydrolysis 90
		3.1.7	C–N Bond Formation: Alkylation Reactions 90
		3.1.8	C–N Bond Formation: Acylation Reactions 93
		3.1.9	C–N Bond Formation: Arylation Reactions 98
		3.1.10	C–N Bond Formation: Azidation Reactions 99
		3.1.11	C–N Bond Formation: Synthesis of Hydroxamic Acids 102
		3.1.12	C–N Bond Formation: Aminolysis of Epoxides 103
		3.1.13	C–F Bond Formation 109
	3.2	Electrophilic Substitution 113	
		3.2.1	C–C Bond Formation 113
		3.2.2	C–N Bond-Forming Reactions: Nitration Reactions 126
		3.2.3	C-Hetero Bond-Forming Reactions: Halogenations under Flow 130

Contents   vii

		3.2.4	C-Hetero Bond-Forming Reactions: Diazotizations under Flow 133
		3.2.5	C-Hetero Bond-Forming Reactions: Sulfonations under Flow 136
	3.3	Nucleophilic Addition 136	
		3.3.1	C–C Bond Formation: Aldol Reaction/Condensation............ 136
		3.3.2	C–C Bond Formation: Knoevenagel Condensation 139
		3.3.3	C–C Bond Formation: Michael Addition............ 142
		3.3.4	C–C Bond Formation: Diels–Alder Reaction 144
		3.3.5	C–C Bond Formation: Horner–Wadsworth–Emmons............ 146
		3.3.6	C–C Bond Formation: Enantioselective Examples............ 148
		3.3.7	C-Hetero Bond Formation: Aza-Michael Addition............ 149
		3.3.8	C-Hetero Bond Formation: Alkylation of Amines............ 151
		3.3.9	C-Hetero Bond Formation: Synthesis of Triazoles............ 152
		3.3.10	C-Hetero Bond Formation: Addition of Hydrazine to Carbonyl Compounds 157
	3.4	Elimination Reactions 158	
		3.4.1	Dehydration Reactions 158
		3.4.2	Dehalogenations: Tris(Trimethylsilyl)Silane-Mediated Reductions............ 160
	3.5	Oxidations............ 163	
		3.5.1	Oxidations: Inorganic Oxidants 163
		3.5.2	Oxidations: Swern–Moffat Oxidation............ 164
		3.5.3	Oxidations: TEMPO-Mediated Oxidations 167
		3.5.4	Oxidations Using Oxone 168
		3.5.5	Oxidations: Epoxidations under Flow Conditions ... 168
		3.5.6	Oxidation: Deprotection of Amines............ 172
	3.6	Reductions 173	
		3.6.1	Transition Metal Free Reductions 173
		3.6.2	Dibal-H Reductions............ 173
	3.7	Metal-Catalyzed Cross-Coupling Reactions............ 175	
		3.7.1	Suzuki–Miyaura Reaction............ 175
		3.7.2	Heck Reaction 177
		3.7.3	Sonogashira Reaction............ 183
		3.7.4	Other Metal-Catalyzed Coupling Reactions 185
	3.8	Rearrangements............ 187	
		3.8.1	Claisen Rearrangement 187
		3.8.2	Newman–Kwart Rearrangement............ 191
		3.8.3	Hofmann Rearrangement............ 193

		3.8.4	Fisher Indolization 195
		3.8.5	Curtius Rearrangement 196
		3.8.6	Dimroth Rearrangement 197
	3.9	Multistep/Multicomponent Liquid-Phase Reactions 198	
		3.9.1	Multicomponent Synthesis of Heterocycles 198
		3.9.2	Multistep Synthesis of 1,2,4-Oxadiazoles 199
		3.9.3	Continuous Flow Synthesis of Ibuprofen 201
		3.9.4	Cation-Mediated Sialylation Reactions 203
		3.9.5	Oligosaccharide Synthesis 205
		3.9.6	Synthesis of Indole Alkaloids Using Metal-Coated Capillary Reactors 206
		3.9.7	Iododeamination under Flow 206
		3.9.8	Radical Additions under Flow 209
	3.10	Summary 211	
	References 211		

Chapter 4 Multi-Phase Micro Reactions 221

	4.1	Nucleophilic Substitution 221	
		4.1.1	C–O Bond-Forming Reactions: Esterifications 221
		4.1.2	C–N Bond-Forming Reactions: Azidations 224
	4.2	Electrophilic Substitution 225	
		4.2.1	Brominations 225
		4.2.2	Phosgene Synthesis 226
	4.3	Nucleophilic Addition 227	
		4.3.1	C–C Bond-Forming Reactions: Knoevenagel Condensation 227
		4.3.2	C–C Bond-Forming Reactions: Michael Additions 229
		4.3.3	C–C Bond-Forming Reactions: Henry Reaction 230
		4.3.4	C–C Bond-Forming Reactions: Diels–Alder 232
		4.3.5	C–C Bond-Forming Reactions: Benzoin Condensation 232
		4.3.6	C–C Bond-Forming Reactions: Trifluoromethylation under Continuous Flow 234
		4.3.7	C–C Bond Formation: Aldol Reaction 234
		4.3.8	C–N Bond Formation: Cycloaddition Reactions 237
		4.3.9	C–O Bond-Forming Reactions: Acetalizations 240
		4.3.10	C–S Bond-Forming Reactions: Thioacetalizations 241
	4.4	Elimination Reactions 241	
		4.4.1	Dehydration Reactions 241
		4.4.2	Dehydration Reactions 242
	4.5	Oxidation Reactions 242	
		4.5.1	Catalytic Oxidations 243
		4.5.2	Epoxidations 247

	4.6	Metal-Catalyzed Cross-Coupling Reactions	250
		4.6.1 Suzuki–Miyaura Reaction	250
		4.6.2 Heck Coupling Reactions	252
		4.6.3 Other Metal-Catalyzed Coupling Reactions	255
	4.7	Rearrangements	261
	4.8	Enantioselective Reactions	262
		4.8.1 Chemically Promoted Reactions	262
		4.8.2 Enzymatic Enantioselective Micro Reactions	266
	4.9	Multistep/Multicomponent Reactions	272
		4.9.1 Independent Multistep Flow Reactions	272
		4.9.2 Integrated Multistep Sequences	272
		4.9.3 Reagents and Scavengers in Series	274
		4.9.4 Combined Chemical and Biochemical Catalysis	274
		4.9.5 "Catch and Release" Strategies under Continuous Flow	277
		4.9.6 Casein Kinase I Inhibitor Synthesis	280
	4.10	Summary	281
	References		283
Chapter 5	Electrochemical and Photochemical Applications of Micro Reaction Technology		289
	5.1	Electrochemical Synthesis under Continuous Flow	289
		5.1.1 Electrochemical Oxidations	290
		5.1.2 Electrolyte-Free Electroorganic Synthesis	296
		5.1.3 Electrochemical Reductions	298
		5.1.4 Electrolyte-Free Reductions under Flow	299
		5.1.5 Summary	301
	5.2	Photochemical Synthesis under Continuous Flow	302
		5.2.1 Photocycloadditions under Continuous Flow	303
		5.2.2 Photodecarboxylative Addition	307
		5.2.3 Photocyanation	307
		5.2.4 Photochemical Halogenations	308
		5.2.5 Nitrite Photolysis under Flow Conditions	309
		5.2.6 Photochemical Dimerization	311
		5.2.7 Photosensitized Diastereo Differentiation	312
	5.3	Multiphase Photochemical Reactions	312
		5.3.1 Photocatalytic Reductions	313
		5.3.2 Photocatalytic Oxidation Reactions	314
		5.3.3 Photocatalytic Alkylation Reactions	315
		5.3.4 Photocatalytic Cyclizations	316
		5.3.5 Gas–Liquid Transformations	317
		5.3.6 Gas–Liquid–Solid Reactions	321
		5.3.7 Summary	321
	References		321

Chapter 6 The Use of Microfluidic Devices for the Preparation and Manipulation of Droplets and Inorganic/Organic Particles 325

    6.1   Droplet Formation Using Continuous Flow Methodology 325
            6.1.1   Polymerization of Droplets under Flow 328
    6.2   Preparation of Inorganic Nanoparticles under Continuous Processing Conditions 330
    6.3   Formation of Organic Particles within Continuous Flow Devices 332
    6.4   The Use of Micro Reactors for the Postsynthetic Manipulation of Organic Compounds 336
    6.5   Mixed Particle Formation 338
            6.5.1   Microencapsulation of Active Pharmaceuticals 338
    6.6   Summary 341
    References 343

Chapter 7 Industrial Interest in Micro Reaction Technology 347

    7.1   MRT in Production Environments 347
    7.2   Synthesis of Fine Chemicals Using Micro Reactors 349
            7.2.1   Synthesis of Carbamates under Continuous Flow Conditions 350
            7.2.2   Production-Scale Synthesis of Ionic Liquids 351
            7.2.3   Scalable Technique for the Synthesis of Diarylethenes 353
            7.2.4   Continuous Flow Synthesis of Light-Emitting Materials 354
            7.2.5   2-(2,5-Dimethyl-1H-Pyrrol-1-yl)Ethanol Synthesis ... 356
            7.2.6   Synthesis of Pigments under Flow Conditions 357
            7.2.7   Production of Thermally Labile Compounds under Flow Conditions 358
            7.2.8   Peracetic Acid Production Using an On-Site Microprocess 360
            7.2.9   The *In Situ* Synthesis and Use of Diazomethane 362
    7.3   Synthesis of Pharmaceuticals and Natural Products Using Continuous Flow Methodology 363
            7.3.1   Ciprofloxacin and Its Analogs 363
            7.3.2   Synthesis of Pristane 363
            7.3.3   Synthesis of Imatinib under Flow Conditions 366
            7.3.4   Synthesis of Aspirin and Vanisal Sodium 368
            7.3.5   Synthesis of Suberoylanilide Hydroxamic Acid 369
            7.3.6   Synthesis of Rimonabant and Efaproxiral Using AlMe$_3$ 370
            7.3.7   Continuous Flow Synthesis of Sildenafil 372
            7.3.8   Synthesis of 6-Hydroxybuspirone 372
            7.3.9   A Key Step in the Synthesis of (*rac*)-Tramadol 373

		7.3.10	Claisen Rearrangement to Afford 2,2-Dimethyl-2H-1-Benzopyrans 373
		7.3.11	Synthesis of a 5HT$_{1B}$ Antagonist 375
		7.3.12	Serial Approach to a Novel Anticancer Agent Using Flow Reactors 378
		7.3.13	Synthesis of Grossamide under Flow Conditions 378
		7.3.14	Synthesis of the Natural Product (±)-Oxomaritidine 378
		7.3.15	Synthesis of Furofuran Ligans 381
	7.4	Synthesis of Small Doses of Radiopharmaceuticals 382	
	7.5	Summary 384	
	References 384		
Chapter 8	Microscale Continuous Separations and Purifications 387		
	8.1	Introduction 387	
	8.2	Liquid–Liquid Extractions 387	
		8.2.1	Side-by-Side (Stratified) Contacting 388
		8.2.2	Three-Phase Microextractions 392
		8.2.3	Segmented Flow 395
		8.2.4	Fluorous Phase Extractions 398
		8.2.5	Comparison of Liquid–Liquid Extraction Efficiencies 401
	8.3	Gas–Liquid Separation 403	
		8.3.1	Membrane Separators 403
		8.3.2	Microfluidic Distillations 403
	8.4	Solvent Exchange and Solvent Removal 407	
	8.5	The Use of Scavenger Resins for Product Purification under Flow 410	
		8.5.1	Trace Metal Removal 410
		8.5.2	Removal of Unreacted Starting Materials 411
	8.6	Continuous Flow Resolutions 413	
		8.6.1	Biocatalytic Resolutions 415
		8.6.2	Chemical Racemization 416
	8.7	Product Isolation 418	
		8.7.1	Antisolvent Precipitation 418
		8.7.2	Lysozyme Crystallization 420
		8.7.3	Solution Crystallization 420
	8.8	Summary 421	
	References 421		
Index 427			

Preface

In spite of the fact that continuous processes have, for many years, found widespread application within chemical production, in research and development the advantages associated with this mode of operation have, until now, not been widely acknowledged; with chemists favoring the centuries-old technique of iterative batch reactions. With the exception of combinatorial and microwave chemistry, little has been done to change the way that synthetic chemists conduct their research. When a synthetic chemist steps away from the batch mindset, and embarks upon an investigation under continuous flow, the advantages of efficient fluidic and thermal control become undeniable; affording the researcher's access to previously forbidden reaction conditions and new ways of investigating synthetic challenges.

With an ever-increasing number of commercially available flow reaction platforms available, it is the aim of this text to highlight the current state of the technology with the vision that more synthetic chemists will embark upon flow chemistry programs of research; facilitating the identification of novel and interesting synthetic methodologies that possess the potential to be scaled directly to production.

Acknowledgments

The authors would like to acknowledge the following contemporaries—Dr. Kasper Koch, Managing Director at Future Chemistry Holding BV, The Netherlands; Paul Pergande, Commercial Director at Uniqsis Ltd., UK; Hugo Delissen, CEO of Chemtrix BV, The Netherlands; Paul Griffin, Applications Manager at Vapourtec Ltd., UK; Dr Ildikó Kovacs at ThalesNano Inc., Hungary; Jan K. Hughes, CEO of Accendo Corporation, USA; Andy Holley, European Marketing Manager at Advion, USA; and Mike Hawes, CEO Syrris Ltd., UK—for their assistance in compiling the section on commercially available flow-reactor systems and kindly providing images to enable the reader to gain an appreciation of progress made with respect to the commercialization of the research described herein.

Authors

Dr. Charlotte Wiles is the Chief Technology Officer at Chemtrix BV, The Netherlands, and has been actively researching within the area of micro reactor synthesis for ten years, starting with a PhD in micro reactors in organic chemistry, which she obtained from the University of Hull in 2003. In the past decade, she has authored and coauthored many scientific papers and review articles on the subject of micro reaction technology and has also contributed to numerous books. More recently, she has tailored her experience to the development and evaluation of commercially available continuous flow reactors, systems, and peripheral equipment.

Dr. Paul Watts is a reader in organic chemistry at the University of Hull and since graduating from the University of Bristol, where he completed a PhD in bio-organic natural product chemistry, he has led the Micro Reactor group at Hull. In this role he has published more than 70 papers, and he regularly contributes to the field of micro reaction technology by way of invited book chapters, review articles, and keynote lectures on the subject of micro reaction technology in organic synthesis.

Abbreviations

Ac	Acetyl
Acac	Acetylacetonate
AIBN	Azobisisobutyronitrile
aq.	Aqueous
BEMP	2-*tert*-Butylimino-2-diethylamino-1,3-dimethylperhydro-1,3,2-diazaphosphorane
Bn	Benzyl
Boc	*t*-Butoxycarbonyl
Bu	Butyl
Bz	Benzoyl
°C	Temperature in degrees Centigrade
cat.	Catalyst
Cbz	Carbobenzyloxy
Cp	Cyclopentadienyl
CPC	Cellular process chemistry
CTAB	Cetyltrimethylammonium bromide
DABCO	1,4-Diazobicyclo[2.2.2]octane
DAST	Diethylaminosulfur trifluoride
dba	Dibenzylideneacetone
DBE	1,2-Dibromoethane
DBU	1,8-Diazabicyclo[5.4.0]undec-7-ene
DCC	1,3-Dicyclohexylcarbodiimide
DCE	1,2-Dichloroethane
% de	Percentage diastereomeric excess
DEAD	Diethylazodicarboxylate
DI	Deionized
Diazald®	*N*-Methyl-*N*-nitroso-*p*-toluenesulfonamide
Dibal-H	Diisobutylaluminum hydride
DMA	Dimethylacetamide
DMAD	Dimethylacetylene dicarboxylate
DMAP	4-Dimethylaminopyridine
DME	Dimethoxyethane
DMF	*N*,*N*-Dimethylformamide
DMSO	Dimethylsulfoxide
dvb	Divinylbenzene
EDDA	Ethylenediamine acetate
EDTA	Ethylenediaminetetraacetic acid
% ee	Percentage enantiomeric excess
Et	Ethyl
FSPE	Fluorous solid-phase extraction
GC	Gas chromatography

GC-MS	Gas chromatography–mass spectrometry
h	Hour(s)
hυ	Irradiation with light
HMPA	Hexamethylphosphoramide
HOBt	1-Hydroxybenzotriazole
HPLC	High-performance liquid chromatography
HPLC-MS	High-performance liquid chromatography–mass spectrometry
K	Kelvin
L	Liter(s)
μL	Microliter(s)
LA	Lewis acid
LDA	Lithium diisopropylamide
LiHMDS	Lithium *bis*(trimethylsilyl)amide
LLDPE	Linear low-density polyethylene
MACOS	Microwave-assisted continuous-flow organic synthesis
max.	Maximum
m-CPBA	*meta*-Chloroperoxybenzoic acid
mL	Milliliter(s)
Me	Methyl
Mes	Mesityl
MOM	Methoxymethyl
Ms	Methanesulfonyl
MS	Mass spectrometry
MTBE	Methyl *tert*-butyl ether
MVK	Methyl vinyl ketone
NAD	Nicotinamide adenine dinucleotide
Napth	Napthyl
NBS	*N*-Bromosuccinimide
NCS	*N*-Chlorosuccinimide
NMP	*N*-Methyl-2-pyrrolidinone
ODS	Octadecylsilane
OLED	Organic light-emitting diode
p	Pressure
PCC	Pyridinium chlorochromate
PCTFE	Polychlorotrifluoroethylene
PDC	Pyridinium dichromate
PEG	Polyethylene glycol
PEPPSI	Pyridine-enhanced precatalyst preparation stabilization and initiation
PET	Positron emission tomography
PFMD	Perfluoromethyldecalin
Ph	Phenyl
PhMe	Toluene
PMP	4-Methoxyphenyl
Pr	Propyl
PTFE	Polytetrafluoroethylene

PVA	Polyvinyl alcohol
PVC	Polyvinyl chloride
PVSZ	Polyvinylsilazane
Py	Pyridine
quant.	Quantitative
SDS	Sodium dodecyl sulfate
SEM	Scanning electron microscopy
SLM	Selective laser melting
T	Temperature
TBAF	Tetrabutylammonium fluoride
TBD	1,5,7-Triazabicylco[4.4.0]dec-1-ene
TBDMS	t-Butyldimethylsilyl
TBS	Tributylsilane
TBTU	O-(Benzotriazol-1yl)-N,N,N',N'-tetramethyluronium tetrafluoroborate
TEM	Transmission electron microscopy
TEMPO	2,2,6,6-Tetramethyl-piperidin-1-oxyl
Tf (OTf)	Triflate
TFA	Trifluoroacetic acid
TFAA	Trifluoroacetic anhydride
THF	Tetrahydrofuran
THP	Tetrahydropyran
TLC	Thin-layer chromatography
TM	Trade mark
TMG	1,1,3,3-Tetramethylguanidine
TMP	2,2,6,6-Tetramethylpiperidine
TMS	Trimethylsilyl
Tr	Trityl
Ts	Tosyl/p-toluenesulfonyl chloride
TTMSS	Tris(trimethylsilyl)silane
UPLC	Ultra-performance liquid chromatography
wrt	With respect to
wt.	Weight
XRD	X-ray diffraction

1 Introduction to Micro Reaction Technology

Today's synthetic chemist is under increasing pressure to discover and deliver compounds quickly, with an eye on devising synthetic methodology that can be readily scaled to enable the next stage of development to be performed rapidly. While emerging technologies such as microwave chemistry, and combinatorial applications have over the decades been implemented within research laboratories, a bottleneck still remained at the scale-up step. Micro reaction technology attempts to solve the problem from the perspective of developing reaction methodology within the laboratory that can be used directly for the performance of reactions at a production-scale, thus driving down the timescales required to put compounds into production. This goal is achieved by basing the technique on the rapid generation of high-quality chemical information, which can be readily scaled once a compound of interest is identified.

1.1 WHAT IS MICRO REACTION TECHNOLOGY?

Micro reaction technology is a term widely used to describe the performance of reactions in a continuous manner, within well-defined reaction channels, where typical dimensions are of the order <1000 µm and volumes, at a research level, span the microliter to milliliter range. At this length scale, laminar flow dominates, due to high viscous forces and surface effects, and mixing occurs by diffusion only. The impact of this on reactions is that uniform reaction conditions are obtained, with each molecule passing through the device experiencing the same reaction times, temperatures, and pressures: the result being the development of operator independent synthetic techniques. A particularly attractive feature of micro reaction technology is therefore the ease with which reaction conditions can be transferred from laboratory-scale reactors to production sites without the need for reoptimization and redesign of the synthetic process. As such, novel processing windows are accessible with researchers who are now able to safely perform reactions well above the boiling point of common organic solvents with ease.

Other benefits include reduced exposure to hazardous chemicals, increased atom efficiency, and the ability to integrate in-line analytics into processes. Upon combining these advantages, researchers have observed an overwhelming trend, whereby increased reaction selectivities, yields, and purities are obtained when comparing micro reaction technology to stirred batch reactions.

While the roots of micro chemical processing lie with chemical engineers, it is the aim of this chapter to approach the developments made in the field from that of a synthetic organic chemist. Consequently, should the reader require additional physical information about micro reactors, where relevant, additional reading has been provided in the form of recommended articles and texts.

1.2 FABRICATION/CONSTRUCTION OF MICRO REACTORS

Since the late 1990s, researchers have reported a wide array of techniques for the fabrication of microstructured reactors, which have been comprehensively discussed in titles such as *Microreactors: New Technology for Modern Chemistry* by Ehrfeld et al. [1], *Micro Process Engineering, A Comprehensive Handbook Volume 2: Devices, Reactions and Applications* edited by Hessel et al. [2], and *Micro Systems and Devices for (Bio)chemical Processes* edited by Schouten [3].

When considering the performance of organic transformations within such reaction systems, the first step to take is to identify the most suitable material from which to fabricate the reactor. Due to the wide-ranging demands of synthetic chemistry, micro reactors have been fabricated from an array of substrates, with researchers reporting the use of glass [4], quartz [5], diamond [6], polymethylmethacrylate (PMMA) [7], polydimethylsiloxane (PDMS) [8,9], SU-8 [10], polyimide [11], allylhydridopolycarbosilane (AHPCS) or polyvinylsilazane (PVSZ) [12], ceramic [13], silicon [14], stainless steel [15], nickel [16], and Hastelloy [17], all tailor-made to suit their end-use or available fabrication techniques. In addition to reactant compatibility, when selecting the reactor material it is also important to consider the operating conditions that will be used, that is, reduced, ambient or high temperatures, and atmospheric or elevated pressures [18] as these parameters will impact greatly on the materials available for selection, along with the techniques used to seal the microchannels and make interconnections to the real world [19].

Once the substrate is selected, a method for incorporating microstructures within the bulk material is required, with each material having its own associated fabrication technique as described below.

1.2.1 GLASS DEVICES

Owing to its compatibility with organic solvents, and the synthetic chemists' familiarity with its use, a large proportion of research within micro reactors has been conducted using glass devices. Microstructures are fabricated within glass substrates such as Pyrex®, Borofloat®, and FOTURAN® using a combination of photolithography and wet etching [20] or powder blasting [21], the result being the formation of reaction channels that are wider than they are deep [22]. Upon covering the microchannel network with a top plate, also glass, the devices are thermally annealed to afford a single integrated device that cannot be opened. Using this approach, robust devices can be fabricated that are suitable for the manipulation of organic reactants at high temperatures and pressures, with fluidic interconnects made using adhesives, gaskets, and commercially available fittings. A limitation of glass is the inability to employ hydrofluoric acid as either a reagent or a by-product of a reaction, as this can lead to additional etching of the channels, resulting in changes to the critical reactor dimensions.

Introduction to Micro Reaction Technology

1.2.2 Silicon Devices

Silicon devices can be fabricated using either wet or dry techniques depending on the channel features required [23]. Using wet-etching channels, grooves and cantilevers have been produced; however, if more complex features are required, such as high aspect ratio channels or mixing elements, dry techniques such as reactive ion etching (RIE) are required. Once the features are formed, the reactor can be fusion bonded to a second silicon layer or anodically bonded to a glass plate. Reactors fabricated for chemical research routinely employ a glass cover plate as this enables the microchannels to be viewed while in use. In order to increase the chemical stability of the resulting devices, a thermal oxide layer is grown on the surface of the microchannels to afford a chemically resistant device. Again fluidic connections are made using commercially available couplings, adhesives, or soldering [24].

1.2.3 Polymeric Devices

Depending on the polymer selected, a number of fabrication techniques exist for the preparation of micro reactor devices, including injection molding [25,26], casting [27], hot embossing [28], and more recently, gas-entrained polymer extrusion [29,30]. While at a research level the preparation of PDMS devices is relatively simple, the polymer is known to swell in some organic solvents and as such is more suited to aqueous chemistries [31]. Due to the excellent heat and chemical resistance demonstrated by polytetrafluoroethylene (PTFE), researchers have begun to develop techniques for the fabrication of microstructured devices using hot embossing without vacuum [32], and more recently Kim and coworkers [33] have reported PVSZ devices that open up the possibility of mass producing chemically tolerant polymeric devices.

1.2.4 Metallic Devices

Along with the vast array of metals available for the production of microfabricated devices, there are also numerous techniques with which to prepare such devices. Wet chemical etching is widely employed for the introduction of grooves and channels into substrates; however, as with its use for glass substrates, the features generated have a minimum width of two times the feature depth [34]. If higher aspect ratio features are required, then techniques such as dry etching using a laser enables rectangular channels to be formed, machining and selective laser melting can be used to realize complex structures, and plates can also be punched from thin sheets of metal and stacked to afford the target structure. Irrespective of the structuring technique employed, the devices need to be assembled in order to afford a liquid/gas-tight device for use. As with all other stages, a series of assembly methods are available ranging from welding, soldering, brazing, clamping, gluing, and diffusion bonding [35]. In addition to microstructured devices, researchers frequently report the use of stainless steel and Hastelloy tubular reactors constructed using commercially available components (see Chapters 3 and 4 for examples).

1.2.5 CERAMIC DEVICES

Ceramics are an interesting material for the fabrication of micro reactors as they are suited to high operating temperatures that may present a problem with other substrates [36]. Such devices are fabricated using a 3D mold, which is filled with a plasticized ceramic powder and then sintered to afford the microstructured device as a replicate of the mold [37]. Alternatively, a process of low-pressure injection molding can be used, which is suitable for the fabrication of ceramic devices from submicron powders [38]. Upon sintering, shrinkage occurs and as such must be taken into account when designing the mold. In addition to fabricating solid devices, these techniques can also be applied for the fabrication of porous membranes, examples illustrating wall coating with catalysts and surface derivatization whereby stable gas–liquid interfaces are obtained have been reported [39].

1.2.6 REACTOR TO WORLD INTERFACE

Once fabricated, the device must be connected to a means of delivering reagents, be they liquids, gases, or slurries. At a research level, authors have reported numerous techniques for the connection of gas/liquid feed lines ranging from gluing (epoxy resins) [40], commercially available Swagelock® connectors [41], adhesive ports [42,43], NanoPort™ assemblies [44], gaskets [45], and soldering [24], with examples illustrated in Figure 1.1.

1.3 MANIPULATION OF REACTANTS AND PRODUCTS WITHIN FLOW REACTORS

In order to perform synthetic reactions within micro flow reactors, reactant solutions and/or gases must be brought together within the microchannels of the device, where they mix and react prior to exiting the reactor. To achieve this, there are two main types of pumping mechanism that have been employed: (a) pressure-driven flow and (b) electroosmotic flow (EOF).

Prior to the development of robust interconnection strategies, such as those illustrated in Figure 1.1, EOF was employed widely as it enabled the manipulation of reagents without the generation of pressure; synthetic examples of its use can be found in Chapters 3 and 4. With significant advances made with the commercial availability of adhesive ports and the array of adhesive and clamping strategies featured within the literature, the majority of researchers now employ pressure-driven flow.

Compared with EOF, pressure-driven flow is advantageous as it enables the propulsion of fluids at preset flow rates within microchannels, with flow rate independent of the fluid properties. Pressure-driven flow is most widely attained at a research level by the use of commercially available syringe, displacement, or high-performance liquid chromatography (HPLC) pumps, with larger capacity pumps employed when looking at scaling reactors for production purposes (see Chapter 7). For details on EOF as a pumping mechanism refer to discussions by Chen et al. [46], Dasgupta and Liu [47], and Banerjee et al. [48].

Introduction to Micro Reaction Technology

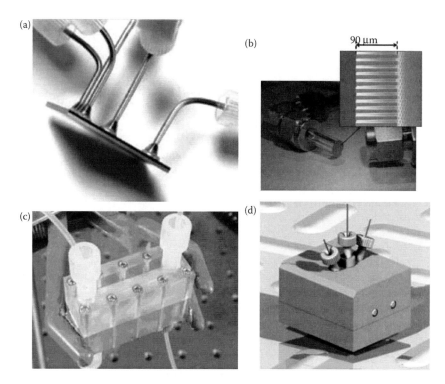

FIGURE 1.1 Illustrations of various "reactor-to-world" interfaces utilized by the research community. [(a) Murphy, E. R. et al. 2007. Solder-based chip-to-tune and chip-to-chip packaging for microfluidic devices, *Lab Chip* 7: 1309–1314; (b) Unnikrishnan, S. et al. 2009. MEMS within a Swagelok®: A new platform for microfluidic devices, *Lab Chip* 9: 1966–1969; (c) Kralj, J. G., Sahoo, H. R., and Jensen, K. F. 2007. Integrated continuous microfluidic liquid–liquid extraction, *Lab Chip* 7: 256–263; (d) de Mello, A. J. 2001. *Lab Chip* 1: 148–152. Reproduced by permission of the Royal Society of Chemistry.]

1.3.1 Mixing within Micro Reactors

When performing a synthetic micro reaction, the choice of reaction solvent is governed by a range of parameters, including the solubility of reactants, intermediates, and products, along with the stability of the reactant solutions contained within syringes or reagent bottles (depending on the type of pump employed). Unlike batch reactions, where mixing is promoted by the use of stirrers or agitators, within the microspace high viscous forces dominate and mixing occurs by diffusion only [49]; consequently, the time taken for two fluids to mix completely is directly linked to the channel size and the diffusion coefficients of the reactants employed [50]. While these timescales can be of the order of milliseconds to minutes, micromixers have been investigated to increase the speed with which homogeneous reaction mixtures are obtained, reducing the channel volume taken up with mixing and increasing the volume available for reaction.

The simplest micromixer available for use within microfabricated reactors is the T-mixer, which works by bringing two reactant streams together at a T-junction and allowing sufficient time for mixing to take place. As mixing only occurs by diffusion, this type of mixer can be limited in its application to those reactant streams with the correct diffusivity; therefore, in the case of reactions that are either very rapid [51] or posses low diffusivity, the mixing attained with a T-mixer may not be sufficient. In these cases, micromixers have been developed to increase the efficiency of the mixing process; however, their application comes at a price, significantly increasing the complexity of the fabrication process and the costs associated with a single device. Micromixers that employ moving parts are termed active and those that require no additional energy, other than that provided by the pumping technique, are referred to as passive [52].

1.3.2 Flow Types within Biphasic Systems

In addition to considering the time taken to mix two reactant solutions, and ensuring that concentrations are managed, in order to prevent precipitation of intermediates and products within the microchannels, the type of flow obtained should also be considered.

Liquid–Liquid: When miscible liquids are brought together at a T-mixer, coflowing lamellae diffuse into one another until the system is completely mixed. In the case of immiscible fluids, such as an organic and aqueous system, these streams can pass down the entire channel and exit the reactor as two separate streams (Figure 1.2).

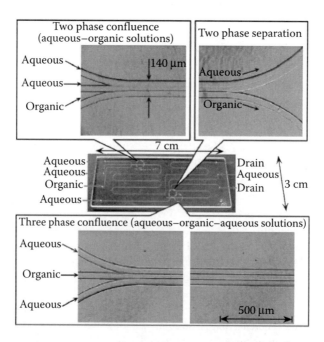

FIGURE 1.2 Illustration of the longitudinal interface obtained between immiscible fluids in a microchannel. (Kitamori, T. 2009. Parallel multiphase microflows: Fundamental physics, stabilisation methods and applications, *Lab Chip* 2470–2476. Reproduced by permission of the Royal Society of Chemistry.)

Depending on the two fluids employed, stable longitudinal flow may however not be possible due to interfacial forces, which act across the fluid, leading to destabilization of the interface. This results in the formation of slugs and or droplets within a continuous phase: often referred to as Taylor flow. While slug flow makes separation of the phases difficult at the microscale (Chapter 8), with products often phase separated in a collection vessel, the internal circulation (10–100 Hz) obtained within the slugs of liquid, due to the combined effects of shear within the channel and interfacial phenomena, has been shown to result in enhanced mixing efficiencies [53]. When compared to longitudinal biphasic systems (Figure 1.2), slug flow has been shown to accelerate some synthetic reactions, details of which can be found in Chapters 3 and 4 [54].

Gas–Liquid: When performing a biphasic reaction where one component is a gas and the other a liquid, several flow regimes can be obtained depending on the relative flow rates employed for either phase [55]. Introducing the reactants into a microchannel, again at a T-mixer, results in the formation of a single line of bubbles with uniform size and frequency within a liquid continuous phase. Depending on the relative flow rates, and the size of the gas bubbles, this can be described as either slug flow or bubbly flow. In the case of high gas flow rates, the liquid phase can be forced towards the walls of the channel or capillary affording what is termed as annular flow [56].

1.4 ADVANTAGES OF MICRO REACTION TECHNOLOGY

1.4.1 Process Safety

When conducting reactions on a laboratory-scale that may one day be scaled to production, the exothermicity of a transformation is one of the most important considerations to make; as a reaction that can be controlled by immersing a flask into an ice bath soon becomes a significant engineering challenge when the volume of the batch reactor is increased. Looking at the issue from both a safety perspective and one of cost, industry finds it is desirable to scale reactions that do not require large amounts of heat management. Consequently, there are lists of reactions that are not used when transferring synthetic routes from the research laboratory through to process scale-up. If heat management could be improved then these "forbidden" reactions could be applied to production-scale syntheses, thus increasing the portfolio of reactions available to the operator.

When using micro reactors, the high surface-to-volume ratio which can be in the range of 10,000–50,000 $m^2\ m^{-3}$ *cf.* 100 to <1000 $m^2\ m^{-3}$ for a batch reactor affords excellent heat transfer, which is useful in the performance of both highly exothermic and endothermic reactions [57]. Owing to the size of microchannels, the convective heat transfer coefficient is large, typically 6×10^4 W m^{-2} K^{-1} (depending on the substrate) and affords efficient thermal conductivity when compared to batch reactors (740 W m^{-2} K^{-1}). Therefore, reactions that once would have required cryogenic conditions to maintain them in a safe thermal regime, can now be performed without risk. In addition to the efficient heat management, the fact that only a small amount of material is committed to the reaction at any given time limits the hazards associated with a process, should the worst happen and control be lost: typically milliliters to liters.

1.4.2 Novel Reaction Conditions

While the majority of batch reactions are routinely performed at atmospheric pressure, giving rise to a limited thermal window for each reaction solvent, micro reactors are sealed units and can therefore be readily pressurized [58]. Pressurization has most commonly been employed as a method of increasing the reaction temperatures accessible while maintaining the reaction solvent in the liquid phase; typically pressures of <30 bar have been employed by means of backpressure regulation or via the fabrication of narrow, restrictor channels within devices.

More recently, examples have been reported whereby reaction units have been pressurized in the range of 140–600 bar [59], resulting in dramatic rate enhancements compared to reactions performed using standard temperatures and pressures [60,61]. The ability to access a wide temperature and pressure range means that specialized equipment is no longer required for performing reactions under extreme conditions, expanding the screening window available to chemists when researching synthetic transformations [62]. Whereas previously reactors would have been heated using thermal baths or heated tubes [63], reactors have now been fabricated to contain heating elements [64,65], although the use of heating blocks or Peltier elements are more common place due to reactor flexibility. Microwaves have also been quite widely used as a means of thermally activating flow reactions [66,67].

Supercooled Fluid: In addition to putting thermal energy into reactions, there is also a need to cool reactions particularly with respect to maintaining product selectivity. In 2006, Kitamori and coworkers [68] reported the use of supercooled micro flows for an asymmetric biphasic reaction. Employing Pyrex glass micro reactors (dimensions = 100 μm (wide) × 40 μm (deep)), with a hydrophobic coating (ODS), the authors found that the freezing point of water could be reduced to −28°C, independent of fluid flow rate. With this in mind, the authors applied the technique to an asymmetric biphasic alkylation reaction, as depicted in Scheme 1.1, between *N*-(diphenylmethylene)glycine-*tert*-butyl ester **1** and benzyl bromide **2** in dichloromethane (DCM) and aqueous KOH **3**, in the presence of a phase transfer catalyst ((*S*,*S*)-3,4,5-trifluorophenyl-NAS bromide). Investigating the alkylation reaction at 20°C and −20°C, the authors observed an increase in enantioselectivity with decreasing reactor temperature (43–50%ee). In addition to demonstrating a facile

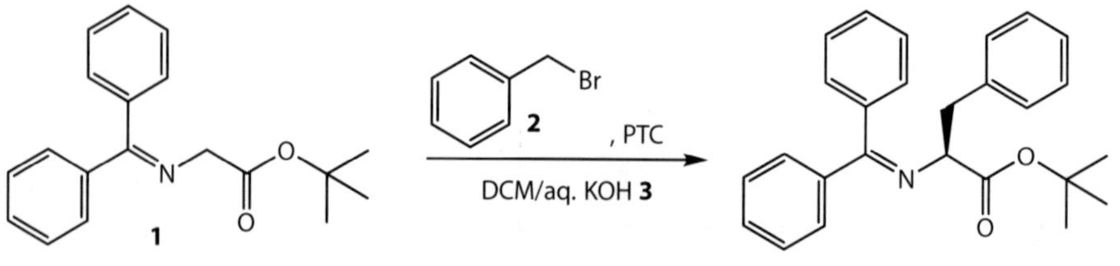

SCHEME 1.1 Illustration of the phase transfer reaction performed utilizing a supercooled micro reactor.

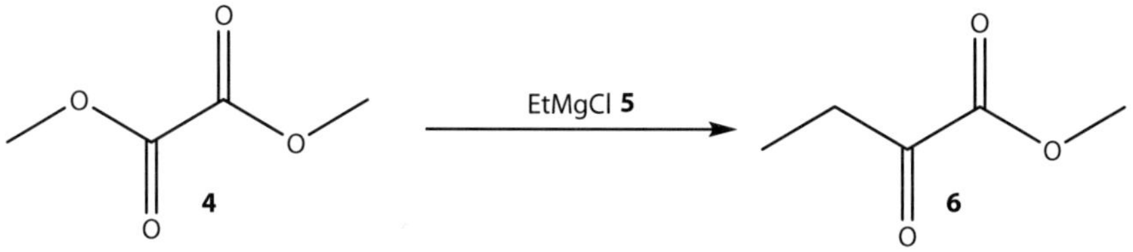

SCHEME 1.2 Illustration of the model reaction employed to determine the heat exchange performance of a series of micro reactors.

method for tuning reaction selectivity, the example illustrates the use of reaction conditions not accessible using conventional stirred flask techniques, owing to the freezing point of aqueous KOH **3** (4°C), thus further increasing the range of reaction conditions accessible to the research chemist.

Hot-Spot Formation: While it is widely commented that micro reactors can provide more uniform thermal conditions than those obtained within stirred batch reactors, the practical assessment of hot-spot formation within such systems is not an easy task. With this in mind, Roberge and coworkers [69,70] developed a Grignard test reaction to characterize heat exchange performance in a series of micro reactors, using commercially available precursors (Scheme 1.2). The first feed solution employed was dimethyl oxalate **4** in DME (various concentrations ranging from 10 to 20 wt.%) and the second, ethyl magnesium chloride **5** in THF (25 wt.%), diluted with DME to avoid precipitate formation at temperatures <0°C. Prior to use, the feedstocks were cooled to −15°C, −5°C, or 5°C and reacted in a series of glass or Hastelloy reactors, with the reaction products quenched upon exiting the reactors with 1.0 M HCl.

Using this approach the authors were able to identify the most effective parameter for isothermal operation as being the multiport injection of dimethyl oxalate **4** (up to 4 points) with Hastelloy affording ~10% higher product **6** yield compared to the glass reactor evaluated, an observation attributed to increased heat conductivity (~10×). Through conducting this investigation, the authors were able to conclude that in order to ensure good local temperature control, multiport injection had a greater effect than concentration, feedstock temperature, and reactor material.

1.4.3 REDUCED WASTE GENERATION

In addition to ensuring process safety, chemical industries are continually met with changing restrictions on the volume and type of waste that they can generate. With pharmaceutical process generating typically 25 kg of water for every kilogram of product prepared, improvements in process are required in order to reduce the environmental burden associated with chemical production. It is envisaged that by developing atom-efficient syntheses that require reduced energy input to maintain, chemical companies can start to reduce the volume of waste generated and move toward production on-demand [71].

In addition to reducing the volume of waste generated, microchemical processing also offers the ability to use new techniques and the implementation of greener reaction solvents. Supercritical fluids are becoming increasingly more popular within

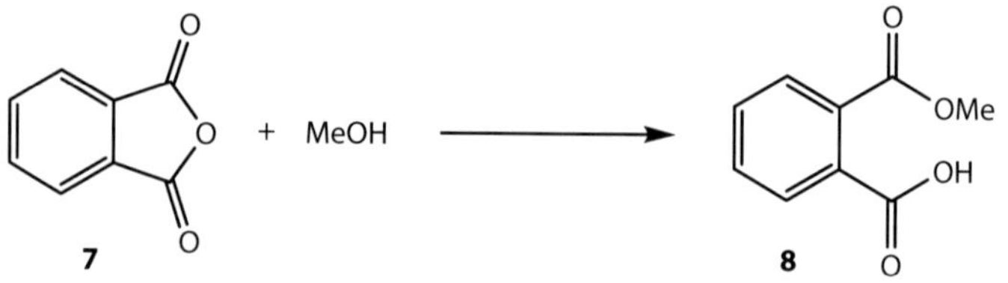

SCHEME 1.3 Illustration of the model reaction used to evaluate the manipulation of $scCO_2$ within a glass micro reactor.

industry as solvents in both separations and extractions. However, due to the need for specialized equipment, their application toward synthetic chemistry is not widely realized at a laboratory level. Due to the moderate conditions required to generate fluids such as supercritical carbon dioxide ($scCO_2$) (pressure = 73.8 bar and temperature = 31.1°C) and the wide-ranging physical properties obtained with slight changes in temperature and pressure, the fluid has the potential to vary the kinetics or reactions, while increasing selectivities and yields. Due to the safe and facile nature associated with heating and pressurizing fluids within microfabricated devices, numerous authors have demonstrated the generation and use of supercritical fluids, removing the perception that these solvents are only available to those research laboratories that have access to specialized rigs.

Verboom and coworkers [72] demonstrated the ease with which T and p could be varied within a glass micro reactor and evaluated the effect on the esterification of phthalic anhydride **7** with MeOH to afford 2-(methoxycarbonyl)benzoic acid **8** (Scheme 1.3). As Figure 1.3 illustrates, the reactor comprised of a heated and cooled

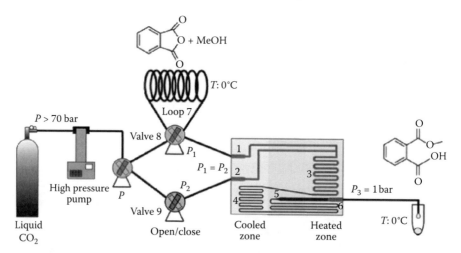

FIGURE 1.3 Schematic illustrating the reaction setup used to evaluate the esterification of phthalic anhydride **7** under high pressures and/or $scCO_2$ conditions. (Benito-Lopez, F. et al. 2007. Substantial rate enhancements of the esterification reaction of phthalic anhydride with methanol at high pressure and using supercritical CO_2 as a co-solvent in a glass microreactor, *Lab Chip* 7: 1345–1351. Reproduced by permission of the Royal Society of Chemistry.)

zone, with the reaction mixture introduced into the reactor at inlet 1, via a pressurized loop and carbon dioxide added via inlet 2, with pressurization occurring along the channel. Upon meeting, the reactants mix (3) prior to entering a fluid restrictor (4), fabricated to ensure constant pressure, where the majority of the reaction takes place, prior to returning to atmospheric conditions in the hydraulic resistor (5–6). Using this approach, authors investigated temperatures of 20–120°C and pressures of 10–100 bar, employing UV/VIS detection to collect the kinetic profiles. This approach enabled the authors to rapidly screen a series of reaction conditions, identifying a significant rate enhancement (5400-fold) when compared to batch experiments performed at 1 bar.

1.4.4 On-Site Chemical Processing

The on-site production of raw materials has the potential to remove the risks currently associated with the ground transportation and storage requirements of hazardous chemicals, while possessing the potential to reduce operating costs. Through conducting a detailed safety study into the synthesis of peracetic acid via a conventional batch process and a microprocess, Kolehmainen and coworkers [73] were able to conclude that micro reactors offer a step toward inherently safer chemical processing not only by reducing the risk associated with catastrophic failure of equipment, process downtime, and cost, but also by enabling process chemists to generate potentially hazardous and explosive raw materials at the site of use.

One such example of this was demonstrated by Jensen and Schmidt [74] who developed a silicon microfabricated reactor suitable for the on-site synthesis of phosgene (see Chapter 4 for details).

1.5 DISADVANTAGES OF MICRO REACTORS

As with all new technologies, there are perceived disadvantages associated with the use of micro reactors and continuous flow operation, none so widely discussed as the issue of "clogging." Owing to the small and intricate nature of microchannel networks, researchers more familiar with standard batch techniques are fearful of the potential to "block" or "clog" microchannels with precipitated material forming over the course of a reaction. For those active within the field it becomes readily apparent that this is not a problem encountered often, although there is potential for it to occur if the synthetic process under investigation is not carefully considered prior to investigation.

When looking to transfer a reaction from batch to flow, the main consideration that should be taken is what reaction solvent to use. This should be viewed from several perspectives, firstly reactant compatibility/stability and secondly reactant, intermediate and product solubility. To ensure long-term operation of a continuous flow micro reaction, it is imperative that a suitable precursor concentration be identified, which enables all reactants and products to remain in solution for the duration of the investigation. Should the concentration be incorrectly selected, this can lead to the precipitation of reaction intermediates and products within the microchannels upon contact. With this in mind, low reactant concentrations should be investigated for reactions that are unfamiliar to the researcher, increasing concentration with experience of the synthetic transformation. Typical concentrations employed, as can be seen throughout, range from 1.0×10^{-3} to 1.0 M, with examples reported at greater concentrations particularly when liquid reactants are employed.

1.5.1 Handling of Insoluble Materials

Although in the main there is a need to maintain the solubility of reactants/intermediates and products, if a product is particularly insoluble this does not instantly mean that it is unsuitable for production within micro flow reactors. As can be seen throughout, there are many modes of operation available with this technology, with reactor design and flow regime adjustable to suit the process under investigation.

With this in mind, McQuade and coworkers [75] demonstrated a facile method for preventing clogging within a polyvinyl chloride (PVC) tube reactor (dimensions = 1.59 mm (i.d.)) utilizing biphasic flow, comprising an immiscible carrier phase and organic droplets containing the reactants. Using this approach, the authors were able to perform the synthesis of two insoluble products, Indigo **9** and N,N'-dicyclohexylethylenediimine **10** without fouling of the reactor, enabling continuous production of solid products to be performed (Scheme 1.4). Even utilizing techniques such as this, it is imperative that the precursor solutions are completely soluble thus enabling dispensation via syringe pumps.

In a second example, Corning SAS (France) [76] has developed glass continuous flow reactors in which slurry hydrogenations can be performed, demonstrating the ability to manipulate three-phase reaction mixtures containing liquid, solid, and gaseous reactants for long periods of time without interruption to the flow process (for more details see Chapter 2) (Figure 1.4).

Authors have also reported the use of micro reaction technology (MRT) for the production of highly viscous polymeric materials [77] and the manipulation of precipitated APIs [78]. With the developments made over the past 5 years, it can be seen that initial concerns over the clogging of reaction channels are unfounded with solutions to many of the scenarios initially perceived as being problematic, being

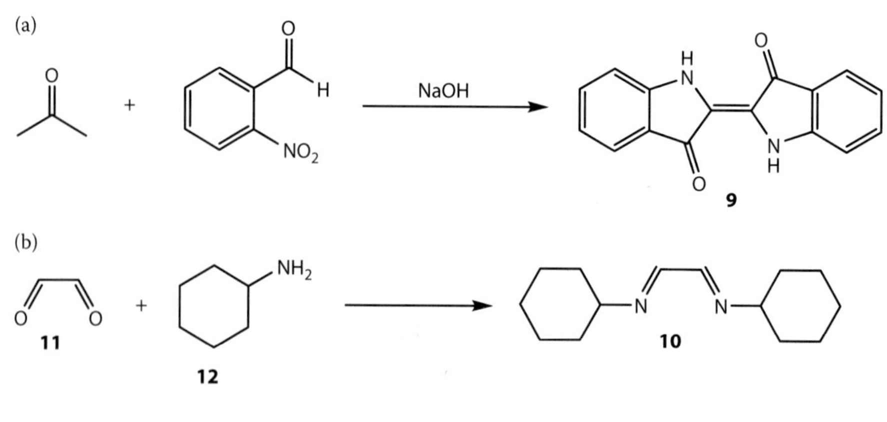

SCHEME 1.4 Illustration of the model reactions used to demonstrate the handling of solids within micro flow reactors: (a) the synthesis of Indigo **9** via an aldol condensation and (b) the reaction of glyoxal **11** and cyclohexylamine **12** to afford N,N'-dicyclohexylethylenediimine **10**.

Introduction to Micro Reaction Technology

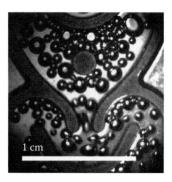

FIGURE 1.4 Illustration of the micromixer employed to ensure efficient mixing of the three-phase system. (Reproduced from Buisson B. et al. 2009. *Chem. Today* 27: 12–14. With permission of Tekno Science Srl.)

overcome with custom reactor design and careful consideration over the synthetic process under investigation.

1.6 PROCESS INTENSIFICATION

The ability to rapidly scale a synthetic reaction without the need for reoptimization of a process is possibly the greatest advantage that micro reaction technology has to offer, when compared to the conventional batch approach of scale-up from the laboratory → pilot plant → production. Unlike stirred tank reactors, in general, near plug flow is obtained within microfabricated reactors, which means that all components passing through the reactor experience the same reaction conditions [79]. It is this uniformity that affords increased process control within micro reactors and the nonuniformity in conventional reactors that can cause problems when the scale of a reaction is altered: highlighted particularly when considering reactions where a range of products can be formed.

While newcomers to the field may be questioning how micron-sized reaction channels can be used for the production-scale synthesis of fine chemicals and pharmaceuticals, the answer lies in obtaining robust chemical information from microchannel devices and then translating this information into higher-throughput continuous flow systems [80]. Being able to determine these system parameters is therefore critical for the design of larger capacity systems, with measurement techniques also required to validate the devices once fabricated [81,82].

Increasing the production capacity of micro reactors can be achieved using several techniques, the first being parallelization—where single reaction channels are operated in parallel; using either internal [83] or external numbering-up [84,85]. Alternatively, the reaction channels can be scaled to enable a higher liquid flow rate to be employed [86]; however, in this case, it is essential that all the reaction characteristics identified on the microscale are maintained, that is, mixing efficiency, residence time, temperature, and pressure. Whichever technique is employed, it is imperative that the intrinsic reaction conditions remain the same to enable successful process intensification [87]. In addition to removing the need to reoptimize processes at each stage, the use of continuous flow technology removes the financial risks associated with failing to safely scale a process as, if the reaction can be performed in a single channel reactor, then it can be performed in multiple reactors [88]. Examples of such systems installed within industrial facilities can be found in Chapter 7.

1.7 IN SITU REACTION MONITORING

During the early stages of micro reactor development, researchers characterized their reactions using offline analytical techniques, identifying a new bottleneck sample analysis. To resolve this, researchers have more recently investigated the coupling of micro reactors with online analytical techniques in order to meet the need for increased sample throughput. This shift has also come at a time when the technology is being taken into industrial situations (see Chapter 7), where a need for constant process evaluation is acknowledged. With this in mind, the following section describes advances made in the area of continuous micro process monitoring using synthetic examples.

1.7.1 RAMAN SPECTROSCOPY

Using the Paal–Knorr synthesis of 2-(2,5-dimethyl-1H-pyrrol-1-yl)ethanol **13** as a model reaction (Scheme 1.5), Mechtilde and coworkers [89] demonstrated the wealth of information attainable through the combination of micro reaction technology and online reaction monitoring via Raman Spectroscopy. Employing either Invar® stainless steel (dimensions = 200–400 μm (wide) × 200 μm (deep) × 5–50 cm (long)) or silicon (dimensions = 50–100 μm (wide) × 50–70 μm (deep) × 5–50 cm (long)) reaction channels and Pyrex glass cover plates the authors investigated the effect of time and channel geometry on the reaction. Taking spectra both across and along the reaction channel, the authors were able to map the consumption of ethanolamine **14** and 2,4-pentanedione **15** along with the formation of 2-(2,5-dimethyl-1H-pyrrol-1-yl)ethanol **13**.

Nitta et al. [90] subsequently demonstrated the use of Raman spectroscopy for the monitoring of reaction intermediates within Pyrex microchannel devices (channel dimensions = 230 μm (wide) × 70 μm (deep)). Using the bromination of cyclohexene **16** as a model reaction, the authors evaluated the use of Raman spectroscopy as a tool for the determination of intermediate **17** concentrations with high spatial resolution.

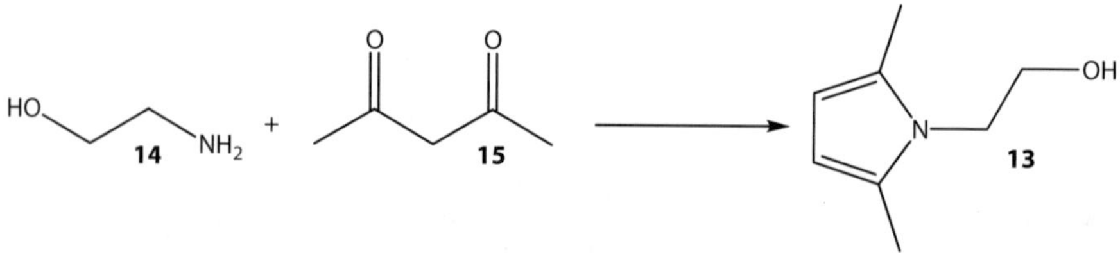

SCHEME 1.5 Schematic illustrating the model Paal–Knorr pyrrole synthesis monitored by *in situ* Raman spectroscopy.

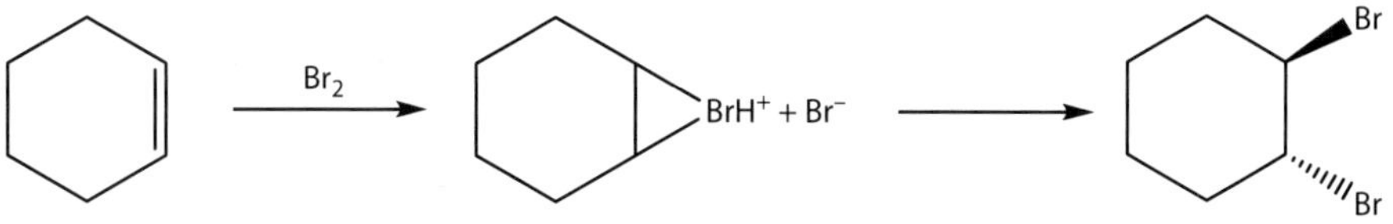

Introduction to Micro Reaction Technology

SCHEME 1.6 Illustration of the model reaction selected to demonstrate the measurement of reactive intermediates using Raman spectroscopy.

Using benzene as the reaction solvent, the authors monitored the Raman spectra 50 μm downstream from the Y-shaped mixing point, whereby they detected a peak at 265 cm^{-1} at the reactant interface, this was attributed to intermediate **17** (Scheme 1.6). By measuring the peak intensity at various positions within the microchannel, the authors were able to derive a relationship between the intermediate **17** and the reactant concentrations of **16**: using benzene as an internal standard. The results of this investigation corresponded well with a theoretical assessment of the reaction, confirming the synthetic utility of this technique.

To date, the *in situ* monitoring of reaction processes within microchannels has been performed using inverted objectives and has been limited to the use of reaction setups that can be located on a microscope stage. Gavriilidis and coworkers [91] devised a technique for the *in situ* monitoring of catalytic reactions within such systems by utilizing a 90° extension arm from a Raman microscope, enabling the reactor and peripheral equipment to be set up alongside the microscope (Figure 1.5).

To demonstrate the utility of the technique, the authors investigated the three-phase oxidation of benzyl alcohol **18** to benzaldehyde **19**, as depicted in Scheme 1.7, using a supported Au–Pd/TiO$_2$ **20** (particle size = 90–125 μm) catalyst packed within a silicon/glass micro reactor (channel dimensions = 600 μm (wide) × 300 μm (deep) × cm (long)) mounted on a heating unit.

Employing 13.7 mg of catalyst **20**, the authors evaluated the effect of O$_2$ **21** on the conversion of benzyl alcohol **18** to the respective aldehyde **19**, using signals at 1000, 1597, and 3066 cm^{-1}. Using this approach, the authors demonstrated that system

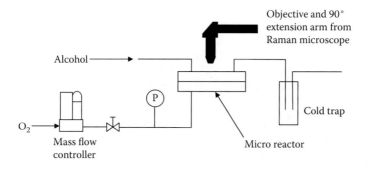

FIGURE 1.5 Schematic illustrating the 90° extension arm used to perform *in situ* Raman monitoring of an oxidation reaction performed in a silicon/glass microchannel.

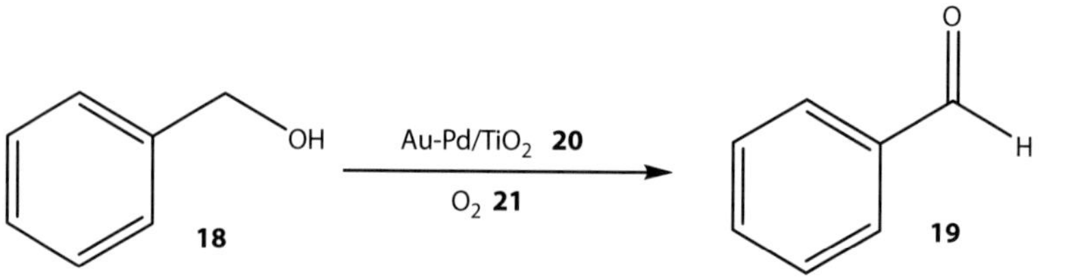

SCHEME 1.7 Model reaction used to demonstrate the Raman monitoring of Au-catalyzed oxidation reaction.

pressure had a dramatic effect on product selectivity, with pressures of 0.7 and 3.7 bar affording 60% and 83% selectivity, respectively (36% and 12% toluene obtained).

Up to now, examples have involved the profiling of liquid-phase reactions performed within microfluidic devices; however, it is possible to apply the technique to monitor catalyst beds, as illustrated by Baiker and coworkers [92] in Figure 1.6.

Fabricated from silicon and glass, the authors investigated the heterogeneously catalyzed, gas–liquid, hydrogenation of cyclohexene within a micro reactor. Using commercially available Pd/Al$_2$O$_3$ (2 wt.%) as the catalyst, the authors were able to monitor the effect of reaction time and temperature on the hydrogenation reaction, obtaining reference spectra at the fluid inlets and reaction spectra at various points before, in and after the catalyst bed. Employing offline gas chromatography (GC) analysis as a reference technique, a trend of increasing reaction with increasing reaction temperature was identified.

1.7.2 IN SITU ANALYSIS BY FTIR SPECTROSCOPY

An early example of *in situ* monitoring using FTIR spectroscopy was reported by Jensen et al. [93] and involved the use of a silicon micromixer placed within the sample holder of a conventional bench-top FTIR spectrometer as depicted in Figure 1.7.

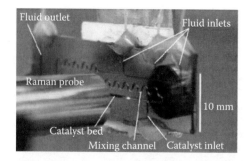

FIGURE 1.6 Illustration of the reaction setup used to obtain *in situ* Raman analysis of a packed-bed reactor. (Urakawa, A. et al. 2008. On-chip Raman analysis of heterogeneous catalytic reaction in supercritical CO$_2$: Phase behavior monitoring and activity profiling, *Analyst* 133: 1352–1354. Reproduced by permission of the Royal Society of Chemistry.)

Introduction to Micro Reaction Technology

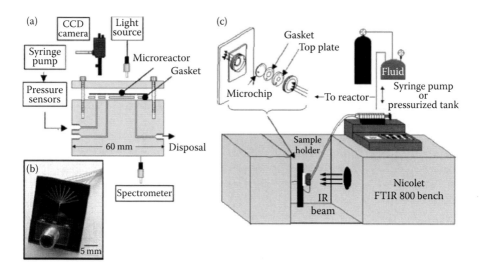

FIGURE 1.7 Schematic illustrating the reaction set-up used to evaluate a continuous flow reaction by FTIR spectroscopy: (a) side view of the micro reactor interface to the fluidic and optical connectors, (b) photograph of the silicone micromixer, and (c) schematic of the FTIR spectrometer/reactor interface. (Reproduced with permission from Floyd, T. M., Schmidt, M. A., and Jensen, K. F. 2005. *Ind. Eng. Chem. Res.* 44: 2351–2358. Copyright (2005) American Chemical Society.)

To evaluate the technique and the micromixer, the authors monitored the acid–base reaction of phenol red whereby a yellow solution at pH 6.4 was mixed with an orange solution at pH 8.0. The reaction was monitored at 560 nm and a calibration revealed that 0% mixing would afford an absorbance of 0.81% and 100% mixing an absorbance of 0.33. Using this approach, the authors evaluated a range of flow rates (1–4000 µL min^{-1}) accessing residence times of 3.2–12,700 ms, identifying complete mixing in 60 ms (in line with model predictions). Having practically evaluated the online detection mechanism, the authors subsequently investigated the reaction setup for the hydrolysis of methyl formate, obtaining a kinetic profile for the reaction which was in line with batch investigations previously reported within the literature.

In a second example, Jähnisch and coworkers [94,95] illustrated the noninvasive *in situ* monitoring of the ozonolysis–reduction sequence illustrated in Scheme 1.8 employing a miniaturized attenuated total reflectance (ATR) sensor and flow cell assembly located close to the outlet of their micro reactor (Figure 1.8).

Employing a glass microstructured gas–liquid contactor and cyclone mixer, the authors initially evaluated the ozonolysis of substrate **22** (0.1 mmol min^{-1}) in MeOH/DCM utilizing O_3 **23** at a flow rate of 1.66 mmol min^{-1} in oxygen (400 mmol min^{-1}). Figure 1.9 illustrates typical IR spectra obtained during evaluation of the micro reactor, with the gray spectra representing substrate **22**, at −8°C, and the black spectrum obtained upon addition of O_3 **23**. The difference spectra can then be used to show the appearance of intermediates/products and, in this case, clearly illustrates the simultaneous formation of hydroperoxide **24** and aldehyde **25**. The signal intensities can then be used to obtain semiquantitative measurements, enabling rapid

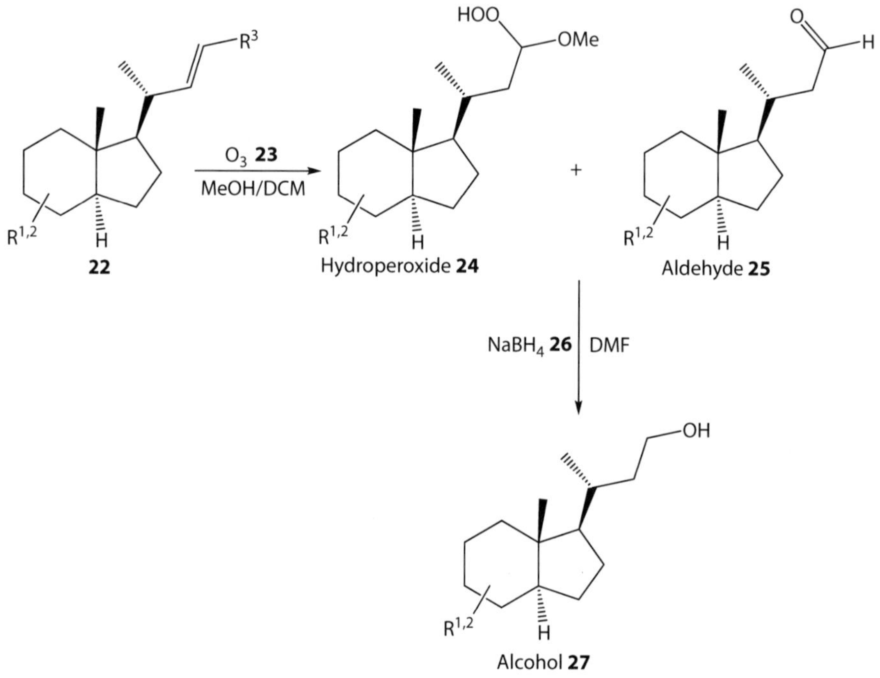

SCHEME 1.8 Illustration of the ozonolysis–reduction sequence monitored by FTIR spectroscopy in a series of continuous flow reactors.

optimization of the ozonolysis step. Using this approach, the authors investigated the effect of O_3 **23** concentration in the O_2 stream, finding that at 0.14 mmol min^{-1}, substrate **22** conversion was optimal.

To demonstrate the synthetic utility of this mode of analysis, the authors subsequently investigated the serial reduction of the ozonolysis mixture (**24** and **25**)

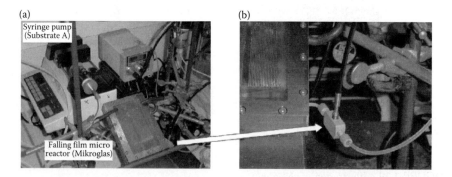

FIGURE 1.8 Photographs illustrating (a) the experimental set-up used for the online analysis of continuous flow reactions by FTIR and (b) the integrated flow cell. (Reproduced with permission from Hübner, S. et al. 2009. *Org. Proc. Res. Dev.* 952–960. Copyright (2009) American Chemical Society.)

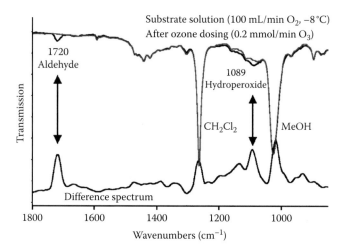

FIGURE 1.9 Spectra obtained through the online analysis of a continuous flow process. (Reproduced with permission from Hübner, S. et al. 2009. *Org. Proc. Res. Dev.* 952–960. Copyright (2009) American Chemical Society.)

utilizing NaBH$_4$ **26** in DMF, whereby quantitative conversion into the alcohol **27** was obtained using a threefold excess of NaBH$_4$ **26**.

Baxendale and coworkers [96] recently demonstrated the incorporation of online IR analysis (Figure 1.10) into a commercially available flow reactor (React-IR™, Mettler Toledo, Figure 1.11) for the synthesis of butane-2,3-tartrates which find application as building blocks in the synthesis of natural products such as (+)-*O*-methylpiscidic acid dimethyl ester, (+)-nephrosteranic acid, and antascomicin B.

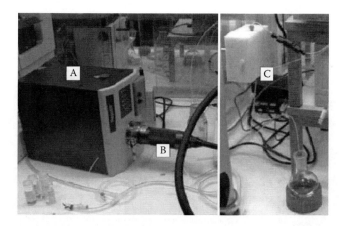

FIGURE 1.10 Illustration of the FTIR diamond probe flow cell incorporated into a flow reactor setup: (A) React-IR 45 m, (B) fiber optic cable, and (C) flow cell. (Carter, C. F. et al. 2009. Synthesis of acetal protected building blocks using flow chemistry with flow I.R. analysis: Preparation of butane-2.3-diacetal tartrates, *Org. Biomol. Chem.* 7: 4594–4597. Reproduced by permission of the Royal Society of Chemistry.)

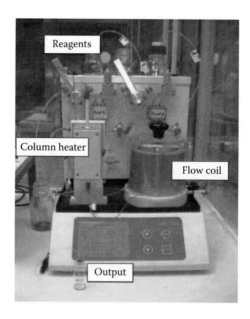

FIGURE 1.11 Continuous flow apparatus used to demonstrate in *in situ* monitoring of synthetic reactions using online spectroscopy. (Carter, C. F. et al. 2009. Synthesis of acetal protected building blocks using flow chemistry with flow I.R. analysis: Preparation of butane-2.3-diacetal tartrates, *Org. Biomol. Chem.* 7: 4594–4597. Reproduced by permission of the Royal Society of Chemistry.)

As Scheme 1.9 illustrates, the model reaction investigated involved the reaction of dimethyl-L-tartrate **28** with butanedione **29** and trimethylorthoformate **30** in the presence of DL-camphorsulfonic acid **31**, with the best reaction conditions found to be 2 M reactant solutions, a residence time of 2 h at a temperature of 90°C. Using this approach, the target butane-2,3-diacetal tartrate **32** was obtained in 78% conversion, representing an 8% increase compared to batch investigations. In order to purify the reaction products generated, an in-line scavenger cartridge was employed, which contained polymer-supported benzylamine **33**, to remove unreacted butandione **29** and periodate **34** to sequester any free diol **28**. Following this procedure, the target compound **32** was isolated by removal of the reaction solvent (MeOH and MeCN).

Having optimized the flow reaction using online IR spectroscopy, the authors subsequently demonstrated continuous operation of the flow reactor for a period of 10 h, generating 40 g of the BDA-protected tartrate **32** with a yield of 75%.

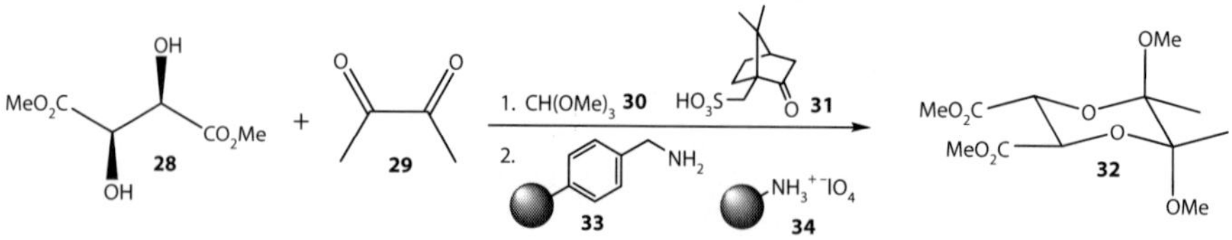

SCHEME 1.9 Schematic illustrating the model reaction used to demonstrate the synthetic value associated with the coupling of online IR spectroscopy to continuous flow reactors.

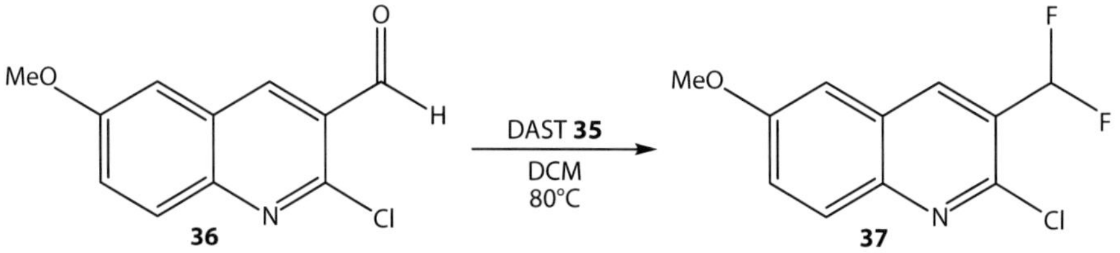

Introduction to Micro Reaction Technology

SCHEME 1.10 Schematic illustrating a model reaction used to demonstrate the synthetic utility of an online flow IR detector.

More recently, Ley et al. [97] reported use of the codeveloped React-IR flow cell for the optimization of a series of flow reactions including the previously reported diethylaminosulfurtrifluoride (DAST) **35** mediated fluorination of 2-chloro-6-methoxyquinoline-3-carbaldehyde **36** to afford 2-chloro-3-(difluoromethyl)-6-methoxyquinoline **37** under continuous flow [98] (Scheme 1.10). To perform a reaction, the aldehyde (0.5 mmol) and DAST **35** (1.0 mmol) in DCM were pumped into the reactor through separate inlets and met at a T-connector where they mixed prior to entering a tubular reactor which was heated to 80°C. Upon exiting the reactor, the reaction mixture was pumped through a cartridge containing calcium carbonate and silica gel, used to quench and clean up the reaction products prior to passing through the in-line detector.

Monitoring the appearance of a C–F band ($\upsilon = 1201$ cm^{-1}), the authors confirmed the formation of 2-chloro-3-(difluoromethyl)-6-methoxyquinoline **37** after 65 min. The effectiveness of the silica plug also noted as no carbonyl stretch, from the unreacted aldehyde **36**, was detected in the reaction products.

The examples cited herein therefore demonstrate the synthetic utility of online IR techniques to monitor the consumption of reactants and formation of products affording rapid optimization and/or monitoring of continuous flow processes. The technique has also been applied to the evaluation of catalytic materials deposited within microchannels and their use under gaseous conditions [99].

1.7.3 Nuclear Magnetic Resonance Detection

In addition to *in situ* optical measurements, the year 2009 saw the publication of a novel high-resolution nuclear magnetic resonance (NMR) spectroscopic probe suitable for the *in situ* monitoring of chemical reactions within microfluidic devices (Figure 1.12), offering structural information unparalleled by other spectroscopic techniques. Unlike previous attempts to develop miniaturized NMR coils, the device described herein employed a stripline "coil" which overcame previous problems with sensitivity, enabling ^{1}H NMR spectral resolutions of 0.7 Hz for ethanol. Using the acetylation of benzyl alcohol **18** as a model reaction (Scheme 1.11), Kentgens and coworkers [100] demonstrated the ability to obtain high-resolution spectra in time frames ranging from several seconds to 30 min, whereby reaction intermediates were observed that were not routinely identified within standard 5 mm sample tubes (Figure 1.13). This observation clearly demonstrates the possibility to study reaction

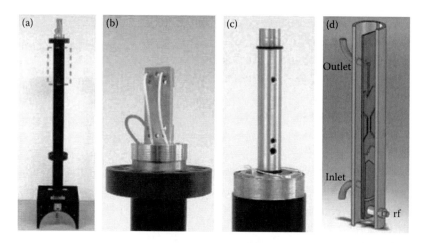

FIGURE 1.12 Illustration of the custom made microfluidic probe used for the *in situ* monitoring of micro reaction by NMR spectroscopy: (a) custom-made microfluidic probe (dashed line represents the position of the NMR chip), (b) micro reactor holder located on the top of the NMR probe, (c) stripline reactor holder, and (d) schematic of the arrangement of the micro reactor within the holder. (Reproduced with permission from Bart, J. et al. 2009, *J. Am. Chem. Soc.* 131: 5014–5015. Copyright (2009) American Chemical Society.)

kinetics within micro reaction channels. As such, a more detailed investigation is therefore underway, the results of which are highly anticipated.

1.7.4 Chromatographic Techniques

Chromatographic techniques have also been demonstrated with Welch and Gong [101] employing a micro flow HPLC system for the online analysis of continuous flow reactant streams at, or near, ambient pressure. Having validated the system using an analyte stream of methyl mandelate in MeOH, the authors investigated the thermal isomerization of *endo* **39** to *exo* **40** as a model reaction (Scheme 1.12).

Employing a reactor coil (volume = 1.0 mL), housed with a programmable GC oven, the effect of reactant residence time (0.2–10 min) was initially evaluated at an isothermal temperature of 230°C. Utilizing a sampling frequency of 4 min, the authors obtained an optimal isomer ratio of 1.8:1 (*exo* **39**: *endo* **40**) with a reactant residence time of 10 min. Under the aforementioned conditions, a series of reaction temperatures (180–250°C) were subsequently evaluated and the results were used to

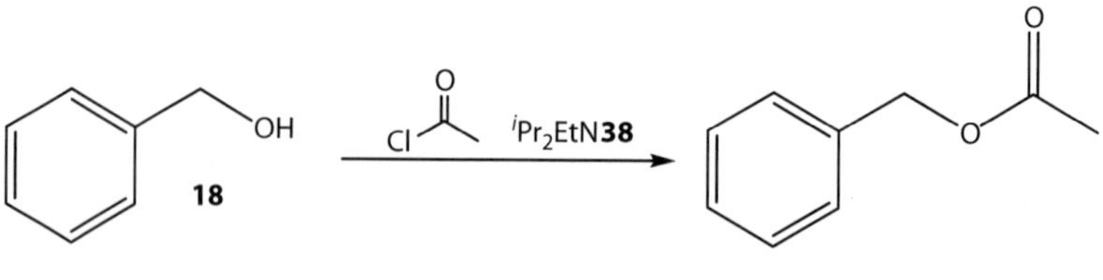

SCHEME 1.11 Schematic illustrating the model reaction used to illustrate the high-resolution NMR spectra attainable within a specially fabricated flow probe.

Introduction to Micro Reaction Technology

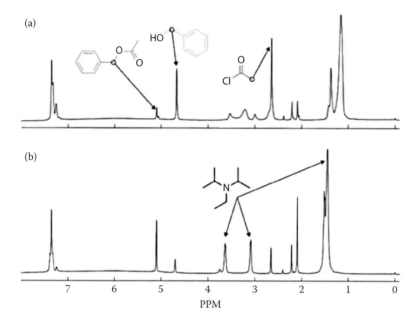

FIGURE 1.13 ^{1}H NMR spectra obtained via the *in situ* monitoring of the acetylation of benzyl alcohol **18** in the presence of the base ^{i}Pr$_2$EtN **38**: (a) reaction time = 9 s and (b) reaction time = 3 min. (Reproduced with permission from Bart, J. et al. 2009, *J. Am. Chem. Soc.* 131: 5014–5015. Copyright (2009) American Chemical Society.)

construct a response surface plot. From this, the authors were able to rapidly identify the optimal conditions to be a residence time of 5 min and a reactor temperature of 240°C. This example is particularly important as it demonstrates the ability to analyze reactants and products which may otherwise be indistinguishable by optical detection alone.

1.7.5 Development of Sensors for Process Monitoring

Along with spectroscopic and chromatographic techniques, the development of inexpensive in-line sensors is an area of significant interest with respect to the application of micro reaction technology within production sites. While at a research level it is

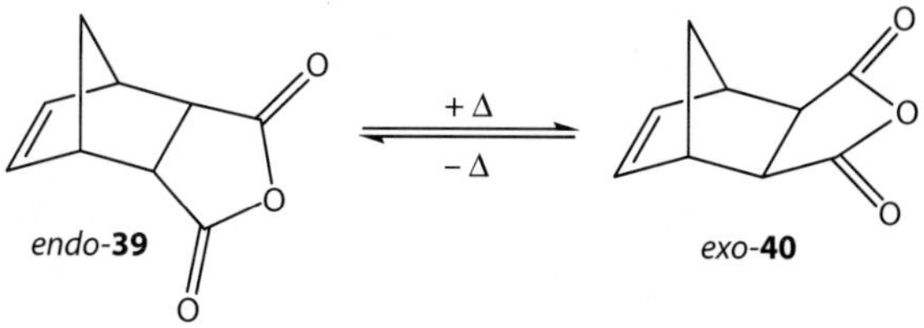

SCHEME 1.12 Illustration of the thermal isomerization of *endo*-bicyclo[2.2.1]hept-5-ene-2,3-dicarboxylic anhydride **39** to *exo*-bicyclo[2.2.1]hept-5-ene-2,3-dicarboxylic anhydride **40**.

important to be able to structurally characterize products and quantify components, at a production level more emphasis is placed on the ability to flag any changes in a process and diagnose when, where, and why these changes occurred.

With this in mind, several research groups have embarked upon the development and evaluation of sensors suitable for incorporation into micro flow processes to enable the assessment of physical parameters such as flow rate, temperature, and pressure.

Thermal Evaluation: Recognizing a need for in-line process monitoring and control, Jacobs and coworkers [102] reported the development of a pressure stable thermoelectric sensor, operated in a nonsteady-state/pulsed mode. Compared to the previous work in the area, the use of pulsed excitation of the sensor heater was found to enable improved distinction between system heat capacity and thermal conductivity; in addition, the proportion of heat transferred into the reaction system was minimized by this mode of operation. Using their thermoelectric flow sensor, the authors were able to discriminate between a series of common organic solvents, based on their measured thermal properties.

Multifaceted Detection: In a second development, Jacobs and coworkers [103] reported the construction of a thermoelectric flow and impedimetric sensor suitable for monitoring chemical conversion within flow reactors. This combination of sensing elements was selected as changes in dielectric/conductive properties, along with variations in the thermal conductivity/isobaric heat capacity of the fluid would enable the authors to monitor chemical conversions based on four physical parameters. To demonstrate the efficacy of their devices, the sensors were placed inside a stainless steel microfluidic chamber (dimensions = $6.0 \times 1.2 \times 0.22$ mm) and the steady-state response of a series of solvents was investigated. Using this approach, the authors were pleased to report the ability to distinguish between 1-butanol, 2-propanol, and EtOH over a flow rate range of 40–120 g h^{-1}. Based on these results, the authors concluded that their in-line sensors were suitable for detecting changes of <1% within a fluid stream.

Using a combination of sensing and chromatographic techniques, Ferstl and coworkers [104] constructed a modular flow micro reactor plant in which they synthesized 2-(4-chloro-3-nitrobenzoyl)benzoic acid via a single-phase nitration reaction. Defining their processing window as being the synthesis of kg-scale quantities of material, the authors developed and incorporated a series of sensing elements, including temperature (200°C), pressure (0–10 bar), and flow, into their system. The use of online Raman and IR spectroscopy enabled the authors to monitor product quality with time and, in addition, the construction of an online HPLC interface enabled the use of the system for process development as well as long-term process monitoring.

1.8 COMMERCIAL AVAILABILITY OF CONTINUOUS FLOW REACTOR TECHNOLOGY

From the examples provided, it can be seen that there are an array of materials, methods, and techniques available for the fabrication and use of micro reactors; however, if it is the chemistry that is of interest to you, rather than the design and

Introduction to Micro Reaction Technology

fabrication of a device, there are now several commercially available flow systems on the market.*

Conjure™ and Propel™: The Conjure system developed by Accedo Corporation (USA) [105] is based on the use of segmented flow and enables the users to automatically screen reaction conditions, in pursuit of faster, cleaner reaction methodology. Coupled with liquid chromatography–mass spectrometry (LC–MS) detection, the system enabled rapid identification of the materials generated (Figure 1.14). In addition, a walk-up multiple user system is also available, called the Propel flow reactor system enabling up to nine conditions to be investigated, and synthesis up to 100 g's (Figure 1.14).

NanoTek®: The NanoTek microfluidic synthesis system from Advion Biosciences Inc. (USA) [106] is a capillary-based platform suitable for the high-purity synthesis of multiple positron emission tomography (PET) compounds with typical reaction times in the range of minutes. In addition, the system has been used for conventional syntheses (see Chapter 3 for an example) demonstrating system versatility (Figure 1.15).

Labtrix-S1® and Plantrix®: The Labtrix system from Chemtrix BV (The Netherlands) [107] is a fully automated "plug and play" platform for the laboratory-based optimization of reactions (within standard or customized glass micro reactors)

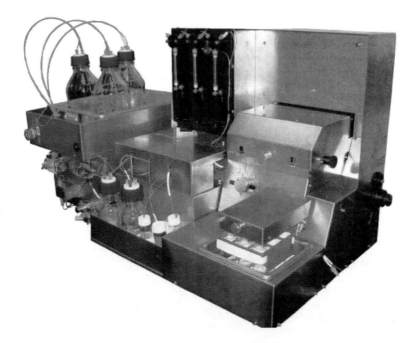

FIGURE 1.14 (See color insert) The Propel system from Accedo Corporation, USA. (Reproduced by permission of Accedo Corporation, 2010, Belmont, CA, USA.)

---

* At the time of going to press the systems described did not present an exhaustive list of those commercially available; the selection was made to provide the reader with an overview of the different types and functions of systems currently on the market.

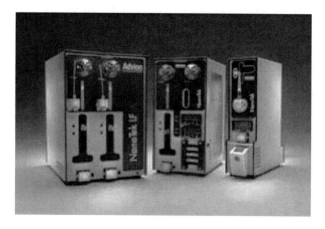

FIGURE 1.15 The NanoTek microfluidic synthesis system from Advion, USA. (Reproduced by permission of Advion Biosciences Inc., 2010, Ithyca, NY, USA.)

(a)
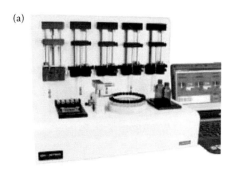

(b)
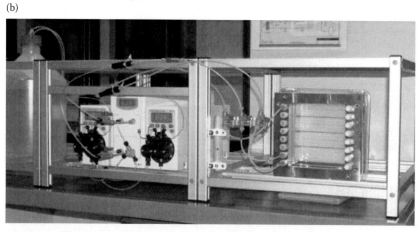

FIGURE 1.16 (**See color insert**) Glass-based flow reactor platforms: (a) Labtrix and (b) Plantrix from Chemtrix BV, The Netherlands. (Reproduced by permission of Chemtrix BV, 2010, Geleen, The Netherlands.)

Introduction to Micro Reaction Technology

generating milligrams of material with an operating range of −15 to 195°C (25 bar). Once reactions are optimized using Labtrix (Figure 1.16), production capacity can be increased to tonnes annum^{-1} using Plantrix, a mesoreactor platform designed for chemical production (Figure 1.16).

FlowStart and FlowScreen: FlowStart, from FutureChemistry (Njimegen, The Netherlands) [108], is a manually operated system for the evaluation of flow reactions within glass micro reactors over an operating range of 0–90°C, enabling the production of mg–g quantities of material per day (Figure 1.17). Combined with data set modeling, the FlowScreen system is an automated platform with which users can perform the rapid optimization of flow reactions over the previously stated thermal range (Figure 1.17).

Africa® and FRX Systems: The Africa flow reactor platform, from Syrris Ltd. (Royston, UK) [109], is a fully automated glass micro reactor flow system that enables the rapid optimization of reaction conditions and with overnight operation has the potential to generate g-scale quantities of materials (Figure 1.18). The FRX products

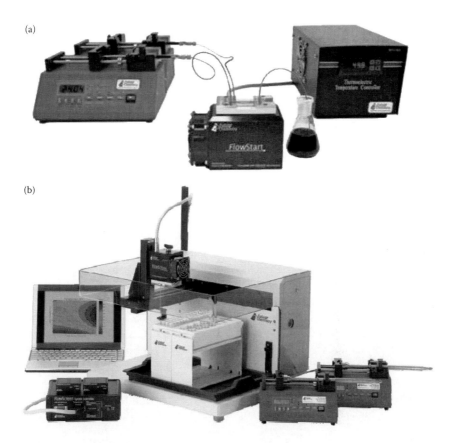

FIGURE 1.17 (See color insert) Glass-based flow reactor platforms: (a) FlowStart and (b) FlowScreen, from FutureChemistry BV, The Netherlands. (Reproduced by permission of FutureChemistry BV, 2010, Njimegen, The Netherlands.)

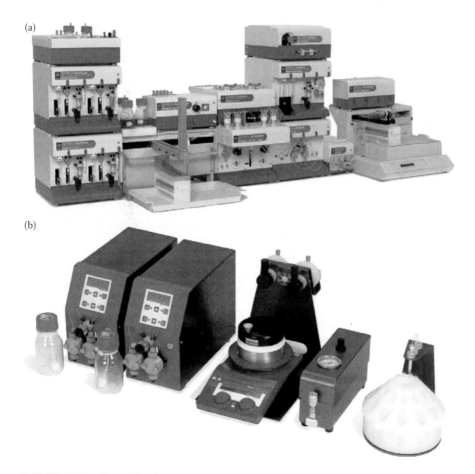

FIGURE 1.18 (See color insert) Glass-based flow reactor platforms: (a) AFRICA and (b) FRX system, from Syrris Ltd., UK. (Reproduced by permission of Syrris Ltd., 2010, Royston, UK.)

are a modular micro reactor system which is manually operated to enable the screening of liquid-phase reactions (Figure 1.18).

H-Cube®, X-Cube™, and O-Cube™: The family of flow reactors from ThalesNano Inc. [110] includes a continuous flow hydrogenator (H-Cube, 100°C and 100 bar), a heterogeneous catalyst flow reactor system (X-Cube, 200°C and 150 bar) and a continuous ozonolysis reactor (O-Cube, −25–25°C) for performing reactions at the lab scale. In addition, the suppliers also have systems suitable for process scale flow hydrogenations (H-Cube Midi™) (Figure 1.19).

FlowSyn: The FlowSyn system from Uniqsis, UK [111] is an integrated continuous flow reactor which enables the user to optimize reactions within a glass mesoreactor and produce chemicals using a series of tubular reactors fabricated from a range of chemically compatible materials. The system has an operating temperature of −40 to 260°C and also enables the user to employ packed-bed cartridges within the system (Figure 1.20).

Introduction to Micro Reaction Technology

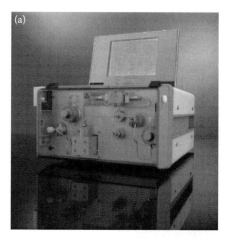
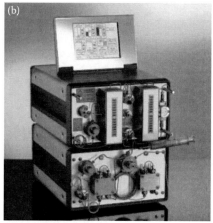
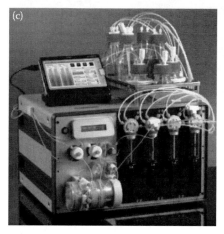

FIGURE 1.19 (See color insert) Tubular reactor systems: (a) H-Cube, (b) X-Cube, and (c) O-Cube from ThalesNano Inc., Hungary. (Reproduced by permission of ThalesNano Inc., 2010, Zahony, Hungary.)

R-Series: The R-series from Vapourtec Ltd., UK [112] is a meso-scale flow chemistry system which offers flexible configurations with an operating range of −70 to 250°C (50 bar). The system offers up to four reaction steps in tube (homogeneous) or column (heterogeneous) reactors, each independently temperature controlled (Figure 1.21).

In addition to the aforementioned system providers, there are also currently several micro reactor vendors specializing in standard glass and polymer devices, with some able to design and fabricate customized devices. Commercialized metal micro reactors are also available with modular systems enabling the user to select the number of fluidic inputs, mixers, and heated zones employed. At a nonresearch level, suppliers are also starting to offer flow reactor solutions for production-scale processes, with examples of the Corning SAS (France) glass mesoreactor platform described in Chapters 2 and 7.

FIGURE 1.20 (See color insert) FlowSyn, a tubular and mesoreactor system from Uniqsis, UK. (Reproduced by permission of Uniqsis Ltd., 2010, Shepreth, UK.)

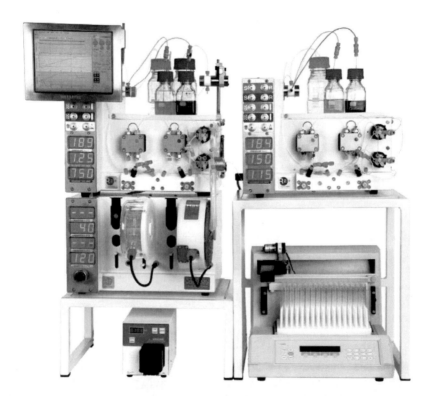

FIGURE 1.21 (See color insert) Vapourtec R-series, a flexible tubular flow reactor system from Vapourtec Ltd., UK. (Reproduced by permission of Vapourtec Ltd., 2010, Bury St Edmonds, UK.)

1.9 OUTLOOK

With the focus of micro reactor research moving from the proof of concept stage to more mainstream research activities, the outlook for the technology is of great interest. With the advent of commercially available flow reactor platforms, the technology can now be implemented in conventional research laboratories and harnessed as a tool for facile reaction optimization and the rapid production of small volumes of material. In addition, its use within academia is becoming more widespread with several academic institutions offering undergraduate degree courses with training in the use of this technology. While examples of production-scale use are not openly discussed within the scientific literature, searches of patents show that the application of MRT in a production environment is an area that is sure to undergo rapid growth in the coming years.

With this in mind, it is projected that MRT will move away from the use of test cases and into novel reaction and process development, finding application in all areas of synthetic chemistry, be it teaching, academic research, industrial research and development, process development to production.

REFERENCES

1. Ehrfeld, W., Hessel, V., and Löwe, H. 2000. *Microreactors New Technology for Modern Chemistry*, Germany: Wiley-VCH.
2. Hessel, V., Renken, A., Schouten, J. C., and Yoshida, J. (Eds.) 2009. *Micro Process Engineering: A Comprehensive Handbook Volume 2: Devices, Reactions and Applications*, Germany: Wiley-VCH.
3. Schouten, J. C. (Ed.) 2010. *Micro Systems and Devices for (Bio)Chemical Processes*, The Netherlands: Academic Press.
4. Hibara, I., Tokeshi, M., Uchiyama, K., Hisamoto, H., and Kitamori, T. 2001. Integrated multilayer flow system on a microchip, *Anal. Sci.* 17: 89–93.
5. Beato, P., Kraehnert, R., Engelschalt, S., Frank, T., and Schlögl, R. 2008. A microstructured quartz reactor for kinetic and *in-situ* spectroscopic studies in heterogeneous catalysis, *Chem. Eng. J.* 135S: S247–S253.
6. Adamschick, M., Hinz, M., Maier, C., Schmid, P., Seliger, H., Hofer, E. P., and Kohn, E. 2001. Diamond micro system for biochemistry, *Diamond Relat. Mater.* 10: 722–730.
7. Becker, H. and Gärtner, C. 2001. Polymer based micro-reactors, *Rev. Mol. Biotechnol.* 82: 89–99.
8. McCreedy, T. 2001. Rapid prototyping of blass and PDMS microstructures for micro total analytical systems and micro chemical reactors by microfabrication in the general laboratory, *Anal. Chim. Acta.* 427: 39–43.
9. Schudel, B. R., Choi, C. J., Cunningham, B. T., and Kenis, P. J. A. 2009. Microfluidic chip for combinatorial mixing and screening of assays, *Lab Chip* 9: 1676–1680.
10. Song, Y., Kumar, C. S. S. R., and Hormes, J. 2004. Fabrication of an SU-8 based microfluidic reactor on a PEEK substrate sealed by a "flexible semi-solid transfer" (FST) process, *J. Micromech. Microeng.* 14: 932–940.
11. Min, K.-I., Lee, T.-H., Park, C. P., Wu, Z.-Y., Girault, H. H., Ryu, I., Fukuyama, T., Mukai, Y., and Kim, D.-P. 2010. Monolithic and flexible polyimide film microreactors for organic microchemical applications fabricated by laser ablation, *Angew. Chem. Int. Edn* 49: 7063–7067.
12. Yoon, T.-H., Park, S.-H., Min, K.-I., Zhang, X., Haswell, S. J., and Kim, D.-P. 2008. Novel inorganic polymer derived microreactors for organic microchemistry applications, *Lab Chip* 8: 1454–1459.

13. Ueno, K., Kim, H., and Kitamura, N. 2003. Characteristic electrochemical responses of polymer microchannel–microelectrode chips, *Anal. Chem.* 75: 2086–2091.
14. Arana, L. R., de Mas, N., Schmidt, R., Franz, A. J., Schmidt, M. A., and Jensen, K. F. 2007. Isotropic etching of silicon in fluorine gas for MEMS micromachining, *J. Micromech. Microeng.* 17: 384–392.
15. De la Iglesia, O., Sebastián, V., Mallada, R., Nikolaidis, G., Coronas, J., Kolb, G., Zapf, R., Hessel, V., and Santamaría, J. 2007. Preparation of Pt/ZSM-5 films on stainless steel microreactors, *Catal. Today* 125: 2–10.
16. De Mas, N., Günther, A., Kraus, T., Schmidt, M. A., and Jensen, K. F. 2005. Scales-out multilayer gas–liquid microreactor with integrated velocimetry sensors, *Ind. Eng. Chem. Res.* 44: 8997–9013.
17. Okafor, O. C., Tadepalli, S., Tampy, G., and Lawal, A. 2010. Microreactor performance studies of the cycloaddition of isoamylene and α-methylstyrene, *Ind. Eng. Chem. Res.* 49: 5549–5560.
18. Tiggelaar, R. M., Benito-Lópex, F., Hermes, D. C., Rathgen, H., Egberink, R. M., Mugele, F. C., Reinhoudt, D. N., van den Berg, A., Verboom, W., and Gardeniers, H. J. G. E. 2007. Fabrication, mechanical testing, and application of high-pressure glass microreactor chips, *Chem. Eng. J.* 131: 163–170.
19. Trachsel, F., Hutter, C., and von Rohr, P. R. 2008. Transparent silicon/glass microreactor for high-pressure and high-temperature reactions, *Chem. Eng. J.* 135S: S309–S316.
20. Rodriguez, I., Spicar-Mihalic, P., Kuyper, C. L., Fiorini, G. S., and Chiu, D. T. 2003. Rapid prototyping of glass microchannels, *Anal. Chim. Acta.* 496: 205–215.
21. Oosterbroek, R. E., Hermes, D. C., Kakuta, M., Benito-Lopez, F., Gardeniers, J. G. E., Verboom, W., Reinhoudt, D. N., and van den Berg, A. 2006. Fabrication and mechanical testing of glass chips for high-pressures synthetic or analytical chemistry, *Microsyst. Technol.* 12: 450–454.
22. Fletcher, P. D. I., Haswell, S. J., Pombo-Villar, E., Warrington, B. H., Watts, P., Wong, S. Y. F., and Zhang, Z. 2002. Micro reactors: Principles and applications in organic synthesis, *Tetrahedron* 58: 4735–4757.
23. Jensen, K. F. 2006. Silicon-based microchemical systems: Characteristics and applications, *MRS Bull.* 31: 101–107.
24. Murphy, E. R., Inoue, T., Sahoo, H. R., Zaborenko, N., and Jensen, K. F. 2007. Solder-based chip-to-tune and chip-to-chip packaging for microfluidic devices, *Lab Chip* 7: 1309–1314.
25. Lee, D.-S., Yang, H., Chung, K.-H., and Pyo, H.-B. 2005. Wafer-scale fabrication of polymer-based microdevices via injection molding and photolithographic micropatterning protocols, *Anal. Chem.* 77: 5414–5420.
26. Attia, U. M., Marson, S., and Alcock, J. R. 2009. Micro-injection moulding of polymer microfluidic devices, *Microfluid. Nanofluid.* 7: 1–28.
27. Zhou, W. X. and Chan-Park, M. B. 2005. Large area UV casting using diverse polyacrylates of microchannels separated by high aspect ratio microwalls, *Lab Chip* 5: 512–518.
28. Alonso-Amigo, M. G. 2000. Polymer microfabrication for microarrays, microreactors and microfluidics, *J. Assoc. Lab. Auto.* 5: 96–101.
29. Hallmark, B., Mackley, M. R., and Gadala-Maria, F. 2005. Hollow microcapillary arrays in thin plastic films, *Adv. Eng. Mater.* 7(6): 545–547.
30. Hornung, C. H., Mackley, M. R., Baxendale, I. R., and Ley, S. V. 2007. A microcapillary flow disk reactor for organic synthesis, *Org. Proc. Res. Dev.* 11: 399–405.
31. Lee, J. N., Park, C., and Whitesides, G. M. 2003. Solvent compatibility of poly(dimethylsiloxane)-based microfluidic devices, *Anal. Chem.* 75: 6544–6554.
32. Wang, Z., Hira, S., and Yoshioka, M. 2010. Effect of process parameters on polymer flow by hot embossing, *11th International Conference on Microreaction Technology*, Kyoto, Japan, pp. 286–287.

33. Yoon, T.-H., Hong, L.-H., Park, S.-H., Min, K.-I., Park, S.-J., and Kim, D.-P. 2008. Novel inorganic polymer derived microreactors for the application of organic microchemical synthesis, *12th International Conference on Miniaturized Systems for Chemistry and Life Sciences*, San Diego, USA, pp. 1242–1244.
34. Brandner, J. J., Bohn, L., Schygulla, U., Wenka, A., and Schubert, K. 2001. *Microreactors: Epoch-making Technology for Synthesis*, J. Yoshida (Ed.). Japan: CMC Publishing Company.
35. Brandner, J. J. 2008. Fabrication of microreactors made from metals and ceramic, *Microreactors in Organic Synthesis and Catalysis*, T. Wirth (Ed.). Germany: Wiley-VCH, pp. 1–17.
36. Jain, K., Wu, C., Atre, S. V., Jovanovic, G., Narayanan, V., Kimura, S., Sprenkle, V., Canfield, N., and Roy, S. 2009. Synthesis of nanoparticles in high temperature ceramic microreactors: Design, fabrication and testing, *Int. J. Appl. Ceram. Technol.* 6: 410–419.
37. Knitter, R., Göhring, D., Risthaus, P., and Haußlet, J. 2001. Microfabrication of ceramic microreactors, *Microsystem Technol.* 7: 85–90.
38. Müller, M., Bauer, W., and Ritzhaupt-Kleissl, H.-J. 2008. Low-pressure injection molding of ceramic micro devices using sub-micron and nanoscaled powders, *Multi-material Micro Manuf.* 1–4.
39. Aran, H. C., Chinthaginjala, J. K., Salamon, D., Lammertink, R. G. H., Lefferts, L., and Wessling, M. 2010. Porous ceramic microreactors for multi-phase reaction systems, *11th International Conference on Microreaction Technology*, Kyoto, Japan, pp. 26–27.
40. Pattekar, A. V. and Kothare, M. V. 2003. Novel microfluidic interconnectors for high temperature and pressure applications, *J. Micromech. Microeng.* 13: 337–345.
41. Unnikrishnan, S., Jansen, H., Berenschot, E., Mogulkoc, B., and Elwenspoek, M. 2009. MEMS within a Swagelok®: A new platform for microfluidic devices, *Lab Chip* 9: 1966–1969.
42. Fredrickson, C. K. and Fan, Z. H. 2004. Macro-to-micro interfaces for microfluidic devices, *Lab Chip* 4: 526–533.
43. Shintani, Y., Hirako, K., Motokawa, M., Takano, Y., Furuno, M., Minakuchi, H., and Ueda, M. 2004. Polydimethylsiloxane connection for quartz microchips in a high-pressure system, *Anal. Sci.* 20: 1721–1723.
44. Upchurch Scientific Inc., Oak Harbor, WA, USA (http://webstore.idex-hs.com).
45. Kralj, J. G., Sahoo, H. R., and Jensen, K. F. 2007. Integrated continuous microfluidic liquid–liquid extraction, *Lab Chip* 7: 256–263.
46. Chen, Z., Wang, P., and Chang, H.-C. 2005. An electroosmotic micro-pump based on monolithic silica for micro-flow analyses and electro-sprays, *Anal. Bioanal. Chem.* 382: 817–824.
47. Dasgupta, P. K. and Liu, S. 1994. Electroosmosis: A reliable fluid propulsion system for flow injection analysis, *Anal. Chem.* 66: 1792–1798.
48. Comandur, K. A., Bhagat, A. A. S., Dasgupta, S., Papautsky, I., and Banerjee, R. K. 2010. Transport and reaction of nanoliter samples in a microfluidic reactor using electro-osmotic flow, *J. Micromech. Microeng.* 20: 1–12.
49. Gervais, T. and Jensen, K. F. 2006. Mass transport and surface reactions in microfluidic systems, *Chem. Eng. Sci.* 61: 1102–1121.
50. Kockmann, N. 2008. Pressure loss and transfer rates in microstructured devices with chemical reactions, *Chem. Eng. Technol.* 31: 1188–1195.
51. Coti, K. K., Wang, Y., Lin, W.-Y., Chen, C.-C., Yu, Z. T. F., Liu, K., Shen, C. K.-F. et al. 2008. A dynamic micromixer for arbitrary control of disguised chemical selectivity, *Chem. Commun.* 3426–3428.
52. Hartman, R. L. and Jensen, K. F. 2009. Microchemical systems for continuous flow synthesis, *Lab Chip* 9: 2495–2507 and references cited herein.

53. Kashid, M. N., Gerlach, I., Goetz, S., Franzke, J., Acker, J. F., Platte, F., Agar, D. W., and Turek, S. 2005. Internal circulation within the liquid slugs of a liquid–liquid slug-flow capillary microreactor, *Ind. Eng. Chem. Res.* 44: 5003–5010.
54. Doku, G. N., Verboon, W., Reinhoudt, D. N., and van den Berg, A. 2005. On-chip multiphase chemistry—A review of microreactor design principles and reagent contacting modes, *Tetrahedron* 61: 2733–2742.
55. Günther, A., Jhunjhunwala, M., Thalmann, M., Schmidt, M. A., and Jensen, K. F. 2005. Micro-mixing of miscible liquids in segmented gas–liquid flow, *Langmuir* 21: 1547–1555.
56. Shui, L., Eijkel, J. C. T., and van den Berg, A. 2007. Multiphase flow in microfluidic systems—Control and applications of droplets and interfaces, *Adv. Coll. Inter. Sci.* 133: 35–49.
57. Illg, T., Lob, P., and Hessel, V. 2010. Flow chemistry using milli- and microstructured reactors—From conventional to novel process windows, *Bioorg. Med. Chem.* 18: 3707–3719.
58. Lin, W.-Y., Wang, Y., Wang, S., and Tseng, H.-R. 2009. Integrated microfluidic reactors, *Nano Today* 4: 470–481.
59. Verboom, W. 2009. Selected examples of high-pressure reactions in glass microreactors, *Chem. Eng. Technol.* 32(11): 1695–1701.
60. Benito-Lopez, F., Verboom, W., Kakuta, M., Gardeniers, J. H. G. E., Egberink, R. J. M., Oosterbroek, E. R., van den Berg, A., and Reinhoudt, D. N. 2005. Optical fiber-based online UV/Vis spectroscopic monitoring of chemical reaction kinetics under high pressure in a capillary microreactor, *Chem. Commun.* 2857–2859.
61. Razzaq, T., Glasnov, T. N., and Kappe, C. O. 2009. Continuous-flow microreactor chemistry under high-temperature/pressure conditions, *Eur. J. Org. Chem.* 1321–1325.
62. Razzaq, T. and Kappe, C. O. 2010. Continuous flow organic synthesis under high temperature/pressure conditions, *Chem. Asian J.* 5: 1274–1289.
63. Darvas, F., Dormán, G., Lengyel, L, Kovács, I., Jones, R., and Ürge, L. 2009. High pressure, high temperature reactions in continuous flow: Merging discovery and process chemistry, *Chimica Oggi* 27(3): 40–43.
64. Kusakabe, K., Morooka, S., and Maeda, H. 2001. Development of a microchannel catalytic reactor system, *Korean J. Chem. Eng.* 18(3): 271–276.
65. Fortt, R., Wootton, R. C. R., and de Mello, A. J. 2003. Continuous-flow generation of anhydrous diazonium species: Monolithic microfluidic reactors for the chemistry of unstable intermediates, *Org. Proc. Res. Dev.* 7(5): 762–768.
66. Glasnov, T. N. and Kappe, C. O. 2007. Microwave-assisted synthesis under continuous flow conditions, *Macromol. Rapid Commun.* 28: 395–410.
67. Moseley, J. D., Lenden, P., Lockwood, M., Ruda, K., Sherlock, J. P., Thomson, A. D., and Gilday, J. P. 2008. A comparison of commercial microwave reactors for scale-up within process chemistry, *Org. Proc. Res. Dev.* 12: 30–40.
68. Matsuoka, S., Hibara, A., Ueno, M., and Kitamori, T. 2006. Supercooled micro flows and application for asymmetric synthesis, *Lab Chip* 6: 1236–1238.
69. Roberge, D. M., Gottsponer, M., and Eyholzer, M. 2010. A test reaction to understand hot-spot formation and heat exchange in microreactors, *11th International Conference on Microreaction Technology*, Kyoto, Japan, pp. 52–53.
70. Roberge, D. M., Bieler, N., Mathier, M., Eyholzer, E., Zimmermann, B., Barthe, P., Guermeur, C., Lobet, O., Moreno, M., and Woehl, P. 2008. Development of an industrial multi-injection microreactor for fast and exothermic reactions—Part II, *Chem. Eng. Technol.* 31(8): 1155–1161.
71. Lerou, J. J., Tonkovich, A. L., Silva, L., Perry, S., and McDaniel, J. 2010. Microchannel reactor architecture enables greener processes, *Chem. Eng. Sci.* 65: 380–385.

72. Benito-Lopez, F., Tiggleaar, R. M., Salbut, K., Huskens, J., Eberbrink, R. J. M., Reinhoudt, D. N., Gardeniers, H. J. G. E., and Verboom, W. 2007. Substantial rate enhancements of the esterification reaction of phthalic anhydride with methanol at high pressure and using supercritical CO_2 as a co-solvent in a glass microreactor, *Lab Chip* 7: 1345–1351.
73. Ebrahimi, F., Kolehmainen, E., and Turunen, I. 2009. Safety advantages of on-site microprocesses, *Org. Proc. Res. Dev.* 13: 965–969.
74. Ajmera, S. K., Losey, M. W., Jensen, K. F., and Schmidt, M. A. 2001. Microfabricated packed-bed reactor for phosgene synthesis, *AIChE J.* 47: 1639–1647.
75. Poe, S. L., Cummings, M. A., Haaf, M. P., and McQuade, D. T. 2006. Solving the clogging problem: Precipitate-forming reactions in flow, *Angew. Chem. Int. Edn.* 45: 1544–1548.
76. Buisson, B., Donegan, S., Wray, D., Parracho, A., Gamble, J., Caze, P., Jorda, J., and Guermeur, C. 2009. Slurry hydrogenation in a continuous flow reactor for pharmaceutical application, *Chem. Today* 27: 12–14.
77. Nagaki, A., Tomida, Y., and Yoshida, J. 2008. Microflow-system-controlled anionic polymerization of styrenes, *Macromolecules* 41: 6322–6330.
78. Zhao, H., Wang, J.-X., Wang, Q.-A., Chen, J.-F., and Yun, J. 2007. Controlled liquid antisolvent precipitation of hydrophobic pharmaceutical nanoparticles in a microchannel reactor, *Industrial Engineering and Chemical Research* 46: 8229–8235.
79. Hornung, C. H. and Mackley, M. R. 2009. The measurement and characterization of residence time distributions for laminar flow in plastic microcapillary arrays, *Chem. Eng. Sci.* 64: 3889–3902.
80. Roberge, D. M., Gottsponer, M., Eyholzer, M., and Kockmann, N. 2009. Industrial design, scale-up, and use of microreactors, *Chem. Today* 27: 8–11.
81. Lohse, S., Kohnen, B. T., Janasek, D., Dittrich, P. S., Franzke, J., and Agar, D. W. 2008. A novel method for determining residence time distribution in intricately structured microreactors, *Lab Chip* 8: 431–438.
82. Saber, M., Commenge, J.-M., and Falk, L. 2010. Experimental measurement of the flow distribution through multi-scale microstructured reactors, *11th International Conference on Microreaction Technology*, Kyoto, Japan, pp. 54–55.
83. Stange, O., Herbstritt, F., Kroschel, M., Jahn, P., Raffa, C., and Schael, F. 2010. Modular microreaction technology—The scale-up from lab to production, *11th International Conference on Microreaction Technology*, Kyoto, Japan, pp. 96–97.
84. Amador, C., Gavriilidis, A., and Angeli, P. 2004. Flow distribution in different microreactor scale-out geometries and the effect of manufacturing tolerances and channel blockage, *Chem. Eng. J.* 101: 379–390.
85. Tonomura, O., Tominari, T., Kano, M., and Hasebe, S. 2008. Operation policy for micro chemical plants with external numbering-up structure, *Chem. Eng. J.* 135S: S131–S137.
86. Knockmann, N., Gottsponer, M., and Roberge, D. 2010. Scale-up concept of single-channel microreactors from process development to industrial production, *11th International Conference on Microreaction Technology*, Kyoto, Japan, pp. 94–95.
87. Sotowa, K.-I., Sugiyama, S., and Nakagawa, K. 2009. Flow uniformity in deep microchannel reactor under high throughput conditions, *Org. Proc. Res. Dev.* 13: 1026–1031.
88. Weiler, A., Wille, G., Kaiser, P., and Wahl, F. 2009. Micro reactors offer improved solutions for liquid multipurpose bulk fine chemical production, *Chem. Today* 27: 38–39.
89. Mechtilde, S., Eduard, S., and Andreas, F. 2006. Controlled chemical micro-reactor, *J. Phys. Conf. Ser.* 28: 115–118.
90. Nitta, K., Tanaka, T., and Okada, Y. 2010. Measurement of reaction intermediates in microreactors, *11th International Conference on Microreaction Technology*, Kyoto, Japan, pp. 232–233.
91. Cao, E., Firth, S., McMillan, P. F., Meenakshisundaram, S., Bethell, D., Hutchings, G. J., and Gavriilidis, A. 2010. Raman and reaction studies of an alcohol oxidation

on gold catalysts in microstructured reactors, *11th International Conference on Microreaction Technology*, Kyoto, Japan, pp. 122–123.
92. Urakawa, A., Trachsel, F., Rudolf von Rohr, P., and Baiker, A. 2008. On-chip Raman analysis of heterogeneous catalytic reaction in supercritical CO_2: Phase behavior monitoring and activity profiling, *Analyst* 133: 1352–1354.
93. Floyd, T. M., Schmidt, M. A., and Jensen, K. F. 2005. Silicon micromixers with infrared detection for studies of liquid-phase reaction, *Ind. Eng. Chem. Res.* 44: 2351–2358.
94. Bentrup, U., Küpper, L., Budde, U., Lovis, K., and Jähnisch, K. 2006. Mid-infrared monitoring of gas–liquid reactions in a vitamin D analogue synthesis with novel fiber optical diamond ATR sensor, *Chem. Eng. Technol.* 29(10): 1216–1220.
95. Hübner, S., Bentrup, U., Budde, U., Lovis, K., Dietrich, T., Freitag, A., Küpper, L., and Jähnisch, K. 2009. An ozonolysis–reduction sequence for the synthesis of pharmaceutical intermediates in microstructured devices, *Org. Proc. Res. Dev.* 952–960.
96. Carter, C. F., Baxendale, I. R., OBrien, M., Pavey, J. B. J., and Ley, S. V. 2009. Synthesis of acetal protected building blocks using flow chemistry with flow I.R. analysis: Preparation of butane-2.3-diacetal tartrates, *Org. Biomol. Chem.* 7: 4594–4597.
97. Carter, C. F., Lange, H., Ley, S. V., Baxendale, I. R., Wittkamp, B., Goode, J. G., and Gaunt, N. L. 2010. ReactIR flow cell: A new analytical tool for continuous flow chemical processing, *Org. Proc. Res. Dev.*, 14: 393–404.
98. Baumann, M., Baxendale, I. R., Martin, L. J., and Ley, S. V. 2009. Development of fluorination methods using continuous-flow microreactors, *Tetrahedron* 65: 6611–6625.
99. Tanaka, M., Isobe, M., Yamada, H., and Tagawa, T. 2010. Evaluation of the catalyst prepared in microchannel with *in-situ* FT-IR microscopy, *11th International Conference on Microreaction Technology*, Kyoto, Japan, pp. 248–249.
100. Bart, J., Kolkman, A. J., Oosthoek-de Vries, A. J., Koch, K., Nieuwland, P. J., Janssen, H. J. W. G., van Bentum, J. P. J. M. et al. 2009. A microfluidic high-resolution NMR flow probe, *J. Am. Chem. Soc.* 131: 5014–5015.
101. Welch, C. J., Gong, X., Cuff, J., Dolman, S., Nyrop, J., Lin, F., and Rogers, H. 2009. Online analysis of flowing streams using microflow HPLC, *Org. Proc. Res. Dev.* 13: 1022–1025.
102. Jacobs, T., Kutzner, C., Kropp, M., Lang, W., Kienle, A., and Huptmann, P. 2009. Novel pressure stable thermoelectric flow sensor in non-steady state operation mode for in-line process analysis in micro reactors, *Procedia Chem.* 1: 148–151.
103. Jacobs, T., Kutzzner, C., Kropp, M., Brokmann, G., Lang,, W., Steinke, A., Kienle, A., and Hauptmann, P. 2009. Combination of a novel perforated thermoelectric flow and impedimetric sensor for monitoring chemical conversion in micro fluidic channels, *Procedia Chem.* 1: 1127–1130.
104. Ferstl, W., Loebbecke, S., Antes, J., Krause, H., Haeberl, M., Schmalz, D., Muntermann, H. et al. 2004. Development of an automated microreaction system with integrated sensorics for process screening and production, *Chem. Eng. J.* 101: 431–438.
105. Accendo Corporation, Northridge, CA (http://www.accendocorporation.com).
106. Advion Biosciences Inc., Ithyca, NY, USA (http://www.advion.com/biosystems/nanotek/nanotek-positron-emission-tomography.php).
107. Chemtrix BV, Geleen, NL (http://www.chemtrix.com).
108. Future Chemistry, Njimegen, NL (http://www.futurechemistry.com).
109. Syrris, Royston, UK (http://www.syrris.com).
110. Thales Nano, Hungary (http://www.thalesnano.com).
111. Uniqsis, Shepreth, UK (http://www.uniqsis.com).
112. Vaportec, Fornham St. Genevieve, UK (http://www.vapourtec.com).
113. de Mello, A. J. 2001. A high-pressure interconnect for chemical microsystem applications, *Lab Chip* 1: 148–152.
114. Kitamori, T. 2009. Parallel multiphase microflows: fundamental physics, stabilisation methods and applications, *Lab Chip* 9: 2470–2476.

2 Micro Reactions Employing a Gaseous Component

With the vast majority of gas-phase reactions requiring the presence of a catalyst, usually supplied in the form of pellets or coatings, it stands to reason that these reactions would be dominated by mass transfer effects. In addition, many heterogeneously catalyzed gas-phase reactions are highly exothermic and it is therefore, imperative that any large releases of heat be effectively managed in order to maintain a safe and controlled process.

When considering conventional reactor methodology, it has been shown that poor mass transport and the presence of thermal gradients can impact the product distribution obtained. With these two factors in mind, it was believed that gas-phase reactions had a lot to gain from micro reaction technology. While gas-phase reactions are largely employed for the preparation of raw materials and feedstocks, from a synthetic organic perspective the chemistries performed using these materials are of greater interest and as such only a brief overview of gas-phase micro reactions is presented.

Observations made by pioneers in the field, however, led to an increase in the number of researchers investigating the use of micro reaction technology, which soon spread from chemical engineers to synthetic chemists who harnessed the advantages shown and applied them toward more routine liquid–liquid-phase reactions (Chapters 3, 4, and 7).

With this in mind, this chapter provides examples of gas-phase transformations however, the majority of examples focus on multiphase reactions using gases as a reactant source. For recent reviews on the subject of gas-phase reactions within micro reactor devices, refer to Jähnisch et al. [1], Kashid and Kiwi-Minsker [2], Kolb and Hessel [3], and Löb et al. [4], along with papers by Ehrfeld and coworkers [5].

2.1 GAS-PHASE MICRO REACTIONS

When considering the fabrication of a gas-phase micro reactor, one of the first challenges is how to incorporate the catalytic material. One of the most widely used approaches involves the use of stacked, coated plates bonded to afford a single reactor unit. Using this arrangement, the plate material itself can be the catalytic surface or a catalyst can be deposited using a range of techniques such as oxidation of the bulk material, coating via deposition or wet impregnation [6,7]. In addition, packed-bed, wall-coated, and tubular fixed-bed reactors can also be applied, with microstructured examples designed to facilitate large-scale production [8]. At the turn of the century, several research groups evaluated these approaches with a view to identifying the practical advantages associated with the use of micro reaction technology in gas-phase transformations.

Controlled Reaction of H_2 and O_2: When hydrogen and oxygen are reacted together at room temperature, ignition is observed within conventional fixed-bed reactors resulting in a risk of explosion. Performing the reaction in a silicon catalytic wire reactor (dimensions = 300 μm (wide) × 525 μm (deep) × 20 mm (long)), Zengerle and coworkers [9] demonstrated the ability to mix hydrogen and oxygen without ignition until the reactor temperature reached 100°C, illustrating the excellent mass and heat transfer properties of such devices. Upon ignition within the micro reactor, the temperature increased to 300°C, which was still able to be controlled, and demonstrated no damage to the structure of the device.

More recently, Jensen et al. [10] demonstrated the use of this approach for the direct synthesis of hydrogen peroxide from hydrogen and oxygen. Using a 10-channel micro reactor fabricated from silicon, illustrated in Figure 2.1, the authors were able to distribute the gases evenly ensuring good contact with the heterogeneous Pd catalysts packed within the device. Using a solution of sulfuric acid, phosphoric acid, and sodium bromide, the authors investigated the role of bromine as a stabilizer during the synthesis of hydrogen peroxide. In order to distinguish water formed during the reaction with that from the reactant solution, hydrogen was added in deuterium and oxygen in nitrogen. Operating within a conventionally explosive regime (2–3 MPa), the authors were able to obtain significantly enhanced mass transfer (3.8) compared to conventional reactors when the inlet zone of the reactor was packed with inert silica particles.

2.1.1 Gas-Phase Oxidations

An early example of a gas-phase oxidation was reported by Hönicke and Wießmeier [11], who demonstrated the partial oxidation of propene **1** to acrolein **2** (Scheme 2.1).

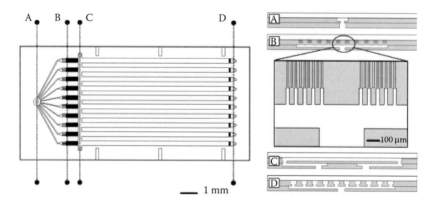

FIGURE 2.1 Schematic illustrating the 10-channel silicon reactor used for the direct synthesis of H_2O_2 from hydrogen and oxygen feeds. (Reproduced with permission from Inoue, T., Schmidt, M. A., and Jensen, K. F. 2007. *Ind. Eng. Chem. Res.* 46: 1153–1160. Copyright (2007) American Chemical Society.)

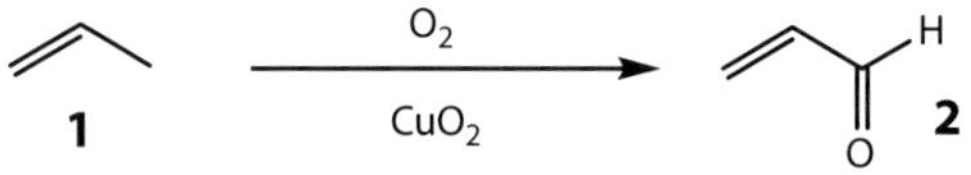

Micro Reactions Employing a Gaseous Component

SCHEME 2.1 Illustration of an early gas-phase reaction performed using a CuO_2 heterogeneous catalyst.

Using a copper plate reactor with Cu_2O as the active surface, the authors evaluated the effect of reaction temperature (350–375°C) on the oxidation of 1% propene **1** in N_2; containing bromomethane as a stabilizer. Supplying the reactor with a cold reactant stream, the authors demonstrated the ability to rapidly heat the reactant mixture to 300°C over a channel length of only 5 µm, enabling the fabrication of short reaction channels reducing the associated pressure drop. While this early example proved to be of limited industrial use due to the sensitive nature of the catalyst surface to O_2, the authors did demonstrate the potential of this mode of operation and helped to open a new area of research.

Oxidation of Ethylene: Using the synthesis of ethylene oxide, selected due to the fact that the process is well characterized on the macroscale, Ehrfeld and Löwe [12] developed a micro reactor comprising of gold-plated mixing foils and silver catalyst layers contained within a stainless-steel housing to afford 50 µm deep channels. Multilamination of the inlet gas streams of oxygen and ethylene ensured efficient mixing by diffusion, which was followed by reaction within the catalyst zone. Preliminary examples illustrated insufficient stability of the interconnections between the layers when reactions were performed at temperatures in excess of 300°C. The investigation did however, bolster interest in this area of process development, with many research groups subsequently investigating this reaction, including Kestenbaum et al. [13,14].

Oxidation of 1-Butene: Maleic anhydride **3** is a synthetically important precursor, used as a potent dienophile in the Diels–Alder reaction, as a ligand for metal complexes and in the preparation of polyimides; which feature in a rich array of material applications. When the oxidation of 1-butene **4** to maleic anhydride **3** is performed using conventional fixed-bed technology, the highly exothermic nature of the reaction poses a safety risk. With this in mind, Hönicke et al. [15] demonstrated the fabrication of a silicon reactor impregnated with anodically oxidized aluminum wires. Using this approach, the authors were able to obtain maleic anhydride **3** with a selectivity of 30% and a total conversion of 87% (Scheme 2.2).

The gas-phase synthesis of maleic anhydride **3** was also investigated by Dalmon and coworkers [16] using *n*-butane as the raw material. Employing vanadium

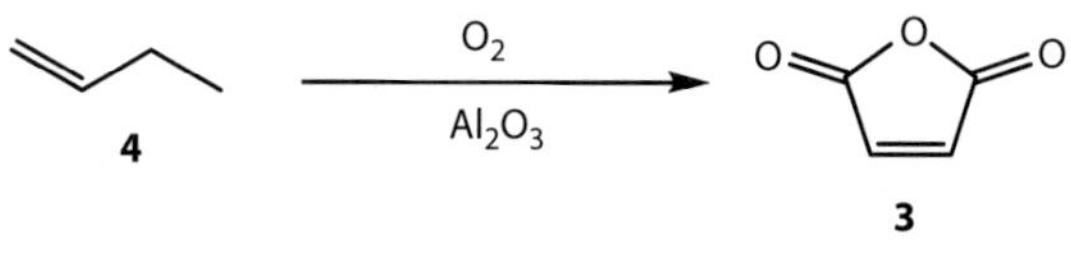

SCHEME 2.2 Schematic illustrating the selective oxidation of 1-butene **4** to maleic anhydride **3**.

phosphorus mixed oxide catalysts within a membrane reactor, the authors investigated reaction parameters such as reactor temperature (397°C) and feed configuration. Using this approach the authors identified that segregating the feeds of oxygen and *n*-butane, but mixing prior to entering the catalytic region of the reactor, afforded 68% selectivity toward the formation of maleic anhydride **3**.

Liu and coworkers [17] subsequently demonstrated the fabrication of a fixed-bed micro reactor, containing Pd-supported catalysts for the selective oxidation of 1-butene **4** to butanone **5**. Investigating a series of supported catalysts, prepared using an impregnation technique, the authors operated their reactor at atmospheric pressure and analyzed the condensed reaction products by GC-FID. As Table 2.1 illustrates, variable product distributions were obtained depending upon the catalyst employed, with Pd–Fe–HCl/Al–Ti **6** proving to be the best catalyst of those evaluated.

Kinetic Studies of Oxidations in Micro Reactors: Using the oxidation of EtOH, *n*-butane and unsymmetrical dimethylhydrazine, Ismagilov and coworkers [18] developed a microstructured catalytic reactor, comprising of 63 catalytically coated plates containing cylindrical micro channels (dimensions = 208 µm (radius)), for an investigation into reaction kinetics. Due to the high temperature and concentration uniformity obtained within the system, the authors reported the ability to function in a kinetic rate controlling regime over a wide temperature range. Using this approach, the authors were able to study the reaction kinetics of the deep oxidation of the target analytes over a thermal range of 150–375°C, identifying all intermediate products formed. Rate constants and apparent activation energies were also derived using kinetic modeling, illustrating the potential that microstructured reactors have as both tools for chemical production and for the gathering of chemical information.

Preferential Carbon Monoxide Oxidation: In addition to the examples discussed, Kim and coworkers [19] recently reported the highly selective preferential carbon monoxide oxidation, within a silicon-based micro reactor. The authors targeted this

TABLE 2.1
Summary of the Results Obtained for the Pd-Catalyzed Oxidation of 1-Butene 4 Performed in a Micro Reactor

Catalyst	Conversion (%)	Product Distribution (%)					
		A	B	C	D 5	E	F
Pd–Fe–HCl/SiO_2	1.2	0	21.5	16.4	48.8	13.3	0
Pd–Fe–HCl/TiO_2	0.8	2.0	11.9	0	61.3	19.7	3.9
Pd–Fe–HCl/Al_2O_3	10.8	4.5	2.7	0	81.1	8.0	3.7
Pd–Fe–HCl/Al-Ti	16.1	2.0	2.0	0	83.0	10.6	2.5
Pd–Fe–HCl/MCM-22	7.8	3.1	1.8	0	7.5	87.6	0

A = CH_3CHO, B = EtOH, AcOH, acetone, C = $CH_3(CH_2)_2COH$, D = $CH_3COCH_2CH_3$ **5**, E = CH_3CH_2COOH, and F = $CH_3(CH_2)_2COH_3$.

reaction as a means of developing a continuous purification method for CO removal from H_2-rich reformed gas, enabling the use of MeOH, EtOH, or methane as storable sources of hydrogen for fuel cell operation. Employing Pt–Al_2O_3 as the catalyst, housed within a packed-bed, the authors investigated the effect of channel geometry and reactor temperature on the CO oxidation of a synthetic feedstock, selected to mimic that formed upon steam reforming of alcohols to H_2. Using this approach, the authors identified 99.4% conversion of CO and 44.1% selectivity for CO at 260°C; finding that channel length, not width, played an important role in the success of the reactors. Other authors to investigate the purification of gaseous feedstock for fuel cell applications include Ye and coworkers [20], who evaluated the hydrogenation of 1-butene **4** using Pd-membrane reactors.

2.1.2 Hydrogenation Reactions within Microstructured Reactors

Partial Hydrogenation of Benzene: Hönicke and Wießmeier [11] also reported the partial hydrogenation of benzene **7** (Scheme 2.3) demonstrating the ability to isolate highly reactive intermediates such as cyclohexene **8** which is not normally possible when using conventional reactor methodology. Employing a Ru/Al_2O_3 catalyst formed within the micro channels (dimensions = 200 μm (wide) × 200 μm (deep) × 50 mm (long) × 450 channels) of a stacked aluminum platelet (15 layers) reactor via anoidic oxidation and Ru impregnation, the authors were able to obtain cyclohexene **8** in 20% selectivity at 13% benzene **7** conversion. Adjusting the MeOH content of the feedstock, they were able to increase selectivities to 38%; however, this was at the expense of conversion (3%).

Hydrogenation of 2-Methyl-3-butyne-2-ol: Employing a wall-coated capillary reactor (coating = 110 nm), Rebrov et al. [21] investigated the selective hydrogenation of 2-methyl-3-butyne-2-ol **9** as a model for the hydrogenation of acetylene alcohols which are widely used in the synthesis of vitamins. Using a coating of polymer stabilized metal nanoparticles ($Pd_{25}Zn_{75}$/TiO_2), the authors investigated the effect of concentration (1.1×10^{-2} to 0.45 M), temperature (55–64°C) and hydrogen partial pressure (0.3–0.8) on the product distribution obtained (Scheme 2.4). Using this approach, the authors were able to identify the optimum conditions for the transformation, obtaining 90% selectivity toward 2-methylbut-3-en-2-ol **10** and 99.9% conversion of the alkyne **9**. This was subsequently increased to 97% selectivity, upon addition of pyridine to the substrate **9**. In addition to demonstrating enhanced selectivity compared to stirred reactors, the authors confirmed stability of the catalytic coating over a 1-month investigation performed at 60°C.

The authors subsequently reported the development of a new method for catalyst deposition within micro structured devices whereby mesoporous thin film catalyst

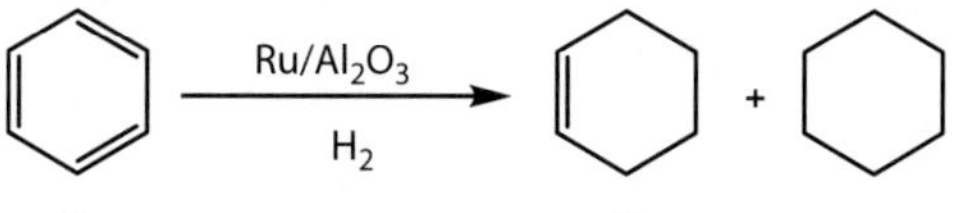

SCHEME 2.3 Schematic illustrating the partial hydrogenation of benzene **7**.

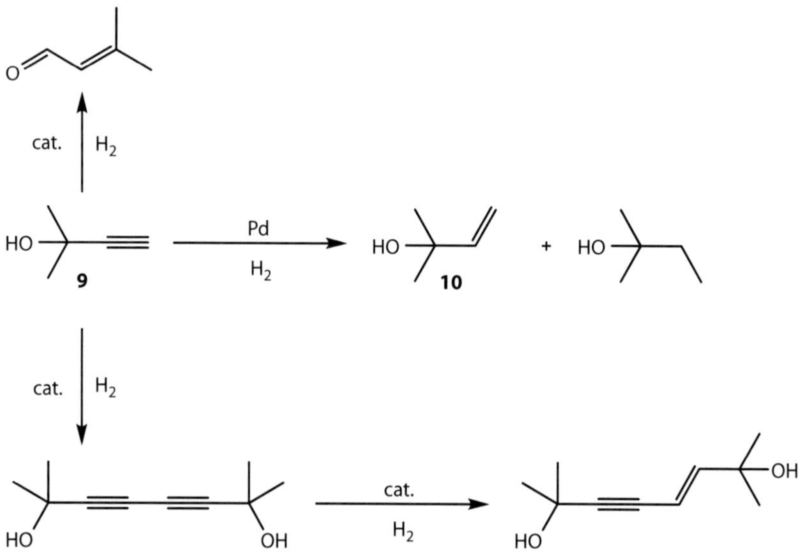

SCHEME 2.4 Illustration of the possible reaction products obtained upon hydrogenation of 2-methyl-3-butyne-2-ol **9**.

supports containing 1 wt% of Pd nanoparticles were used as wall coatings [22]. Utilizing capillary micro reactors, the authors demonstrated the selective hydrogenation of phenyl acetylene obtaining activities and selectivities approaching those obtained with homogeneous Pd catalysts. In addition, the authors demonstrated the robustness of the immobilization technique by operating the reactor for 1000 h without a loss of catalytic activity. Concomitantly, Kiwi-Minsker and coworkers [23] reported the fabrication of a novel compact reactor for three-phase hydrogenations using the aforementioned reaction as a model.

2.1.3 Dehydration and Dehydrogenation Reactions

Oxidative Dehydration of Propane: The dehydration of propanol is an exothermic oxidative reaction which is coupled to an endothermic nonoxidative process; as a result, coke formation accompanies the nonoxidative process. In order to maintain high productivity, this means that periodic operation of the reactor is required, with burning off of the coke performed intermittently to maintain catalytic activity. Employing vanadium oxide as the catalyst and a titanium micro reactor, Creaser et al. [24] investigated the transient kinetics of the oxidation of propanol.

See Chapter 4 for an example of EtOH dehydration using sulfated zirconia within a glass-PDMS micro reactor [25], the work of Koubeck et al. [26] for a discussion on the dehydration of 2-propanol, and a recent report by Yuan et al. [27], describing the catalytic dehydration of bioethanol.

Oxidative Dehydrogenation of Alcohols: Wörz et al. [28,29] reported the use of gold catalysts at 550°C for the oxidative dehydrogenation of alcohols to aldehydes, finding that isothermal operation of the reactor within the laboratory afforded the target compounds in high selectivity (90%). Researchers at BASF have also investigated this transformation, observing a selectivity of 85% within a micro channel reactor compared with 45% in a large-scale pan reactor [30]. In addition to packed-bed reactors, the dehydrogenation of cyclohexane has been reported within a micro channel device containing a Pd membrane [31] and using monoliths as the catalyst support [32].

2.1.4 FISCHER–TROPSCH SYNTHESIS

Using a highly active cobalt/alumina catalyst, Myrstad et al. [33] recently evaluated the Fischer–Tropsch synthesis in a stacked micro reactor owing to its use industrially for the synthesis of long-chain paraffins from natural gas. Employing a stainless-steel reactor, comprising of eight catalyst sections, sandwiched between heat exchange channels, the authors investigated the use of CO-based catalyst foils due to the metals selectivity toward the formation of linear paraffins. Under what are termed severe conditions (high temperature, pressure, and CO conversion), the authors were able to conduct the reaction without temperature runaway which maintained activity of the catalyst and afforded excellent selectivity toward C_5 products.

2.1.5 SYNTHESIS OF METHYLISOCYANATE IN A MICRO REACTOR

Basing their investigations on industrial processes, researchers at DuPont (Willmington, USA) analyzed the technical feasibility associated with the synthesis of methyl isocyanate **11** from methylformamide **12** as illustrated in Scheme 2.5 [34]. The reaction was selected as it falls into the class of a high-temperature, hazardous catalytic reaction, which demands intense cooling due to its exothermicity. Recognizing the potential for process intensification, the researchers evaluated the use of a micro reactor which comprised of three etched silicon wafers and two cover plates. The device contained a heat exchanger for the reactants, one for the products and a catalyst chamber, which was packed with polycrystalline silver particles. Using GC analysis of the reaction products, the authors obtained 95% conversion of **12**. However, due to an inability to maintain isothermal operation, selectivity was lower than that obtained in a conventional laboratory-scale reactor.

Other reactions performed within gas-phase micro reactors include the oxidation of ammonia [35,36], hydrogen cyanide synthesis via the Andrussov process [37], oxidation of carbon monoxide [38], hydrogen production [39], and the oxidation of benzene to phenol [40] to highlight a selection.

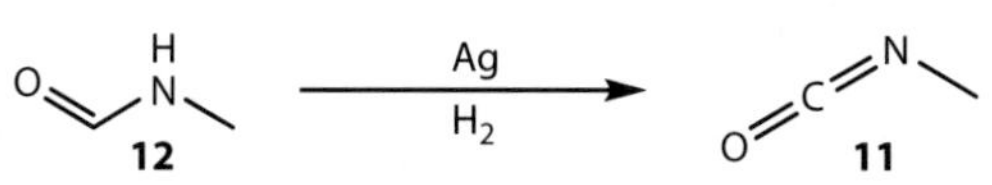

SCHEME 2.5 Illustration of the reaction protocol used for the synthesis of methyl isocyanate **11** within microstructured reactors.

2.2 GAS–LIQUID-PHASE MICRO REACTIONS

When looking at performing gas–liquid-phase reactions, good contacting between the phases is important in order to maintain uniformity within the reactor and high product selectivity [41]. With this in mind, two main techniques are applied to biphasic micro reactors, the first fluid-film contacting and the second gas bubbles within a continuous liquid phase [42], with surface modification techniques sometimes used to guide reactant streams [43,44]. Throughout the following section, examples will be provided in order to demonstrate the significant processing advantages that can be obtained as a result of employing biphasic reaction conditions within micro reactors.

Owing to the hazardous nature of elemental gases such as chlorine **13**, fluorine **14**, and ozone **15**, their use within research laboratories and process environments has been under exploited. The following section however, illustrates how these aggressive chemicals can be handled with ease and applied to the reaction of small organic molecules, for the first time enabling research to be performed without the need for specialized equipment.

2.2.1 Continuous Flow Chlorination Reactions

In 2000, Wehle et al. [45] published a patent detailing the chlorination of acetic acid within a falling film micro reactor. Industrially, the reaction is performed by feeding a solution of acetic acid (38.5%) and acetic anhydride (11.5%) into a bubble column where it contacts with Cl_2 **13** (50% at 3.5 bar) at 115–145°C. The resulting reaction products are then purified by distillation and crystallized to remove any dichlorinated product (3.5%). Utilizing a micro reactor (dimensions = 1500 μm (wide) × 300 μm (deep)), the authors were able to synthesize the target monochlorinated product in high yield (90%) and selectivity with only 0.05% dichloride formed; demonstrating significant advantages compared to the industrial process.

Photochemical Chlorination Reaction: Again using a falling-film micro reactor (film thickness ~100 μm) fabricated from nickel and iron, Jähnisch and coworkers [46] reported an investigation into a series of gas–liquid photochemical transformations, disclosing their findings for the photochlorination of toluene-2,4-diisocyanate **16** (Scheme 2.6). Compared to conventional batch protocols whereby 65% conversion was obtained and 45% selectivity respectively, the micro process afforded 55% conversion of toluene-2,4-diisocyanate **16** into 1-chloromethyl-2,4-diisocyanobenzene **17** and toluene-5-chloro-2,4-diisocyanate **18** in 80% and 5%

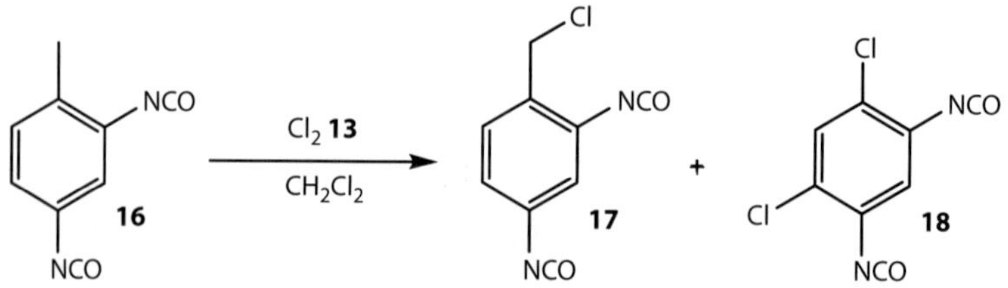

SCHEME 2.6 An example of a photochlorination reaction performed in a falling-film micro reactor.

selectivity, respectively. In addition to the enhanced selectivity obtained, the authors also calculated a 308-fold increase in space–time yield obtaining the chlorinated product **17** in a throughput of 400 mL $L^{-1} h^{-1}$. Details of other photochemical transformations can be found in Chapter 5.

2.2.2 Continuous Flow Fluorination Reactions

With a large number of biologically active compounds possessing fluorine molecules, research chemists have sought to find efficient methods for their incorporation into small organic molecules. In 1999, Chambers and Spink [47] began to report their findings into the use of micro reactors for reactions using elemental fluorine **14** (10% in N_2). Early examples focused on describing the micro reactor set-up, which comprised of a stainless-steel channel plate, closed using nickel and PCTFE plates, and demonstrating the selective fluorination of a series of 1,3-diketones (Scheme 2.7) [48]. Over the coming decade, the authors reported the use of this reactor for the fluorination of compounds such as ethyl-2-chloroacetate and deactivated aromatics [49,50] again demonstrating excellent reaction control. More recently, the authors described the development of a parallel reactor as a means of addressing the need to scale this technology [51], demonstrating the fluorination of *para*- and *meta*-substituted benzaldehyde derivatives, the results of which are as summarized in Table 2.2 [52].

de Mas et al. [53] more recently reported the ability to increase the productivity of fluorination reactions by increasing the superficial gas and liquid velocities within a single channel reactor, converting 63–76% of toluene (Figure 2.2).

2.2.3 Ozonolysis within Micro Reactors

Owing to the hazards associated with conducting ozonolysis reactions on a large-scale, several research groups have investigated the development of micro fabricated devices for the safe and scalable performance of these reactions.

Silicon Flow Distributor: In 2007, Wada et al. [54] reported the design, fabrication, and evaluation of a multichannel micro reactor for the ozonolysis of octylamine **19** to nitrooctane **20** (Scheme 2.8), obtaining 98.9% conversion and 79.7% selectivity with a reaction time of 0.32 s.

Falling-Film Micro Reactor: In the same year, Steinfeldt et al. [55] reported the use of a falling-film micro reactor (film thickness = 100 μm) for the ozonolysis of acetic acid 1-vinyl ester. With conventional processes performed at −60°C, the authors set about investigating the effect of variables such as temperature (−50–0°C),

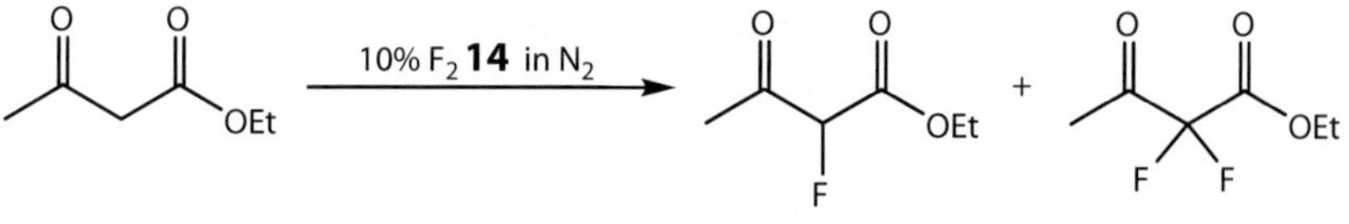

SCHEME 2.7 Schematic illustrating the model reaction used to demonstrate the selective fluorination attainable within a continuous flow reactor.

TABLE 2.2
A Selection of Fluorinated *Meta*-Substituted Benzaldehydes Synthesized Using a 9-Channel Micro Reactor

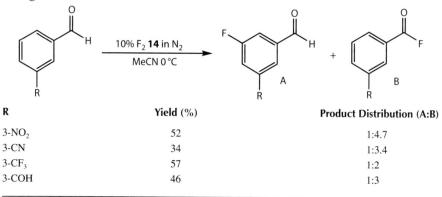

R	Yield (%)	Product Distribution (A:B)
3-NO$_2$	52	1:4.7
3-CN	34	1:3.4
3-CF$_3$	57	1:2
3-COH	46	1:3

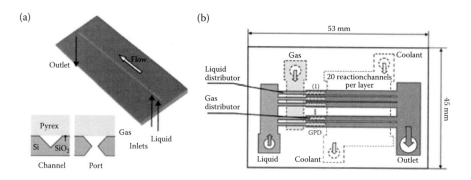

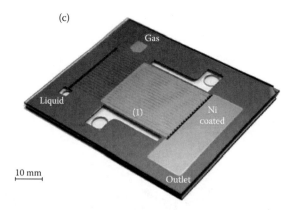

FIGURE 2.2 Schematic illustrating a single-channel reactor used for the direct fluorination of toluene: (a) microchannel, (b) top view of reactor holder and reagent inlets, and (c) illustration of the closed device. (Reproduced with permission from de Mas, N. et al. 2009. *Ind. Eng. Chem. Res.* 48: 1428–1434. Copyright (2009) American Chemical Society.)

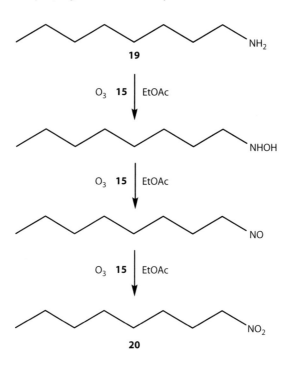

SCHEME 2.8 Illustration of the model reaction used to demonstrate ozonolysis in a silicon micro channel reactor.

olefin concentration (0.03–0.2 M) and ozone **15** partial pressure on the efficiency of the reaction. Using this approach, the authors were able to conclude that the use of a falling-film micro reactor enabled the use of ozone **15** without risk over the temperature range −50 to 0°C with products formed in excellent conversion and equivalent selectivities at concentrations <0.03 M. An unidentified by-product was observed to form at concentrations >0.03 M and up to 0.1 M. Hübner et al. [56] subsequently utilized the knowledge gained from this investigation for the synthesis of Vitamin D intermediate **21**, as illustrated in Scheme 2.9; this time employing glass-based micro reactors (Mikroglass, Germany).

Gas Permeable Tube Reactor: Employing a semipermeable Teflon (AT-2400) tube reactor (dimensions = 0.6 mm (i.d.) × 0.8 mm (o.d.) × 90 cm (long)), housed within a gas-tight vessel, Ley and coworkers [57], recently reported a facile method for the ozonolysis of a series of alkenes to afford the respective carbonyl compound. Whereas conventional gas–liquid flow reactions have been performed by delivering and mixing both gaseous and liquid reactants within microstructured devices, the apparatus described consisted of a semipermeable tube coiled within a glass vessel, into which ozone **15** was pumped directly from an ozone generator (Figure 2.3).

Initial investigations focused on the bleaching of Sudan red 7B in a range of solvents, which served to visually confirm the effective transfer of ozone **15** from the sealed vessel to the flow reactor. A control experiment was also performed utilizing PTFE tubing, whereby no bleaching of the dye was observed. Having established the

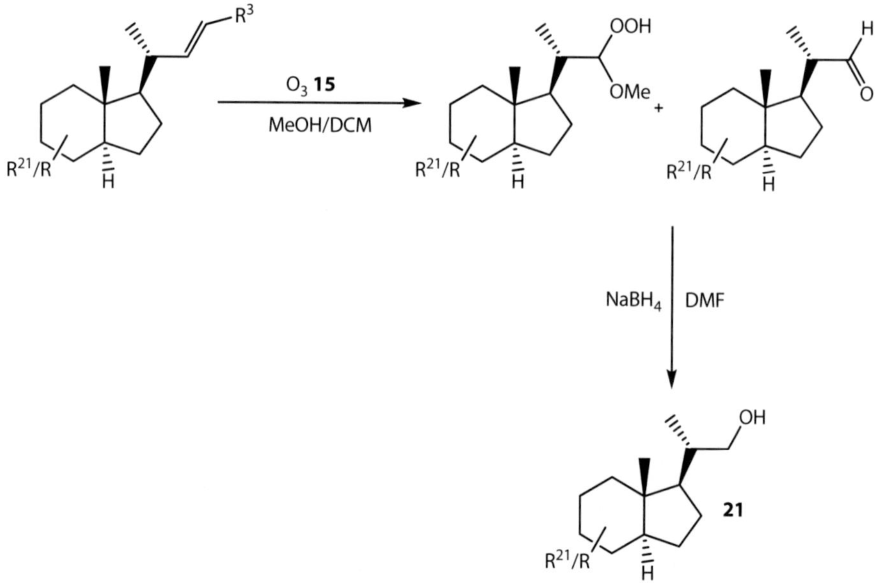

SCHEME 2.9 Illustration of the synthetic sequence employed for the continuous flow synthesis of a Vitamin D precursor **21**.

reproducibility of this process, the authors began their investigation into the ozonolysis of a series of alkenes, employing MeOH as the reaction solvent.

Using a residence time of 1 h, the authors found that they were able to quantitatively convert a 0.1 M solution of 1,1-diphenylethylene **22** into benzophenone **23**; as determined by NMR spectroscopy (Scheme 2.10). To ensure that any unreacted ozone **15** was destroyed, the reaction products were treated with polymer-supported triphenylphosphine in the receiving flask; which facilitated the reaction clean-up. Using this approach, the authors evaluated the reaction of a further 10 alkenes, a selection of the results are summarized in Table 2.3, which illustrate the generality of the technique developed.

On the basis of the facile nature of this technique and ease of reactor construction, the authors propose that this methodology would be suitable for the use of other reactive gases including CO, H_2, ethylene, and NO.

2.2.4 BIPHASIC CARBONYLATIONS

Using a glass micro reactor (channel dimensions = 200 μm (wide) × 75 μm (deep) × 5 m (long)), de Mello and coworkers [58] demonstrated the continuous flow synthesis of a series of secondary amides via a carbonylative coupling reaction. Employing a biphasic reaction stream, comprised of CO and iodobenzene **24**, benzylamine and a palladium–phosphine catalyst, the authors investigated the effect of liquid flow rate with a fixed gas flow of 2 sccm. Using this approach, annular flow dominated and the authors were able to identify a link between increasing reaction time and product

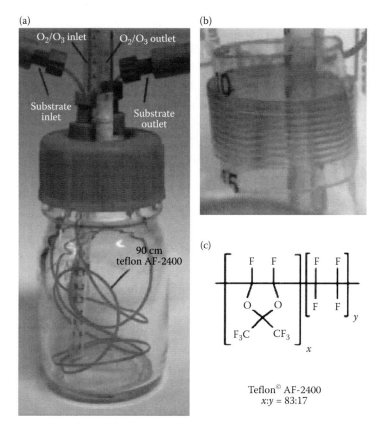

FIGURE 2.3 (See color insert) Photograph of the experimental set-up employed for the continuous flow ozonolysis of alkenes; (a,b) reactor filled with Sudan Red to enable visualization of the process and (c) porous tube structure. (Reproduced with permission from O'Brien, M., Baxendale, I. R., and Ley, S. V. 2010. *Org. Lett.* 12(7): 1596–1598. Copyright (2010) American Chemical Society.)

formation. Under optimal conditions, the authors obtained 46% conversion to the amide and 9% α-ketoamide, compared with 25% in batch.

In order to demonstrate the ability to vary product selectivity as a function of reaction time, temperature, and pressure, Buchwald and coworkers [59] evaluated the Pd-catalyzed Heck aminocarbonylation of a series of aryl halides, as depicted in Scheme 2.11.

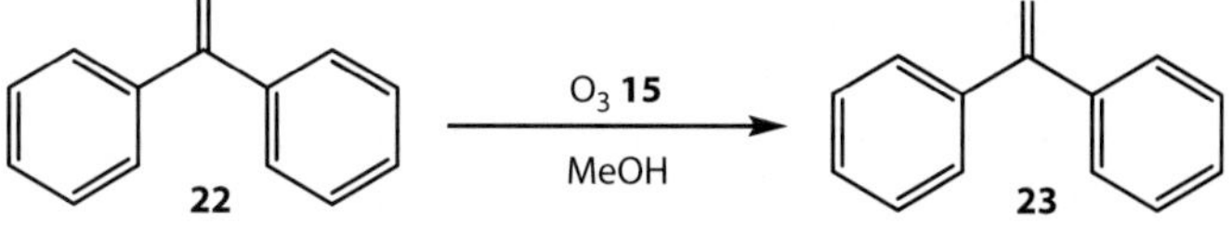

SCHEME 2.10 Schematic illustrating the model reaction used to develop a continuous ozonolysis reactor.

TABLE 2.3
Summary of the Results Obtained for the Continuous Flow Ozonolysis Using a Semipermeable Flow Reactor

Substrate	Product	Conversion (%)[a]	Yield (%)[b]
(styrene)	(acetophenone) **43**	100	83
(2-iodo-α-methylstyrene)	(2-iodoacetophenone)	100	93
(4-fluoro-α-methylstyrene)	(4-fluoroacetophenone)	100	95
(1,3-diphenyl-2-methylenepropane)	(1,3-diphenyl-2-propanone derivative)	100	88
(glycal with OMe groups)	(lactone with OMe groups)	100	73
(OTBS alkene)	(OTBS ketone)	100	93
(TBSO alkene)	(TBSO ketone)	100	76

[a] Determined by TLC and NMR analysis of the crude products.
[b] Isolated yield after purification by column chromatography.

Using a silicon and glass micro reactor (Reaction channel = 400 μm (wide) × 400 μm (deep) × 43.0 cm (long)), with a reaction volume of 68.8 μL, the authors investigated the effect of temperatures ranging from 109°C to 160°C and CO **25** pressures of 2.7–14.8 bar on the product distribution obtained. To perform a reaction, a reactant solution containing the aryl halide (1.0 M) and DBU **26** (3.0 M) in morpholine **27** and a second solution containing the catalyst Pd(OAc)$_2$ **28** (0.02 M) and ligand Xantphos **29** (0.02 M) in toluene, were introduced into the reactor using a high-pressure syringe pump and the CO **25** delivered to the reactor from a cylinder using a needle valve. Employing slug flow, the authors were able to manipulate the amide and α-ketoamide distributions, with high temperatures favoring the amide and high pressures giving rise to the α-ketoamide; quantified using GC-FID with dodecane as an internal standard (Table 2.4). Due to the reduced reactivity observed for aryl bromides, a residence time

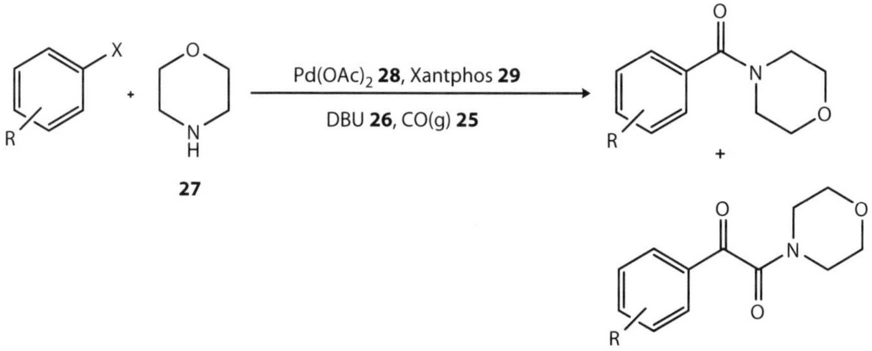

Micro Reactions Employing a Gaseous Component

SCHEME 2.11 Illustration of the Pd-catalyzed Heck aminocarbonylation performed within a silicon/glass micro reactor.

unit was added to the reactor (stainless-steel tube = 1.17 mm (i.d.) × 30 cm (long)) employed to afford a total reactor volume of 400 µL enabling increased reaction times to be investigated. Using this approach, the authors obtained increased conversions, while maintaining moderate productivities.

A more recent example of this transformation was reported by Skoda-Földes and coworkers [60,61], who described the use of the X-cube™ (ThalesNano Inc., Hungary) flow reactor for the double carbonylation of aryl halides in the presence of amines (Table 2.5). While the previous example had highlighted a link between pressure and α-ketoamide formation, the current investigation performed a detailed investigation into the reaction conditions that promote the formation of α-ketoamides. Screening a series of solvents, temperatures, bases, and reactant stoichiometries, the authors

TABLE 2.4
Summary of the Results Obtained for the Heck Aminocarbonylation of a Series of Aryl Halides Using a Gas–Liquid Micro Reactor

Aryl Halide	Pressure (bar)	Temperature (°C)	Time (min)	Conversion (%)	Product Distribution Amide (%)	α-Ketoamide (%)
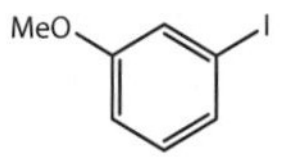 MeO–C₆H₄–I	7.9	146	3.3	100	68	28
	7.9	116	4.2	100	35	65
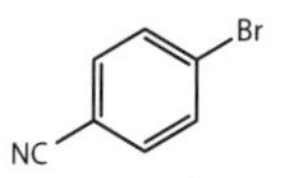 NC–C₆H₄–Br	2.7	160	7.1	100	83	0
	14.8	109	6.6	99	32	57
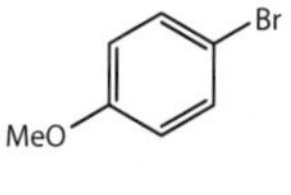 MeO–C₆H₄–Br	2.7	150	12.7	48	35	0

TABLE 2.5
A Selection of the Results Obtained for the Double Carbonylation of Iodobenzene 24

$$\text{24} \xrightarrow[\text{Base, Toluene}]{\text{Pd(pph}_3)_4\text{, CO 25}} \text{PhC(O)NRR}^1 + \text{PhC(O)C(O)NRR}^1$$

Amine	Eq. Amine	Base	Temperature (°C)	Conversion (%)	Product Distribution (%)	
					Amide	α-Ketoamide
Morpholine	1	Et₃N	80	31	79	21
	2	Et₃N	80	35	58	42
	2	DABCO	80	80	44	56
	2	DBU **26**	80	61	30	70
Cyclohexylamine	1	Et₃N	80	22	42	58
	2	Et₃N	80	61	38	62
	2	DABCO	80	96	24	76
	2	DBU **26**	80	92	13	87
Allylamine	2	Et₃N	80	25	25	75
	2	DBU **26**	80	64	9	91
n-Butylamine	2	Et₃N	80	40	39	61
	2	DBU **26**	80	76	4	96

found that employing excess of the nucleophile the target α-ketoamides could be generated in high selectivity at 80°C (Table 2.5).

Employing a closed-loop system, such as the ones described, was found to be advantageous for the safe manipulation of a toxic gas and air sensitive catalysts. Coupled with the increased gas–liquid contacting obtained within microfluidic systems, operating conditions unattainable using conventional reactor methodology were readily accessed.

Tin-Mediated Radical Carbonylations: A more recent example illustrating the advantages of exploiting gas–liquid-phase reactions under continuous flow was reported by the group of Ryu [62,63] and focused on a series of radical-based reactions including formylations, carbonylative cyclizations, and a multicomponent coupling reaction. Employing a stainless-steel reactor, comprising of a micromixer (dimensions = 1000 μm (i.d.)) and a tube reactor (dimensions = 1000 μm (i.d.) × 8 or 18 m (long)) capable of being pressurized to 80 atm, the authors evaluated the tin-mediated radical carbonylation of a series of alkyl halides, investigating the ability to suppress the competing reduction associated with this transformation. As Scheme 2.12 depicts, the thermally induced radical formylation of bromododecane **30** was

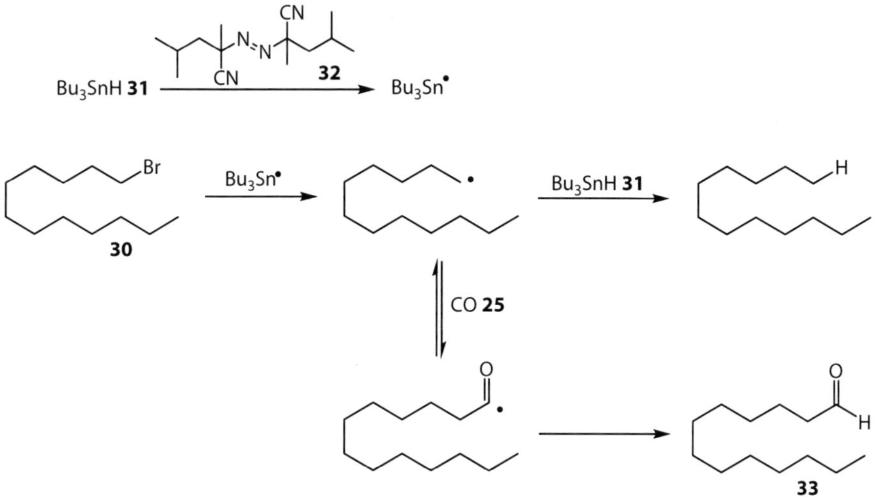

SCHEME 2.12 Schematic illustrating the tin-mediated radical carbonylation, and competing reduction, conducted under continuous flow.

selected as a model reaction, with tributyltin hydride **31** as the mediator. Using 10 mol% of 2,2'-azobis(2,4-dimethylvaleronitrile) (V 65) **32** as the initiator (half-lifetime at 80°C = 12 min), bromododecane **30** (0.02 M) in toluene, a CO **25** pressure of 83 atm and a residence time of 12 min, the authors obtained the target compound, tridecanal **33**, in 77% yield (Table 2.6).

The authors subsequently extended the investigation to the carbonylative cyclisation of 1,6-azaenyne, whereby a residence time of 27 min afforded the six-membered ring lactam in 85% yield. Using continuous flow methodology, the authors have developed a facile technique for the radical-mediated carbonylation of alkyl halides with reaction times ranging from 12 to 27 min.

Reductive Carbonylation of Nitro Groups: More recently, Takebayashi et al. [64] reported an efficient method for the direct conversion of nitrobenzene **34** into synthetically useful isocyanates such as phenylisocyanate **35**, based on their importance as intermediates in the manufacture of thermoplastic foams, elastomers and agrochemicals. As illustrated in Scheme 2.13a the conventional route to these compounds involved a two-step process comprising of a reduction step followed by phosgenation. This approach is however problematic, due to (1) the use of toxic gas, (2) the formation of HCl as a by-product, and (3) contamination of the reaction product with chlorine. With these factors in mind, the authors were interested in the development of a safe and efficient alternative for the direct production of isocyanates from nitroaromatic compounds. As Scheme 2.13b illustrates, the reductive carbonylation of nitro-compounds represents a potential direct route, however when performed in a batch environment, the technique's harsh reaction conditions ($\geq$200°C and $\geq$10 MPa CO **25**) and long reaction times ($\geq$1 h) led to the formation of undesirable polymerization products.

TABLE 2.6
Summary of the Results Obtained for a Series of Radical Carbonylations Conducted under Continuous Flow

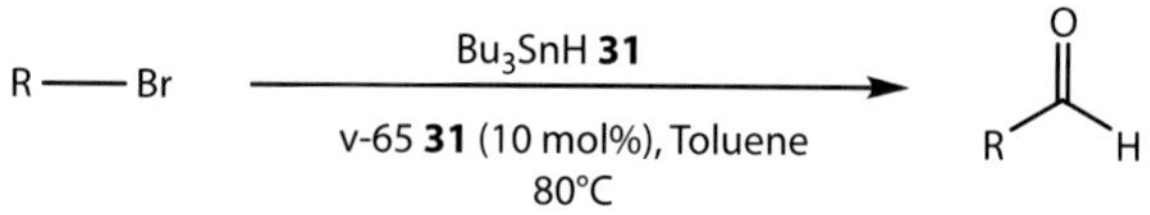

R—Br $\xrightarrow[\text{80°C}]{\text{Bu}_3\text{SnH 31} \\ \text{v-65 31 (10 mol\%), Toluene}}$ R—C(=O)H

Substrate	Product	CO Pressure (atm)	Residence Time (min)	Yield (%)
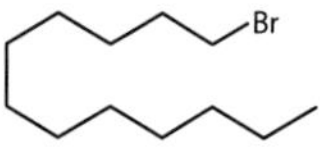 30	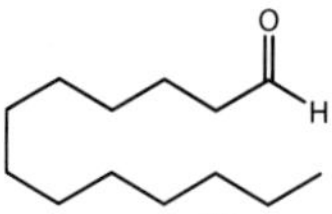 33	83	12	77
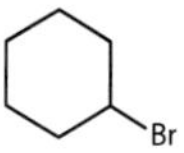 (cyclohexyl bromide)	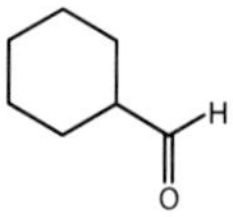 (cyclohexanecarbaldehyde)	85	12	68
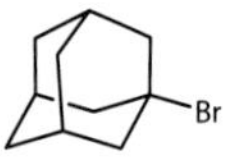 (1-bromoadamantane)	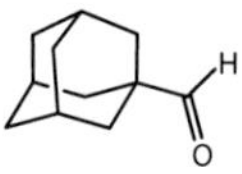 (adamantane-1-carbaldehyde)	85	12	86
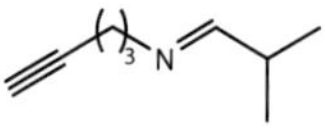 (alkyne-enamide substrate)	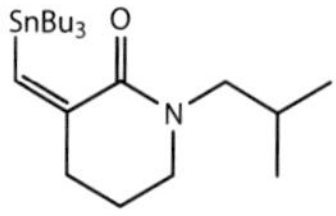 (SnBu₃ piperidinone product)	85	29	85 (E/Z = 0/100)

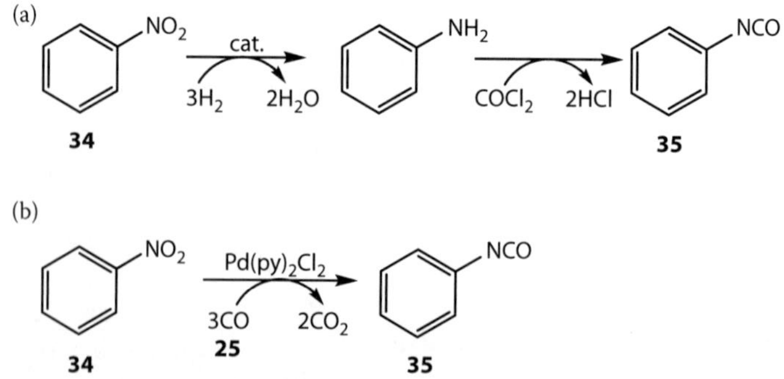

SCHEME 2.13 Illustration of the (a) conventional route and (b) direct route to the synthesis of organic isocyanates.

(a) 34 (Ph-NO₂) $\xrightarrow[3H_2 \quad 2H_2O]{\text{cat.}}$ (Ph-NH₂) $\xrightarrow[COCl_2 \quad 2HCl]{}$ 35 (Ph-NCO)

(b) 34 (Ph-NO₂) $\xrightarrow[3CO \quad 2CO_2]{\text{Pd(py)}_2\text{Cl}_2 \\ \mathbf{25}}$ 35 (Ph-NCO)

Owing to the improved mass transfer across gas–liquid interfaces, the authors were able to efficiently perform the reductive carbonylation of nitrobenzene **34** to phenyl isocyanate **35** employing CO **25** pressures of <1 MPa; affording a direct and efficient route to synthetically useful isocyanates.

2.2.5 Transfer Hydrogenations under Continuous Flow Conditions

Using a microstructured single channel, falling film reactor constructed from stainless steel (channel dimensions = 300 µm (wide) × 100 µm (deep), volume = 14 µL), de Bellefon et al. [65] developed a device suitable for the screening of rhodium diphosphine catalysts toward the asymmetric hydrogenation of methyl-Z-α-acetamidocinnamate **36** (Scheme 2.14).

To conduct a reaction, the authors preformed the catalyst in a Schlenck tube (metal:ligand, 1:1) containing MeOH at 28°C over a period of 15 min, prior to the addition of methyl-Z-α-acetamidocinnamate **36** (0.8 M); affording a catalyst concentration of 10 mol%. An aliquot (20 µL) of the resulting reaction mixture was transferred into a Rheodyne injection loop, connected to a syringe pump and pumped through the helicoidal reactor at a flow rate of 12 µL min^{-1} and a hydrogen flow rate of 5 scc min^{-1}. Using this approach, the authors screened 17 chiral phosphines at a reaction time of 3 min. As Table 2.7 illustrates, (R,R)-Diop **37** afforded the most encouraging results under the aforementioned conditions, with (R,S)-cy-cy-Josiphos **38** providing better enantioselectivity (75.1%). Using the (R,R)-Diop **37** system, the reproducibility of the device was investigated over five separate runs, whereby the authors obtained a deviation of 2% for conversions and <1% for the enatiomeric excess; confirming the fabrication of a reliable screening device. Compared to batch investigations, the use of a helicoidal falling film reactor proved advantageous as each reaction only employed 20 nmol of Rh and 22 nmol of ligand; an important feature when evaluating often precious ligands.

In a subsequent experiment, the authors evaluated the ability to screen a single catalyst $((R)$-[Rh(Binap)COD]BF$_4$ **39**) toward a series of prochiral alkenes, using a reaction time of 8 min, as illustrated in Figure 2.4. Of the substrates evaluated, methylacetamidoacrylate **40** and **36** were observed to be the most active toward the catalyst **39** affording 99% (56% ee) and 49% (65% ee) respectively, while dimethylitaconate **41** was recovered unreacted and methone **42** in <5% conversion (25% ee). In addition to the reduced reactant/catalyst consumption within the falling film reactor, the device afforded increased reactivities compared to a conventional carousel-type batch reactor.

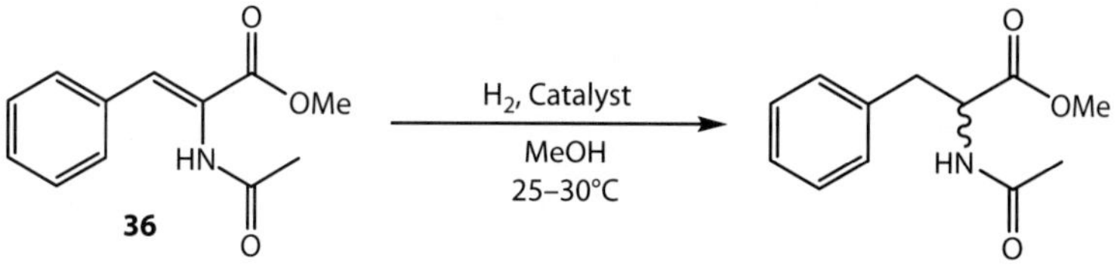

SCHEME 2.14 Illustration of the model reaction used to demonstrate the gas–liquid hydrogenation of methyl-Z-α-acetamidocinnamate **36** in a falling film reactor.

TABLE 2.7
Summary of the Data Obtained from the Helicoidal Falling Film Reactor for the Hydrogenation of Methyl-Z-α-acetamidocinnamate 36

Ligand/Catalyst	Temperature (°C)	Conversion (%)	ee (%)	Ligand (µg)
(S)-NMPP	30.3	<1.0	13	7.13
(R,R)-Et-Duphos	30.1	1.6	>99	7.97
(R,R)-Me-Duphos	30.0	26.0	>99	6.74
(R,R)-Et-BPE	31.5	0	—	6.92
(R,R)-Me-BPE	31.4	1	84.6	5.68
(R,R)-Norphos	31.5	<1.0	—	10.18
(S,S)-Chiraphos	31.5	0	—	9.38
(R)-Prophos	31.5	<1.0	—	9.07
(R)-Trost ligand	27.5	0	—	15.20
(R)-Quinap	27.4	0	—	9.67
(R,R)-Diop 37	33.5	94	63.5	10.97
(R,S)-Josiphos	33.5	26.5	86.2	13.08
(R,S)-t-Bu-P-Josiphos	33.5	3.0	48.6	11.93
(R,S)-Pcy-Josiphos	33.5	3.7	−7.6	13.08
(R,S)-cy-cy-Josiphos 38	33.5	88.1	75.1	13.35
(R)-Binap	35.0	<1.0	23.7	12.45
(R)-[Rh(Binap)COD]BF$_4$ 39	35.0	<1.0	—	12.45
(S,S)-BPPM	35.0	22.5	91.0	11.07

In order to efficiently design scalable synthetic processes for the pharmaceutical and fine chemical industries, Gavriilidis and coworkers [66] described the need to firstly identify any potential scale-up obstacles. Using this thought process, the authors were able to design a novel gas–liquid reactor in which the catalytic asymmetric transfer hydrogenation of acetophenone **43** was performed in the presence of 1R,2S-amino-indanol/pentamethylcyclopentadienylrhodium; with ^{i}PrOH as the proton donor (Scheme 2.15). During their initial investigations Zanfir and Gavriilidis [67] identified a reaction dependence on the proportion of acetone within the system,

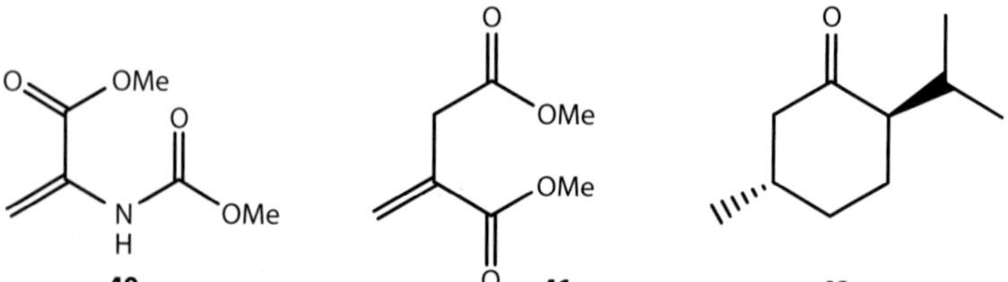

FIGURE 2.4 Illustration of the pro-chiral substrates evaluated within the gas–liquid falling film reactor.

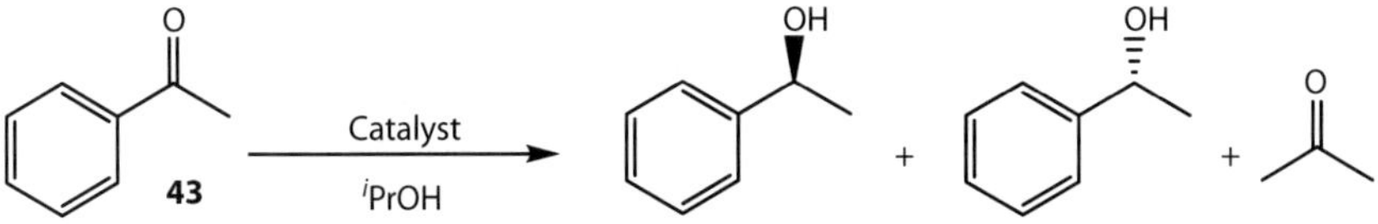

Micro Reactions Employing a Gaseous Component

SCHEME 2.15 Model reaction used to demonstrate the scalability of transfer hydrogenations conducted in micro fabricated reactors.

resulting in retardation of the hydrogenation, coupled with erosion of product enantioselectivity at >0.5 M.

With this information in hand, Sun and Gavriilidis [68] subsequently designed a microstructured reactor capable of removing acetone from the reactor as it formed, thus preventing retardation of the transfer hydrogenation and maintaining high enantioselectivity.

Exploiting the acetone vapor–liquid equilibrium coupled with the principle of solvent stripping, the authors developed a micromesh reactor, depicted in Figure 2.5a, consisting of a copper gas–liquid flow channel sealed with stainless-steel plates (Figure 2.5b). A micromesh (76 μm pore size, 50 μm thick and 26% open area) was placed between the gas and liquid channels to afford a defined gas–liquid interface (Figure 2.5c), which was stabilized by exerting pressure on both phases.

To perform a reaction, the micro reactor was placed in a water bath (30°C) to provide a stable reaction temperature, the liquid phase was introduced into the reactor, from two separate syringes, the first containing the catalyst (2.8×10^{-4} M) in ^{i}PrOH and the second acetophenone **43** (0.28 M) and sodium isopropoxide (2.24×10^{-3} M) and mixed using a standard slit inter-digital micromixer. A total liquid flow rate of 26–400 μL min^{-1} was employed and the gas used to strip acetone from the reaction mixture was nitrogen at a constant flow rate of 70 mL min^{-1}. Employing a residence time of 15 min, the authors were able to quantitatively hydrogenate acetophenone **43** compared to 88% in batch after 1 h; however, along with evaporation of acetone, the authors also noted that ^{i}PrOH was lost leading to erosion of enantioselectivity (*ee* 85%). Therefore, in order to maintain a constant concentration and prevent the observed erosion of enantioselectivity, the N_2 stream was pumped through a solution of ^{i}PrOH prior to entering the micro reactor, affording efficient acetone removal (90% in 5 min) and enhanced enantioselectivities.

2.2.6 Miscellaneous Biphasic Micro Reactions

Epoxidation Using HOF:MeCN: Employing a nickel/PCTFE single-channel flow reactor (dimensions = 500 μm (wide) × 500 μm (deep)), Murray and coworkers [69] reported the development of a continuous flow technique for the epoxidation of alkenes using HOF:MeCN **44**. In order to reduce the risks associated with the reaction, the authors generated HOF:MeCN **44** *in situ* from elemental fluorine and MeCN. The alkene (in DCM) under investigation was subsequently reacted (1:1) at room temperature to afford the target epoxide. As an additional safety precaution, the reactor effluent was quenched in a solution of aq. $NaHCO_3$ and purged with N_2 prior to extraction of the reaction products into DCM. Using this approach, the authors

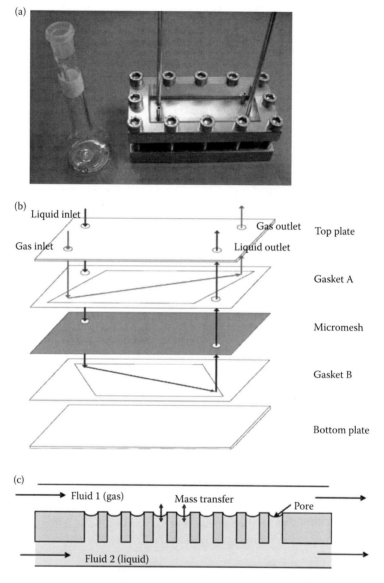

FIGURE 2.5 Schematic illustrating (a) the microstructured reactor, (b) an exploded view of the reactor components, and (c) an illustration of the gas–liquid interface obtained using a micromesh. (Reproduced with permission from Sun, X. and Gavriilidis, A. 2008. *Org. Proc. Res. Dev.* 12: 1218–1222. Copyright (2008) American Chemical Society.)

were able to synthesize a wide array of epoxides in good to excellent yield depending on the alkene used (Table 2.8).

Biphasic Heck Coupling Reactions: Using a series of PDMS microchannel reactors, Kim and coworkers [70] evaluated the biphasic boron-Heck reaction under continuous flow, as summarized in Table 2.9. When performed in batch, H_2O_2 generated

TABLE 2.8
Summary of the Results Obtained for the Epoxidation of Alkenes Using *In Situ* Generated HOF:MeCN 44

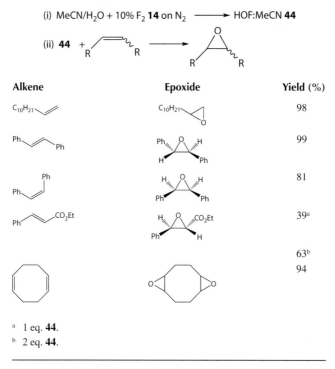

[a] 1 eq. **44**.
[b] 2 eq. **44**.

as a by-product of the reaction was found to lead to decomposition of the boronic acid and ligand **45**, leading to the need for an excess to be employed [71]. Conducting the reactions under continuous flow, the authors obtained increased conversions as a result of increased contact area between the gas and liquid phases. In their first reactor, single channel (400 µm (wide)), slug flow dominated with an 800 µm liquid gap between slugs of gas, affording 51–59% conversion to the target alkene compared with 58–70% in batch. Attempts to reduce the slug spacing and increase mixing efficiency were unsuccessful, as were attempts to obtain steady flow within the reactor.

To address this, the authors fabricated a double-channel reactor, where liquid flows and gas flows were separated by a gas permeable PDMS membrane (20 µm). Using this configuration, the authors were able to obtain stable and reproducible reaction times, which led to increases in alkene conversion compared to both the single channel and batch reactors (Table 2.9). The improvements in conversion are attributed to the efficient mixing obtained between phases within the dual channel micro reactor, coupled with a decrease in boronic acid decomposition due to separation of any H_2O_2 formed. As an extension to this investigation, research into the application of this reactor configuration toward hydrogenations and carbonylations is currently underway.

TABLE 2.9
Summary of the Results Obtained for the Gas–Liquid Boron-Heck Performed in Batch and Continuous Flow

$$\text{ArBH(OH)}_2 + \text{alkene} \xrightarrow[\text{O}_2, \text{DMF}]{\substack{\text{Pd(OAc)}_2 \; 28 \\ [2,2']\text{-Bipyridyl } 45}} \text{product} + \text{ArOH}$$

Boronic Acid	Alkene	Batch[a]	Product Distribution Single Channel[b]	Dual Channel
Ph-B(OH)$_2$	ethyl acrylate (OEt)	69:24	59:6	82:8
Ph-B(OH)$_2$	acrylonitrile (CN)	67:25	53:8	79:9
Ph-B(OH)$_2$	styrene	70:25	57:6	85:9
4-acetylphenyl-B(OH)$_2$	ethyl acrylate (OEt)	58:21	51:6	76:6
3-MeO-phenyl-B(OH)$_2$	ethyl acrylate (OEt)	64:24	55:7	79:8

[a] Reaction time 12 h.
[b] Reaction time 10 min.

In addition to the examples described, authors have also reported selective oxidations under explosive conditions [72], the multiphase formation of hydrogen peroxide [73], sulfonations using SO_3 [74] and the high-throughput screening of multiphase reactions [75], to name a few.

2.3 GAS–LIQUID–SOLID MICRO REACTIONS

As observed with gas–liquid reactions, the same is true of gas–liquid–solid reactions where it is imperative to ensure good reactant contacting between the phases. To achieve this, the solid element is conventionally incorporated into the devices using either packed-bed technology [76] or wall-coating of the micro channels; however, examples of monoliths have also been reported [77] (see Chapter 4 for more details of packing techniques). Using these approaches, many authors have demonstrated the synthetic advantages associated with the use of microfabricated devices for this complex mode of operation.

2.3.1 TRIPHASIC OXIDATIONS UNDER FLOW CONDITIONS

As can be seen in subsequent examples (Chapters 3 and 4), the oxidation of carbonyl containing compounds is a fundamental transformation utilized by the synthetic chemist; however the use of inorganic oxidants can lead to the generation of large quantities of chemical waste. In comparison, the use of molecular oxygen in conjunction with a heterogeneous catalyst has been reported as a potential solution to this problem. However, the low reactivity of such gas–liquid–solid systems means that this technique does not currently provide an adequate alternative to the use of inorganic oxidants.

Using the process of surface derivatization, depicted in Scheme 2.16, Kobayashi and coworkers [78] converted a polysiloxane coated capillary (250 mm (i.d.) × 50 cm (long) × 0.25 μm (film thickness)) into a gold-coated micro reactor and evaluated its efficiency toward the oxidation of both 1° and 2° alcohols.

To perform an oxidation reaction, separate solutions of aqueous (aq.) K_2CO_3 (0.3 M, 0.5 μL min^{-1}) and alcohol (0.1 M, 1.7 μL min^{-1}) were introduced into the reactor, via a T-connector, from two separate syringe pumps followed by O_2 at a flow rate of 1.5 mL min^{-1}. Employing the multiphase oxidation of 1-phenylethanol **46** as a model reaction, the authors investigated the effect of reaction temperature on the conversion to acetophenone **43** identifying the optimal temperature as 60°C and the

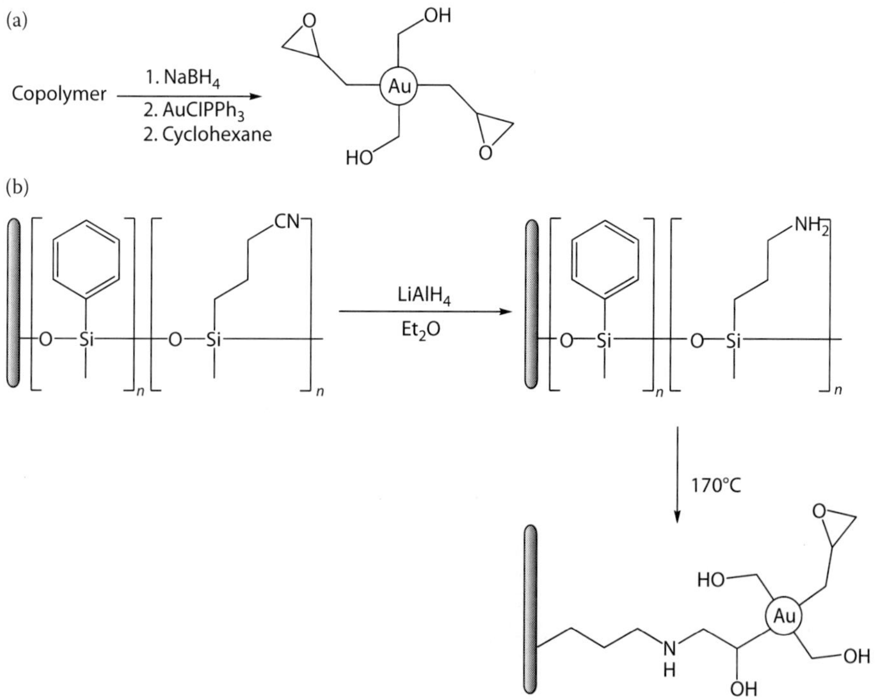

SCHEME 2.16 Summary of the derivatization protocol employed for the preparation of (a) functionalized Au nanoparticles and (b) Au-wall-coated micro channels.

residence time as 90 s (99% conversion). Under the aforementioned conditions, the system was found to be stable for several days with no sign of Au leaching. In comparison, an analogous batch reaction was performed using Au/C whereby only trace amounts of acetophenone **43** were obtained, illustrating the dramatic increase in reaction efficiency attained using a capillary flow reactor.

The generality of the aerobic oxidation was subsequently investigated, and as illustrated in Table 2.10, excellent conversions were obtained for all alcohols evaluated; however, in the case of benzyl alcohol **47** low product selectivity was observed for the aldehyde. Application of an Au/Pd-coated capillary, operated under analogous conditions was found to restore reaction selectivity for the aldehyde.

2.3.2 Carbonylations Using Solid-Supported Catalysts

As an extension to their earlier work into gas–liquid-phase carbonylations, de Mello and coworkers [79] subsequently investigated the incorporation of a silica-Pd catalyst into their flow reactor as a means of simplifying the post reaction work-up and enabling the techniques application to positron emission tomography ^{11}C-radiolabeling. To perform a reaction, the authors prepared a solution of aryl halide and amine in THF, upon entering the reactor it was mixed with CO prior to passing through the tubular packed-bed reactor (dimensions = 1 mm (i.d.) × 45 cm (long)) maintained at 75°C. After a reaction time of 12 min, the reaction products were collected and analyzed using offline GC with six benzamides synthesized in yields ranging from 23% to 99%.

Encouraged by these results, the authors subsequently employed ^{11}CO in order to prepare the respective amides. As Table 2.11 illustrates, the target compounds were

TABLE 2.10
Summary of the Results Obtained for the Oxidation of 1° and 2° Alcohols within a Gold-Coated Micro Reactor

Substrate	T (°C)	Conversion (%)	Yield (%)
PhCH(OH)Me **46**	60	>99	99
p-MeOC$_6$H$_4$CH(OH)Me	60	>99	99
p-FC$_6$H$_4$CH(OH)Me	60	>99	99
p-ClC$_6$H$_4$CH(OH)Me	70	>99	99
$trans$-PhCH=CHCH(OH)Me	70	>99	99
2-Thienyl-CH(OH)Me	60	>99	99
Ph(CH$_2$)$_2$CH(OH)Me[a]	65	>99	89
1-Indanol	60	>99	99
PhCH$_2$OH **47**	60	>99	53
PhCH$_2$OH **47**[a]	50	>99	92
p-MeC$_6$H$_4$CH$_2$OH[a]	60	>99	95

[a] Au/Pd-capillary reactor used in place of the Au-capillary reactor.

Micro Reactions Employing a Gaseous Component

obtained in high radiochemical yield and purity; however, run-to-run reproducibility was poor and the system suffered from reactant carry over between reactions.

2.3.3 Hydrogenations within Continuous Flow Reactors

Owing to the fact that almost 20% of reaction steps employed in fine chemical syntheses are catalytic hydrogenations [80], it comes as no surprise that this transformation is one of the most widely studies in gas–liquid–solid micro reactors, with examples of commercially available equipment dedicated to this transformation available (see Chapter 1).

With this in mind, researchers at ThalesNano Inc. [81] have reported many examples of continuous flow hydrogenations, ranging from the reduction of nitro groups to amines, the deprotection of benzyl/Cbz protecting groups, hydrogenation of alkenes to alkanes, nitriles to amines, and enantioselective hydrogenations [82], using a standalone instrument, that generates H_2 *in situ* via the electrolysis of DI H_2O; with developments such as these enabling any laboratory worker to perform hydrogenation reactions safely.

Using the H-cube™ (ThalesNano Inc., Hungary) system, Lou et al. [83] reported the continuous flow synthesis of tetrahydropyrimidones, citing this equipment as

TABLE 2.11
Summary of the Results Obtained for a Series of Gas–Liquid–Solid Carbonylations

Product	Radiochemical Yield (%)[a]	Radiochemical Purity (%)[b]
(benzamide, N-benzyl)	79, 65	96, 94
(4-cyano benzamide, N-benzyl)	67, 70	95, 95
(4-trifluoromethyl benzamide, N-benzyl)	46, 68	70, 90
(4-methoxy benzamide, N-benzyl)	45, 33	72, 80

[a] Decay-corrected.
[b] Determined by analytical radio HPLC and * = ^{11}C.

enabling faster catalyst screening, ease of reaction optimization and the potential for library generation. As part of a batch led investigation, the authors screened a series of catalysts (Pd/C, Pt/C, Pt/Al$_2$O$_3$, and Raney-Ni), solvent systems and reactor temperatures for the diastereoselective hydrogenation of substituted dihydropyrimidones (Table 2.12). After a detailed study, the authors identified Raney Ni **48** as the catalyst, MeOH as the solvent and a reactor temperature of 45°C as the optimal conditions. Employing a flow rate of 0.5 mL min^{-1} and a system pressure of 90 bar, to prevent boiling of the reaction solvent, the authors were able to readily convert the substrates into the respective *cis,cis*-tetrahydropyrimidone. Analysis of the reaction products by LC–MS–UV–ELSD confirmed complete conversion to the target product was obtained (Table 2.12). To afford the respective *trans,trans*-tetrahydropyrimidones, the reaction products were stirred in a solution of methanolic potassium carbonate for 3 days.

Employing the continuous flow hydrogenator, Desai and Kappe [84] reported the reductive dethionation of 3,4-dihydropyrimidin-2-thiones **49** using MeCN as the reaction solvent. To perform a reaction, the authors dissolved the thione **49** (1.2 × 10^{-2} M) in MeCN and pumped the solution through a prepacked catalyst cartridge containing Raney-Ni **48**, a pressure of 1–2 bar was applied and a flow rate of 1 mL min^{-1} (40°C) employed. The process was conducted over a period of 30 min at

TABLE 2.12
Illustration of a Selection of Tetrahydropyrimidones Prepared Using a Commercially Available Continuous Flow Hydrogenator

R^1	R^2	R^3	Yield (%)[a]	dr[b]	Purity (%)[c]
PhCH$_2$	Ph	OCH$_3$	85	>20:1	>98
EtOCOCH$_2$	Ph	CH$_3$	83	19:1	95
Et	(CH$_2$)$_2$Ph	CH$_3$	82	17:1	>98
EtOCOCH$_2$	Ph	OCH$_3$	86	19:1	95
MeO(CH$_2$)$_2$	(CH$_2$)$_2$Ph	OCH$_3$	87	16:1	>98

[a] Isolated yield.
[b] dr by LC/MS/ELSD.
[c] Purity after preparative RP-HPLC (LC/MS/UV/ELSD).

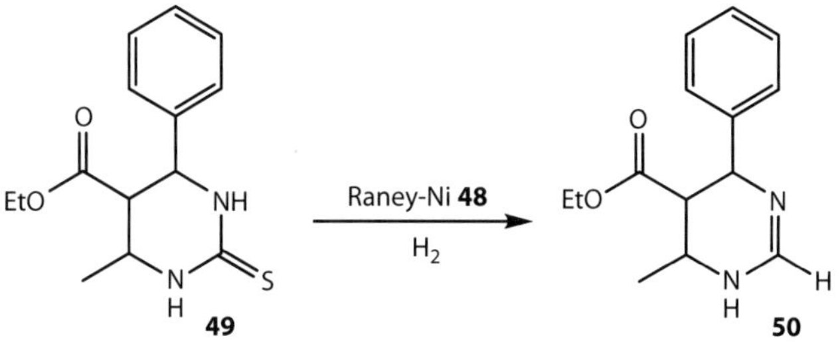

SCHEME 2.17 An example of continuous flow reductive dethionation.

which point the resulting 1,4-dihydropyrimidine **50** was obtained in quantitative yield after evaporation of the reaction solvent (Scheme 2.17).

More recently, the authors have demonstrated the hydrogenation of functionalized pyridines under flow conditions [85] utilizing the H-cube™ (ThalesNano Inc., Hungary) (Scheme 2.18). Conventionally this reaction is performed using Pt, Rh, Ru, or Ni catalysts however, upon screening a series of prepacked catalysts for activity toward the reaction, the authors were surprised to find that the transformation could be performed using Pd/C **51** and water as the reaction solvent. To identify if this was a general observation, the authors employed a reactant concentration of 0.1 M, water as the reaction solvent, a reactor temperature of 80°C and a flow rate of 0.5 mL min^{-1} for a series of pyridine derivatives, as illustrated in Table 2.13. The reactions were also repeated using Pt/C and Rh/C whereby comparable results were obtained-conversions determined by ^{1}H NMR spectroscopy.

In addition, researchers have fabricated their own continuous flow hydrogenation devices, with a recent example from the research group of Plucinski [86] fabricating a compact multichannel reactor whereby the reduction of benzaldehyde to benzyl alcohol and tandem C–C coupling hydrogenation depicted in Scheme 2.19 were performed.

In the case of the sequential reaction, the authors employed a premixed solution of iodobenzene **24** and styrene **54** (0.4 M) in EtOH and a reactor temperature of 120°C. Employing a liquid feed flow rate of 0.25 mL min^{-1} and a gas flow rate of 8 mL min^{-1}, the authors obtained complete consumption of the aryl halide **24** and (*E*)-1,2-diphenylethene **55** obtaining 83% 1,2-diphenylethane **56** with 13% benzene **7** as a by-product; showing potential for the development of large-scale continuous flow hydrogenators.

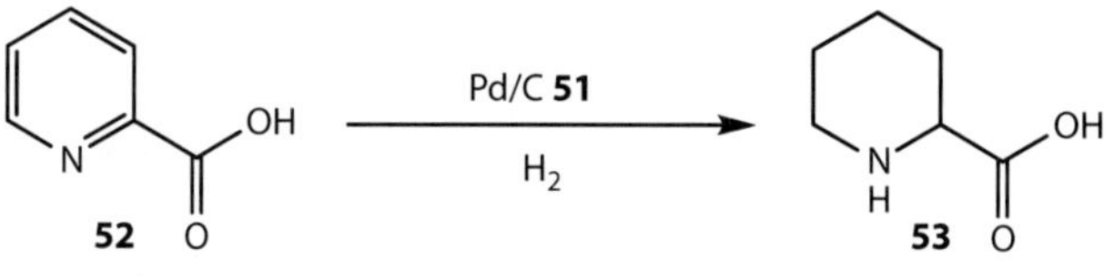

SCHEME 2.18 Illustration of the hydrogenation of picolinic acid **52** to pipecolic acid **53**.

TABLE 2.13
Summary of the Results Obtained for the Continuous Flow Hydrogenation of Pyridines Using Water as the Reaction Solvent

Product	Temperature (°C)	Pressure (bar)	Conversion (%)[a]
methyl piperidine-4-carboxylate	80	30	>99
2-(2-(Boc-amino)ethyl)piperidine	80	30	>99 (74)
2-(2-phenyl-1,3-dioxan-2-yl)piperidine	80	90	>99
ethyl 2-(piperidin-2-yl)acetate	80	90	>99
4-methoxypiperidine	80	30	>99 (91)

[a] The number in parentheses represents the isolated yield.

Kobayashi et al. [87] reported an early example of a microfluidic device for gas–liquid–solid hydrogenations and more recently, the authors [88] demonstrated the use of scCO$_2$ as a reaction solvent for hydrogenations using a Pd-wall-coated micro-channel reactor.

2.3.4 SLURRY-BASED MICRO REACTIONS

While the use of packed-bed and/or wall-coated micro reactors illustrate a novel method for the performance of gas–liquid–solid reactions, the throughputs afforded by such systems typically span the range of 10's mg to 10's g h^{-1}; as such they are not suited to industrial-scale production. With this in mind, Guermeur and coworkers [89] investigated the use of slurries within microfluidic reactors as a means of accessing high efficiencies from reactions that are conventionally mass and heat transfer limited. As a result of poor mass transfer, the rate of reaction can often be masked, with inefficient heat removal leading to the formation of several products and by-products due to localized concentration gradients. Using a slurry-based glass

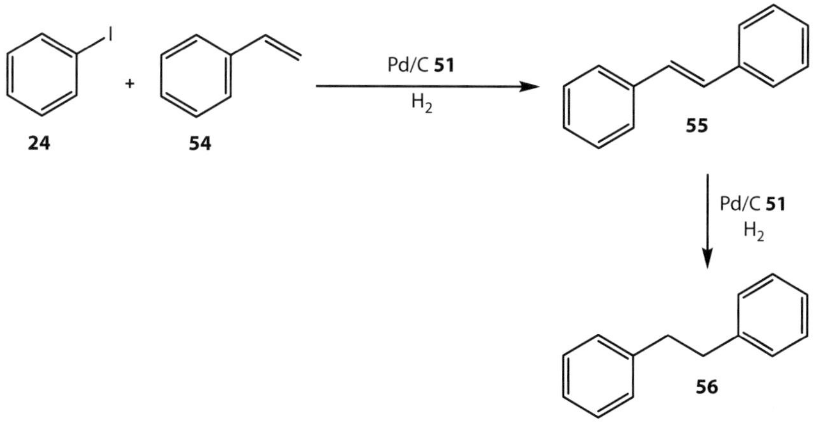

Micro Reactions Employing a Gaseous Component

SCHEME 2.19 Synthesis of 1,2-diphenylethane **56** via Heck coupling and hydrogenation reactions.

reactor, developed by Corning Inc. (USA), the authors devised a fluidic protocol capable of performing multiphase hydrogenation reactions on a scale suitable for industrial production, that is, 1000s tonne annum^{-1}.

Using glass reactors, containing reaction channels with hydraulic diameters of the order of millimeter-sandwiched between heat exchange layers—the authors were able to obtain both efficient mass and heat transfer ($H = 400$–500 kJ mol^{-1}) (Figure 2.6). The reactor developed consisted of 15 glass modules, used to perform steps such as preheating of reactants, gas introduction, reactant mixing, residence time units and

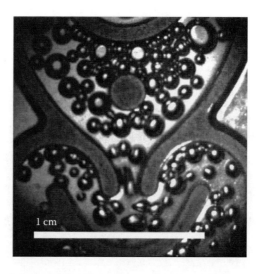

FIGURE 2.6 Illustration of the micro mixer employed to ensure efficient mixing of the three-phase system. (Reproduced from Buisson, B. et al. 2009. *Chem. Today* 27: 12–14. With permission of Tekno Science Srl.)

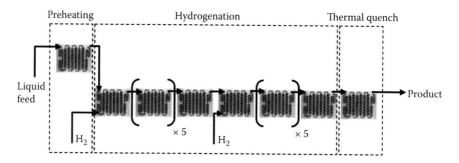

FIGURE 2.7 Schematic illustrating the various glass modules employed to afford a continuous flow process for slurry based hydrogenation reactions. (Reproduced from Buisson, B. et al. 2009. *Chem. Today* 27: 12–14. With permission of Tekno Science Srl.)

thermal quenching (Figure 2.7). To demonstrate the synthetic utility of the reactor, the authors selected the catalytic hydrogenation of an undisclosed substrate using a three phase reaction mixture (solid–liquid–gas), comprising of an alcoholic solvent, hydrogen gas, and a solid noble-metal catalyst (particle size = 30 µm). Using this approach, the authors investigated the effect of reactor temperature (30–140°C), H_2 molar ratio (3.0–4.7), dosing ratio (from two inlets) and catalyst content (0.1–0.4 wt.%) on the conversion and selectivity of the reaction. As Table 2.14 illustrates, the reaction responded well to an increase in reactor temperature, obtaining 98.9% conversion (93.0% selectivity) at 140°C; analysis performed by off-line HPLC with UV detection.

In addition to enabling the use of higher operating temperatures, compared to standard batch hydrogenation equipment, the authors also comment on the ability to recover unused H_2 in a downstream process, enabling recovery, and recycle. Using the aforementioned experimental set-up, the authors synthesized the target compound in 0.43 kg h^{-1}; therefore, by operating 20 of the reactors in parallel, the technique would be suitable for production-scale synthesis (200 tonne annum^{-1}) of the undisclosed product. In addition to the large throughput of the system, the development of a

TABLE 2.14
Summary of the Results Obtained for the Optimization of a Three-Phase Hydrogenation Reaction Performed in a Glass-Based Micro Reactor

Temperature (°C)	Pressure (bar)	Liquid Flow Rate (g min^{-1})	H_2 Molar Ratio	Reactor Volume (mL)	Conversion (%)	Selectivity (%)
30	11	16	4.0	64	<20	99.0
70	11	16	4.0	64	61.9	97.3
140	11	16	4.0	64	85.0	97.0
140	17	16	4.0	112	98.9	93.0

Employing a substrate concentration of 30 wt% and a catalyst concentration of 0.4 wt%.

slurry-based continuous process has the potential to be applied to a significant proportion of synthetic transformations that would have previously been overlooked as candidates to benefit from continuous processing due to precipitation of intermediates/products or poor solubility of precursors (see Chapter 1).

Another example of the manipulation of slurries within continuous flow reactors was reported by Stitt and coworkers [90] demonstrating the hydrogenation of resorcinol.

2.3.5 MISCELLANEOUS TRIPHASIC MICRO REACTIONS

To demonstrate the ability to perform multistep three-phase reactions under continuous flow, Abahamane et al. [91] developed a method for the synthesis of α-substituted pyridines using Montmorillonite **57** and nanoparticle-impregnated Alumina **58**.

As Scheme 2.20 illustrates, the reaction comprises of three stages, firstly enamine formations, followed by a pericyclic reaction (6-endo-dig ring closure) of the enamine to afford a dihydropyrimidine and finally oxidation to afford the target pyridine derivative. When conducted in batches, numerous side reactions result, largely due to the polymerization of propargylamine **59**, even when performed in the presence of activating catalysts. With this in mind, the authors split the reaction into two steps, first, the imine formation in the presence of Montmorillonite **57** and second, the oxidation in the presence of nanoparticle-impregnated alumina and air, as a means of improving the cleanliness of the process.

At a T-junction, solutions of propargylamine **59** and the methyl ketone were brought together prior to entering a packed-bed (dimensions = 1 mm (i.d.) × 50 cm (long)) fabricated from PEEK tubing (coiled around a heater). At the channel outlet, air was introduced via a T-mixer before the reaction mixture entered the second packed bed.

A series of ketones were investigated with the aim to span a series of boiling points, in the case of aliphatic ketones, stoichiometric ketone to amine **59** ratio was found to afford moderate to high conversions of the desired imine at 75°C, whereas aromatic ketones were employed in 2:1 ratio and a reactor temperature of 125°C.

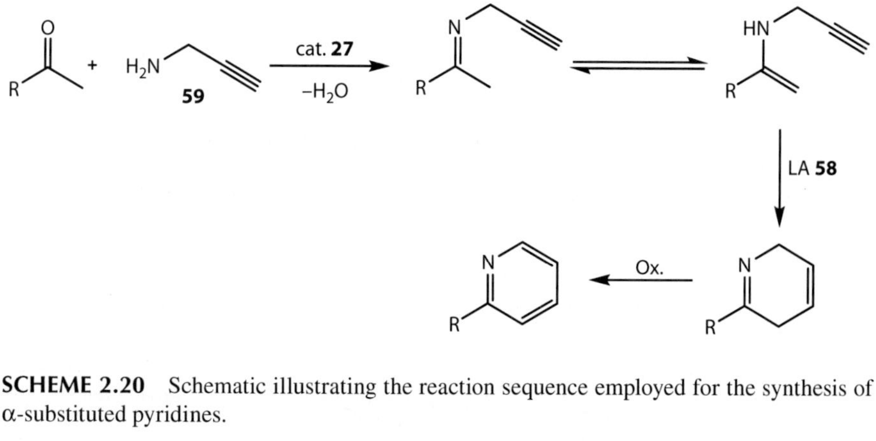

SCHEME 2.20 Schematic illustrating the reaction sequence employed for the synthesis of α-substituted pyridines.

TABLE 2.15
Comparison of the Results Obtained in Batch and a Two-Step PEEK Flow Reactor

Ketone	α-Substituted Pyridine	Conversion (%)	
		One-Step Batch	Two-Step Flow
(phenyl methyl ketone) 43	2-phenylpyridine	17	94
(butan-2-one)	2-ethylpyridine	21	97
(acetone)	2-methylpyridine	61	100

In the second-stage reactor, air was employed as the oxidizing agent and Au-impregnated Alumina 58 (2.5 wt.%) as the catalyst. Employing a reactor temperature of 125°C, excellent conversions to the target α-substituted pyridine were obtained with a reaction time of 25 s. As Table 2.15 illustrates, using this two-step approach dramatic improvements in yield were obtained when compared to one-pot batch reactions.

REFERENCES

1. Jähnisch, K., Hessel, V., Löwe, H., and Baerns, M. 2004. Chemistry in microstructured reactors, *Angew. Chem. Int. Ed.* 43: 406–446.
2. Kashid, M. V. and Kiwi-Minsker, L. 2009. Microstructured reactors for multiphase reactions: State of the art, *Ind. Eng. Chem. Res.* 48: 6465–6485.
3. Kolb, G. and Hessel, V. 2004. Micro-structured reactors for gas phase reactions, *Chem. Eng. J.* 98: 1–38.
4. Löb, P., Löwe, H., and Hessel, V. 2004. Fluorinations, chlorinations and brominations of organic compounds in micro reactors, *J. Fluor. Chem.* 125: 1677–1694.
5. Ehrfeld, W., Hessel, V., and Löwe, H. 2000. *Microreactors New Technology for Modern Chemistry*, Wiley-VCH, Germany.
6. Schwarz, O., Duong, P.-Q., Schäfer, G., and Schomäcker, R. 2009. Development of a microstructured reactor for heterogeneously catalyzed gas phase reactions: Part 1. Reactor fabrication and catalyst coatings', *Chem. Eng. J.* 145: 402–428.
7. Kiwi-Minsker, L. and Renken, A. 2005. Microstructured reactors for catalytic reactions, *Catal. Today* 110: 2–14.
8. Vervloet, D., Kamali, M. R., Gillissen, J. J. J., Nijenhuis, J., van den Akker, H. E. A., Kapteijn, F., and van Ommen, J. J. 2009. Intensification of co-current gas–liquid reactors using structured catalytic packings: A multi-scale approach, *Catal. Today* 147S: S138–S143.
9. Veser, G, Friedrich, G., Freyganag, M., and Zengerle, R. 2000. A modular microreactor design for high temperature catalytic oxidation reactions, *3rd International Conference on Microreaction Technology*, Frankfurt, Germany, 678–686.

10. Inoue, T., Schmidt, M. A., and Jensen, K. F. 2007. Microfabricated multiphase reactors for the direct synthesis of hydrogen peroxide from hydrogen and oxygen, *Ind. Eng. Chem. Res.* 46: 1153–1160.
11. Hönicke, D. and Wießmeier, G. 1996. Heterogeneously catalyzed reactions in a microreactor, *Microsystem Technology for Chemical and Biochemical Microreactors*: DECHEMA Monologue, 132: 93–107.
12. Löwe, H. and Ehrfeld, W. 1999. State-of-the-art in microreaction technology: Concepts, manufacturing and applications, *Electrochimica Acta.* 44: 3679–3689.
13. Kestenbaum, H., de Oliveria, A. L., Schmidt, W., Schüth, F, Ehrfeld, W., Gebauer, K., Löwe, H., and Richter, Th. 2000. Synthesis of ethylene oxide in a microreaction system, *Microreaction Technology, Industrial Prospects*, 207.
14. Kestenbaum, H., de Oliveira, A. L., Schmidt, W., and Schüth, F. 2002. Silver-catalyzed oxidation of ethylene to ethylene oxide in a microreaction system, *Ind. Eng. Chem. Res.* 41: 710–719.
15. Kursawe, A., Dietzsch, E., Kah, S, Hönicke, D., Fichtner, M., Schubert, K., and Wießmeier, G. 2000. Selective reactions in microchannels, *3rd International Conference on Microreaction Technology*, Frankfurt, Germany, 213–223.
16. Mota, S., Miachon, S., Volta, J.-C., and Dalmon, J.-A. 2001. Membrane reactor for selective oxidation of butane to maleic anhydride, *Catal. Today* 67: 169–176.
17. Liu, S., Xu, L., Xie, S., and Wang, Q. 2003. Performance of palladium-containing supported catalysts in the oxidation of 1-butene, *React. Kinet. Catal. Lett.* 79: 281–286.
18. Ismagilov, I. Z., Michurin, E. M., Sukhova, O. B., Tsykoza, L. T., Matus, E. V., Kerzhentsev, M. A., Ismagilov, Z. R. et al. 2008. Oxidation of organic compounds in a microstructured catalytic reactor, *Chem. Eng. J.* 135S: S57–S65.
19. Hwang, S., Kwon, O. J., Ahn, S. H., and Kim. J. J. 2009. Silicon-based micro-reactor for preferential CO oxidation, *Chem. Eng. J.* 146: 105–111.
20. Y. S., Tanaka, S., Esashi, M., Hamakawa, S., Hanaoka, T., and Mizukami, F. 2005. Thin palladium membrane microreactors with oxidized porous silicon supports and their application, *J. Micromech. Microeng.* 15: 2011–2018.
21. Rebrov, E. V., Klinger, E. A., Berenguer-Murcia, A., Sulman, E. S., and Schouten, J. C. 2009. Selective hydrogenation of 2-methyl-3-butyne-2-ol in a wall-coated capillary microreactor with a $Pd_{25}Zn_{75}/TiO_2$ catalyst, *Org. Proc. Res. Dev.* 13: 991–998.
22. Rebrov, E. V., Berenguer-Murcia, A., Skelton, H. E., Johnson, B. F. G., Wheatley, A. E. H., and Schouten, J. C. 2009. Capillary microreactors wall-coated with mesoporous titania thin film catalyst supports, *Lab Chip* 9: 503–506.
23. Grasemann, M., Renken, A., Kashid, M., and Kiwi-Minsker, L. 2010. A novel compact reactor for three-phase hydrogenations, *Chem. Eng. Sci.* 65: 364–371.
24. Creaser, D., Andersson, B., Hudgins, R. R., and Silverston, P. L. 1999. Transient kinetic analysis of the oxidative dehydrogenation of propane, *J. Catal.* 182: 264–269.
25. Wilson, N. G. and McCreedy, T. 2000. On-chip catalysis using a lithographically fabricated glass microreactor—The dehydration of alcohols using sulfated zirconia, *Chem. Commun.* 733–734.
26. Koubeck, J., Pasek, J., and Ruzicka, V. 1980. Exploitation of a non-stationary kinetic phenomena for the elucidation of surface processes in a catalytic reaction, *New Horizons Catal.* 853–562.
27. Chen, G., Li, S., Jiao, F., and Yuan, Q. 2007. Catalytic dehydration of bioethanol to ethylene over $TiO_2.\gamma\text{-}Al_2O_3$ catalysts in microchannel reactors, *Catal. Today* 125: 111–119.
28. Wörz, O., Jäckel, K.-P., Richter, T., and Wolf, A. 2001. Microreactors—A new efficient tool for reactor development, *Chem. Eng. Technol.* 24: 138–142.
29. Wörz, O., Jäckel, K.-P., Richter, T., and Wolf, A. 2001. Microreactors—A new efficient tool for optimum reactor design, *Chem. Eng. Sci.* 56: 1029–1033.

30. Jäckel, K. P. 1996. Microtechnology: Application opportunities in the chemical industry, *Microsystem Technol. Chem. Biochem. Microreactors* 132: 29–50.
31. Yamamto, S., Hanaoka, T., Hamakawa, S., Sato, K., and Mizukami, F. 2006. Application of a microchannel to catalytic dehydrogenation of cyclohexane on Pd membrane, *Catal. Today* 118: 2–6.
32. Sadykov, V. A., Pavlova, S. N., Saputina, N. F., Zolotarskii, I. A., Pakhomov, N. A., Moroz, E. M., Kuzmin, V. A., and Kalinkin, A. V. 2000. Oxidative dehydrogenation of propane over monoliths at short contact times, *Catal. Today* 61: 93–99.
33. Myrstad, R., Eri, S., Pfeifer, P., Rytter, E., and Holmen, A. 2009. Fischer-Tropsch synthesis in a microstructured reactor, *Catal. Today* 147S: S301–S304.
34. Ehrfeld, W., Hessel, V., and Löwe, H. 2000. *Microreactors New Technology for Modern Chemistry*, Wiley-VCH, Germany.
35. Srinivasan, R., Hsing, I.-M., Berger, P., Jensen, K. F., Firebaugh, S. L., Schmidt, M. A., Harold, M. P., Lerou, J. J., and Ryley, J. F. 1997. Micromachined reactors for catalytic partial oxidation reactions, *AIChE J.* 43: 3059–3069.
36. Rebrov, E. V., de Croon, M. H. J. M., and Schouten, J. C. 2002. Development of a cooled microreactor for platinum catalyzed ammonia oxidation, *5th International Conference on Microreaction Technology*, Berlin, 45–59.
37. Hessel, V., Ehrfeld, W., Golbig, K., Hofmann, C., Jungwirth, S., Löwe, H., Richter, T. et al. 2000. High temperature HCN generation in an integrated microreaction system, *3rd International Conference on Microreaction Technology*, Frankfurt, Germany, 151–164.
38. Ajmera, S. K., Delattre, C., Schmidt, M. A., and Jensen, K. F. 2002. Microfabricated differential reactor for heterogeneous gas phase catalyst testing, *J. Catal.* 209: 401–412.
39. Reuse, P., Renken, A., Haas-Santo, K., Görke, O., and Schubert, K. 2004. Hydrogen production for fuel cell application in an auto-thermal micro-channel reactor, *Chem. Eng. J.* 101: 133–141.
40. Louis, B., Reuse, P., Kiwi-Minsker, L., and Renken, A. 1993. Synthesis of ZSM-5 coatings on stainless steel grids and their catalytic performance for partial oxidation of benzene by N_2O, *Appl. Catal. A: Gen.* 96: 3–13.
41. Günther, A., Jhunjhunwala, M., Thalmann, M., Schmidt, M. A., and Jensen, K. F. 2005. Micromixing of miscible liquids in segmented gas–liquid flow, *Langmuir* 21: 1547–1555.
42. Günther, A., Khan, S. A., Thalmann, M., Trachsel, F., and Jensen, K. F. 2004. Transport and reaction in microscale segmented gas–liquid flow, *Lab Chip* 4: 278–286.
43. Hibara, A., Iwayama, S., Matsuoka, S., Ueno, M., Kikutani, Y., Tokeshi, M., and Kitamori, T. 2005. Surface modification method of microchannels for gas–liquid two-phase flow in microchips, *Anal. Chem.* 77: 943–947.
44. Oskooei, S. A. K. and Sinton, D. 2010. Partial wetting gas–liquid segmented flow microreactor, 2010. *Lab Chip* 10: 1732–1734.
45. Wehle, D., Dejmek, M., Rosenthal, J., Ernst, H., Kampmann, D., Trautschold, S., and Pechatschek, R. 2000. Verfahren zur Herstellung von Monochloressihaure in mikroreaktoren, DE 10036603 A1.
46. Ehrich, H., Linke, D., Morgenschweis, K., Baerns, M., and Jähnisch, K. 2002. Application of microstructured reactor technology for the photochemical chlorination of alkylaromatics, *Chimia* 56(11): 647–653.
47. Chambers, R. D. and Spink, R. C. H. 1999. Micro reactors for elemental fluorine, *Chem. Commun.* 883–884.
48. Chambers, R. D., Holling, D., Rees, A. J., and Sandford, G. 2003. Microreactors for oxidations using fluorine, *J. Fluor. Chem.* 119: 81–82.
49. Chambers, R. D., Fox, M. A., Sandford, G., Trmcic, J., and Goeta, A. 2007. Elemental fluorine: Part 20, direct fluorination of deactivated aromatic systems using microreactor techniques, *J. Fluor. Chem.* 1228: 29–33.

50. Chambers, R. D., Holling, D., Spink, R. C. H., and Sandford, G. 2001. Elemental fluorine. Part 13. Gas–liquid thin film microreactors for selective direct fluorination, *Lab Chip* 1L 132–137.
51. Chambers, R. D., Fox, M. A., Holling, D., Nakano, T., Okaze, T., and Sandford, G. 2005. Elemental fluorine. Part 16. Versatile thin-film gas–liquid multi-channel microreactors for effective scale-out, *Lab Chip* 5: 191–198.
52. Chambers, R. D., Sandford, G., Trmcic, J., and Okazoe, T. 2008. Elemental fluorine. Part 21. Direct fluorination of benzaldehyde derivatives, *Org. Proc. Res. Dev.* 12: 339–344.
53. de Mas, N., Günther, A., Schmidt, M. A., and Jensen, K. F. 2009. Increasing productivity of microreactors for fast gas–liquid reactions: The case of direct fluorination of toluene, *Ind. Eng. Chem. Res.* 48: 1428–1434.
54. Wada, Y., Schmidt, M. A., and Jensen, K. F. 2006. Flow distribution and ozonolysis in gas–liquid multichannel microreactors, *Ind. Eng. Chem. Res.* 45: 8036–8042.
55. Steinfeldt, N., Abdallah, R., Dingerdissen, U., and Jähnisch, K. 2007. Ozonolysis of acetic acid 1-vinyl-hexyl ester in a falling film microreactor, *Org. Proc. Res. Dev.* 11: 1025–1031.
56. Hübner, S., Bentrup, U., Budde, U., Lovis, K., Dietrich, T., Freitag, A., Küpper, L., and Jähnisch, J. 2009. An ozonolysis-reduction sequence for the synthesis of pharmaceutical intermediates in microstructured devices, *Org. Proc. Res. Dev.* 13: 952–960.
57. O'Brien, M., Baxendale, I. R., and Ley, S. V. 2010. Flow ozonolysis using a semipermeable Teflon AF-2400 membrane to effect gas–liquid contact, *Org. Lett.* 12(7): 1596–1598.
58. Miller, P. W., Long, N. J., de Mello, A. J., Vilar, R., Passchier, J., and Gee, A. 2006. Rapid formation of amides via carbonylative coupling reactions using a microfluidic device, *Chem. Commun.* 546–548.
59. Murphy, E. R., Martinelli, J. R., Zaborenko, N., Buchwald, S. L., and Jensen, K. F. 2007. Accelerating Reaction within microreactors at elevated temperatures and pressures: Profiling aminocarbonylation reactions, *Angew. Chem. Int. Ed.* 46: 1734–1737.
60. Balogh, J., Kuik, A., Urge, L., Darvas, F., Bakos, J., and Skoda-Földes, R. 2009. Double carbonylation of iodobenzene in a microfluidics-based high throughput flow reactor, *J. Mol. Catal. A: Chem.* 302: 76–79.
61. Csajági, C., Borcsek, B., Niesz, K., Kovás, I., Székelyhidi, Z., Bajkó, Z., Urge, L., and Darvas, F. 2008. High-efficiency aminocarbonylation by introducing CO to a pressurized continuous flow reactor, *Org. Lett.* 10: 1589–1592.
62. Fukuyama, T., Rahman, M. T., Kamata, N., and Ryu, I. 2009. Radical carbonylations using a continuous flow microflow system. *Beilstein J. Org. Chem.* 5(34).
63. Rahman, M. T., Fukuyama, T., Kamata, N., Sato, M., and Ryu, I. 2006. Low pressure Pd-catalyzed carbonylation in an ionic liquid using a multiphase microflow system, *Chem. Commun.* 2236–2238.
64. Takebayashi, Y., Sue, K., Yoda, S., Furuya, T., and Mae, K. 2010. Direct carbonylation of nitrobenzene to phenylisocyanate with microreaction system, *11th International Conference on Microreaction Technology*, Kyoto, Japan, 188–189.
65. de Bellefon, C., Lamouille, T., Pestre, N., Bornette, F., Pennemann, H., Neumann, F., and Hessel, V. 2005. Asymmetric catalytic hydrogenations at micro-litre scale in a helicoidal single channel falling film micro-reactor, *Catal. Today* 110: 179–187.
66. Caygill, G., Zanfir, M., and Gavriilidis, A. 2006. Scalable reactor design for pharmaceuticals and fine chemicals production 1: Potential scale-up obstacles, *Org. Proc. Res. Dev.* 10: 539–552.
67. Zanfir, M. and Gavriilidis, A. 2007. Scalable reactor design for pharmaceuticals and fine chemicals production 2: Evaluation of potential scale-up obstacles for asymmetric transfer hydrogenation, 11: 966–971.

68. Sun. X. and Gavriilidis, A. 2008. Scalable reactor design for pharmaceuticals and fine chemicals production 3. A novel gas–liquid reactor for catalytic asymmetric transfer hydrogenation with simultaneous acetone stripping, *Org. Proc. Res. Dev.* 12: 1218–1222.
69. McPake, C. B., Murray, C. B., and Sandford, G. 2009. Epoxidation of alkenes using HOF: MeCN by a continuous process, *Tetrahedron Lett.* 50: 1674–1676.
70. Park, C. P., Fang, Q., Li, X., and Kim, D. 2010. Boron-Heck reaction microreactor with thin PDMS membrane layer, *11th International Conference on Microreaction Technology*, Kyoto, Japan, 186–187.
71. Gligorich, K. M., and Sigman, M. S. 2009. Recent advancements and challenges of palladium^{II}-catalyzed oxidation reactions with molecular oxygen as the sole oxidant, *Chem. Commun.* 3854–3867.
72. Leclerc, A., Alamé, M., Schweich, D., Pouteau, P., Selattre, C., and de Bellefon, C. 2008. Gas–liquid selective oxidations with oxygen under explosive conditions in a microstructured reactor, *Lab Chip* 8: 814–817.
73. Wang, X., Nie, Y., Lee, J. l. C., and Jaenicke, S. 2007. Evaluation of multiphase microreactors for the direct formation of hydrogen peroxide, *Appl. Catal. A: Gen.* 317: 258–265.
74. Müller, A., Cominos, V., Hessel, V., Horn, B., Schürer, J., Ziogas, A., Jähnisch, K. et al. 2005. Fluidic bus system for chemical process engineering in the laboratory and for small-scale production, *Chem. Eng. J.* 107: 205–214.
75. de Bellefon, C., Tanchoux, N., Caravieilhes, S., Grenouillet, P., and Hessel, V. 2000. Microreactors for dynamic, high-throughput screening of fluid/liquid molecular catalysis, *Angew. Chem. Int. Ed.* 39: 3442–3445.
76. Losey, M. W., Schmidt, M. A., and Jensen, K. F. 2001. Microfabricated multiphase packed-bed reactors: Characterization of mass transfer and reactions, *Ind. Eng. Chem. Res.* 40: 2555–2562.
77. Nijhuis, T. A., Kreutzer, M. T., Romijn, A. C. Kapteijn, F., and Moulijn, J. A. 2001. Monolithic catalysts as more efficient three-phase reactors, *Catal. Today* 66: 157–165.
78. Wang. N., Matsumoto, T., Ueno, M., Miyamura, H., and Kobayashi, S. 2009. A gold-immobilized microchannel flow reactor for oxidation of alcohols with molecular oxygen, *Angew. Chem. Int. Ed.* 48: 4744–4746.
79. Miller, P. W., Long, N. J., de Mello, A. J., Vilar, R., Audrain, H., Bender, D., Passchier, J., and Gee, A. 2007. Rapid multiphase carbonylation reactions by using a microtube reactor: Application in positron emission tomography ^{11}C-radiolabeling, *Angew. Chem. Int. Ed.* 46: 2875–2878.
80. Hessel, V., Ehrfeld, W., Golbig, K., Haverkamp, V., Löwe, H., Storz, M., Wille, Ch., Guber, A., Jahnisch, K., and Baerns, M., 2000. In *Microreaction Technology, Industrial Prospects*, Springer, Berlin, p. 526.
81. Jones, R. V., Godorhazy, L., Varga, N., Szalay, D., Urge, L., and Darvas, F. 2006. Continuous flow high pressure hydrogenation reactor for optimization and high-throughput synthesis, *J. Comb. Chem.* 8: 110–116.
82. Szöllösi, G., Cserényi, S., Fülöp, F., and Bartók, M. 2008. New data to the origin of rate enhancement on the Pt-cinchona catalyzed enantioselective hydrogenation of activated ketones using continuous-flow fixed-bed reactor system, *J. Catal.* 260: 245–253.
83. Lou, S., Dai, P., and Schaus, S. E. 2007. Asymmetric Mannich reaction of dicarbonyl compounds with α-amido sulfones catalyzed by cinchona alkaloids and synthesis of chiral dihydropyrimidones, *J. Org. Chem.* 72: 9998–10008.
84. Desai, B. and Kappe, C. O. 2005. Heterogeneous hydrogenation reactions using a continuous flow high-pressure device, *J. Comb. Chem.* 7: 641–643.
85. Irfan, M., Petricci, E., Glasnov, T. N., Taddei, M., and Kappe, C. O. 2009. Continuous flow hydrogenation of functionalized pyridines, *Eur. J. Org. Chem.* 1327–1334.

86. Fan, X., Lapkin, A. A., and Plucinski, P. K. 2009. Liquid phase hydrogenation in a structured multichannel reactor, *Catal. Today* 147S: S313–S318.
87. Kobayashi, J., Mori, Y., Okamoto, K., Akiyama, R., Ueno, M., Kitamori, T., and Kobayashi, S. 2000. A microfluidic device for conducting gas–liquid–solid hydrogenation reactions, *Science* 304: 1305–1308.
88. Kobayashi, J., Mori, Y., and Kobayashi, S. 2005. Hydrogenation reactions using $scCO_2$ as a solvent in microchannel reactors, *Chem. Commun.* 2567–2568.
89. Buisson, B., Donegan, S., Parracho, A., Gamble, J., Caze, P., Jorda, J., and Guermeur, C. 2009. Slurry hydrogenation in a continuous flow reactor for pharmaceutical application, *Chem. Today* 27: 12–14.
90. Enache, D. I., Hutchings, G. J., Taylor, S. H., Raymahasay, S., Winterbottom, J. M., Mantle, M. D., Sederman, A. J. et al. 2007. Multiphase hydrogenation of resorcinol in structured and heat exchange reactor systems influence of the catalyst and the reactor configuration, *Catal. Today* 128: 26–35.
91. Abahamane, L., Knauer, A., Ritter, U., Köhler, J. M., and Groß, G. A. 2009. Heterogeneous catalyzed pyridine synthesis using montmorillionite and nanoparticle-impregnated alumina in a continuous micro flow system, *Chem. Eng. Technol.* 32(11): 1799–1805.

3 Liquid-Phase Micro Reactions

With a vast majority of synthetic reactions performed within research laboratories centering on liquid-phase reaction conditions, due to the ease of operation and availability of equipment, it is not surprising that this area of continuous flow chemistry has received significant interest from researchers over the past 15 years. With this in mind for ease of use, the following chapter has been split into classes of reaction (Sections 3.1 through 3.8), walking those new to the field through developments made and directing those actively researching flow chemistry to reactions of particular interest. The chapter then concludes with a discussion of multistep/multicomponent reactions (Section 3.9), all of which have been selected to demonstrate the ability to perform complex and synthetically challenging reactions within liquid-phase flow reactors.

For an in-depth coverage of recent developments made in the field of liquid-phase micro reactions, see reviews by Hartman and Jensen [1], Jähnisch et al. [2], Steven and coworkers [3], Wiles and Watts [4,5], books written by Hessel et al. [6] and Yoshida [7] or those titles edited by Hessel et al. [8], Schouten [9], and Wirth [10].

3.1 NUCLEOPHILIC SUBSTITUTION

3.1.1 C–C Bond Formation: Acylation Reactions

With the enolate often referred to as the most important synthetic intermediate used in the formation of C–C bonds, several authors have investigated their formation and subsequent reaction within microflow reactors. As observed under standard batch conditions, a significant proportion of work has been performed in order to gain further understanding of the reaction conditions that promote the clean and efficient preparation and acylation of ambident enolates; including reaction solvent, temperature, and reactant stoichiometry.

Regioselective Acylation of Enol Ethers: Exploiting the fluidic control obtained using EOF as a pumping mechanism, Wiles et al. [11] evaluated the regioselective acylation of silyl enol ethers using a range of acylating agents, as illustrated in Scheme 3.1 ($X = $ Cl **1**, F **2**, and CN **3**). Building on their experience gained in a previous batch investigation [12], the authors employed an in-house fabricated borosilicate glass micro reactor (channel dimensions = 100 μm (wide) × 50 μm (deep)) exploring the use of preformed enolates, in the form of silyl enol ethers.

By independently introducing solutions of acylating agent (1.0 M), anhydrous tetrabutylammonium fluoride **4** (TBAF) (0.1 M) and the enol ether **5** (1.0 M), prepared in anhydrous THF, the authors were able to convert the enol ether **5** into the

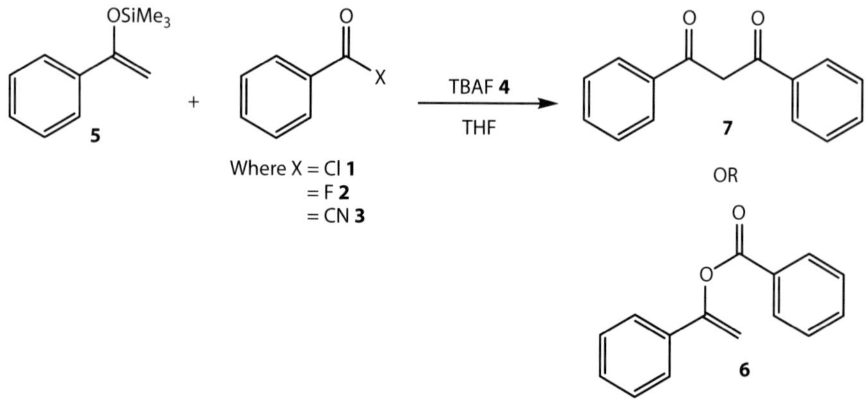

SCHEME 3.1 Illustration of the model reaction selected to enable the evaluation of regioselectivity under continuous flow conditions.

respective enolate, *in situ*, and react it immediately with the acylating agent to afford the respective *O*- **6** or *C*- **7** product. Using offline GC–MS analysis, with retention times determined via the analysis of fully characterized synthetic standards, the authors were able to compare the effect of leaving group on the product formed.

In the case of benzoyl fluoride **2**, applied fields of 333, 455, 333, and 0 V cm^{-1} afforded quantitative conversion of the enol ether **5** to 1-phenylvinyl benzoate **6**, with no competing *C*-acylation **7** detected; analogous results were observed for benzoyl chloride **1**. Exchanging the acylating agent for benzoyl cyanide **3**, the authors were able to readily alter the product distribution, this time obtaining dibenzoylmethane **7** in quantitative conversion. In the case of α-substituted enolates, *C*-acylation was obtained independent of the acylating agent employed and in no case was diacylation of the precursor observed.

The authors concluded that silyl enol ethers were advantageous as an enolate precursor as they remove the effect of a metal counter-ion on the reaction, circumventing by-product formation conventionally observed between acylating agents and organic base residues. The technique also removed the need to use an inorganic base, which due to their sparing solubility in etherated solvents had the potential to precipitate and cause problems with blockage formation within micro reaction channels (see Chapter 1). Although useful as a research tool, due to the simplicity of the equipment required, EOF is not suitable for the large-scale production of chemicals under continuous flow.

For details of a continuous enolate formation applied to multistep syntheses, see the synthesis of Rimonabant (Chapter 7, Section 7.3.6) and refer to Section 3.17 for an example of diasteroselective enolate alkylation.

3.1.2 C–C Bond Formation: Alkylation Reactions

Using the alkylation of ethyl 2-oxocyclopentanecarboxylate **8** with benzyl bromide **9**, to afford **10** (Scheme 3.2) Kobayashi and coworkers [13] demonstrated the ability to perform phase transfer C–C bond forming reactions under continuous flow.

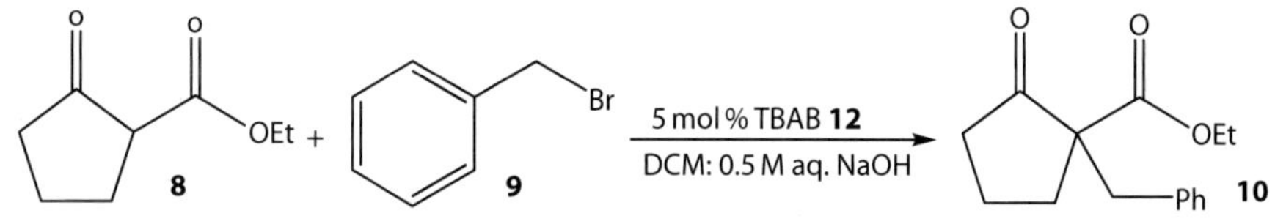

Liquid-Phase Micro Reactions

SCHEME 3.2 Schematic illustrating the phase-transfer alkylation of ethyl 2-oxocyclopentanecarboxylate **8** performed within a glass micro reactor (channel dimensions = 200 μm (wide) × 100 μm (deep) × 45.0 cm (long)).

To perform a reaction, the authors employed a glass micro reactor (channel dimensions = 200 μm (wide) × 100 μm (deep) × 45.0 cm (long)) and as illustrated in Figure 3.1, the organic reactants, ethyl 2-oxopentanecarboxylate **8** (0.3 M), and benzyl bromide **9** (0.45 M) in DCM, were introduced from one inlet and the aqueous reactants, sodium hydroxide **11** (0.5 M) and tetrabutylammonium bromide **12** (1.5×10^{-2} M), were introduced from a second inlet. The reactant streams met at a Y-shaped channel where a segmented flow resulted and the authors evaluated the effect of residence time with respect to the volumetric flow rate, with analysis of the organic fraction (by HPLC) used to quantify the proportion of product **8** formed.

The reaction proceeded well within the microchannel, affording 57% conversion to **8** in 60s, increasing to 96% after 10 min; compared to macroscale reactions, this represented a dramatic increase in reaction efficiency, as depicted in Figure 3.2.

Having demonstrated superior results within a continuous flow system for the alkylation of ethyl 2-oxopentanecarboxylate **8**, the authors investigated the reaction of a series of ketonic substrates with a range of alkyl bromides (Table 3.1) obtaining

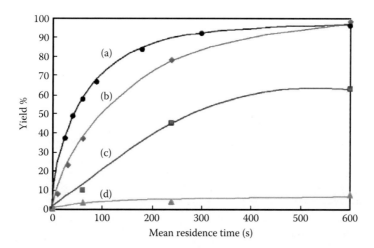

FIGURE 3.1 Summary of the results obtained for the alkylation of ethyl 2-oxocyclopentanecarboxylate **8** under (a) micro reactor, (b) round-bottomed flask (1350 rpm), (c) round-bottomed flask (400 rpm), and (d) round-bottomed flask (0 rpm) conditions. (Ueno, M. et al., 2003. Phase-transfer alkylation reactions using microreactors, *Chem. Commun.* 936–937. Reproduced by permission of the Royal Society of Chemistry.)

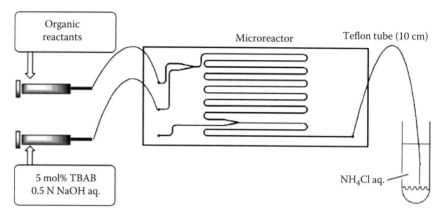

FIGURE 3.2 Schematic illustrating the reaction setup employed for the phase transfer alkylations performed under continuous flow. (Ueno, M. et al., 2003. Phase-transfer alkylation reactions using microreactors, *Chem. Commun.* 936–937. Reproduced by permission of the Royal Society of Chemistry.)

enhanced isolated yields in all cases; a finding that is due to the increased interfacial area obtained through the use of segmented flow (see Chapter 1).

Generation and Reaction of N-(t-Butylsulfonyl)aziridinyllithiums: Despite their versatility as building blocks in organic synthesis, the production of α-substituted aziridines is often problematic due to the instability of aziridinyllithiums; which readily undergo side reactions such as C–H insertion reactions and reductive alkylations; consequently, reduced processing temperatures are necessary. When looking at compound production, the use of reduced reaction temperatures is both costly and difficult to achieve; as such, synthetic routes such as this are often overlooked. With this in mind, Yoshida and coworkers [14] recently investigated the generation and subsequent reaction of α-lithiated aziridines under continuous flow in an effort toward developing methods for the efficient synthesis of α-substituted aziridines at scale. Having previously illustrated the effectiveness of microflow systems for the generation of unstable intermediates, the authors evaluated a stainless-steel (SUS316) reactor for the generation, and subsequent reaction of, N-(t-butylsulfonyl)-α-phenylaziridinyllithium **13** (Scheme 3.3).

Employing a tubular flow system comprising two T-shaped micromixers (mixer 1 = 250 μm (i.d.) and mixer 2 = 500 μm (i.d.)) and two microtube reactors (dimensions = 1000 μm (i.d.) and 50 cm (long)), the authors evaluated the deprotonation of N-Bus-α-phenylaziridine **14** (0.05 M, 6.0 mL min^{-1}) using n-BuLi **15** (0.24 M, 1.5 mL min^{-1}) in THF. The N-Bus-α-phenylaziridinyllithium **13** was subsequently trapped using an electrophile (0.14 M, 3.0 mL min^{-1}) to afford the respective α-substituted N-Bus-2-phenylaziridine. Using the synthesis of N-Bus-2-methyl-2-phenylaziridine **16** as a model reaction, the effect of residence time (0.4 to 12.6 s) and reactor temperature (−48 to −18°C) was investigated.

As Table 3.2 illustrates, the yield of aziridine **16** was found to depend on both reaction time and temperature, with the optimal conditions identified as a residence time of 1.57 s (reactor 1 = 1000 μm (i.d.) × 25 cm (long)) and a reactor temperature

TABLE 3.1
Illustration of the Synthetic Scope of the Phase-Transfer Alkylation Methodology

Product	Isolated Yield (%)		
	Batch Reaction (2 min)	Micro Reaction (2 min)	Micro Reaction (10 min)
8	49	75	96
	48	73	85
	18	45	87
	18	35	92
	87	91	97
	20	65	91
	17	44	90

of −28°C affording the target compound **16** in 82% yield. Under the aforementioned conditions, the authors subsequently investigated the reactivity of **13** toward a series of electrophiles, observing that chlorotrimethylsilane **17**, tributylchlorostannane **18**, benzaldehyde **19**, and acetone afforded the corresponding α-substituted *N*-Bus-2-phenylaziridines in moderate-to-excellent yields (Table 3.3). In addition to enabling the generation of *N*-Bus-α-phenylaziridinyllithium **13** at −28°C, compared to −78°C in batch, the use of a microflow system also enabled the intermediate **13** to be generated in the absence of TMEDA; affording a practical and scalable route to

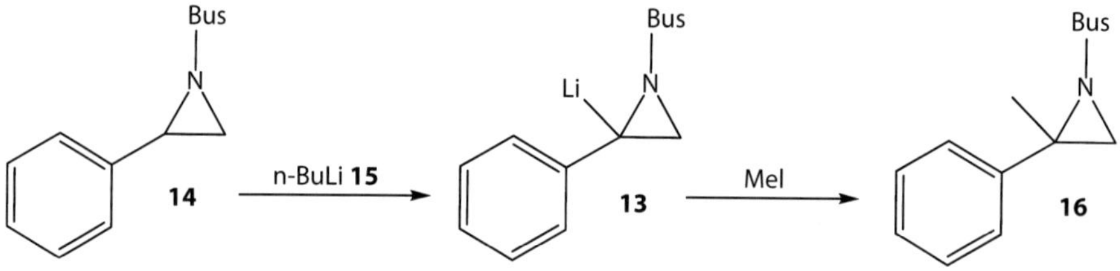
Bus
N
14
n-BuLi 15
Li
13
MeI
16
Bus = t-butylsulfonyl

SCHEME 3.3 Schematic illustrating the preparation, and reaction of, *N*-Bus-α-phenylaziridinyllithium **14**.

TABLE 3.2
Summary of the Results Obtained for the Optimization of *N*-Bus-α-Phenylaziridine 14 Deprotonation Performed under Continuous Flow Conditions

Residence Time (s)	Reactor Temperature (°C)	Conversion (%)	Yield (%)
0.38	−18	85	71
0.79		93	79
1.57		96	74
3.14		97	52
6.28		93	46
12.60		98	32
0.38	−28	71	60
0.79		84	76
1.57		93	82
3.14		95	83
6.28		96	68
12.60		96	58
0.38	−38	57	47
0.79		73	64
1.57		86	78
3.14		92	84
6.28		94	83
12.60		95	80
0.38	−48	36	31
0.79		54	44
1.57		71	64
3.14		83	77
6.28		89	82
12.60		93	83

TABLE 3.3
Summary of the Results Obtained for the Optimization of Reaction N-Bus-α-Phenylaziridine 14 with a Range of Nucleophiles under Continuous Flow Conditions

Electrophile	Product	Yield (%)
MeI		82
Me$_3$SiCl 17		59
Bu$_3$SnCl 18		81
PhCHO 19		78[a]
Me$_2$CO		74

[a] Diastereomer ratio 56:44 as determined by ^{1}H NMR spectroscopy.

these synthetically useful precursors; removing the need for postreaction purification steps.

3.1.3 ENANTIOSELECTIVE C–C BOND-FORMING REACTIONS

In addition to those examples discussed, Odedra and Seeberger [15] reported the organo-catalyzed aldol reaction within a glass micro reactor (mixing zone = 161 μm (wide) × 1240 μm (deep) × 53.6 cm (long) and reaction zone = 391 μm (wide) × 1240 μm (deep) × 18.4 cm (long)). Initial reaction screening identified that the reactions proceeded, with maintained selectivity, at higher temperatures within the micro reactor compared to batch, an observation that was attributed to a more uniform thermal profile within the micro reactor. To perform a reaction, the authors introduced the

TABLE 3.4
Summary of the Results Obtained for a Selection of 5-(Pyrrolidin-2-Yl) Tetrazole 20 Catalyzed Aldol Reactions Performed at 60°C

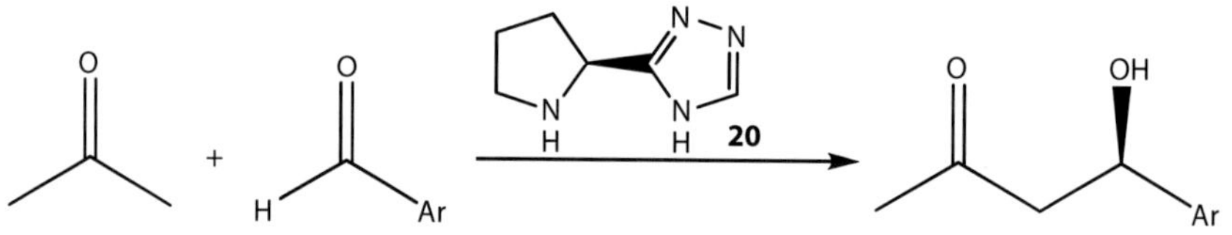

Ar	20 (mol%)	Reaction Time (min)	Yield (%)	ee (%)[a]
4-F$_3$CC$_6$H$_4$	10	20	71	74
4-F$_3$CC$_6$H$_4$	5	30	77	68
4-CNC$_6$H$_4$	5	30	77	70
4-BrC$_6$H$_4$	10	20	53	74
4-BrC$_6$H$_4$	5	30	78	75
C$_6$H$_5$ 19	15	30	36 (25)[b]	62
C$_6$H$_5$ 19	10	30	38 (18)[b]	63
2-Naphthyl	10	30	44 (19)[b]	57

[a] Determined by chiral HPLC analysis.
[b] Corresponding dehydration product.

reactants into the micro reactor as two separate solutions, the first contained the aldehyde (1.0 M) in acetone and the second a solution of 5-(pyrrolidin-2-yl)tetrazole **20** (0.05–0.10 M) in DMSO. Employing a reactor temperature of 60°C, it can be seen from Table 3.4 that the use of a micro reactor affords the target β-hydroxyketones in high yield and enantioselectivity with reaction times as low as 20 min; compared to 40 h in an analogous batch reaction.

In an extension to this, the authors evaluated the Mannich reaction between an α-iminoglyoxylate **21** and cyclohexanone **22** to afford the respective β-amino ketone **23** (Scheme 3.4); again the authors employed a glass micro reactor and DMSO as the

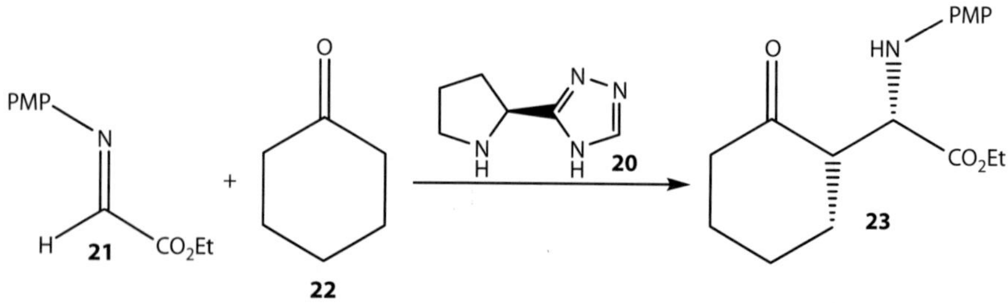

SCHEME 3.4 Schematic illustrating the organo-catalyzed **20** Mannich reaction performed in a glass micro reactor.

reaction solvent. Utilizing the organocatalyst **20** in 5 mol%, the authors were able to synthesize the β-amino ketone **23** in 91% yield and >95% *de* at a reaction temperature of 60°C.

The performance of these two examples enabled the authors to conclude that micro reactors have the potential to accelerate aldol and Mannich reactions, while reducing the proportion of organocatalyst needed to obtain equivalent/higher yields and enantioselectivities than conventional batch techniques.

3.1.4 C–O BOND FORMATION: ESTERIFICATION REACTIONS

Esters are widely employed within the chemical industry, with many low molecular weight derivatives finding application in flavors, fragrances, and cosmetics. In addition, *tert*-butyl esters find widespread application as protecting groups in multistep syntheses. The reversible nature of the reaction can however, prove problematic, especially when reactions are performed on an industrial scale. With these considerations in mind, several authors have reported the synthesis of esters under continuous flow, using both chemical and enzymatic catalysts, as a means of obtaining simple esters in high yield and purity.

Acid Catalyzed: Employing an in-house fabricated glass tubular flow reactor (dimensions = 1.07 cm (i.d.) × 42 cm (long)), Pipus et al. [16] reported an investigation into the acid-catalyzed esterification of benzoic acid **24** to afford the respective ethyl ester **25**. Utilizing microwave irradiation as a means of heating the reactor and a backpressure of 7 atm, the authors investigated the effect of reactor temperatures, up to 140°C, at a fixed flow rate of 1 L h^{-1}. Employing EtOH in a 10-fold excess, with respect to benzoic acid **24**, and 2.5 wt. % H_2SO_4 **26**, the authors quantified product **25** formation by offline HPLC analysis. Using this approach, the authors were able to significantly increase the rate of reaction, which typically takes several days to reach equilibrium in batch (at 80°C).

Mixed Anhydrides: Utilizing electroosmotic flow as a pumping mechanism, Wiles et al. [17] demonstrated the catalytic synthesis of esters derived from a series of *in situ* prepared mixed anhydrides. Employing a borosilicate glass micro reactor, fabricated in-house (channel dimensions = 350 μm (wide) × 52 μm (deep) × 2.5 cm (long)), the authors investigated the synthesis of methyl-, ethyl-, and benzyl-esters (Scheme 3.5).

To perform a reaction, three solutions comprising Et_3N **27** (1.00 M), premixed Boc-glycine **28** and alkylchloroformate (1.00 M), and 4-dimethylaminopyridine

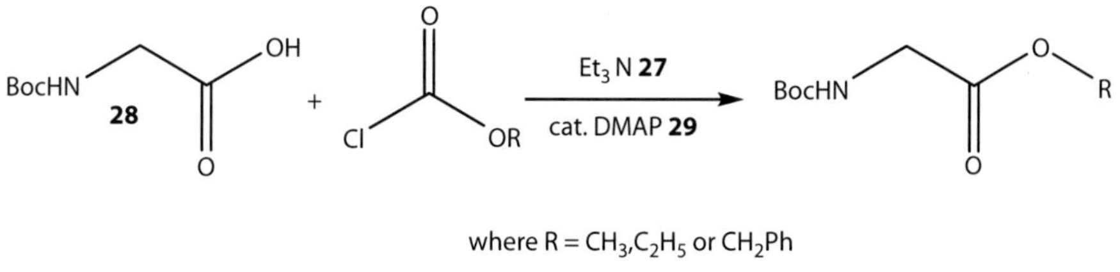

where R = CH_3, C_2H_5 or CH_2Ph

SCHEME 3.5 Schematic illustrating the reaction protocol employed for the esterification of Boc-glycine **28** under continuous flow.

(DMAP) **29** (0.50 M) in anhydrous MeCN, were mobilized through the reactor from inlets A, B, and C using 385, 417, and 364 V cm^{-1}, respectively. Reactants met at T mixers and reactions were conducted for periods of 20 min, with the reaction products collected in MeCN (0 V cm^{-1}) and analyzed offline by GC–MS in order to determine the percentage conversion to the target ester. Using this approach, the authors obtained the target esters in quantitative conversion and excellent product purity; detecting no residual mixed anhydride, or starting materials. The authors also extended their investigation to incorporate the base catalyzed esterification of a series of aromatic carboxylic acids and phenolic derivatives, again obtaining the target aromatic esters in excellent yield and purity at room temperature.

Base-Catalyzed Esterifications: In a more applied example, Lu et al. [18] reported the esterification of a carboxylic acid **30** (0.01 M), with $^{11}CH_3I$ **31** (0.01 M) in the presence of tetra-*n*-butylammonium hydroxide **32** (0.01 M), to afford a ^{11}C-labeled peripheral benzodiazepine ligand **33** (Scheme 3.6). Employing DMF as the reaction solvent, the authors employed an in-house fabricated borosilicate glass T-micro reactor (channel dimensions = 220 μm (wide) × 60 μm (deep) × 1.4 cm (long)), analyzing the reaction products generated by offline HPLC. Using an optimized residence time of 12s, the authors were able to obtain the target ester **33** with a radiochemical yield of 65% (*n* = 2) in an overall processing time of 10 min; comparable to current synthetic methodology for the preparation of PET tracer molecules.

Catalyst-Free Esterifications: Accessing reaction conditions unattainable within conventional reactors, Sato et al. [19] demonstrated the ability to perform the rapid acylation of a series of alcohols using subcritical water (sub-H$_2$O) as the reaction solvent. Unlike conventional esterifications, the authors were able to perform this transformation in the absence of a base or acid catalyst, using water as the reaction solvent; with no competing ester hydrolysis observed.

Using an SUS-316 tube reactor (volume = 49 μL, dimensions = 0.5 mm (i.d.) × 24.7 cm (long)), housed within a furnace, the authors constructed a continuous flow

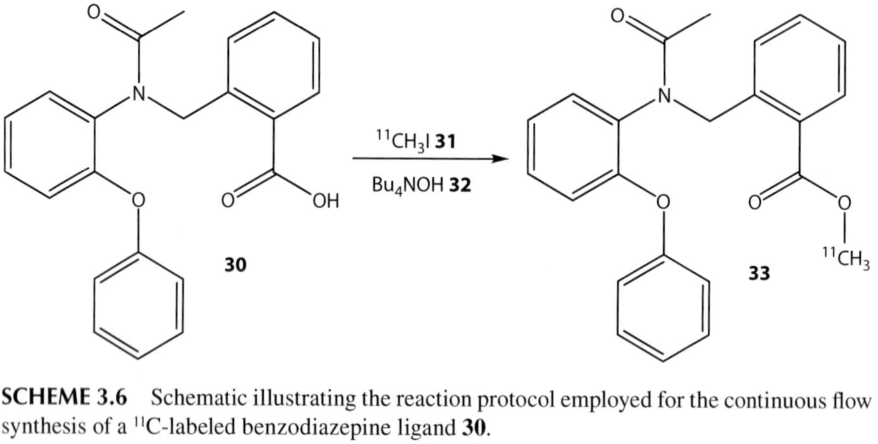

SCHEME 3.6 Schematic illustrating the reaction protocol employed for the continuous flow synthesis of a ^{11}C-labeled benzodiazepine ligand **30**.

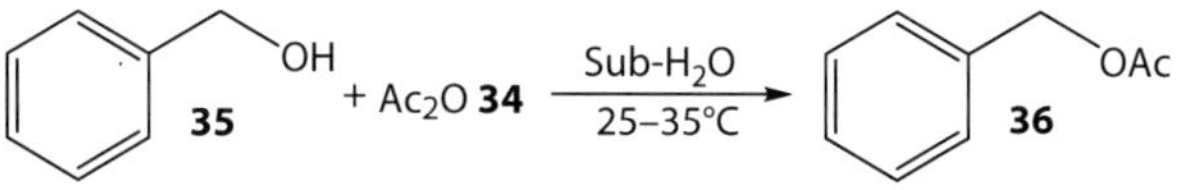

Liquid-Phase Micro Reactions

SCHEME 3.7 Schematic illustrating the model reaction used to demonstrate the O-acylation of alcohols in sub-H_2O.

set-up capable of instantaneously heating reactants to 250°C, with a cooling coil (dimensions = 0.5 mm (i.d.) × 46.0 cm (long)) affording efficient termination of reactions (10 s). To perform a reaction, the authors pumped a solution of alcohol and acetic anhydride **34** into the reactor at a linear velocity of 4.2 cm s^{-1} (5 MPa), where at a T-junction this solution collided with sub-H_2O (42 cm s^{-1}, 5 MPa). The resulting particle dispersion rapidly underwent heating within the tube reactor to induce a reaction which upon cooling afforded a binary mixture; enabling facile separation of the water from the reaction products.

To demonstrate the synthetic utility of this technique, the authors employed the acetylation of benzyl alcohol **35** as a model reaction (Scheme 3.7), evaluating the effect of reactor temperature on the product formed at a fixed residence time of 9.9 s. Using this approach, the authors identified a link between the reactor temperature and benzyl acetate **36** formation, obtaining 99.9% conversion at a reactor temperature of 200°C. In comparison, performing the reaction in a batch system utilizing sub-H_2O, the authors obtained only 17 % benzyl acetate **36**; an observation which is attributed to the strong hydrolytic properties of the reaction medium when heated for a prolonged period of time.

In order to evaluate the scope of their system, the authors evaluated a diverse array of alcohols, ranging from phenolic to tertiary aliphatic derivatives. As Table 3.5 illustrates, employing 1.1 equivalents of Ac$_2$O **34** the authors were able to acetylate the alcohols in high yield and, where relevant, high selectivity. In the case of 1-hydroxyisobutyric acid **37**, acetylation occurred rapidly; however, a large proportion of olefinic by-products resulted due to a dehydration side reaction, performing the acetylation in the presence of 30 equivalents of Ac$_2$O **34** under the aforementioned conditions, however, did enable the reaction to proceed with increased selectivity and isolated yield.

The authors noted that this technique compares favorably, in terms of scope and reactivity, with those reactions utilizing volatile organic solvents and catalysts; however, the methodology employs an environmentally benign solvent and requires no catalyst to afford selective, hydrolysis-free O-acylation of alcohols; demonstrating its potential for use on a production scale.

Elevated Reaction Temperatures: Kappe and coworkers [20] also recently demonstrated the synthetic advantages associated with the performance of homogeneous reactions at elevated temperatures (350°C) and pressures (50–200 bar) within stainless-steel tubular reactors (dimensions = 1 mm (i.d.), volumes = 4, 8, and 16 mL).

Such equipment enabled the authors to readily evaluate the use of supercritical solvents allowing high-temperature reactions to be performed in low-boiling organic solvents facilitating postreaction removal and product isolation. Additionally, due to

TABLE 3.5
Illustration of the Reaction Scope of Sub-H_2O as a Reaction Solvent for the Esterification of Alcohols

Substrate[a]	Molality (mol kg^{-1})	Conversion (%)	Selectivity (%)	Yield (%)
[benzyl alcohol]	0.38	97	—	97
[2-methylbenzyl alcohol]	0.29	97	97	97
[2-methoxy-4-allylphenol derivative]	0.25	95	97	95
[tert-butanol]	0.37	100	94	94
[1-ethynylcyclohexanol]	0.33	100	86	86
[compound 37]	0.28	18	31	6
	0.02	77	98	75[b]

[a] Unless otherwise stated, all reaction conducted at 1.1 eq. Ac_2O, 5 MPa, 200°C.
[b] 30 eq. Ac_2O.

the high ionic product of supercritical alcohols, the authors found it possible to conduct esterifications (Scheme 3.8a) and transesterifications (Scheme 3.8b) in the absence of an acid catalyst.

In the case of benzoic acid **24**, the authors employed a reactant concentration of 0.33 M in EtOH and found that a reaction temperature of 300°C afforded excellent conversion (87%) to ethyl benzoate **25**, at 120 bar, with a residence time of 12 min. When considering the transesterification of ethyl 3-phenylpropanoate **38** (0.05 M), the authors employed supercritical MeOH at 350°C and 180 bar obtaining 85% conversion to methyl 3-phenylpropanoate **39** with a reaction time of only 8 min; demonstrating significant processing advantages when compared with conventionally catalyzed systems.

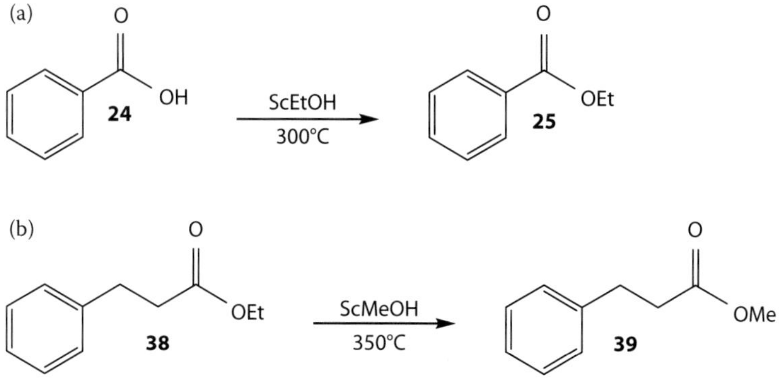

Liquid-Phase Micro Reactions

SCHEME 3.8 Schematic illustrating the (a) esterification and (b) trans-esterification reactions performed under high temperature and pressure within a flow reactor.

3.1.5 C–O BOND FORMATION: ETHERIFICATION REACTIONS

Using a microcapillary flow disc (MFD) reactor developed by Mackley and coworkers [21] as a tool for organic synthesis, Ley and coworkers [22] demonstrated the reactors utility toward a series of chemical reactions including the base-catalyzed synthesis of allylic ethers, as illustrated in Scheme 3.9. The reactors in question comprised of an extruded linear low-density polyethylene (LLDPE) plastic film (dimensions = 0.58 mm (deep) × 1.38 mm (wide)) containing 19 cylindrical capillaries (dimensions = 180 or 220 µm), coiled to afford a reactor of the desired volume. Reactants were introduced into the MFD reactor using HPLC pumps and where necessary, the system is thermostated via immersion into a warming or cooling bath.

To perform the etherification reaction illustrated in Scheme 3.9, selected due to the exothermic nature of the alkylation, two feed solutions were prepared, the first contained salicylaldehyde **40** (2.0 M) and DBU **41** (4.0 M) in MeCN and the second, allyl bromide **42** (4.0 M) in MeCN. Using a T-mixer, the reagents were brought together before entering the MFD reactor where the reaction was evaluated at room temperature.

Employing a residence time of 57 min, the authors obtained the target allylic ether, salicylaldehyde allyl ether **43** at 0.53 kg day^{-1}, following an aqueous extraction and solvent removal. The simplicity of the reactor design makes it suitable for mass

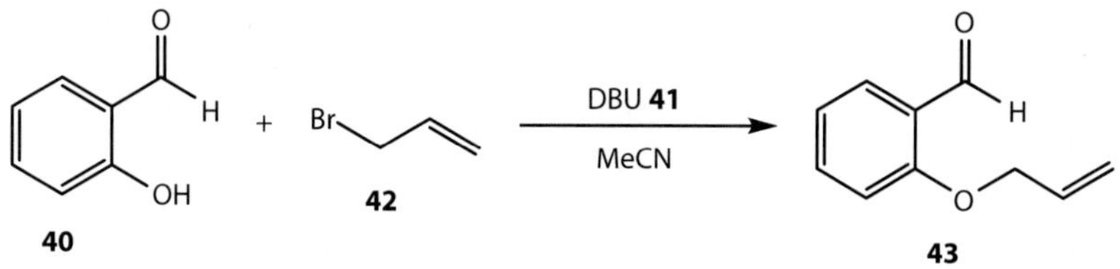

SCHEME 3.9 Illustration of the model reaction utilized to demonstrate the parallel synthesis of allylic ethers in a microcapillary disk flow reactor.

replication and the parallel nature of the reaction capillaries provides a facile method for numbering-up a synthetic process.

3.1.6 C–O BOND FORMATION: EPOXIDE HYDROLYSIS

The hydrolysis of epoxides is a synthetically useful technique for the preparation of *vic*-diols, although conventionally catalyzed by acids or bases more recently biocatalysts have been employed to resolve racemic epoxides as a route to enantio-enriched diols.

Biocatalytic Epoxide Hydrolysis: In a noteworthy example of biocatalysis within micro reactors, Belder et al. [23] fabricated an integrated fused-silica device capable of performing synthetic, separation, and detection steps. To demonstrate the advantages of their miniaturized system, the authors employed the enantioselective synthesis of 3-phenoxypropane-1,2-diol **44** via the hydrolysis of 2-phenoxymethyloxirane **45** using a series of epoxide hydrolase enzymes (wild type, LW086, LW144, and LW202) as the biocatalyst (Scheme 3.10). To perform a reaction, the substrate and catalysts were placed in separate vials on the device and delivered to the reaction channel by the application of pressure or an electric field. The reaction was performed within a meandering channel, prior to guiding the reaction mixture into the separation channel, by means of a voltage-controlled pinched injection. Electrophoretic separation of the products and educts then occurred within the separation channel in the presence of an electric field; native fluorescence detection, using a deep-UV laser (Md:YAG, $\lambda = 266$ nm) was employed to perform on-chip detection of the analytes. Using this approach, the authors were able to evaluate three mutants of the epoxide hydrolase *Aspergillus niger*, obtaining conversions ranging from 33% to 43% and *ee*'s of 49–95%; obtaining enhanced selectivity factors (E) compared to conventional bench-top reactors.

3.1.7 C–N BOND FORMATION: ALKYLATION REACTIONS

With *N*-alkyl products frequently possessing lower pKas than the parent amine, it can prove challenging to selectively monoalkylate aliphatic and aromatic amines. To overcome this, researchers have investigated the transformation under flow conditions with a view to reduce the product distribution by employing efficient mixing and carefully controlling the reaction times used.

Uncatalyzed Amination Reactions: Based on the synthetic utility of the aminopyridine structural motif and the difficulties associated with their preparation

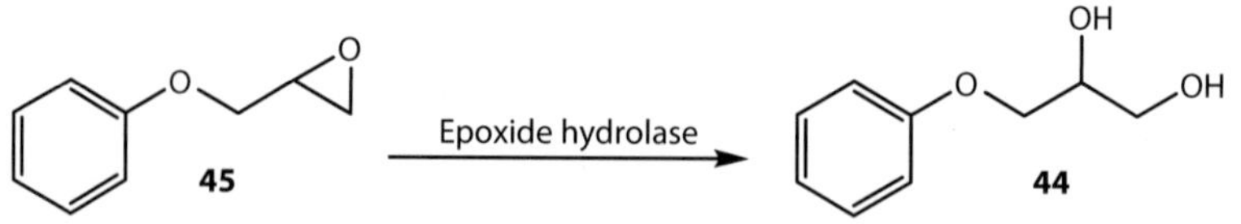

SCHEME 3.10 Schematic of the model reaction used to illustrate the enantioselective catalysis and analysis performed in an integrated microfluidic device.

TABLE 3.6
Illustration of the Model Reaction Used to Develop a Direct Method for the Synthesis of Aminopyridines

Solvent	Residence Time (min)	Piperidine (eq.)	Temperature (°C)	Yield (%)
DMF	4	1	240	9
DMF	4	1	240	47
DMF	4	2	240	52
DMF	20	2	240	63
DMA	20	2.2	260	76
NMP	20	2.2	260	100

utilizing conventional synthetic methodology, Hamper and Tesfu [24] embarked upon an investigation into the nucleophilic substitutions using a continuous flow reactor.

Employing the synthesis of 2-piperidinylpyridine **46** as a model reaction (Table 3.6), the authors investigated the effect of reaction solvent, stoichiometry, reaction time, and reactor temperature. Utilizing a tubular stainless-steel reactor (dimensions = 0.020 in. (i.d.) × 10 m (long), volume = 2.03 mL) housed within an aluminum block, reactions were performed under pressure (69 bar), enabling the authors to routinely access reactor temperatures of 260°C by placing the reactor on a standard laboratory hotplate. Reactant solutions were delivered using a HPLC pump (2-chloropyridine **47** = 0.5 M) and reaction progress monitored by GC–MS, using an internal standard.

Using this approach, the authors were able to readily identify NMP as the best solvent for this transformation, affording quantitative conversion of 2-chloropyridine **47** to 2-piperidinylpyridine **46** with a reaction time of 20 min and a reactor temperature of 260°C.

Under the aforementioned optimized conditions, the authors subsequently investigated the reaction of 2-chloropyridine **47** with a series of secondary amines, finding that they could overcome the activation barrier for this reaction, even when employing inactivated substrates, enabling the catalyst-free synthesis of aminopyridines in throughputs of typically 0.5 g h^{-1}.

Microwave Synthesis: Using a commercially available single-mode microwave and an in-house fabricated glass flow cell (volume = 4 mL), Wilson et al. [25] demonstrated the use of their continuous microwave reactor toward the S$_N$Ar of 1-fluoro-2-nitrobenzene **48** to afford 2-nitro-*N*-phenethylaniline **49** (Scheme 3.11).

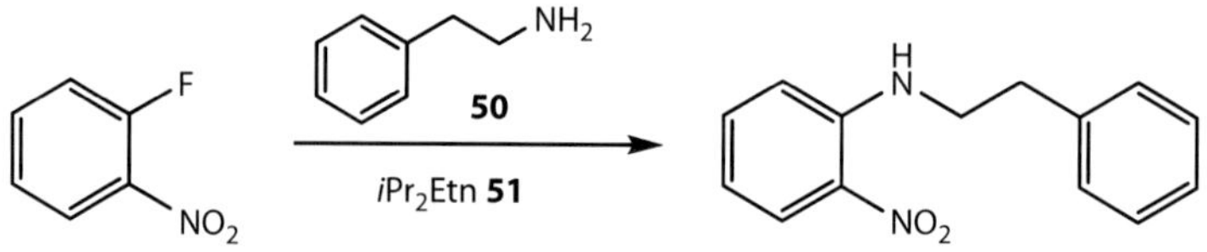

SCHEME 3.11 Illustration of the nucleophilic aromatic substitution of 1-fluoro-2-nitrobenzene **48**, to afford 2-nitro-N-phenethylaniline **49**, performed in a continuous microwave reactor.

Utilizing a stock solution of 1-fluoro-2-nitrobenzene **48** (47.4 mmol), phenylethylamine **50** (94.8 mmol) and diisopropylethylamine **51** in EtOH (67 mL), the reaction mixture was cycled through the flow reactor at a rate of 1 mL min^{-1} over a period of 5 h. Employing an irradiation time of 24 min mL^{-1}, the authors obtained a reactor temperature of 120°C and synthesized 2-nitro-N-phenethylaniline **49** in 81% yield.

For comparison purposes, the authors performed the reaction in batch, using standard heating techniques whereby only 54% conversion to **49** was obtained in 5 h (100°C); highlighting a processing advantage associated with continuous flow microwave reactors.

The authors did, however, acknowledge a limitation of the technique which was the clogging of the reactor when high conversions of the target compound **49** were obtained. With this in mind, the authors found it necessary to terminate the reactions before completion in order to maintain a free-flowing system; however, alternative solutions would have been to employ a different solvent system or more dilute reactant solutions.

Ionic Liquid Synthesis: While the efficient heat transfer obtained within microstructured reactors is very much viewed as an advantage, overcooling of reactors can sometimes occur and lead to a kinetic slow down, resulting in a perceived need to increase reaction time. To address this, Löwe et al. [26] evaluated the effect of passive reactor cooling (with heat pipe capillary pumped loops) for the exothermic synthesis of imidazole **52** based ionic liquids (Scheme 3.12).

In the case of ionic liquid **53**, the reaction was found to be instantaneous, with an adiabatic temperature rise of >250°C making the process unsuitable for performance on a batch scale. Performing the reaction in a steel, microstructured reactor, equipped with heat pipes for cooling, the authors were able to process the ionic liquid at flow rates of 2 L h^{-1}; equivalent to 10 mol h^{-1} **53**. Using this approach, heat transportation with near sonic velocity was obtained, removing the need for recirculation of the cooling fluid.

To exemplify the effect of overcooling a reaction, the authors investigated the synthesis of ionic liquid **54**, identifying that after an induction period of 10 s the reaction slows down dramatically if the temperature falls below ambient; therefore, by controlling heat removal the material can be effectively synthesized under flow conditions. In addition to this example, several authors have investigated the continuous flow synthesis of ionic liquids [27], including Renken et al. [28]; see Chapter 7 for an example of industrial-scale production of ionic liquids using MRT.

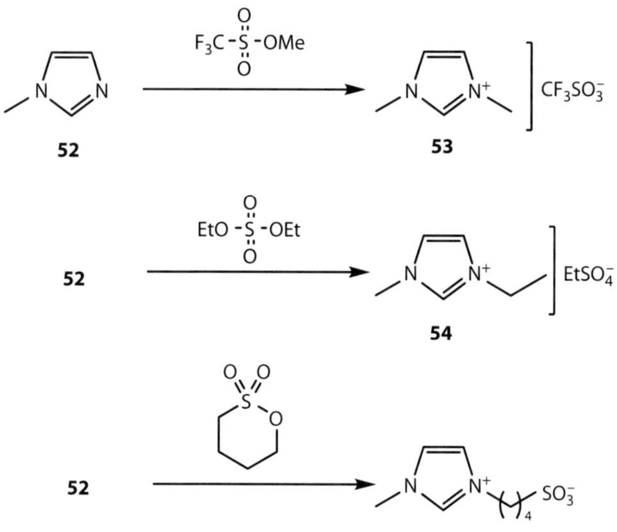

Liquid-Phase Micro Reactions

SCHEME 3.12 Examples of the imidazole **53** based ionic liquids synthesized using passive or intrinsic cooling methods.

Diastereoselective Synthesis: In 2004, Wiles et al. [29] compared the alkylation of an Evans auxiliary derivative performed in batch and a micro reactor, conducted at a series of reduced temperatures (0–100°C). Using a borosilicate glass micro reactor (channel dimensions = 152 µm (wide) × 51 µm (deep) × 2.3 cm (long)), submerged in a solid CO_2/diethyl ether bath (−100°C), the authors initially deprotonated 4-methyl-5-phenyl-3-propionyloxazolidin-2-one **55** using NaHMDS **56** in anhydrous THF, the enolate **57** was subsequently reacted with benzyl bromide **9** and the reaction products collected at room temperature with immediate quenching (DI H_2O at 25°C). The reaction products were subsequently analyzed offline by GC–MS, and the conversion of the *N*-acyl oxazolidinone **55** to diastereomers **58** and **59** determined, along with the ratio of diastereomers obtained (Scheme 3.13).

At a total flow rate of 30 µL min^{-1}, the authors obtained a conversion of 41% **55** with the products formed in a 91:9 ratio (**58:59**) with 59% residual oxazolidinone **55**. In an analogous batch reaction, performed at −100°C, the isomers were obtained in a ratio of 85:15 (**58:59**), with an accompanying 10% decomposition to the *N*-alkyl by-product **60**. This observation is attributed to the efficient reaction of any anion **57** formed and the use of short reaction times, preventing decomposition from occurring. Owing to the presence of a significant proportion of residual *N*-acyl oxazolidinone **55**, this suggests that the use of a greater residence time for the enolate **57** formation would enable the reaction efficiency to be increased further.

3.1.8 C–N Bond Formation: Acylation Reactions

Of the homogeneous liquid-phase reactions performed under continuous flow conditions, the acylation of amines, to afford amides, is one of the most widely studied

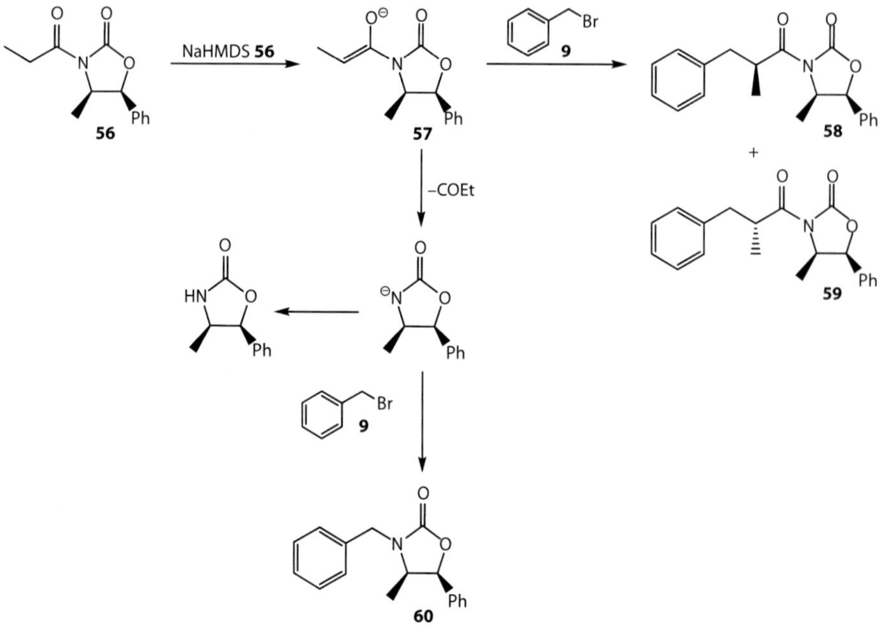

SCHEME 3.13 Schematic illustrating the alkylation of *N*-acyl oxazolidinone **55** investigated under continuous flow.

transformations. Schwalbe et al. [30] reported an early example of continuous acylations, using an organic base (Et₃N **27**) and DMF as the reaction solvent, demonstrating the acylation of aniline **61**, with a residence time of 42 min affording *N*-phenylacetamide in 94% yield. Chevalier et al. [31] subsequently demonstrated the use of Schotten–Baumann conditions for the amidation of α-methylbenzylamine employing a glass micro reactor containing static micromixers, describing its suitability for industrial-scale production.

Atom-efficient Acylations: More recently, Hooper and Watts [32] extended this principle illustrating the use of micro reactors for the incorporation of deuterium labels into an array of small organic compounds, selecting the technique due to its previously demonstrated atom efficiency. Using the base-mediated acylation of primary amines, such as benzylamine **62**, as depicted in Scheme 3.14, the authors

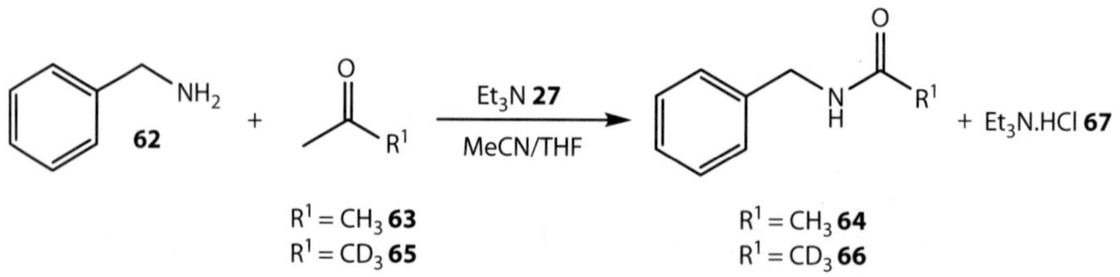

SCHEME 3.14 Schematic illustrating a model reaction selected to illustrate the advantages of performing deuterium labeling under continuous flow conditions.

initially demonstrated optimization of the amidation using unlabeled precursors. Substitution with a labeled reagent followed with no reoptimization, illustrating the facile and cost-effective technique for the synthesis of labeled analogs.

To conduct the reactions depicted in Scheme 3.14, two borosilicate glass micro reactors were employed in series (Reactor 1 = 201 µm (wide) × 75 µm (deep) × 2.0 cm (long) and Reactor 2 = 158 µm (wide) × 60 µm (deep) × 1.5 cm (long)) and reactants were manipulated within the reactors using pressure-driven flow. In order to enable the use of the developed protocol for long-term operation, the authors found it necessary to employ a mixed solvent system which enabled dissolution of the reaction by-product ($Et_3N.HCl$), while maintaining stability of the acylating agent.

Introduction of solutions of benzylamine **62** (0.1 M) and triethylamine **27** (0.1 M) ensured complete mixing in the first reactor, prior to the addition of acetyl chloride **63** (0.05 M, 1.0 eq.) to the reaction mixture, using the second micro reactor. Employing a total flow rate of 40 µL min^{-1}, which equates to a reaction time of 2.6 s, coupled with offline HPLC analysis, the authors obtained the target N-benzamide **64** in 95% conversion. To demonstrate the transferability of the methodology developed, acetyl chloride **63** was substituted with a solution of acetyl [D_3] chloride **65** (0.05 M) and operating under the aforementioned conditions, the authors confirmed the formation of the deuterated N-benzamide **66** derivative in 98% conversion.

Although this application represents a niche area of synthetic chemistry, the development provides important confirmation that reactions can be optimized using readily available compounds and once appropriate methodology is identified, exchanging reactants for deuterated analogs affords a facile, cost-effective route to the synthesis of labeled analogs (see Chapter 7 for additional examples).

Mediator-free Acylations: In addition to the examples described, Sato et al. [33] recently demonstrated the efficient N-acylation of amines without the use of an added catalyst or mediator; employing water as either a Lewis acid or a coolant in the case of endothermic and exothermic reactions, respectively.

Employing a tubular reactor (dimensions = 500 µm (i.d.) × 50 cm (long)), reactions were performed by mixing a solution of amine (1° or 2°) and acetic anhydride **34** (1.1 eq.), the solution was then collided (at 4.2 cm s^{-1}) at a second T-mixer with H_2O at 42.0 cm s^{-1} to induce a reaction. In order to prevent hydrolysis of Ac_2O **34** and/or the target amide, the authors found it imperative to employ reaction times of <10 s. As Table 3.7 illustrates, using this approach the authors were able to synthesize amides in high yield and purity employing ambient H_2O as the reaction solvent. In the case of sterically hindered or resonance-stabilized compounds, it was found necessary to heat the aqueous stream in order to promote the acetylation, achieved via catalyst-like proton transfer. Excellent yields were once again obtained with residence times ranging from 1.1 to 9.9 s, with the authors concluding that the N-acylation of amines was dependent on stereoelectronic effects.

Having previously demonstrated O-acylation using this approach (see Section 3.2.1), finding steric hindrance-influenced reactions, the authors proposed that the technique would be suitable for N-acylation in the presence of aliphatic and aromatic alcohols. Evaluating a series of precursors, as summarized in Table 3.8, the authors

TABLE 3.7
Illustration of a Selection of Amides Formed Using Sub-H_2O as a Reaction Solvent, a 1:1 Stoichiometry of Ac_2O 34, and a Residence Time of 1.1 s

Substrate	Product	Temperature (°C)	Conversion (%)	Selectivity (%)	Yield (%)
isopropylamine	N-isopropylacetamide	26	100	99	99
tert-butylamine	N-tert-butylacetamide	35	95	100	95
piperidine	N-acetylpiperidine	25	100	100	100
2,6-dimethylpiperidine	N-acetyl-2,6-dimethylpiperidine	100	95	100	95
2,6-dimethylaniline	N-(2,6-dimethylphenyl)acetamide	22	98	100	98
4-aminobenzoic acid	4-acetamidobenzoic acid	75	96	99	95

were able to demonstrate the development of a facile and efficient method for the chemoselective acylation of amines, all performed in the absence of a catalyst.

Dendrimers: These highly branched molecules, with core–shell structures, have a unique architecture that enables them to encapsulate small guest molecules within their interiors and conjugate molecules onto their surfaces. Consequently, these macromolecules are finding application in a diverse array of areas such as catalysis, drug delivery, sensing, and gene transfer, putting great pressure on the need to develop methodologies suitable for their large volume of production. Based on a convergent method, Chang and coworkers [34] evaluated the use of micro reactors for the synthesis of polyamide dendrimers.

TABLE 3.8
Illustration of the Types of Bi-Functional Compounds that Undergo Chemoselective N-Acylation Using H_2O as a Reaction Solvent

Substrate	Product	Temperature (°C)	Conversion (%)	Selectivity (%)
$H_2N{\sim}NH_2$	$H_2N{\sim}NHC(O)-$	38	30	100
$H_2N{\sim}NH{-}cyclohexyl$	acylated diamine w/ cyclohexyl	25	95	85
$HO{\sim}NH_2$	$HO{\sim}NHC(O)-$	19	100	100
4-aminophenol (NH_2, OH)	4-hydroxyacetanilide (NH-C(O)-, OH)	50	100	99.5[a]

[a] Residence time = 9.9 s.

Recognizing the importance of mixing and heat removal, the authors investigated the use of an interdigital micromixer (IMM, Germany) with microfabricated components derived from thermally grown silicon dioxide. Reactions were performed by mixing two reactant solutions, delivered at 1.56 mL min^{-1}, with the reaction products collected in water; whereby the product was observed to precipitate and was subsequently dried at 120°C.

In the case of a G1 dendrimer, solution 1 contained the G1 dendron (0.2 M) and solution 2 thionyl chloride (1.5 eq.), with a reactor temperature of 30°C and a residence time of 17 s employed for the initial formation of the intermediate acyl chloride. Upon collection at the reactor outlet, the acyl chloride solution was introduced into a second device where it was reacted with 0.5 eq. of a core molecule. Under the aforementioned conditions, the authors were able to collect the G1 dendrimer as a precipitate from receiver vessel. The dried material was then analyzed by ^1H NMR spectroscopy and MS, confirming the formation of the G1 dendron in high yield and purity. Compared to batch reactions, where limitations such as poor mass transfer lead to incomplete reaction after 3 h, the use of a micromixer afforded access to short reaction times and high throughputs. With iterative processes, it is imperative that each step is efficient in order to obtain the target materials in high yield and purity; as such, this methodology presents a potential route to the large-scale production of these synthetically useful materials.

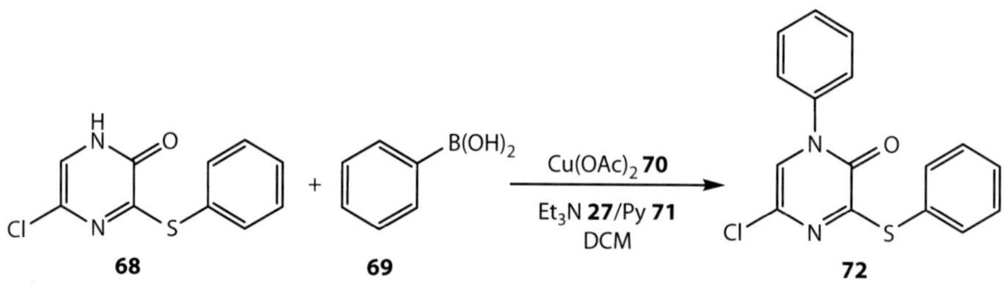

SCHEME 3.15 Schematic illustrating the reaction sequence employed for the continuous flow *N*-arylation of 2(1*H*)-pyrazinone **68**.

3.1.9 C–N BOND FORMATION: ARYLATION REACTIONS

Employing a CYTOS® College System (CPC, Germany), comprising a stacked plate micro reactor and a tubular residence time unit, Stevens and van der Eyken and coworkers [35] reported the development of a scalable technique for the *N*-arylation of 2(1*H*)-pyrazinone **68** (Scheme 3.15), selected due to their application in biologically interesting molecules and compound libraries.

Initial investigations focused on assessing the effect of residence time on the *N*-arylation of 2(1*H*)-pyrazinone **68** (2.5×10^{-2} M), with boronic acid **69** (5.0×10^{-2} M), in the presence of copper acetate **70** (2.5×10^{-2} M), Et$_3$N **27** (2.5×10^{-2} M) and pyridine **71** (5.0×10^{-2} M). Employing DCM as the reaction solvent, the authors investigated residence times ranging from 15 to 120 min, observing an increase in conversion with increasing reaction time, as summarized in Table 3.9. Under the optimal conditions of 2 h, at room temperature, the authors isolated 5-chloro-1-phenyl-3-(phenylthio)pyrazin-2-(1*H*)-one **72** in 100% conversion, as determined by quantitative TLC, and 79% isolated yield; after passing through a plug of silica gel.

TABLE 3.9
Illustration of the Optimization Protocol Employed for the Copper-Mediated *N*-Arylation of 2(1*H*)-Pyrazinone 72

Residence Time (min)	Flow Rate (mL min^{-1})	Conversion (%)[a]
15	1.54	30
45	0.52	50
60	0.39	70
90	0.26	90
120	0.20	100 (79)[b]

[a] Determined using quantitative TLC.
[b] Isolated yield.

TABLE 3.10
Summary of the Results Obtained for the N- and O-Arylations Performed under Continuous Flow

				Product	
Solvent	X	R¹	R²	Yield (%)	Isolated Yield (g)
DCM	N	H	H	71	1.2
DCM	N	H	3-OEt	73	1.5
DCM	N	4-Cl	H	67	1.3
DCM	N	4-Cl	3-OEt	69	1.7
DCM	N	2,4,6-Cl	3-OEt	No reaction	N/A
DCM	N	2,4,6-NO₂	3-OEt	56	1.7
DMF	O	3-OCH₃	H	72	1.4
DMF	O	2-I	4-OMe	69	2.3
DMF	O	4-Et	4-OMe	70	1.6

Note: DCM = Reaction temperature and DMF = 130°C.

In an extension to this, the authors evaluated the *N*-arylation of a series of substituted anilines and phenols, again obtaining the target biaryls in moderate to high yield. Due to an increase in substrate and product solubility, the authors were able to perform these reactions on a 10 mmol scale, demonstrating the ease of synthesizing such compounds on a gram-scale (Table 3.10).

In the case of *O*-arylations, the authors found it necessary to heat the reactor in order to increase the conversion beyond 60%, as such DMF was employed as the reaction solvent which enabled the target compounds to be isolated in yields ranging from 69% to 72%.

3.1.10 C–N BOND FORMATION: AZIDATION REACTIONS

While azides represent one of the most versatile groups of reagents employed in modern synthetic chemistry, their widespread application is not exploited due to the hazards associated with their preparation, purification, and handling at scale [36].

Iodide Azide: To demonstrate the increased reaction safety obtained as a result of performing reactions under continuous flow, Brandt and Wirth [37] conducted the synthesis of a series of carbamoyl azides **73** via the treatment of aromatic aldehydes with an excess of an inorganic azide (Table 3.11). As sodium azide is sparingly soluble in organic solvents, the authors investigated the *in situ* generation of iodine azide **74** from tetrabutylammonium azide and iodine monochloride, finding that within a capillary reactor iodine azide **74** was formed readily.

Using this approach, the authors employed a tubular PTFE reactor (dimensions = 250 μm (i.d.), volume = 196 μL) in which the aldehyde was reacted with the *in-situ* generated iodine azide **74** (at 80°C) to afford the respective acyl azide **75**; which underwent thermal rearrangement to afford the isocyanate **76**, and in the presence of excess iodine azide **74** yielded the target carbamoyl azide **73**. Reactions were terminated at the reactor outlet using an aqueous solution of sodium thiosulfate and the reaction products isolated by organic extraction into DCM followed by purification via column chromatography. Interestingly, in the case of 4-bromobenzaldehyde **77** no carbamoyl azide **78** was obtained, only the respective acyl azide; however, when the aldehyde **77** was distilled, the carbamoyl azide **78** was obtained in 27% yield. As Table 3.11 illustrates, using this approach, the carbamoyl azides of benzaldehyde **19** and a series of substituted aromatic aldehydes, affording yields ranging from 21% to 44%.

Trimethylsilyl Azide: Nieuwland et al. [38] subsequently reported the use of trimethylsilyl azide **79**, after themselves finding the reaction of sodium azide too slow for use within micro reactors, hampering their aim of process intensification. To perform a reaction, the authors added TMS-N$_3$ **79** (0.4 M in THF) to a solution of TBAF **4** (1.0 M in THF) and benzyl bromide **9** (0.74 M in THF) within a heated

TABLE 3.11
Summary of the Results Obtained for the Continuous Flow Synthesis of Carbamoyl Azides Using *In Situ* Generated Iodine Azide 74

Aldehyde	Product	Yield (%)
Benzaldehyde **19**		44
4-Bromobenzaldehyde **77**	**78**	27
4-Chlorobenzaldehyde		32
2,4-Dimethoxybenzaldehyde		21

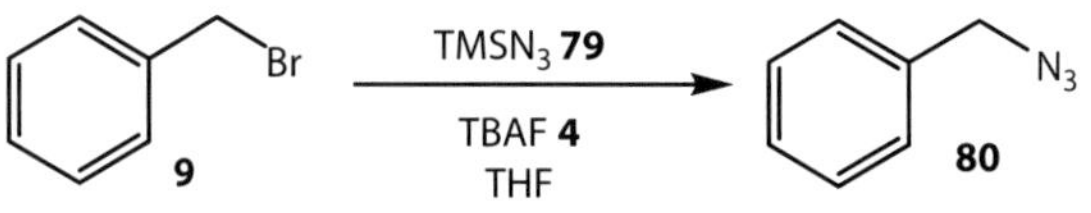

SCHEME 3.16 Schematic illustrating the azidation of benzyl bromide **9** utilizing trimethylsilyl azide **79** under continuous flow.

micro reactor and employed HCl (1.0 M) in EtOAc/acetone (1:1) as a quench solvent. Using this approach, the authors evaluated the effect of reaction time (5–30 s), reactor temperature (60–80°C), and TMS-N_3 **79**/benzyl bromide **9** ratio on the formation of (azidomethyl)benzene **80** (Scheme 3.16); finding that a ratio of 0.8, reaction time of 10 s, and temperature of 80°C afforded the optimal synthetic conditions.

Large-Scale Synthesis: Owing to their interest in the synthesis of potent NK1-II antagonists (Figure 3.3), Kopach et al. [39] required a safe and efficient method for the preparation of 1-(azidomethyl)-3,5-bis-trifluoromethyl)benzene **81** in sufficient quantities to enable the synthesis of a phase 2 investigational drug candidate. Due to the risks associated with the build-up of hydrazoic acid in the headspace of batch reaction vessels, the authors investigated the development of a continuous flow method for the synthesis of 1-(azidomethyl)-3,5-bis-trifluoromethyl)benzene **81** from 3,5-bis-(trifluoromethyl)benzyl chloride **82** (Scheme 3.17). Compared to batch protocols, the advantages of using continuous flow methodology lies in the fact that reactors are completely filled; as such no headspace exists for HN_3 build-up, removing the risk of explosion often associated with the large-scale synthesis of organic azides.

With this in mind, the authors constructed a flow reactor comprising a 316 stainless-steel tubular reactor (dimensions = 1.59 mm (o.d.) × 0.64 mm (i.d.) × 63.1 m (long), volume = 20 mL), with reactants introduced into the system using two high-pressure ISCO syringe pumps and a T-mixer. To enable the investigation of elevated reaction temperatures, the system was housed within a GC oven and a back-pressure regulator (200 psi) placed at the reactor outlet.

Investigating the effect of reactor temperature on the azidation, the authors observed increasing azide **81** formation with increasing reactor temperature, identifying 90°C and a residence time of 20 min as optimal; affording 97.3% azide **81** and

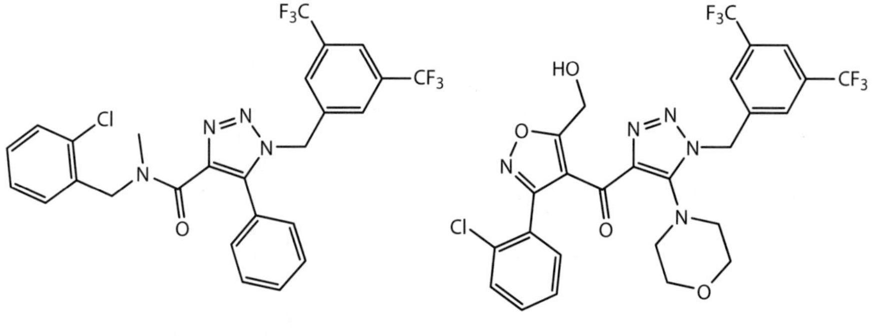

FIGURE 3.3 Illustration of a selection of NK1-II antagonists.

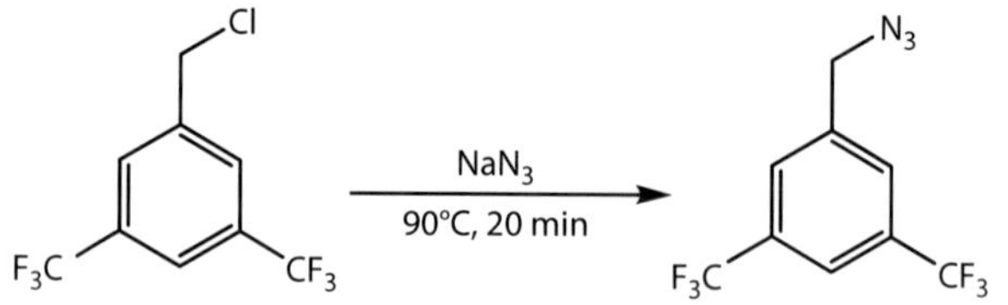

SCHEME 3.17 Schematic illustrating the model reaction used to evaluate the advantages associated with the synthesis of organic azides under continuous flow.

0.39% residual benzyl chloride **82**. Operating the reactor continuously for a period of 2.8 h, under the aforementioned conditions, the authors were able to synthesize 25 g of 1-(azidomethyl)-3,5-bis-trifluoromethyl)benzene **81** in 94% yield, after partitioning between heptane and water.

In addition to increasing process safety, the use of a continuous flow reactor enabled the authors to access higher reaction temperatures than attainable in batch systems, which led to a reduction in reaction time; affording an efficient method for the preparation of organic azides. Further work is currently underway to investigate the use of alternative solvents as a means of reducing the volume of solvent required and hence, the *e*-factor.

3.1.11 C–N Bond Formation: Synthesis of Hydroxamic Acids

Forming part of a medicinal chemistry project, Martinelli and coworkers [40] investigated the conversion of esters into hydroxamic acids, as a means of developing a synthetic protocol for the preparation of a diverse array of compounds of this type.

Using the synthesis of *N*-hydroxybenzamide **83** as a model reaction, the authors evaluated the effect of residence time (5–30 min) and reaction temperature (50–80°C) on the conversion of the ester, as determined by offline LC–MS analysis. Employing PTFE flow reactor (Reactor 1 = 10 mL), a methanolic solution of ester (0.5 M) and hydroxylamine **82** (10 eq.) was mixed with a solution of sodium methoxide (0.5 M), where it reacted prior to collection and analysis. Using this approach, the authors systematically screened a series of potential reaction conditions, identifying a reaction time of 30 min and temperature of 70°C as optimal; finding temperatures >70°C promoted the formation of benzoic acid **24**.

Comparison of the flow reaction (80% **83**) with conventional methodologies such as a stirred batch reactor (58% **83**) and a microwave-irradiated sealed tube (72% **83**), the authors were pleased to see that the uniform temperature distribution obtained within the flow reactor resulted in increased product yield and purity, due to the suppression of the corresponding carboxylic acid **24** formation.

With this information in hand, the authors investigated the conversion of a series of esters, obtaining the target hydroxamic acids in moderate-to-excellent conversion depending upon the functionality selected. As Table 3.12 illustrates, the technique was found to be applicable to aromatic, aliphatic, and heterocyclic derivatives, along with methyl and ethyl esters, affording isolated yields of between 52% to 100%

TABLE 3.12
Summary of the Results Obtained for the Conversion of Esters to Hydroxamic Acids under Flow Conditions

Ester	Hydroxamic Acid	Yield (%)
PhCH₂C(O)OEt	PhCH₂C(O)NHOH	96
PhCH₂NHCH₂C(O)OEt	PhCH₂NHCH₂C(O)NHOH	96
4-MeO-C₆H₄-SO₂-NH-CH₂-C(O)OMe	4-MeO-C₆H₄-SO₂-NH-CH₂-C(O)NHOH	95
(3-pyridyl)CH₂NHCH₂C(O)OEt	(3-pyridyl)CH₂NHCH₂C(O)NHOH	97
2-methylfuran-3-C(O)OMe	2-methylfuran-3-C(O)NHOH	52
Val-NHBoc-OMe	Val-NHBoc-NHOH	100
Ala-NHBoc-OMe	Ala-NHBoc-NHOH	81
Phe-NHBoc-OMe	Phe-NHBoc-NHOH	83

and purities of >95% (LC–MS). In addition to enhancing product yield through the suppression of carboxylic acid formation, the flow reactor methodology was also applicable to the reaction of enantiomerically pure esters, without loss of stereochemical integrity (Scheme 3.18).

3.1.12 C–N Bond Formation: Aminolysis of Epoxides

The aminolysis of epoxides is a classic synthetic route to β-amino alcohols, which find application in both organic and medicinal chemistry; with drugs such as

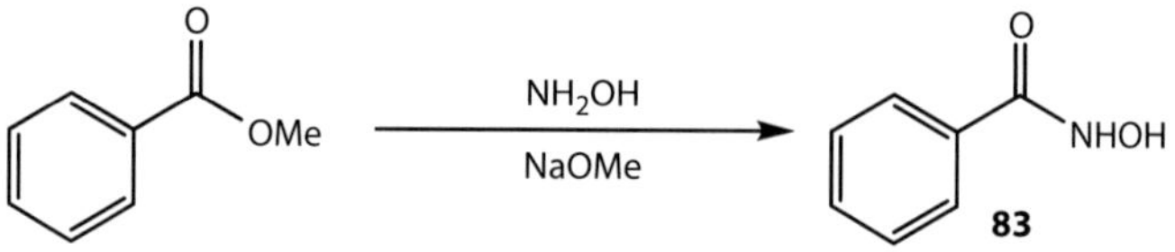

SCHEME 3.18 Model reaction selected to investigate the conversion of esters to synthetically useful hydroxamic acids.

Oxycontin and Toprol-XL, along with the phase III candidate Indacaterol **84** (Figure 3.4), displaying the functional group. While this method has been shown to be useful for simple amines, there remains a need for methodology to enable high-temperature reactions to be performed affording access to a more diverse array of products.

With this in mind, Jensen and coworkers [41] evaluated the aminolysis of epoxides within a silicon micro reactor (Figure 3.5) comprising silicon nitride-coated microchannels (reactor volume = 120 µL) and a borosilicate glass cover plate. In order to evaluate the scope of the technique, the authors investigated the aminolysis of a range of epoxides and substrates, as summarized in Table 3.13, using a back pressure regulator (250 psi) to enable "super-heating" of the reaction mixture. Employing an epoxide concentration of 1.0 M (in EtOH) and 1.2 equivalents of amine (in EtOH), the authors investigated the effect of reactor temperature (150–245°C) and reaction time (1–30 min) on the aminolysis reaction and subsequently on the product distribution. Conducting the reactions under pressure, within a sealed micro reactor, afforded several advantages over conventional batch reactions; the most notable being the ability to employ volatile amines without modification of the reactor setup, that is, *t*-butylamine **85** (boiling point = 46°C).

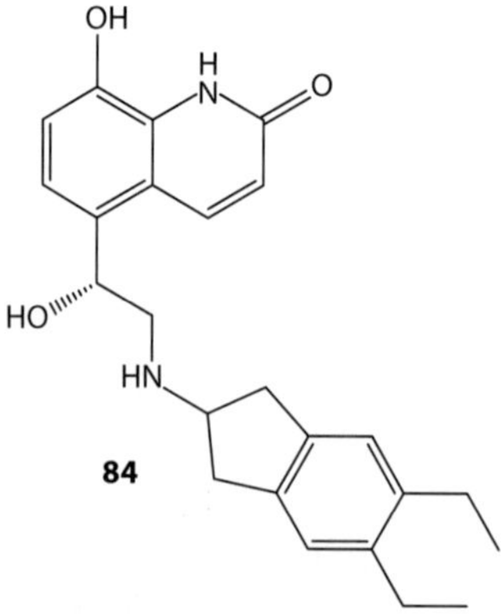

FIGURE 3.4 Indacaterol **84** a drug candidate in phase III clinical trials for the treatment of chronic obstruction pulmonary disease (COPD).

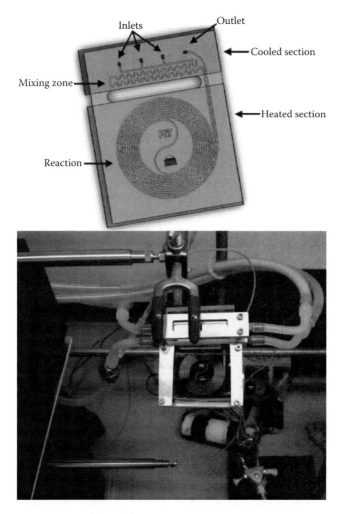

FIGURE 3.5 (**See color insert**) Illustration of the silicon micro reactor used to evaluate the continuous flow synthesis of β-amino alcohols which utilized water cooling to maintain low temperatures at the chuck. (Reproduced with permission from Bedore, M. et al. 2010. *Org. Proc. Res. Dev.* 14: 432–440. Copyright (2010) American Chemical Society.)

Following their successful screening of reaction conditions, the authors extended their investigation to the synthesis of pharmaceutically important β-amino alcohols, Metoprolol **86** (Scheme 3.19) and Indacaterol **84** (Figure 3.4, Scheme 3.20); demonstrating the synthetic utility of the technique developed.

In the first instance, the authors evaluated the synthesis of Metoprolol **86** due to its use as a selective $β_1$-adrenoreceptor blocking agent and current shortage. As Scheme 3.19 illustrates, the protocol involved the aminolysis of 2-((4-(2-methoxyethyl)phenoxy)methyl)oxirane **87** and employing a series of reaction conditions,

TABLE 3.13
Summary of the Results Obtained for the Continuous Flow Synthesis of β-Amino Alcohols under Continuous Flow

Epoxide	Amine[a]	Temperature (°C)	Residence Time (min)	Product Distribution[c]			Conversion (%)
				α-Opened (%)	β-Opened (%)	Bis-Alkyl (%)	
α,β-epoxide-OPh	indanyl-NH₂	150	3	73	—	26	>99
		195	2	72	—	24	98
		195	1	71	—	21	93
α,β-epoxide-OPh	t-Bu-NH₂ (85)	150	30	82	—	16	>99
		195	3	82	—	13	98
		195	1	81	—	6	92
α,β-epoxide-Ph	t-Bu-NH₂ (85)	150	30	62	7	16	94
		195	5	60	8	14	91
		195	5	66	9	8	91

Amine	Epoxide	T (°C)	t (min)				
61 NH₂–Ph	α,β-epoxybutane	195	5	63	—	18	82
		195	5	81	—	13	95
Indoline (NH)	naphthalene epoxide	150	30	39	—	—	40
		195	30	66	—	—	72
		245[d]	30	71	—	—	93
NH₂ [b] (n-butylamine)	Ph-substituted cyclohexene oxide (α,β)	150	30	15	2	—	17
		240[d]	30	68	6	—	78

[a] Unless otherwise stated 1.2 eq. of amine were employed.
[b] 5.0 eq. of amine.
[c] Determined by HPLC analysis using internal standardization.
[d] A back pressure regulator of 500 psi was employed.

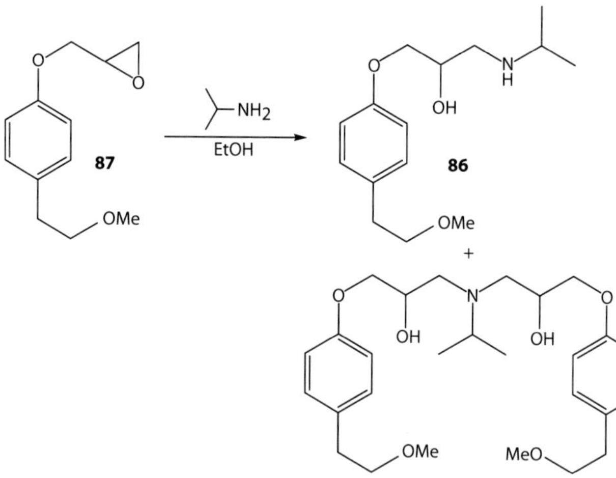

SCHEME 3.19 Synthesis of Metoprolol **86** via the aminolysis of epoxide **87**.

detailed in Table 3.14, the authors identified that the selectivity of Metoprolol **86** synthesis could be improved significantly when performed within a micro reactor, compared to batch reactions conducted using microwave irradiation. Using this approach, the authors comment that the continuous use of a single 120 μL micro reactor with a reaction time of 15 s could generate 7.0 g h^{-1} of Metoprolol **86** which equates to 61 kg annum^{-1}; scaling the system to 17 micro reactors would therefore afford a production throughput of 1 tonne annum^{-1}.

With selectivity issues dominating the synthesis of such molecules, the authors also explored the synthesis of an Indacaterol **84** intermediate **88** using the aminolysis of epoxide **89** with amine **90** as a key step. Conducting the reaction under industrial conditions, the authors obtained the target compound **84** in 69%, a regioisomer in 8% and the bis-adduct in 12% (Scheme 3.20). Employing the alternative solvent system

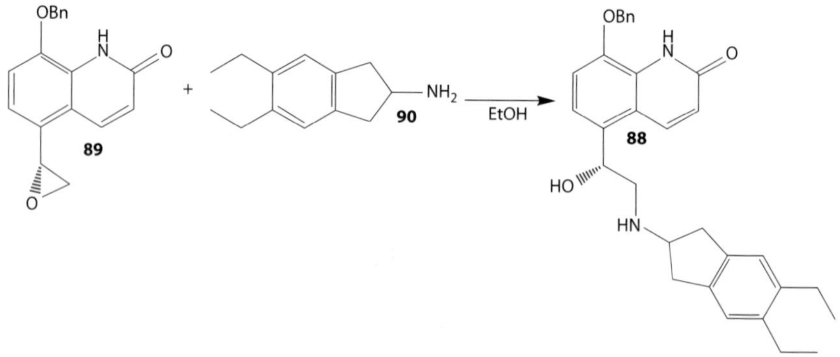

SCHEME 3.20 Schematic illustrating the key aminolysis step used in the synthesis of Indacaterol **84**.

TABLE 3.14
Summary of the Optimization Process Used for the Continuous Flow Synthesis of Metoprolol 86

Condition	Amine (Eq.)	Temperature (°C)	Residence Time (min)	Product 86 (%)[a]	By-Product (%)	Conversion (%)
Batch (microwave)[b]	1.2	150	30	65	31	100
Micro Reactor[c]	1.2	240	0.25	61	14	76
Micro Reactor	1.2	240	0.50	69	21	92
Micro Reactor	1.2	240	1	72	24	99
Micro Reactor	2.0	240	0.25	80	8	89
Micro Reactor	2.0	240	0.50	86	12	99
Micro Reactor	4.0	240	0.25	91	6	98

[a] Determined by HPLC with an internal standard.
[b] At 100 psi.
[c] At 500 psi using a back pressure regulator.

of NMP/H$_2$O (9:1), the authors were pleased to find that the product selectivity could be dramatically improved within the micro reactor, to afford Indacaterol intermediate **88** in 72% yield (Table 3.15).

3.1.13 C–F Bond Formation

Although organofluorine compounds are rare in nature, due to their metabolic stability compared to protonated analogs, such molecules feature widely in both pharmaceuticals and agrochemicals; safe, efficient, and scalable methods are therefore, required for the incorporation of fluorine into organic compounds.

Fluorination Reactions Using DAST: While diethylaminosulfur trifluoride (DAST) **91** has been shown to be a powerful fluorinating agent within the research laboratory, it is widely viewed as being too hazardous to employ on a production-scale due to its propensity to detonate (>90°C) and as such has yet to fulfill its synthetic potential at scale.

With this in mind, in 2008, Gustafsson et al. [42] evaluated the use of DAST **91** within a PTFE flow reactor (reactor volume = 16 mL), reporting the development of a facile approach for the fluorination of alcohols, aldehydes, and carboxylic acids. To perform a reaction, the authors mixed the substrate (0.2 M) of interest and DAST **91** (0.2 M for alcohols/carboxylic acids and 0.4 M for aldehydes) at a T-mixer prior to entering a PTFE tube reactor, heated to 70°C and held under 5 bar of pressure. Employing a residence time of 16 min, the authors obtained the corresponding deoxyfluorination product in moderate-to-excellent yield (40–100%), as summarized in Table 3.16.

More recently, Baumann et al. [43] demonstrated the use of PEEK/PFA/PTFE continuous flow reactors (Vapourtec R2 + /R4, UK) for the fluorination of organic

TABLE 3.15
Summary of the Results Obtained for the Synthesis of Indacaterol Intermediate 88 under Continuous Flow

Condition[a]	Concentration 89 (M)	Temperature (°C)	Residence Time (min)	Yield 88 (%)[b]	Isomer (%)	Bis-Adduct (%)	Conversion (%)
Batch (microwave)[c]	0.5	185	15	68.1	6.3	7.7	95.4
Micro Reactor[d]	0.4	185	15	67.8	8.6	9.1	97.0
Micro Reactor	0.38	185	15	70.0	8.0	7.1	92.8
Micro Reactor	0.38	185	15	68.3	8.2	7.5	95.1
Micro Reactor	0.38	185	15	72.1	8.6	7.9	92.4
Micro Reactor	0.37	165	30	60.7	6.8	6.4	92.3

[a] All reactions conducted using 1 eq. of amine **90**.
[b] Determined by HPLC with an internal standard.
[c] At 100 psi.
[d] At 250 psi using a back pressure regulator.

TABLE 3.16
A Selection of the Results Obtained for the Fluorination of a Series of Alcohols, Aldehydes, Ketones, and Carboxylic Acids Conducted under Continuous Flow

Substrate	Product	Yield (%)[a]
(menthol)	(fluoromenthol)	70[b]
(cholesterol)	(fluorocholestene)	61[c]
AcO-sugar-OAc with OAc, OH	AcO-sugar-OAc with OAc, F (α:β 5:4)	89
PhCH₂CH(Me)C(O)OH	PhCH₂CH(Me)C(O)F	81
4-MeO-C₆H₄-CHO	4-MeO-C₆H₄-CHF₂	89

[a] Isolated yield, determined after purification by column chromatography.
[b] 5:1 mixture of diastereomers.
[c] 6:1 mixture of diastereomers.

substrates using DAST **91**. To perform such reactions, the authors introduced solutions of DAST **91** (0.5–1.0 M) and substrate (0.5 M), in anhydrous DCM, into the flow reactor where they mixed at a T-connector prior to entering the heated reaction channel (reactor volume = 9 mL, Reaction channels = 1000 µm (i.d.)). Upon exiting the flow reactor, the reaction products were passed through a packed-bed reactor containing powdered $CaCO_3$ (~2 g) and a plug of silica gel (~2 g), thus quenching any residual DAST **91** and removing any side products. The final product was then obtained by evaporation of the reaction solvent and characterized by LC–MS and ^{1}H NMR spectroscopy.

As Table 3.17 illustrates, in the case of primary alcohols, the target monofluorides were obtained in high to excellent yield with a residence time of 27 min and a reactor temperature of 70–80°C; demonstrating tolerance toward sensitive functionalities such as vinyl iodides, epoxides, and ethers. In the case of aldehydic precursors, two equivalents of DAST **91** were required in conjunction with a reactor temperature of 80°C. Under the aforementioned conditions, electron-deficient aldehydes were fluorinated readily; however, electron-donating substrates required increased reaction times (45 min) to afford the respective difluoride in moderate-to-excellent yield.

TABLE 3.17
A Selection of the Fluorinated Compounds Prepared Using DAST in a PEEK/PTFE/PFA Flow Reactor

Starting Material	Product	Yield (%)
chloroimidazopyridine-CH2OH	chloroimidazopyridine-CH2F	73
4-Cl-3-NO2-C6H3-CH2OH	4-Cl-3-NO2-C6H3-CH2F	97
phenyl epoxide-CH2OH	phenyl epoxide-CH2F	96
vinyl iodide allylic alcohol	vinyl iodide allylic fluoride	82
bis-tetrahydropyranyl alcohol	bis-tetrahydropyranyl fluoride	83
alkenyl aldehyde	alkenyl difluoride	83
MeO-quinoline-CHO with Cl	MeO-quinoline-CHF2 with Cl	75

TABLE 3.17 (continued)
A Selection of the Fluorinated Compounds Prepared Using DAST in a PEEK/PTFE/PFA Flow Reactor

Starting Material	Product	Yield (%)
		89
		87
		73

See Chapter 7 for additional examples of fluorinations performed under flow conditions.

3.2 ELECTROPHILIC SUBSTITUTION

3.2.1 C–C Bond Formation

The reaction of organolithium compounds with aryl halides represents one of the most synthetically useful methods for the formation of C–C bonds, the need for cryogenic reaction conditions, due to the exothermic nature of reactions, however, often prevents successful scale-up of such reactions and has led to several research groups investigating the transformation under continuous flow.

Bromo-lithium Exchange: Employing two CYTOS Lab Systems (CPC, Germany) in series, thermostatted to 0°C, Schwalbe et al. [44,45] were early pioneers in this field and evaluated the bromo-lithium exchange reaction of 3-bromoanisole **92** with n-BuLi **15**, followed by the addition to DMF, to afford 3-methoxybenzaldehyde **93**; as depicted in Scheme 3.21.

By performing the reaction in two stages, the authors were able to optimize the reaction times for each step, ensuring that the unstable intermediate 3-methoxyphenyllithium **94** underwent formylation efficiently without undergoing decomposition, to anisole **95**, as observed in batch reactions.

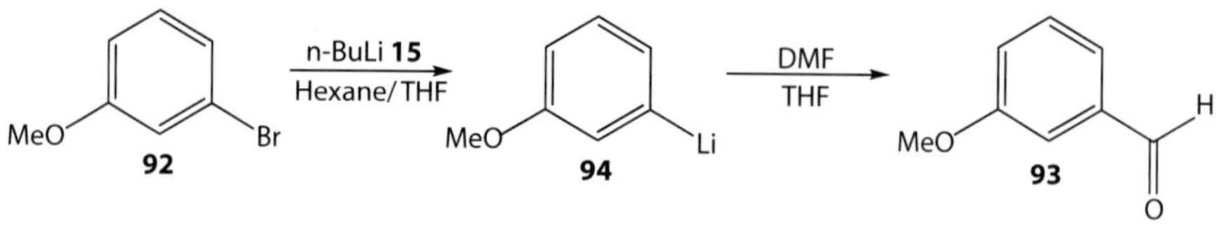

SCHEME 3.21 Illustration of the formylation reaction performed in a thermostatted flow reactor.

To perform the reaction under continuous flow, the authors pumped a solution of n-BuLi **15** (1.6 M, 6.0 mL min^{-1}), in hexanes, into the reactor from one inlet and a solution of 3-bromoanisole **92** (1.9 M, 4.7 mL min^{-1}) from a second inlet. Maintaining the reactor at 0°C, this afforded a reaction time of 11.4 s and generated the lithiated intermediate **94** in quantitative conversion. The output from the first reactor was subsequently pumped into a second reactor, where it mixed with a solution of DMF (5.0 M, 2.5 mL min^{-1}) in THF to afford 3-methoxybenzaldehyde **93** in 9 s. Owing to the reaction control obtained in the fluidic system, the authors were also able to perform the reaction at higher temperatures than in batch, with no degradation of product quality. As Table 3.18 illustrates, this enabled the authors to generate the compound **93** on a kilogram-scale, affording higher yields and purities than obtained in comparable batch reactors maintained under cryogenic conditions.

In 2008, Goto et al. [46] exploited the rapid rate of halogen–lithium exchange as a key step in the coupling of fenchone **96** and 2-bromopyridine **97** reducing the conventional two-step process to a single reaction step. Employing this reaction protocol, it was envisaged that the bromo-lithium exchange could be performed selectively in the presence of a trapping agent, fenchone **96**. To evaluate their strategy, the authors employed a stainless-steel flow reactor, comprising a 2 mL micro reactor cell and a 15 mL residence time unit. To perform a reaction, n-BuLi **15** (0.5–1.0 M) in hexane was added to a solution of fenchone **96** (0.5 to 1.0 M) and 2-bromopyridine **97** (0.46 M) in anhydrous THF at a flow rate of 5 mL min^{-1}. Maintaining the reactor at −25°C, the authors evaluated the effect of reactant stoichiometry at a fixed reaction time of 3 min, collecting the reaction products in ice-H$_2$O, to quench the reaction.

TABLE 3.18
Comparison of Batch and Flow Techniques for the Preparation of 3-Methoxybenzaldehyde 93 on a Kilogram-Scale

Reactor Type	Temperature (°C)	Reaction Scale (mol)	Yield (%)
Batch	−65	0.04	Quant.
Batch	−65	0.8	85 (76)
Batch	−50	4.8	60
Batch	−40	4.8	24
Flow	0	10.3	88

TABLE 3.19
Summary of the Optimization Protocol Employed for the One-Step Synthesis of Pyridine Derivative 98

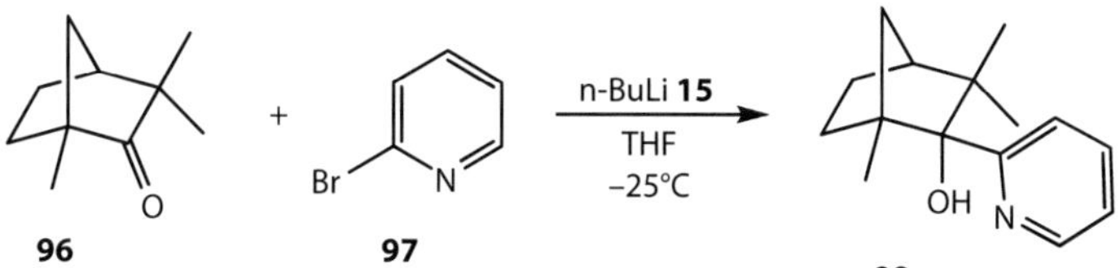

Substrate:Halide:n-BuLi Ratio	Temperature (°C)	Yield (%)[a]
1:1:1	−25	56
2:1:1.5	−25	71
2:1:2	−25	91
2:1:1.5	0	68

[a] Isolated yield.

The organic material was then extracted into ether prior to analysis by GC–MS and ^{1}H NMR spectroscopy. As Table 3.19 illustrates, this approach enabled the authors to rapidly optimize the reaction, obtaining the target pyridine derivative **98** in an isolated yield of 91%.

Yoshida et al. [47] and Yoshida and coworkers [48] have also performed extensive studies into the generation and reaction of unstable aryllithiums under continuous flow, attaining access to compounds that would otherwise have proved difficult to synthesize using the conventional reactor methodology. The control and selectivity of such techniques were demonstrated by Nagaki et al. [49], Nagaki and coworkers [50] for the bromo-lithium exchange reaction of alkyl *o*-bromobenzoates, a reaction commonly observed to suffer from functional group incompatibilities, with 1° and 2° alcohols found to dramatically reduce reaction yields (Scheme 3.22).

When performed in a microflow reactor, comprising of two T-shaped micromixers and two tubular reactors (1 mm i.d.), the authors were able to successfully lithiate

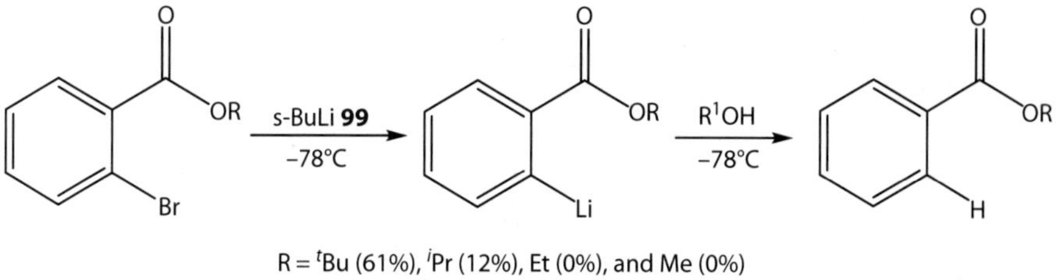

R = ^{t}Bu (61%), ^{i}Pr (12%), Et (0%), and Me (0%)

SCHEME 3.22 Schematic illustrating the effect of alkoxy substituents on bromo-lithium exchange reactions conducted in batch.

TABLE 3.20
Summary of the Results Obtained for the Br–Li Exchange Reaction of Alkyl o-Bromobenzoates and Benzaldehyde 19 Performed under Continuous Flow

Substrate	Product	Temperature (°C)	Residence Time (s)	Yield (%)[a]
o-Br-C₆H₄-CO-O^{t}Bu	3-Ph-isobenzofuranone **100**	0	0.01	82
o-Br-C₆H₄-CO-O^{i}Pr	3-Ph-isobenzofuranone **100**	−28	0.01	66
o-Br-C₆H₄-CO-OEt **101**	3-Ph-isobenzofuranone **100**	−48	0.06	70
o-Br-C₆H₄-CO-OMe **102**	3-Ph-isobenzofuranone **100**	−48	0.02	85

[a] o-Bromobenzoates in THF (0.1 M), s-BuLi **99** in hexane/cyclohexane (0.42 M), and benzaldehyde **19** (0.60 M, 3 eq.) in THF.

and react substituted alkyl o-bromobenzoates in high to excellent yield. Employing benzaldehyde **19** as the electrophile (0.60 M, 3 eq.) and s-BuLi **99** (0.42 M, 1.05 eq.) as the lithiating agent, the authors evaluated the effect of alkoxy substitution on the formation of 3-phenylisobenzofuran-1(3H)-one **100**. As Table 3.20 illustrates, unlike comparable batch reactions, high yields are even obtained for the less sterically demanding ethoxy- **101** and methoxy-derivatives **102**, an observation that the authors attribute to the use of short reaction times (0.02–0.06 s) compared to batch (10 min), which prevent decomposition of the lithiated intermediate prior to electrophilic substitution.

Additional examples utilizing electrophiles such as TMSCl (trimethylsilyl chloride) **17**, EtOH and methyl triflate **103** can be found within the original manuscript; in all cases dramatic improvements in product yield were obtained compared to reactions performed using batch reactor technology.

More recently, the authors have looked at the sequential introduction of two electrophiles, as a means of trapping the unstable aryllithium intermediates and generating substituted aryl derivatives [51]. As illustrated in Table 3.21, the authors evaluated the reaction of o-dibromobenzene **104** with n-BuLi **15** at −78°C prior to the addition of the first electrophile (E¹) also at −78°C. In a third micromixer, the reaction mixture

TABLE 3.21
Illustration of the Sequential Bromo-Lithium Exchange Reactions Performed under Continuous Flow

E^1	E^2	Yield (%)
MeOTf **103**	PhCHO **19**	61
MeOTf **103**	TMSCl **17**	67
PhCHO **19**	TMSCl **17**	74

was warmed to 0°C prior to the second bromo-lithium exchange reaction and addition of electrophile 2 (E^2). Using this approach, the authors were able to demonstrate the ease with which aryllithiums can be trapped with electrophiles, such as methyl triflate **103** and TMSCl **17**, affording the target compounds in moderate-to-high yield.

As an extension to this process, Tomida et al. [52] evaluated the carbolithiation of unsaturated compounds as a synthetically useful C–C bond forming reaction that generates a second organo-lithium intermediate; available for subsequent reaction with an array of electrophiles. To date however, the yields of such reactions have been low due to the difficulties associated with the addition of aryllithiums to unsaturated compounds. Based on the experience gained through the bromo-lithium exchange reactions previously described, the authors envisaged increasing product yield and purity by performing carbolithiations followed by reaction with an electrophile under continuous flow.

Employing the carbolithiation of 4-phenyl-but-1-en-3-yne **105** with 4-methylphenyllithium **106**, as a model reaction, the authors investigated the effect of reactant residence time and temperature on the resulting product distribution (Scheme 3.23). Using this approach, the authors conducted the bromo-lithium exchange reaction at 0°C, whereby they noted incomplete lithiation with a residence time of 60 s. Increasing the reactor temperature to 25°C, again with a reaction time of 60 s, the authors observed complete carbolithiation, which coupled with a reaction time of 25 s in the second reactor, afforded high yields of **107** and **108** while suppressing the formation of butyl bromide derivatives **109** and **110**. Compared to macroscale experiments, this technique affords a dramatic increase in reactor temperature, from −78 to 25°C, and enabled the authors to synthesize a series of allenylsilanes in moderate to high yield, independent of any substituent effect; as depicted in Table 3.22.

Further examples of the synthetic utility of this technique have been reported whereby lithiobenzonitriles [53] and oxiranyl anions [54] have been generated and reacted at elevated temperatures compared with −100 to −78°C in batch.

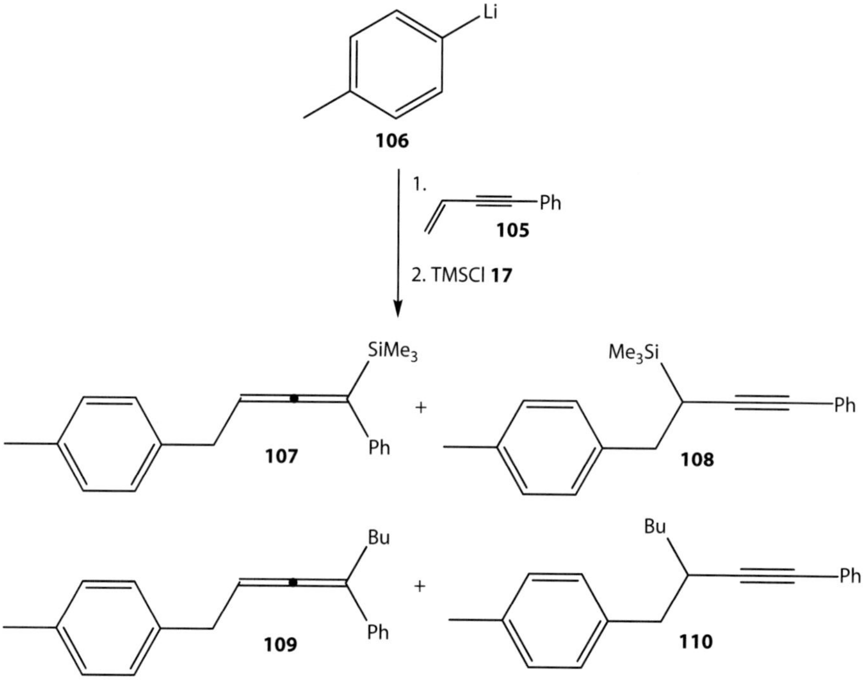

SCHEME 3.23 Schematic illustrating the potential reaction products obtained for the carbolithiation of 4-phenylbut-1-en-3-yne **105**.

Iodo-lithium Exchange: Using a similar approach to that demonstrated for the bromo-lithium exchange reactions, Yoshida and coworkers [55] incorporated iodo-lithium exchange reactions into their toolbox of synthetic transformations. Once again, the micro reactor comprised two T-shaped micromixers and two tube reactors, with aryllithium reactants generated *in situ*, this time via the treatment of *o*-iodonitrobenzene **111**, *m*-iodonitrobenzene **112**, and *p*-iodonitrobenzene **113** with phenyllithium **114**. Using this approach, the authors investigated the effect of temperature and residence time on the formation of the lithiated intermediate, identifying 0°C and 0.01 s as the optimum for **111** and −28°C and 0.01 s for substrates **112** and **113** (Table 3.23). The authors also identified the ability to selectively form either the kinetic or thermodynamic organo-lithiated intermediate as a function of reactant residence time (Figure 3.6).

Grignard Reactions: In addition to demonstrating the advantages associated with the manipulation of short-lived lithiated intermediates, Schwalbe et al. [45] also reported the ability to prepare Grignard reagents from alkyl halides within continuous flow reactors (Scheme 3.24). Due to problems observed in batch with poor metallation of pentafluoroethyliodide **115** and subsequent addition to carbonyl compounds, such as benzophenone **116**, the authors investigated the formation of a Grignard reagent **117** by reaction with methyl magnesium chloride **118**. In batch only 16% of the desired adduct, 2,2,3,3,3-pentafluoro-1,1-diphenylpropan-1-ol **119**, was formed due to decomposition via β-elimination.

TABLE 3.22
Illustration of the Substituent Effect on the Carbolithiation of 4-Aryl-But-1-En-3-Ynes in an Integrated Microflow System

R	Ar	Yield (%)[a]
Me **107**	Ph	80 (91/9)
H	Ph	75 (91/9)
OMe	Ph	78 (90/10)
F	Ph	47 (91/9)
Ph	Ph	73 (90/10)
Me	p-MePh	82 (83/17)
Me	p-OMePh	81 (80/20)
Me	p-FPh	80 (94/6)
Me	Thienyl	62 (98/2)

[a] The numbers in parentheses represent the ratio of products obtained (a/b).

Employing a two-stage micro reactor, the authors were able to efficiently form the Grignard reagent **117** and perform the nucleophilic addition steps at different reaction temperatures, using independent cooling baths (−6 and −4°C). Flow reactions were performed using the following procedure; solutions of MeMgCl **118** (0.82 M in hexane) and C_2F_5I **115** (0.75 M in DCM) were pumped through the reactor to afford a residence time of 0.9 min. Benzophenone **116** (0.62 M in DCM) was then introduced into the second reactor where a residence time of 8 min afforded the target compound in 86% yield demonstrating a dramatic increase of 61% compared to batch.

Exploring the reaction of commercially available Grignard reagents, Rencurosi and coworkers [56] demonstrated the continuous flow synthesis of a series of alcohols using a PTFE tube reactor (Vapourtec, UK). By conducting a brief reaction optimization, the authors were able to explore the effect of reaction time (33–66 min), temperature (−78 to 25°C), and stoichiometry (1.2–2.0 eq.) on the Grignard reaction between 4-isopropylbenzaldehyde **120** and (2-methylallyl)magnesium chloride **121** (Table 3.24). Using this approach, the authors were able to identify a reaction time of 33 min, reactor temperature of 25°C, and a 1.2 eq. excess of Grignard reagent **121** as being optimal, affording 1-(4-isopropylphenyl)-3-methylbut-3-en-1-ol **122** in 98% yield; after purification with polymer-supported benzaldehyde, used to remove residual **121**.

TABLE 3.23
Summary of the Results Obtained for the Iodo-Lithium Exchange Reactions of Iodonitrobenzene Derivatives Performed under Continuous Flow

Substrate	Electrophile	Product	Yield (%)
111 (0°C) *ortho-I, NO₂*	MeOH	*ortho-H, NO₂*	87
	MeI	*ortho-Me, NO₂*	36
	TMSCl **17**	*ortho-SiMe₃, NO₂*	62
	PhCHO **19**	*ortho-CH(OH)Ph, NO₂*	93
112 (−28°C) *meta-I, NO₂*	MeOH	*meta-H, NO₂*	87
	MeI	*meta-Me, NO₂*	44
	TMSCl **17**	*meta-SiMe₃, NO₂*	85
	PhCHO **19**	*meta-CH(OH)Ph, NO₂*	93

Liquid-Phase Micro Reactions

TABLE 3.23 (continued)
Summary of the Results Obtained for the Iodo-Lithium Exchange Reactions of Iodonitrobenzene Derivatives Performed under Continuous Flow

Substrate	Electrophile	Product	Yield (%)
O₂N-C₆H₄-I **113** (−28°C)	MeOH	O₂N-C₆H₄-H	91
	MeI	O₂N-C₆H₄-Me	46
	TMSCl **17**	O₂N-C₆H₄-SiMe₃	70
	PhCHO **19**	O₂N-C₆H₄-CH(OH)Ph	86

Having established a protocol for the continuous flow Grignard reaction, the authors evaluated the scope of the methodology, investigating the effect of Grignard reagent and carbonyl compound on the reaction. As Table 3.24 illustrates, using the continuous flow protocol the authors were able to readily synthesize a series of secondary alcohols in high to excellent yield, with purities >95%, as determined by UPLC-MS. Compared to batch reactions (−20°C), the authors were able to perform reactions utilizing Grignard reagents at room temperature without observing degradation of the reagent or the undesirable formation of by-products. In addition, the authors were able to selectively bring about reaction among carbonyl moieties in the presence of nitriles, affording a facile route to nitrile-substituted secondary alcohols.

Friedel–Crafts Acylation: Owing to their application in catalysis and materials science, methods have been sought for the selective synthesis of ferrocene derivatives. While the Friedel–Crafts acylation of ferrocene **123** has been widely employed as a means of accessing compounds suitable for further activation or derivatization, many of the catalyst and solvent combinations evaluated have resulted in the formation of diacylated products. Based on a previous example whereby product selectivity has been improved as a result of performing reactions under continuous flow, Lei and coworkers [57] evaluated the acetylation of ferrocene **123** in a microfluidic reactor. Using a soda-lime glass device (channel dimensions = 500 μm (wide) × 100 μm (deep) × 10 cm (long)), with two reactant inputs and a single output, the authors investigated the effect of reaction time (37–440 s) and temperature (15–25°C) on the acetylation of ferrocene **123** (Scheme 3.25).

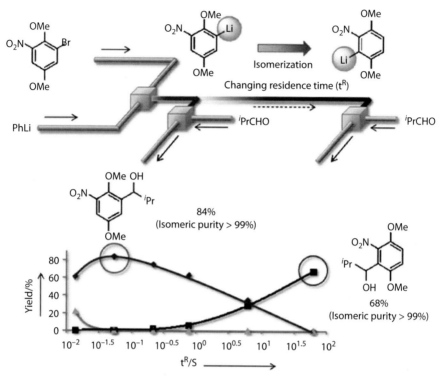

FIGURE 3.6 Illustration of the effect of kinetic and thermodynamic control obtained within a microflow reactor. (Nagaki, A., Kim, H., and Yoshida, J. Nitro-substituted aryl lithium compounds in microreactor synthesis: Switch between kinetic and thermodynamic control, *Angew. Chem. Int. Ed.* 2009. 48: 8063–8065. Copyright Wiley-VCH Verlag GmbH & Co. KGaA. Reproduced with permission.)

Employing ferrocene **123** (0.15 M) and phosphoric acid **124** (1.5 M) in acetic anhydride **34**, the authors identified that high conversions to the monoacetylated product **125** were attainable, with a flow rate of 60 µL min^{-1} affording 98% conversion (at 25°C); product purity was also confirmed by analysis of the resulting reaction products by NMR spectroscopy and MS. In comparison to batch techniques previously reported, where reactor temperatures of 55–60°C were required, the authors were able to effectively demonstrate the selective acylation of ferrocene **123** by performing the reaction in a microfluidic reactor at room temperature.

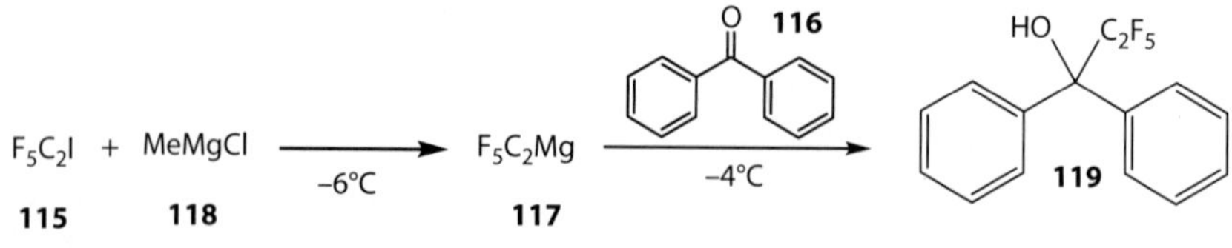

SCHEME 3.24 Illustration of the *in situ* formation of a short-lived organometallic species **117** and its subsequent nucleophilic addition to benzophenone **119**.

Liquid-Phase Micro Reactions

TABLE 3.24
A Selection of Flow Reactions Performed Utilizing Grignard Reagents at Room Temperature

Carbonyl Compound	Grignard Reagent	Product	Yield (%)
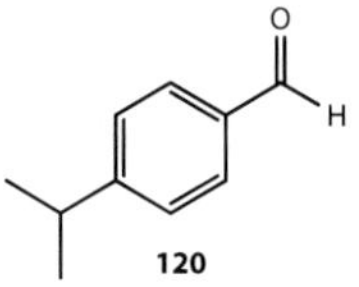 4-isopropylbenzaldehyde **120**	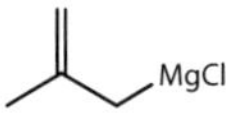 isobutenyl-MgCl **121**	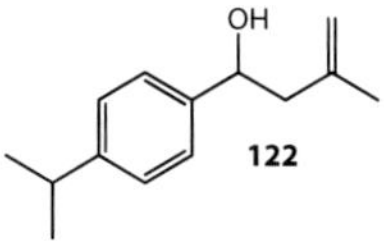 **122**	
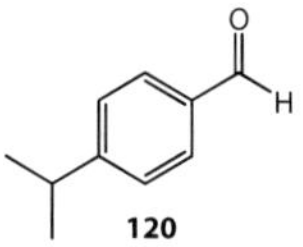 4-isopropylbenzaldehyde **120**	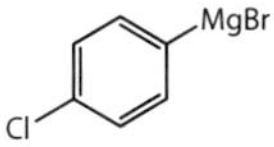 4-chlorophenyl-MgBr	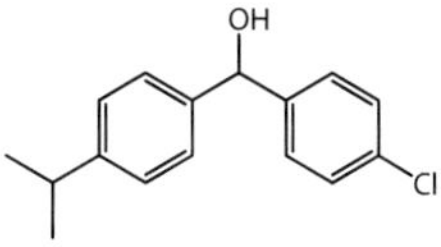	93
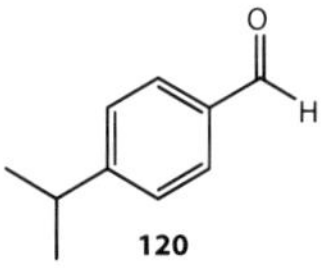 4-isopropylbenzaldehyde **120**	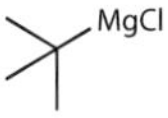 tert-butyl-MgCl	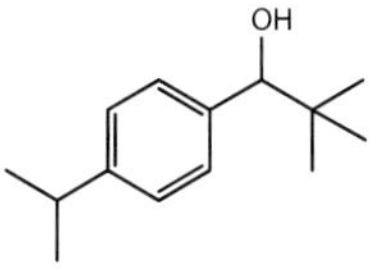	87
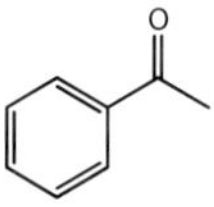 acetophenone	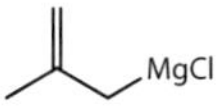 isobutenyl-MgCl **121**	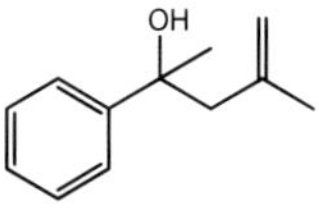	94
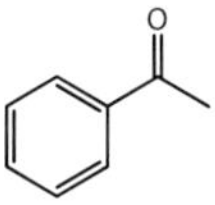 acetophenone	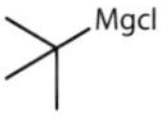 tert-butyl-MgCl	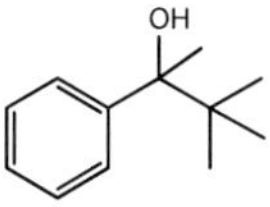	90
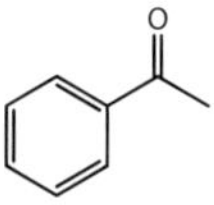 acetophenone	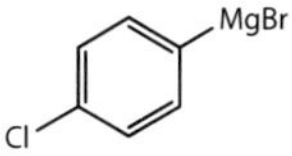 4-chlorophenyl-MgBr	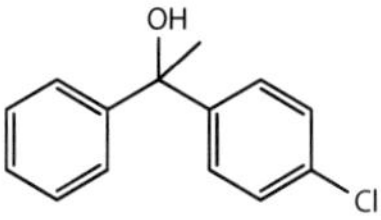	95
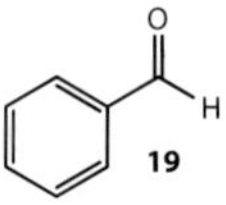 benzaldehyde **19**	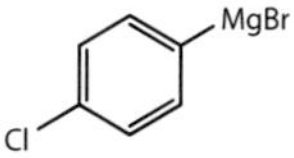 4-chlorophenyl-MgBr	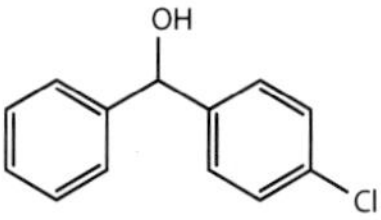	90

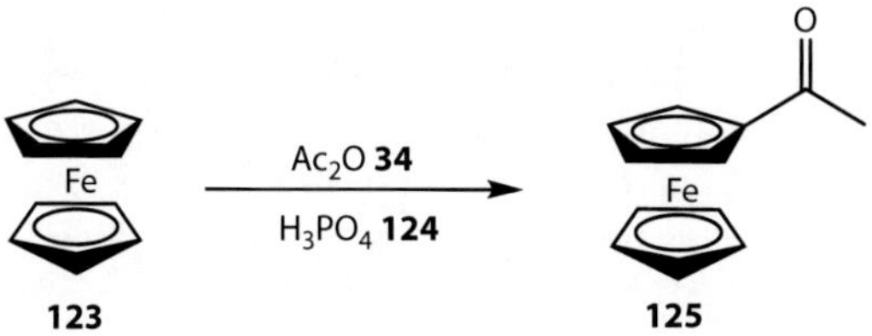

SCHEME 3.25 Illustration of the reaction conditions employed for the selective acetylation of ferrocene **123**.

Ferrocene **123** → (Ac$_2$O **34**, H$_3$PO$_4$ **124**) → acetylferrocene **125**

TABLE 3.25
A Selection of Monoacylated Ferrocene Derivatives Prepared under Continuous Flow

Anhydride	Flow Rate (µL min⁻¹)	Conversion (%)
Propionic	40	95
n-Butyric	40	95
n-Hexanoic	20	93
n-Octanoic	10	92

Gratified by their findings, the authors subsequently investigated the generality of the methodology and by substituting acetic anhydride for an array of alternative anhydrides, four substituted ferrocenes were synthesized, as illustrated in Table 3.25.

In all cases, excellent conversions of ferrocene **123** to the monoacylated derivative were obtained, with a trend observed whereby increasing reaction time was required with increasing acid anhydride chain length. It must be noted, however, that at no point did the authors detect the formation of any diacylferrocene by-products.

Friedel–Crafts Alkylation: The alkylation of aromatic rings, Friedel–Crafts, is of great synthetic importance due to its broad scope and tolerance, the technique is however, prone to the formation of by-products, via polyalkylation (Scheme 3.26). With this in mind, Suga et al. [58] evaluated the alkylation of 1,3,5-trimethoxybenzene **126** using the *N*-acyliminium ion **127** generated by the anodic oxidation of methyl butyl((trimethylsilyl)methyl)carbamate. When performed in batch, at −78°C, the authors found the reaction afforded a 1:1 mixture of the mono- **128** and dialkylated **129** products (54:46), regardless of the order of reactant addition employed; attributing the observation to inefficient mixing.

The importance of efficient mixing was subsequently evaluated by performing the reaction two microfluidic devices, the first a T-shaped tube reactor (dimensions = 500 µm (i.d.)) and the second was an interdigital micromixer, containing 25 µm channels (IMM, GmbH). As Table 3.26 illustrates, the use of a relatively large tube reactor, analogous results to those in batch are obtained, however, by decreasing the mixing time through the use of an interdigital micromixer, the authors were able to dramatically increase the percentage of methybutyl(2,4,6-trimethoxybenzyl)carbamate **128**, while simultaneously decreasing the formation of methyl-2-(3-((butyl(methoxycarbonyl)amino)methyl)-2,4,6-trimethoxybenzyl) hexanoate **129**.

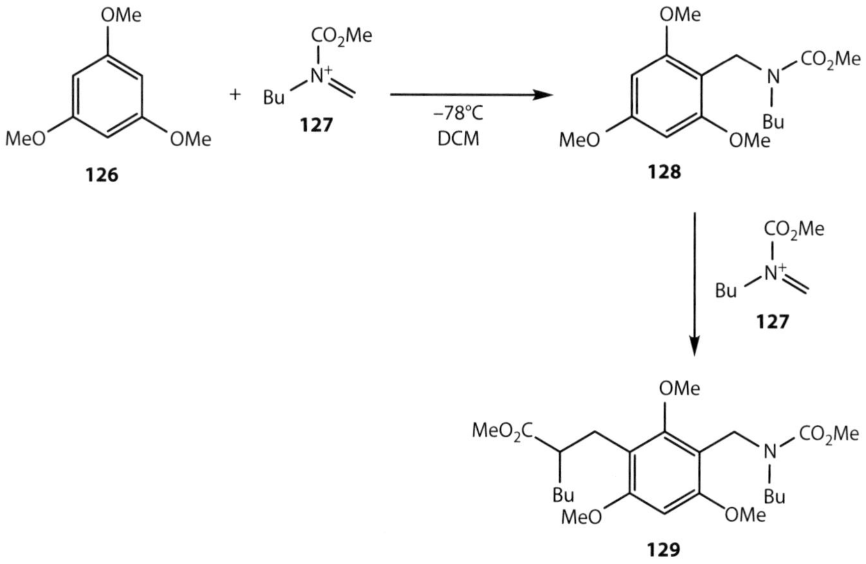

SCHEME 3.26 Schematic illustrating the distribution of products obtained when performing the Friedel–Crafts alkylation of 1,3,5-trimethoxybenzene **126** under conventional conditions.

In addition to improved mixing, the enhanced selectivity can be attributed to the efficient distribution of heat when reactions are performed using microstructured devices, and this was demonstrated by the decrease in yield (**128**) when the reactor temperature was increased from −78 to 0°C (92–30%).

In a second example, the authors demonstrated the selective alkylation of a series of heteroaromatic compounds and as summarized in Table 3.27, the use of a micromixer again afforded a straightforward method for the high-yielding monoalkylation. To demonstrate the synthetic versatility of the methodology developed, the authors concluded their investigation with the controlled dialkylation thiophene **130**, using two electrophiles. As Scheme 3.27 illustrates, the authors employed a two-step protocol, firstly introducing the butyl carbamates moiety **131**, followed by the cyclohexyl derivative to afford the dialkyl derivative **132** in an overall yield of 64%.

TABLE 3.26
Illustration of the Effect of Mixing Efficiency on the Selective Alkylation of 1,3,5-Trimethoxybenzene 126

Reactor Type	Mono-128 (%)	Di-129 (%)	Product Ratio (128:129)
Batch	37	32	54:46
T-shaped tube	36	31	54:46
Micro mixer	92	4	96:4

TABLE 3.27
Comparison of Product Distributions Obtained for the Friedel–Crafts Alkylation of Heterocycles under Batch and Flow Conditions

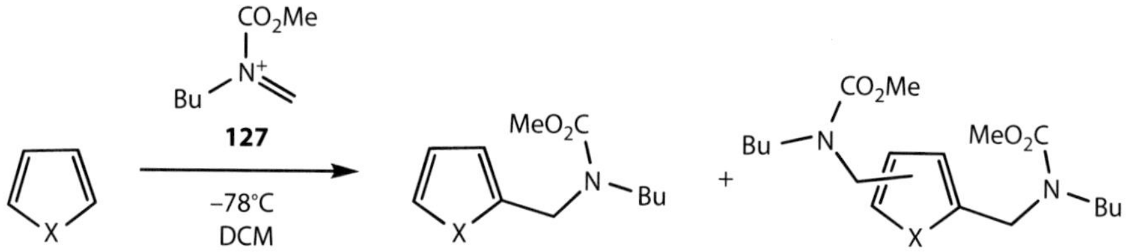

X	Reactor Type	Mono-(%)	Di-(%)
S 130	Batch	14	27
	Micromixer	84	0
O	Batch	11	5
	Micromixer	39	Trace
N-Me	Batch	33	28
	Micromixer	60	6

3.2.2 C–N Bond-Forming Reactions: Nitration Reactions

Due to the hazardous nature associated with nitration reactions, many research groups from both academia and industry have evaluated the performance of this class of reaction under continuous flow with examples reported using electroosmotic flow [59], biphasic liquid–liquid systems [60], and sodium dihydrogen phosphate-catalyzed

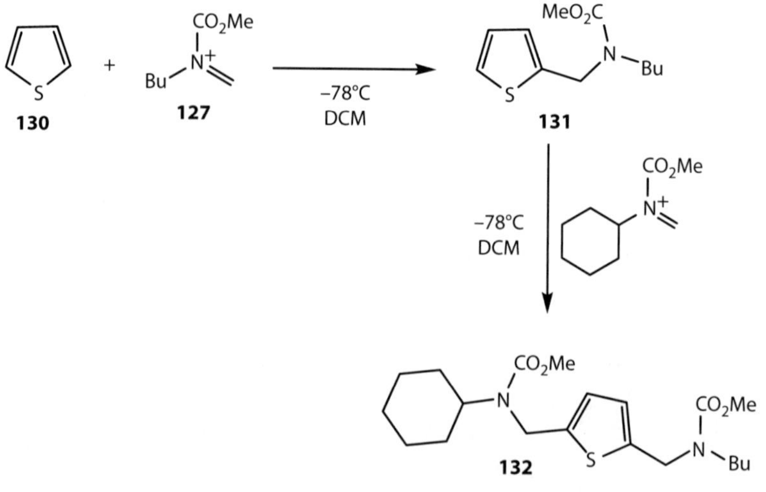

SCHEME 3.27 Schematic illustrating the sequential Freidel–Crafts alkylations performed, exploiting the efficient mixing obtained in microstructured devices.

reactions [61], with many of these examples illustrating the high product selectivity that results from the controlled addition of reactants and efficient thermal control.

Aromatic Nitrations: An early investigation into nitration reactions performed within capillary flow reactors was reported by Burns and Ramshaw [62], who reported that "slug flow" enabled efficient contacting of immiscible liquid phases within narrow channels. This is attributed to internal circulation obtained within the slugs of liquid due to the combined effects of shear within the channel and interfacial phenomena; the authors noted that typical circulation frequencies of 10–100 Hz may be obtained in submillimeter slugs (cm s^{-1}). In addition to affording a rapid method of mixing, the technique also provided a facile method of separating two immiscible phases, at the outlet of the reaction channel.

With this in mind, the authors evaluated the nitration of benzene and toluene in a range of reactors including stainless steel (127, 178, and 254 µm (i.d.)) and PTFE (300 µm (i.d.)); both utilizing a T-mixer as a liquid contactor. These reactions were selected as the nitration of aromatic compounds, using mixed acids, is known to be exothermic and mass transfer limited; as such it was proposed that the heat removal and efficient mixing obtained in microfluidic systems would offer processing advantages over conventional batch methodology.

Using offline GC-FID analysis of the organic fraction, the authors quantified the effect of acid ratio, temperature, flow rate, and organic/aqueous flow ratios on the product distribution obtained. Under the aforementioned conditions, the authors identified that the application of narrow bore tubing afforded a rate enhancement when compared with wider bore tubing and observed an exponential relationship between the rate constant and flow velocity; it must be noted, however, that no attempts were made to optimize the reaction conditions to afford high conversions and/or reduced polynitration.

In a rare example, Ferstl et al. [63] described the single-phase nitration of 2-(4-chlorobenzoyl)benzoic acid **133** to 2-(4-chloro-3-nitrobenzoyl)benzoic acid **134**, a precursor used in the synthesis of a pharmaceutical compound, within a custom built silicon reactor (channel dimensions = 300 µm (wide)), containing split and recombine micromixers and integrated sensors.

Utilizing on-line HPLC analysis, the authors investigated the effect of reactant residence time on the formation of **134** and the dinitrated by-product, for a nitrating solution of H_2SO_4 **26** (97%, 0.4 mL min^{-1}) and nitric acid **135** (0.024 mL min^{-1}), achieved by varying the number of micro reactors employed in series. Using this approach, the authors were able to increase the conversion of **133** to **134** from 42% to 75%, by increasing the number of micro reactors used from 1 to 3, while suppressing the formation of dinitrated by-products. Although not reported, it can be seen from the recovery of unreacted 2-(4-chlorobenzoyl)benzoic acid **133**, that increasing the residence time further would enable more efficient synthesis of 2-(4-chloro-3-nitrobenzoyl)benzoic acid **134** to be performed (Scheme 3.28).

Nitration of Heterocycles: In order to demonstrate the ability to perform challenging reactions under flow conditions, Schwalbe and coworkers [64] investigated the synthesis of 2-methyl-4-nitro-5-propyl-2*H*-pyrazole-3-carboxylic acid, a key intermediate in the synthesis of Sildenafil®. The reaction was selected as it currently proves problematic when performed in large batches due to the thermal degradation of the target compound at temperatures above 100°C. To circumvent this, process

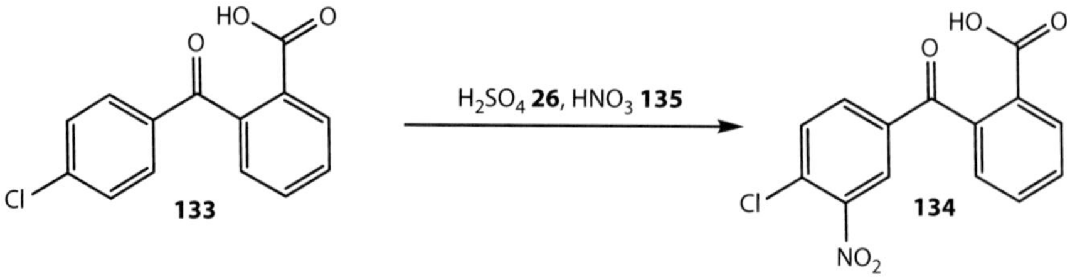

SCHEME 3.28 Illustration of the model nitration used to evaluate a silicon micro reaction system with integrated sensors and on-line HPLC capabilities.

chemists have found that adding the nitrating solution in three aliquots, temperature rises could be reduced (71°C); however, this does lead to an undesirable increase in reaction time from 8 to 10 h.

Utilizing a CYTOS stainless-steel flow reactor, the authors investigated the ability to perform the nitration of 2-methyl-5-propyl-2H-pyrazole-3-carboxylic acid in a controlled and isothermal manner as a means of increasing process safety and productivity. Employing a residence time of 35 min and a reactor temperature of 90°C, the authors were able to chemoselectively synthesize 2-methyl-4-nitro-5-propyl-2H-pyrazole-3-carboxylic acid in 73% yield and a throughput of 5.5 g h^{-1}, therefore reducing the hazards associated with the process and providing a scalable method for the intermediates preparation (see Chapter 7 for details).

Alternative Nitrating Agents: In addition, Panke et al. [64] also reported the nitration of 2-methylindole **136** utilizing sulfuric acid and sodium nitrate **137** as the nitrating agents. Modifying a conventional batch protocol, the authors prepared two reactant solutions, the first contained 2-methylindole **136** (41.2 mmol) in H$_2$SO$_4$ **26** and the second contained sodium nitrate **137** (41.2 mmol) in H$_2$SO$_4$ **26**. The reactant solutions were brought together in a CYTOS (CPC, Germany) micro at a total flow rate of 1.44 mL min^{-1}, a reactor temperature of 3°C was employed and the reaction products collected in ice-H$_2$O at the reaction outlet. Upon standing overnight, the 2-methyl-5-nitroindole **138** was obtained as a yellow solid (70% yield) in >99% purity (GC) (Scheme 3.29).

Aliphatic Nitro Esters: Another example of the safe performance of exothermic nitration reactions was the synthesis of 2-ethylhexylnitrate **139** reported by Chen and coworkers [65]. In addition to being a temperature sensitive reaction, the synthesis of 2-ethylhexylnitrate **139** (Scheme 3.30) is of industrial interest as it is used in the petrochemical sector as a diesel additive; acting to increase the cetane number, reducing hydrocarbon emissions and NO formation. While the nitration of alcohols

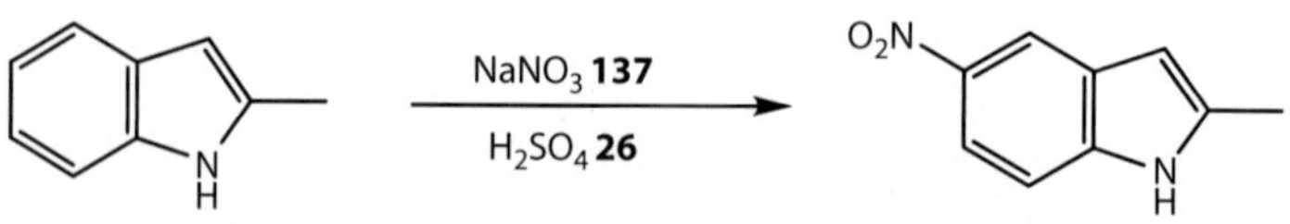

SCHEME 3.29 Illustration of the reaction protocol employed for the synthesis of 2-methyl-5-nitroindole **138** under continuous flow.

Liquid-Phase Micro Reactions

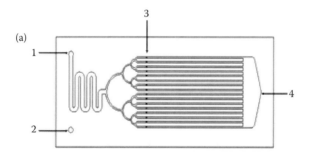

SCHEME 3.30 Illustration of the reaction protocol employed for the synthesis of 2-ethylhexylnitrate **139** under continuous flow.

is an industrially relevant route, the resulting nitrates can readily volatilize and undergo self-sustaining decomposition. Due to the accompanied release of heat, the reactions are conventionally limited to a low thermal regime, and hence slow reaction, to avoid the risk of thermal runaway. While additives such as alkoxyalcohols have been used to avoid the accumulation of heat, their presence also leads to the formation of undesirable by-products resulting in the need for more complex and lengthy purification steps.

Based on the advantages that micro reaction technology has brought to previous nitration reactions, the authors investigated the effect of reaction temperature, reaction time, and mixed acid composition on the selective nitration of 2-ethylhexanol **140** to afford 2-ethylhexylnitrate **139**. To achieve this, the authors fabricated a stainless-steel micro reactor consisting of sixteen parallel microchannels (dimensions = 500 μm (wide), 500 μm (deep) × 7.8 cm (long)), each fed from a single inlet stream per reactant, and used it to investigate the nitration of 2-ethylhexanol **140** under continuous flow (Figure 3.7).

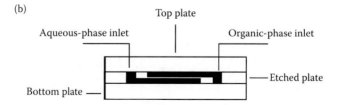

FIGURE 3.7 Schematic illustrating the stainless-steel micro reactor used to investigate the nitration of 2-ethylhexanol **140** (a) top view of the channel network and (b) lateral view of the inlet channels. (Reproduced from Chin. J. Chem. Eng., 17(3), Shen, J. et al., Investigation of nitration processes of *iso*-octanol with mixed acid in a microreactor, 412–418, Copyright (2009), with permission from Elsevier.)

Employing a mixed acid solution comprising 74% H_2SO_4 **26**, 24% HNO_3 **135**, a biphasic reaction mixture resulted upon the introduction of 2-ethylhexanol **140** into the micro reactor and under this liquid–liquid regime, the authors noted that the reaction took place within the aqueous phase (close to the interface) and was controlled by mass transfer. To ensure that any reaction trends observed were due to the nitration reaction occurring within the micro reactor, the reaction products were diluted and cooled to 0°C upon exiting the micro reactor in order to terminate the reaction prior to analysis by GC-FID.

While the commercial process is performed at 15°C, the ability to control fast, exothermic reactions within micro reactors led the authors to investigate the effect of reaction temperature on the nitration reaction. Employing electrical heating, the reactor was evaluated at temperatures ranging from 25 to 40°C, where an increase in conversion from 60% to 82% **139** was observed. The authors noted that a combination of the rapid heat transfer, with the small internal reactor volume, enabled control of the reaction temperature, ensuring process safety and removing the risks associated with thermal decomposition of the target compound **139**.

In addition, the authors observed an increase in the nitration of 2-ethylhexanol **140** with increasing HNO_3 **135**, without the generation of by-products conventionally associated with this approach. Combining all of these effects, a reaction time of 7.2 s, a reaction temperature of 35°C, and a HNO_3 **135**:2-ethylhexanol **140** of 1.5, enabled the authors to obtain 2-ethylhexylnitrate **139** in 97.2% conversion, with no associated by-product formation; affording a liquid hourly space velocity of 500 h^{-1}.

Using a series of commercially available glass micro reactors (Corning Incorporated, Germany), Reintjens and coworkers [66] also demonstrated the ability to perform a selective organic nitration reactions, using neat HNO_3 **135** under cGMP conditions (Scheme 3.31). Using this approach, a reactor volume of 150 mL allowed the undisclosed target at a production rate of 13 kg h^{-1} with intrinsically high levels of safety unattainable in conventional batch techniques. In order to increase productivity further, the authors employed a production unit consisting of 8 micro reactors (two banks of 4 micro reactors) and operated the unit under the previously determined conditions. Using this approach, the authors were able to produce the target in 100 kg h^{-1}, which extrapolates to an annual capacity of 800 tons if operated continuously. This investigation illustrates the ability to safely operate hazardous reactions and design a predictable/reliable production unit for the rapid preparation of high-quality chemicals [67].

3.2.3 C-Hetero Bond-Forming Reactions: Halogenations under Flow

Due to their widespread use in coupling reactions, to name but one example, efficient methodology for the selective synthesis and large-scale preparation of halogenated

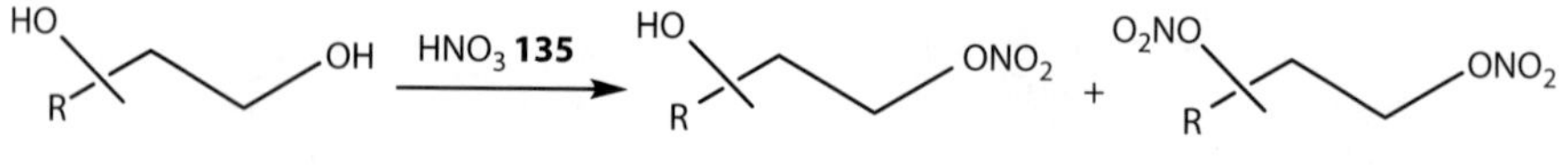

SCHEME 3.31 General scheme illustrating the selective nitration reaction performed under continuous flow conditions.

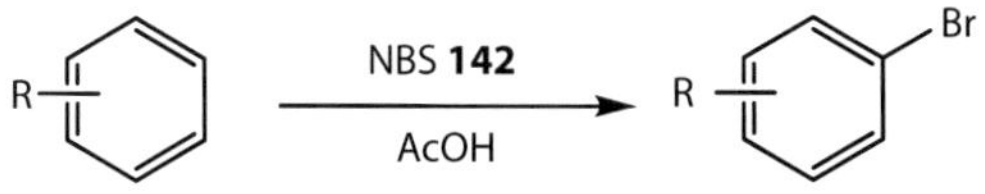

SCHEME 3.32 Generalized scheme illustrating the bromination of aromatic substrates used in a GlaxoSmithKline drug discovery program.

compounds are sought. While examples of such reactions performed using gases have been described elsewhere (see Chapter 2), this section focuses on the liquid-phase reactions investigated as a means of developing efficient halogenation strategies for the preparation of brominated and iodinated compounds.

Brominations: By coupling fluorous spacer technology with a continuous flow microwave reactor, Benali et al. [68] demonstrated the ability to generate and process plugs of reagents within a continuous fluorous phase (perfluoromethyldecalin-PFMD) as a means of using small amounts of reagents (300 µL) for parameter and reactant screening.

Among other reactions investigated, the authors utilized this technique for the bromination of an aromatic derivative **141** using NBS **142**, to afford **143** an intermediate required in large quantities for use in a drug discovery program (Scheme 3.32).

Using the CEM Voyager® (CEM, USA) continuous flow reactor, the authors investigated the bromination reaction, identifying 650 µL min^{-1}, 300 W, and a reaction temperature of 120°C as optimal conditions. Running the system continuously for 1 h, the authors processed 37 mL of reaction mixture which upon work up afforded 1.4 g of the bromination product **143** in 89% yield and 91% purity.

Using elemental bromine **144**, Löb et al. [69] investigated the bromination of toluene (Scheme 3.33) within a tubular flow reactor (dimensions = 2.5 m (long), volume = 4.9 mL), coupled with either a triangular interdigital or caterpillar micromixer. Employing a bromine **144** molar ratio of 1.0, the authors investigated the effect of reaction temperature (0–120°C) and residence time (from 48 ms to 3.9 min) on the formation of benzyl bromide **9** and the 3 isomers of monobromotoluene. Using this approach, the authors identified 80°C, in the triangular interdigital mixer, as affording quantitative consumption of toluene, obtaining 20% selectivity toward benzyl bromide **9**.

More recently, Wahab [70] reported the use of elemental bromine **144** under flow conditions for the bromination of a series of indoles (Table 3.28), forming intermediates useful in the synthesis of pharmaceutically relevant compounds. While this

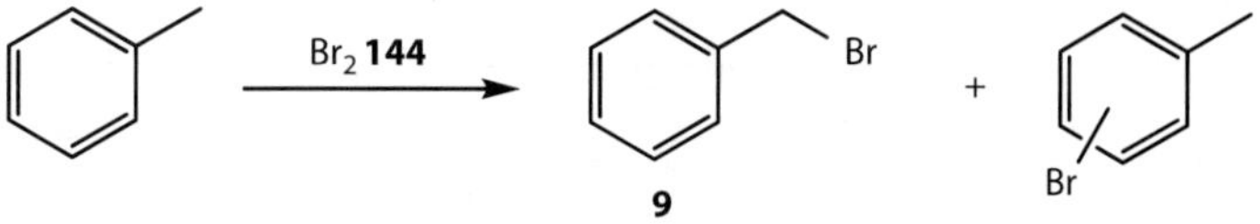

SCHEME 3.33 Schematic illustrating the potential reaction products obtained when brominating toluene.

TABLE 3.28
Summary of the Results Obtained for the Continuous Flow Synthesis of 3-Bromoindoles

Starting Material	Product	Yield (%) Batch	Yield (%) Micro Reactor	Throughput (mg h^{-1})
145	146	77	99.9	12.8
		72	99.9	11.7
		65	97.2	12.2
		54	92.5	15.1

transformation has been readily performed in batch utilizing pyridinium bromide perbromide, the heterogeneous nature of the reaction mixture made the procedure unsuitable for transfer to a microfluidic system. With this in mind, the authors investigated the use of elemental bromine **144** within an inhouse fabricated glass reactor (channel dimensions = 244 μm (wide) × 97 μm (deep) × 45.1 cm (long)). Employing a 0.2 M solution of the indole **145**, the authors investigated a range of reaction conditions including bromine **144** stoichiometry, solvent (MeCN, DMF, and DCM), reaction temperature (25–125°C), and flow rate (2–10 μL min^{-1}). Using this approach, 2 eq. of Br$_2$ **144**, MeCN as solvent, a flow rate of 10 μL min^{-1} and room temperature were identified as the optimal conditions for the continuous flow synthesis of the target 3-bromoindole derivative **146** (99.9% conversion, 12.8 mg h^{-1}). In order to explore the generality of the devised protocol, the bromination of a series of substituted indoles was investigated and as Table 3.28 illustrates, in all cases excellent conversions were obtained with productivities ranging from 11.7 to 15.1 mg h^{-1}).

Solvent-Free Brominations: Löb et al. [71] subsequently demonstrated the bromination of thiophene **130** under solvent and catalyst-free conditions, obtaining 2,5-dibromothiophene **147**, a compound of significance within the OLED manufacturing industry, in typical yields of 85% and high selectivity, representing a dramatic improvement the 50% yields typically reported within the literature. In addition to an improvement in selectivity, the authors found that by performing the reaction at

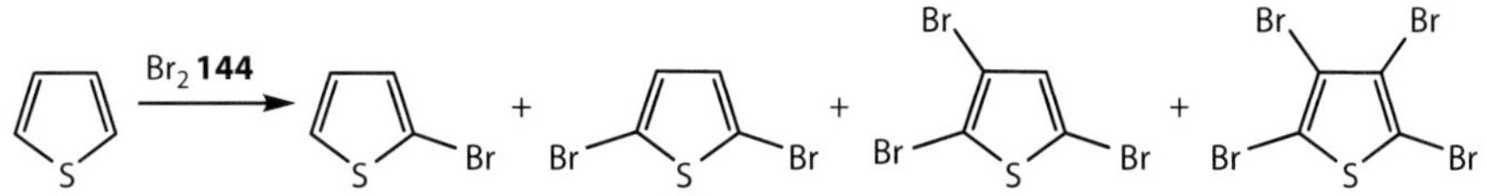

SCHEME 3.34 Illustration of the potential products obtained during the direct bromination of thiophene **130**.

50°C, under flow conditions, the reaction time could be reduced from 2 h to 1 s (Scheme 3.34).

Iodinations: For examples of the selective monoiodination of aromatic compounds, using electrochemically generated I⁺ under liquid-phase conditions, see Chapter 5.

3.2.4 C-Hetero Bond-Forming Reactions: Diazotizations under Flow

A pioneering paper in the field of micro reaction technology was reported in 1997 by Harrison and coworkers [72], who described their findings with respect to the EOF-based synthesis of a red azo-dye **148** from the manipulation, mixing, and reaction of *p*-nitrobenzenediazonium tetrafluoroborate **149** and *N,N*-dimethylaniline **150** within a Pyrex® micro reactor (Scheme 3.35).

Based on this investigation, many research groups have subsequently investigated the synthesis of azo-dyes under continuous flow; however, all have favored the use of pressure-driven pumping mechanisms due to the magnitude of flow obtained being independent of the reactant and solvent compositions employed. De Mello and coworkers [73] report one such example where a glass micro reactor (channel dimensions = 150 µm (wide) × 50 µm (deep) × 8 cm (long)) was used to perform the synthesis of three naphthol-derived azo-dyes. Unlike Harrison's example, these researchers investigated both the synthesis and the reaction of the diazonium salt, demonstrating the ability to not only generate, but also handle and react unstable, reactive intermediates within such devices. Reactions were performed by introducing an acidified solution of aryl amine and sodium nitrite (in aq. DMF) into the reactor, each at a flow rate of 3.5 µL min⁻¹. Over a channel length of 4 cm, the reagents mixed

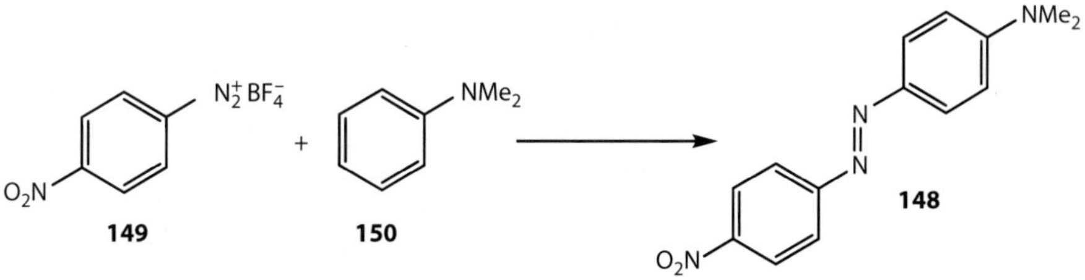

SCHEME 3.35 Illustration of an early example of azo-dye synthesis utilizing electro-osmotic pumping.

by diffusion and reacted to afford the respective diazonium salt, prior to the addition of a basic solution of 2-naphthol at a flow rate of 7 µL min⁻¹. With initial product formation inferred via the red coloration of the reaction channel, the authors subsequently confirmed the azo-dye formation using offline spectroscopic analysis. Under the aforementioned conditions, the authors were able to synthesize a series of three dyes, with yields ranging from 9% to 52%; depending on the diazonium salt formed.

Biphasic Synthesis of Azo-dyes: Using a double injector reactor, fabricated from silicon, Köhler and coworkers [74] reported the azo-coupling of 2-naphthol and cresol novolaks as an example of microsegmented organic reactions. At the first injection point (inlet), the authors introduced a solution containing the coupling component (2-naphthol or an equivalent) into a carrier phase (tetradecane) and the diazonium salt under investigation from a second injector (inlet). Coupling the reactor to a compact spectrometer and stereomicroscope, the authors were able to continuously monitor processes including segment formation, injection, mixing, and dye formation; with dye characterization corroborated by offline spectrophotometric measurements.

Using this approach, the authors were able to perform microliter volume reactions in an addressable manner affording rapid reaction screening and process optimization using small volumes of reactant solutions (Figure 3.8).

Production of Pigments: From a nonresearch and development perspective, Wille et al. [75] reported an early micro reactor plant, configured for the synthesis of an undisclosed azo dye. Using a three-stage pilot plant, the authors reported an experimental technical report detailing the use of MRT at Clariant, for azo-chemistry; including the formation of the diazonium salt and subsequent reaction to afford the target dyes in throughputs of 30 L h⁻¹. In a later report, Pennemann et al. [76] reported significant processing advantages associated with the synthesis of an azo pigment, Yellow 12 **151** (Figure 3.9), finding that micromixing afforded a product of smaller particle size, reduced size distribution, increased glossiness, and increased transparency (see Chapter 7 for more details).

Sandmeyer Reaction: In 2003, Fortt et al. [77] reported the design and fabrication of a soda-lime glass micro reactor for the preparation and reaction of diazonium

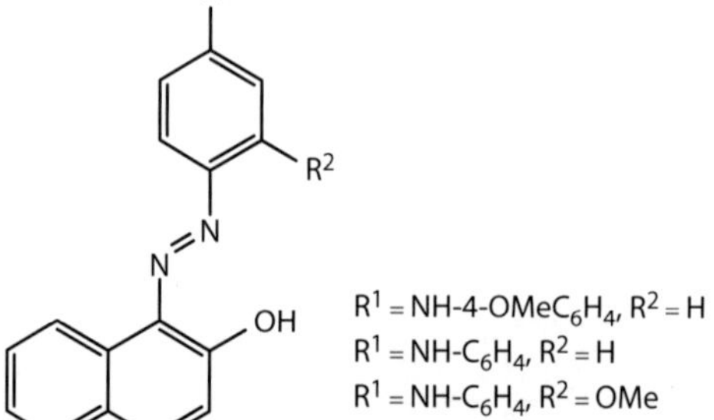

R^1 = NH-4-OMeC$_6$H$_4$, R^2 = H
R^1 = NH-C$_6$H$_4$, R^2 = H
R^1 = NH-C$_6$H$_4$, R^2 = OMe
R^1 = OMe, R^2 = H

FIGURE 3.8 Illustration of a selection of azo-dyes synthesized using a double-injector micro reactor.

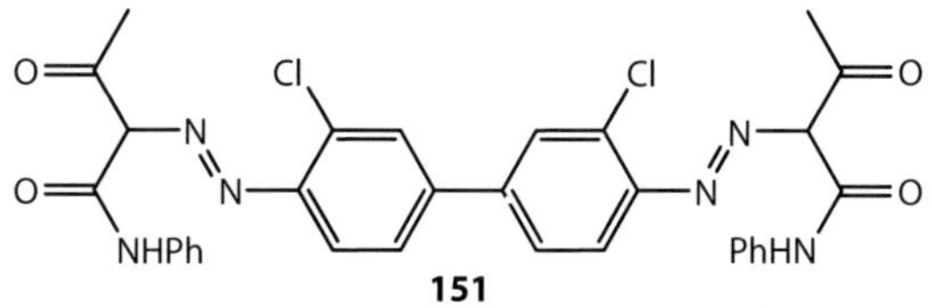

Liquid-Phase Micro Reactions

FIGURE 3.9 Illustration of the azo-pigment synthesized using a micromixer-based continuous process.

species. The reactor in question comprised a T-mixer coupled to a serpentine channel (channel dimensions = 150 μm (wide) × 50 μm (deep) × 8.0 cm (long)), in which the diazonium intermediate was formed, and a second inlet channel (channel dimensions = 150 μm (wide) × 50 μm (deep) × 28.0 cm (long)) in which the chloro-dediazotization was performed. Using the aforementioned reactor configuration the authors evaluated the effect of amine (2–10 mM) and nitrite concentration (1.5–10 mM) on the formation of chlorobenzene **152**; employing anhydrous DMF as the reaction solvent.

Under pressure-driven flow, the authors investigated the effect of reactant residence time on a reaction between aniline **61** and nitrite **153**, to form the diazonium intermediate **154** and subsequent chloro-dediazotization via *in situ* Raman microscopy; for additional examples, see Chapter 1. Using this approach, the authors were able to confirm that a reaction time of 600 s was optimal to afford chlorobenzene **153** in 50.5% yield; whereby quantitative conversion of the diazonium intermediate **154** was obtained (Scheme 3.36).

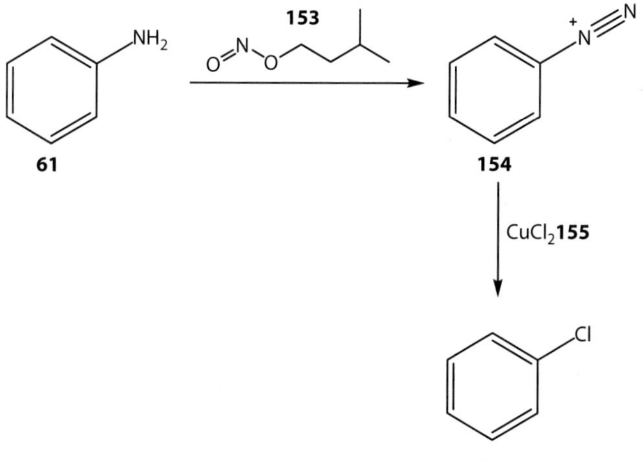

SCHEME 3.36 Illustration of the Sandmeyer reaction selected to demonstrate the performance of radical additions under continuous flow.

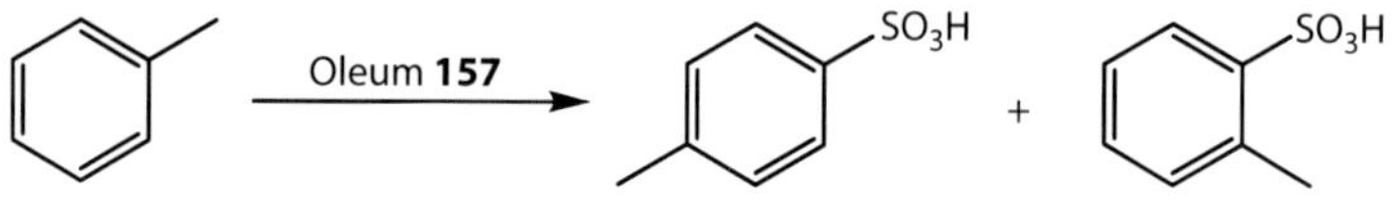

SCHEME 3.37 Illustration of the exothermic sulfonation reaction performed under flow conditions.

More recently, Asano et al. [78] reported the use of quartz glass micro reactors containing split flow microchannels as a means of increasing mixing efficiency at low flow rates. Once again, the synthesis of chlorobenzene **152** was used as a model reaction whereby solutions of aniline **61** (0.55 M) and isopentyl nitrite **153** (1.25 M) in DMF were mixed in a micro reactor at room temperature, prior to the introduction of copper(II) chloride **155** (0.32 M) in anhydrous DMF in a second reactor; maintained at 65°C. Using this approach, the authors obtained chlorobenzene **152** in 80% yield, with a reaction time of 20 s, representing almost a 30% increase over that obtained previously in an open-channel reactor and a 30-fold reduction in reaction time.

3.2.5 C-Hetero Bond-Forming Reactions: Sulfonations under Flow

Forming part of a development project into the construction of an automated reaction system, Löbbecke et al. [79] investigated the sulfonation of toluene to afford the industrially useful p-toluene sulfonic acid **156**. As Scheme 3.37 illustrates, the reaction involving the treatment of toluene with oleum **157** (2:1) has two potential products, the target *para*-substituted product **156** (thermodynamic) and the undesirable *ortho*-product **158** (kinetic).

Operating under thermodynamic control (80°C), reactions were performed in a PTFE reactor (dimensions = 2 m (long)) and the reaction monitored using inline IR spectroscopy (see Chapter 1). Upon completion of the reaction, the products were transferred to a second reactor (20°C) where hydrolysis of the anhydride intermediate was performed, affording the target compound **156**. Using this approach, the authors were able to demonstrate the performance of a highly exothermic reaction under excellent process control giving rise to quantitative conversion of toluene and the formation of p-toluenesulfonic acid **156** in 93% selectivity.

3.3 NUCLEOPHILIC ADDITION

3.3.1 C–C Bond Formation: Aldol Reaction/Condensation

As part of an investigation into the synthesis of natural products, Tanaka and Fukase and [80], Fukase and coworkers [81] evaluated the aldol reaction as a key reaction step. Due to the inefficiencies associated with mixing biphasic reaction mixtures in batch and the propensity for aliphatic aldehydes to polymerize, the authors proposed that employing efficient mixing under continuous flow conditions would enable them to prepare aliphatic β-keto derivatives in higher yield and purity.

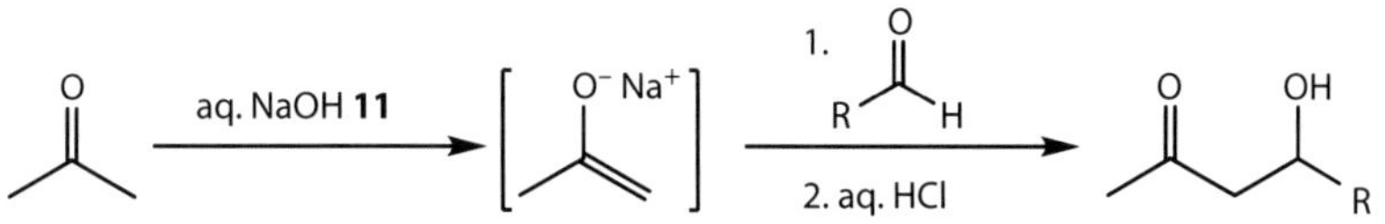

Liquid-Phase Micro Reactions

SCHEME 3.38 Schematic illustrating the aldol condensation of acetone with a series of aldehydes to afford conventionally problematic β-hydroxyketones in high yield.

Investigating a series of acetone-based aldol reactions, the authors employed a biphasic reaction mixture comprising aq. NaOH **11** and the aldehyde (5.5 M) in acetone (Scheme 3.38); quenching the reaction mixture with aq. HCl. Utilizing a reactor comprising of two micromixers (Comet X-01) and tubular reaction zones (dimensions = 1.0 mm (i.d.) × 0.9 or 0.3 m (long)), the authors evaluated the effect of micromixing on the reaction efficiency (Figure 3.10). Employing a residence time of 15 s for the reaction of the enolate with the aldehyde, the authors obtained significant improvements in product yield when compared with analogous batch reactions. These observations are attributed to the efficient mixing of the aldehyde and enolate, enabling the aldol reaction to proceed in preference to the self-aldol reaction of the aldehyde (where relevant); as such, undesired polymerization was suppressed and product quality increased.

Biphasic: Using the Claisen–Schmidt reaction as a model, Yin and coworkers [82] compared the reaction efficiency obtained in batch with that of two microflow

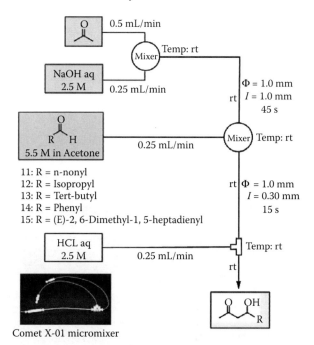

FIGURE 3.10 Illustration of the microfluidic system used to evaluate the biphasic aldol reaction under continuous flow. (Reproduced with permission from Tanaka et al. 2009. *Org. Proc. Res. Dev.* 13(5): 983–990. Copyright (2009) American Chemical Society.)

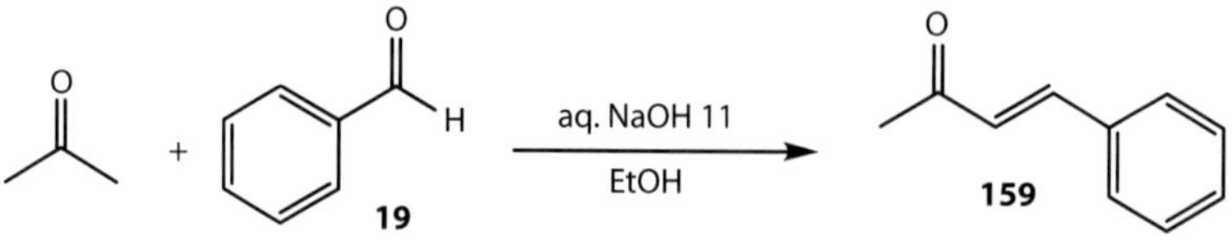

SCHEME 3.39 Illustration of the biphasic Claisen–Schmidt reaction used to determine the effect of various contacting methods within a continuous flow micro reactor.

processes, the first employing side-by-side contacting of the biphasic reactants and the second using Taylor, or slug flow. Reactions were performed using two stock solutions, the first containing acetone and benzaldehyde **19** in EtOH (4.0 M and 1.0 M, respectively) and the second a solution of aqueous NaOH **11** (0.5 M) (Scheme 3.39). Employing a soda-lime glass micro reactor (channel dimensions = 300 or 500 µm (wide) × 80 µm (deep) × 12, 16, 24, 32, 40, 48, 72, and 96 cm (long)), the authors obtained (E)-4-phenylbut-3-en-2-one **159** in higher conversions (93%) than in a comparable batch reaction (60%) performed under vigorous stirring (1250 rpm). Slug flow afforded the highest product yield due to a dramatic increase in interfacial surface area between the reactant solutions.

Fluorous Nanoflow: Coupling fluorous biphasic catalysis with micro reaction technology, Mikami et al. [83] demonstrated a dramatic increase in reactivity of the Mukaiyama aldol reaction performed using the fluorous solvent perfluoromethylcyclohexane and the lanthanide Lewis acid catalyst, scandium bis(perfluorooctanesulfonyl)amine ($Sc[N(SO_2C_8F_{17})_2]_3$) **160** (Scheme 3.40).

Employing a borosilicate glass micro reactor (channel dimensions = 60 µm (wide) × 30 µm (deep)), supplied by Fuji Electric Co. (Japan), the authors evaluated the effect of contact time (5.4–43.2 s) on the reaction of benzaldehyde **19** (0.1 M) and trimethylsilyl enol ether **161** (0.2 M) in toluene, with $Sc[N(SO_2C_8F_{17})_2]_3$ **160** (6.25 mol%) introduced in $CF_3C_6F_{11}$; utilizing side-by-side contacting. Employing a reactor temperature of 55°C and increasing the contact time from 5.4 to 16.2 s resulted in an increase in conversion to the aldol products **162** and **163**, with 43.2 s affording near quantitative conversion (97% by GC). For comparative purposes, the

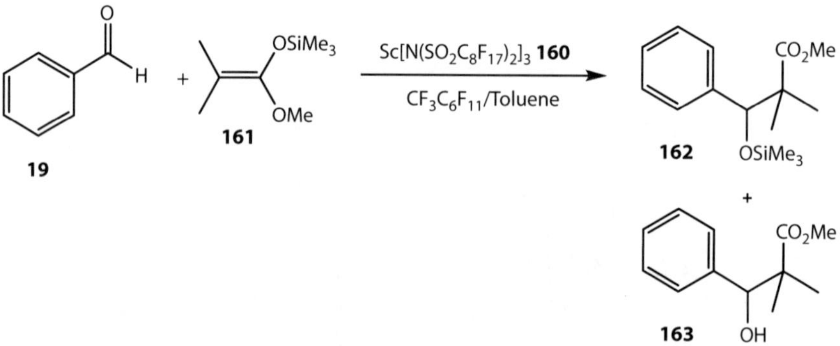

SCHEME 3.40 Schematic illustrating the Mukaiyama aldol reaction performed in a fluorous biphasic micro reaction.

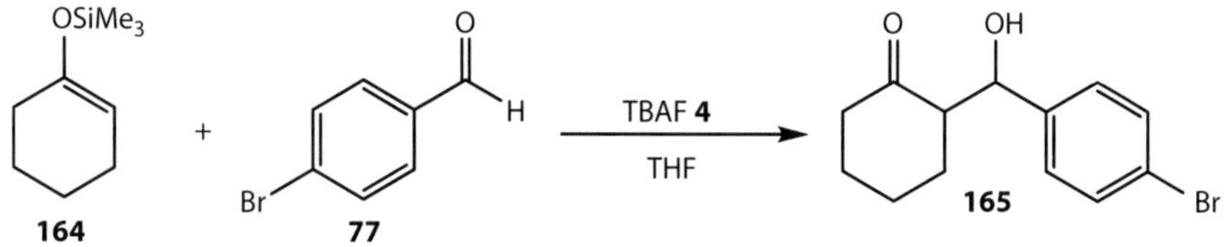

Liquid-Phase Micro Reactions

SCHEME 3.41 Schematic illustrating the synthesis of 2-((4-bromophenyl)(hydroxyl)methyl)cyclohexanone **165** using an EOF-based micro reaction.

authors also performed the biphasic reaction in batch, under analogous conditions, whereby only 11% product **162** was obtained in 2 h (55°C).

Preformed Enolates: Utilizing silyl enol ethers as preformed enolates, Wiles et al. [84] investigated the aldol reaction under continuous flow conditions employed EOF as the pumping mechanism. Within an inhouse fabricated borosilicate glass micro reactor, the authors investigated the aldol reaction between silyl enol ether of cyclohexanone **164** (0.1 M) and 4-bromobenzaldehyde **77** (0.1 M) in anhydrous THF, as illustrated in Scheme 3.41.

Using catalytic TBAF **4**, the authors demonstrated the *in situ* formation of the ammonium enolate and subsequent reaction with 4-bromobenzaldehyde **77** to afford 2-((4-bromophenyl)(hydroxyl)methyl)cyclohexanone **165**. Using GC–MS and a series of fully characterized synthetic standards, the authors quantified the conversion of the enol ether **164** to the target aldol product **165**. Employing applied fields of 417, 341, 333, and 0 V cm^{-1}, the authors were gratified to find that the enol ether **164** was quantitatively converted into the aldol product **165** with no competing condensation products detected.

3.3.2 C–C Bond Formation: Knoevenagel Condensation

An early example of the Knoevenagel condensation performed under continuous flow was reported by Fernandez-Suarez et al. [85] and formed part of a domino reaction used toward the synthesis of a series of pharmaceutically interesting cycloadducts, as illustrated in Scheme 3.42.

To perform the Knoevenagel condensation, the authors employed a soda lime glass reactor (channel dimensions = 74 µm (wide)), fabricated by Caliper Technologies Corp. (USA) and employed pressure-driven flow to manipulate the reactants within the device. Solutions of aldehyde (0.1 M) in 80:20 MeOH:H$_2$O and premixed

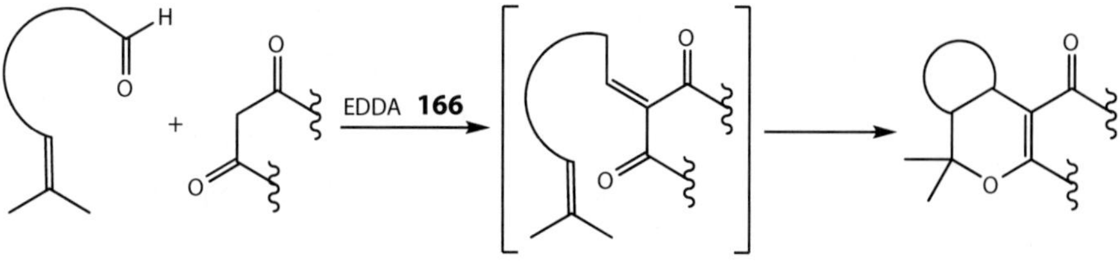

SCHEME 3.42 Illustration of the domino reaction, consisting of a Knoevenagel condensation followed by an intramolecular Diels–Alder reaction, performed under continuous flow.

ethylenediamine acetate (EDDA) **166** (10 mol%) and 1,3-diketone in 80:20 MeOH:H_2O were placed into the respective inlet reservoir of the reactor and 80:20 MeOH:H_2O placed in the collection reservoir. Pressure was applied to the inlet solutions, to afford the desired residence times and the reaction products collected at the outlet over a period of 30 min and analyzed offline by LC–MS. Employing two aldehydes, *rac*-citronellal **167** and 2-(3-methylbut-2-enyloxy)benzaldehyde **168**, and two 1,3-diketones, 1,3-dimethylbarbituric acid **169**, and Meldrum's acid **170**, the authors set about constructing a 2 × 2 library of cycloadducts, investigating three residence times in each case, 120, 240, and 360 s; with all reactions performed at room temperature.

As Table 3.29 illustrates, in all cases 120 s afforded moderate conversions to the target cycloadduct, with increases in conversion obtained as a result of employing longer reaction times. Using this approach, the authors were able to investigate the synthesis of a complex scaffold using only nmol quantities of material.

Using a multichannel quench-flow reactor, fabricated from glass and silicon via deep-reactive ion etching, Bula et al. [86] demonstrated the parallel investigation of reaction kinetics using a single micro reactor containing four different volume reaction channels (channel dimensions = 50 μm (wide) × 53 μm (deep), volumes = 0.35, 0.53, 0.70, and 0.88 μL) (Figure 3.11). Employing the Knoevenagel

TABLE 3.29
Summary of the Materials Synthesized Using the Knoevenagel Condensation as a Key Step and the Conversions Obtained over a Range of Reaction Times

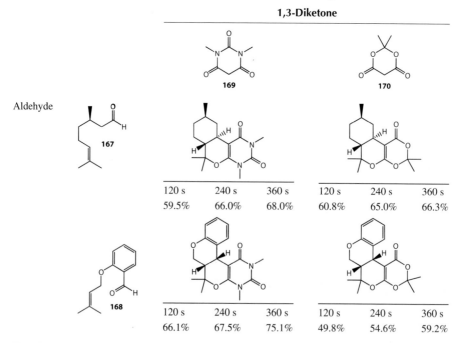

Liquid-Phase Micro Reactions

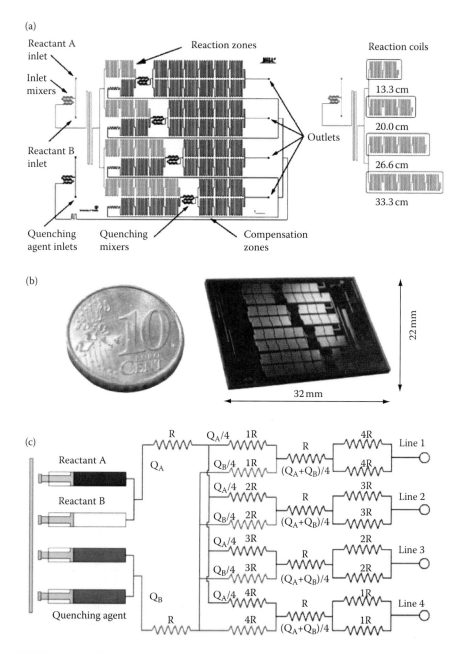

FIGURE 3.11 Illustration of the microstructure used to perform kinetic evaluation of organic reactions under continuous flow conditions: (a) schematic, (b) image, and (c) depiction of regions of variable resistance of the device. (Bula, W. P. et al. 2007. Multi-channel quench-flow microreactor chip for parallel reaction monitoring, *Lab Chip* 7: 1717–1722. Reproduced by permission of the Royal Society of Chemistry.)

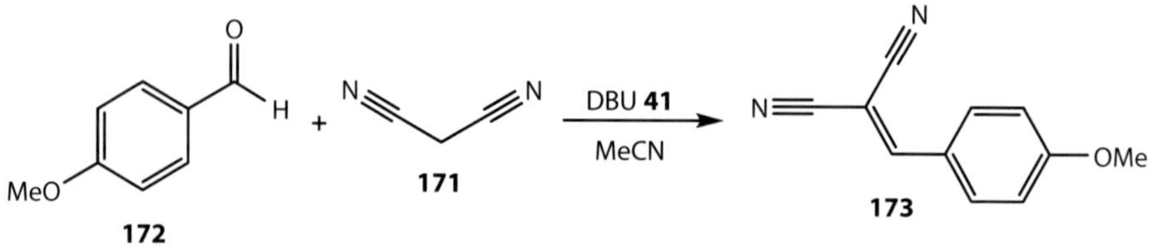

SCHEME 3.43 Illustration of the model Knoevenagel condensation reaction used to demonstrate the principle of parallel reaction monitoring under continuous flow conditions.

condensation reaction between malononitrile **171** (0.04 M) and 4-methoxybenzaldehyde **172** (0.04 M), performed in MeCN in the presence of DBU **41** (0.04 to 0.16 M), the authors evaluated the effect of reaction time and temperature on the formation of 2-(4-methoxybenzylidene)malononitrile **173**; the reaction products were quenched using TFA, prior to offline analysis of the UV–VIS absorbance at $\lambda = 354$ nm.

Using the Levenberg–Marquardt error minimization algorithm, the authors were able to fit kinetic curves to experimentally determined data, from which they concluded that there was a directly proportional link between catalyst concentration and the reaction constant. Furthermore, the investigation also illustrated a dramatic increase in reaction rate within the micro reactor when compared with a batch reaction performed under analogous conditions.

While the authors did not focus on the optimization of the reaction illustrated in Scheme 3.43, the investigation did illustrate the wealth of physical information that can be readily obtained from processes conducted within micro reactors.

3.3.3 C–C Bond Formation: Michael Addition

Using a borosilicate glass micro reactor (channel dimensions = 100 µm (wide) × 50 µm (deep)), Wiles et al. [87] utilized electroosmotic flow (EOF) as the pumping mechanism for an investigation into the Michael addition under continuous flow conditions.

Using the synthesis of (*E*)-ethyl-4-acetyl-5-oxohex-2-enoate **174** as a model reaction, the authors initially investigated the reaction of 2,4-pentanedione **175** (5.0 M) with ethyl propiolate **176** (5.0 M) in the presence of the organic base, diisopropylethylamine **51** (5.0 M) in EtOH. Manipulating the reactants through the reactor by application of the following applied fields, 147, 318, 333, and 0 V cm^{-1}, the authors obtained 56% conversion of diketone **175** to the *trans*-product **174**; as determined by GC and ^{1}H NMR analysis.

As significant proportions of unreacted starting materials were detected, the authors evaluated the effect of increased residence time on the reaction; owing to the use of EOF this was best investigated by the use of stopped flow, that is, reagents pumped for several seconds and then the flow paused, followed by pumping and so on. As Table 3.30 illustrates, using a regime of 2.5 s on and 5 s off an increase of 39% conversion was obtained affording (*E*)-ethyl-4-acetyl-5-oxohex-2-enoate

TABLE 3.30
Summary of the Results Obtained for a Series of Michael Addition Reactions Conducting Using a Range of EOF Regime

		Conversion (%)		
Reaction Type	Flow Regime	Product 174	Product 179	Product 180
Batch	N/A	89	78	91
Micro Reactor	Continuous	56	15	40
Micro Reactor	2.5 s on/5.0 s off	95	34	100
Micro Reactor	5.0 s on/10 s off	N/A	100	N/A

174 in 95% conversion. To extend the investigation a further two diketones were investigated, benzoyl acetone **177** and diethyl malonate **178**, affording (E)-ethyl-4-benzoyl-5-oxohex-2-enoate **179** and (E)-triethyl prop-2-ene-1,1,3-tricarboxylate **180** in quantitative conversion, respectively (Scheme 3.44).

In an additional example, the authors investigated the reaction of 2,4-pentanedione **175** with methyl vinyl ketone (MVK), obtaining the target compound in 95% conversion (stopped flow regime = 2.5 s on/5.0 s off), demonstrating the generality of the technique. In this early example, the residence times required exceeded those accessible within a short micro reaction channel; consequently, the stopped flow regime enabled the authors to increase the residence time without needing to fabricate an additional reactor. The investigation could, however, be repeated using pressure-driven flow, with the respective residence times obtained via the use of a larger reactor volume.

Fluorinated Adducts: Miyake and Kitazume [88] subsequently demonstrated the use of the Michael addition as a means of readily preparing a series of fluorinated organic compounds. Employing a micro reactor, with channel dimensions of 100 μm (wide) × 40 μm (deep) × 80 cm (long), the authors investigated the reaction of ethyl-3-(trifluoromethyl)acrylate **181** (1.38 M) with a series of substituted nitropropanes

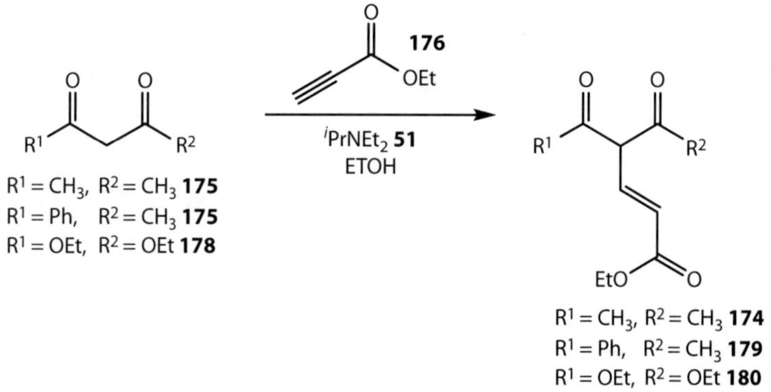

SCHEME 3.44 Illustration of a selection of Michael additions conducted using EOF pumping.

TABLE 3.31
Summary of the Michael Addition Reaction Performed in a Glass Micro Reactor

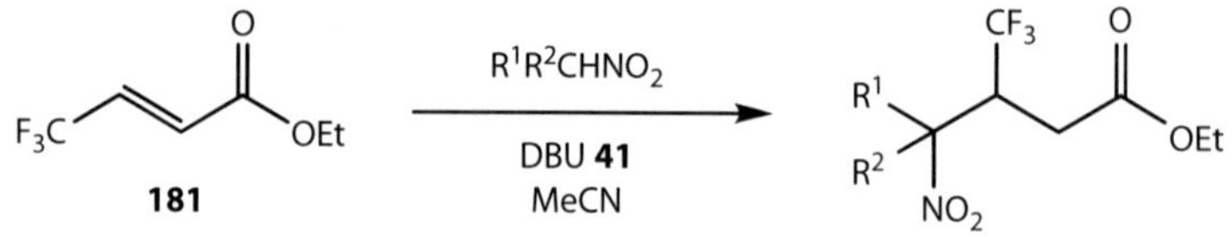

$F_3C-CH=CH-C(=O)-OEt$ **181** $\xrightarrow[\text{MeCN}]{R^1R^2CHNO_2, \text{ DBU } \mathbf{41}}$ $R^1R^2C(NO_2)-CH(CF_3)-CH_2-C(=O)-OEt$

R^1	R^2	Yield (%)[a]
H	H	80 (90)
Me	Me	93 (92)
H	CO_2Et	90 (99)

[a] The numbers in parentheses represent the yield obtained in batch after 30 min.

(2.08 M, 1.5 eq.) in the presence of the organic base DBU **41** (1.95 M) in MeCN; in all cases, a flow rate of 1 µL min⁻¹ was employed and a reactor temperature of 25°C. As Table 3.31 illustrates, using this approach the authors successfully obtained the corresponding 1,4-adducts in high yield, with no polymerization products detected, affording a facile route to the synthesis of fluorinated alkanes.

3.3.4 C–C Bond Formation: Diels–Alder Reaction

Utilizing a modular micro reactor setup comprising functional stainless-steel blocks (Ehrfeld Microtechnik AG), Seeberger and coworkers [89] investigated the cyclo-addition reaction between 2,3-dimethylbuta-1,3-diene **182** (2.0 M) and maleic anhydride **183** (0.95 M) in NMP. Employing a residence time of 30 min and a reactor temperature of 60°C, 3a,4,7,7a-tetrahydro-5,6-dimethylisobenzofuran-1,3-dione **184** was obtained in 98% yield and 98% selectivity; as determined by GC-FID analysis in $CHCl_3$. Using this approach, the authors were able to produce the adduct **184** at a throughput of 17 g h⁻¹ (Scheme 3.45).

Using a microcapillary flow disk (MFD) reactor, comprising eight parallel channels (dimensions = 180–220 µm), described previously, Hallmark et al. [90] investigated the Diels–Alder reaction under flow. Employing the reaction of isoprene **185**

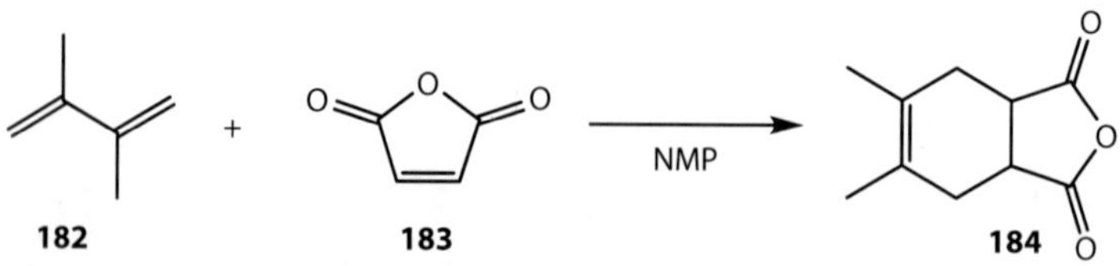

SCHEME 3.45 Illustration of the Diels–Alder cycloaddition performed in a stainless-steel micro reactor.

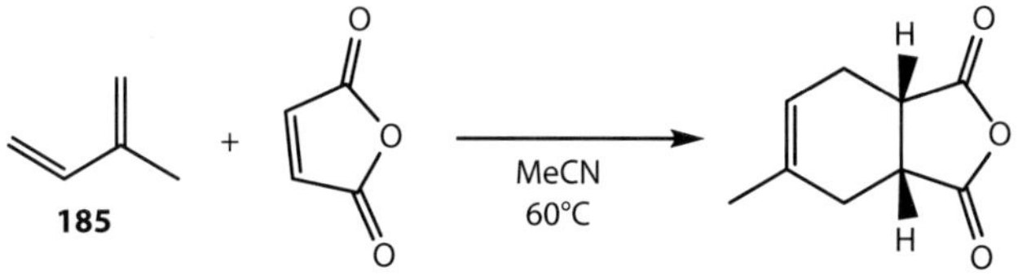

SCHEME 3.46 Illustration of the model Diels–Alder reaction performed in a microcapillary flow disk reactor.

and maleic anhydride **183** to afford cycloadduct 3a,5,7a-trimethyl-3a,4,7,7a-tetrahydro-isobenzofuran-1,3-dione **186**, the authors evaluated the feasibility of constructing this pharmaceutically important scaffold (Scheme 3.46).

To perform a reaction, the reactants were introduced into the MFD reactor from separate inlets, employing isoprene **185** (18.0 M) in a twofold excess with respect to maleic anhydride (9.0 M) and using MeCN as the reaction solvent. Employing a reactor temperature of 60°C, achieved via immersion of the MFD reactor in a water bath, the authors investigated the effect of reactant residence time on the progress of the reaction, with yields ranging from 85% to 98% between 28 and 113 min residence times. Under the optimal conditions described, this methodology was capable of producing the cycloadduct **186** at a rate of 1.05 kg day^{-1}.

Compared to analogous batch protocols, the authors found the use of a flow reactor to be advantageous due to the ease of reactant manipulation, particularly with respect to the volatile isoprene **185**.

More recently, Kappe and coworkers [20] demonstrated the use of a high-temperature (300°C) and high-pressure (200 bar) stainless steel, tubular (1000 μm (i.d.)), reactor for the Diels–Alder cycloaddition of 2,3-dimethylbutadiene **187** and acrylonitrile **188** (Scheme 3.47).

Employing toluene as the reaction solvent, the authors identified a reactant residence time of 5 min and a reactor temperature of 250°C (200 bar) afforded quantitative conversion to 3,4-dimethylcyclohex-3-enecarbonitrile **189**, with lower temperatures and residence times affording incomplete reactant conversion. In a subsequent publication, Kappe and coworkers [91] demonstrated the ability to reduce the reaction time from 5 to 2 min for the reaction of butadiene with acrylonitrile **188**, by simply increasing the reactor temperature to 280°C. Under the aforementioned conditions, the authors were able to synthesize the target compound in 82% yield; equating to a throughput of 80.4 g h^{-1} and a space–time yield of 1.4 kg m^{-3} s.

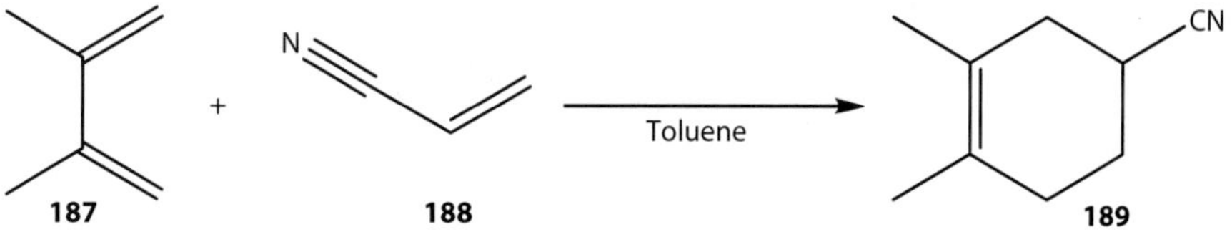

SCHEME 3.47 Schematic illustrating the model reaction used to demonstrate the use of high temperatures and pressure under continuous flow.

Further increases in reactant concentration were investigated as a means of increasing the productivity of each reactor unit; however, the use of a 5 M stock solution led to a 50°C increase in reactor temperature. Should greater throughputs be required, an alternative to diluting reactant streams would be to employ a more efficient heat exchanger, enabling highly exothermic reactions to be performed safely at moderate to high reactant concentrations.

In addition, utilizing a pressurized reactor, the authors were able to substitute the high-boiling solvent toluene for MeCN, THF, and DME which afforded analogous results and facile product isolation.

3.3.5 C–C BOND FORMATION: HORNER–WADSWORTH–EMMONS

In addition to demonstrating the synthesis of fluorinated alkanes under continuous flow (Table 3.31), Miyake and Kitazume [92] also reported the performance of the Horner–Wadsworth–Emmons (HWE) reaction (Scheme 3.48) in a micro reactor (channel dimensions = 100 μm (wide) × 40 μm (deep) × 8 cm (long)) as a means of synthesizing α-fluoro-α,β-unsaturated esters.

To perform a reaction, the authors introduced a premixed solution of aldehyde and triethyl-2-fluoro-2-phosphonoacetate **190** (1.69 and 2.54 M, respectively) in DME into the reactor from inlet one and DBU **41** (1.95 M) in DME from the second inlet and the reaction products collected in aqueous HCl (1 N). The reaction products were isolated via an organic extraction into diethyl ether, with removal of the solvent *in vacuo* affording the target α-fluoro-α,β-unsaturated ester. The reaction products were subsequently analyzed by ^{19}F NMR spectroscopy, using benzotrifluoride as an internal standard, in order to determine the yield and E:Z ratio. As Table 3.32 illustrates, the major product in all cases was the E-isomer (based on coupling constants), with the E:Z ratios obtained under flow conditions comparable to those obtained from reactions performed under batch conditions.

In addition to numerous other reactions, Seeberger and coworkers [93] reported the use of a modular micro reactor (Ehrfeld Microtechnik AG, Germany) for performing the 1,5,7-triazabicyclo[4.4.0]dec-1-ene (TBD) **191** base promoted Horner–Wadsworth–Emmons olefination, as depicted Scheme 3.49.

Employing a reactor temperature of 50°C and MeCN as the reaction solvent, the authors reacted benzaldehyde **19** (0.21 M) with triethylphosphonoacetate **192** (1.2 eq.) in the presence of an equivalent of TBD **191**. The reaction products were subsequently quenched with ammonium acetate (10%) and extracted with EtOAc; analysis

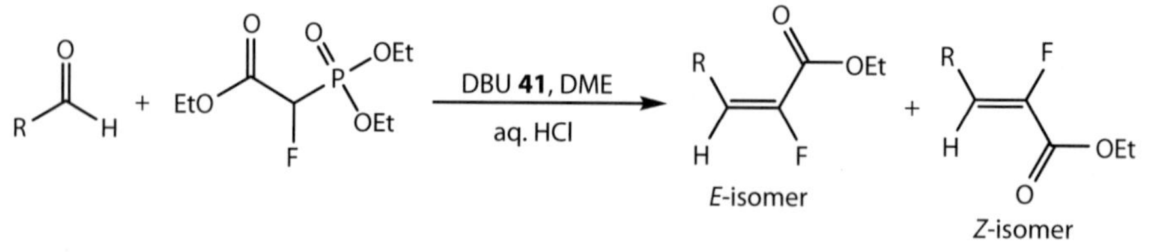

SCHEME 3.48 General reaction scheme illustrating the Horner–Wadsworth–Emmons reaction for the synthesis of fluorinated alkenes.

TABLE 3.32
Summary of the Results Obtained for the Synthesis of α-Fluoro-α,β-Unsaturated Esters under Continuous Flow

R	Method	Yield (%)	Z:E[a]
C_6H_5	Batch[b]	>99	70:30
	Micro reactor	78	77:23
4-$CF_3C_6H_4$	Batch[b]	86	64:36
	Micro reactor	88	68:32
4-$MeOC_6H_4$	Batch[b]	>99	76:24
	Micro reactor	58	74:26
n-Nonyl	Batch[b]	66	64:36
	Micro reactor	81	64:36

[a] Ratios were determined by ^{19}F NMR spectroscopy.
[b] Reactions performed for 30 min.

of the organic portion by GC-FID enabled quantification of the proportion of ethyl cinnamate **193** formed. Using this approach, the authors identified a reaction time of 10 min and a temperature of 50°C as the optimum conditions, forming the title compound **193** in 95% yield, at a throughput of 5.2 g h^{-1}.

Aminonaphthalene Intermediate: More recently, Tietze and Liu [94] reported a continuous flow HWE olefination as a key step in the synthesis of an aminonaphthalene derivative **194** employed in the synthesis of the duocarmycin based prodrug **195** illustrated in Figure 3.12. Utilizing a CYTOS College System (CPC Systems, Germany), comprising a micromixer (volume = 2 mL) and a tubular residence time unit (volume = 45 mL), the authors prepared the anion of 4-*tert*-butyl-1-ethyl-2-diethoxyphosphorylsuccinate **196** *in situ*, upon reaction with sodium ethoxide **197** in EtOH, prior to the addition of benzaldehyde **19**. Employing a residence time of 47 min, (*E*)-*tert*-butyl-1-ethyl-2-benzylidenesuccinate **198** was obtained in a yield of 89% at room temperature. Analysis of the reaction products by NMR spectroscopy confirmed the formation of the olefin **198** in excellent selectivity, with none of the Z-isomer detected. Compared to an analogous batch reaction, the flow protocol represented an acceleration factor (*F*) of 6, where typically 5 h were required to conduct the batch protocol (Scheme 3.50).

SCHEME 3.49 Illustration of the HWE-olefination performed in a modular micro reactor system.

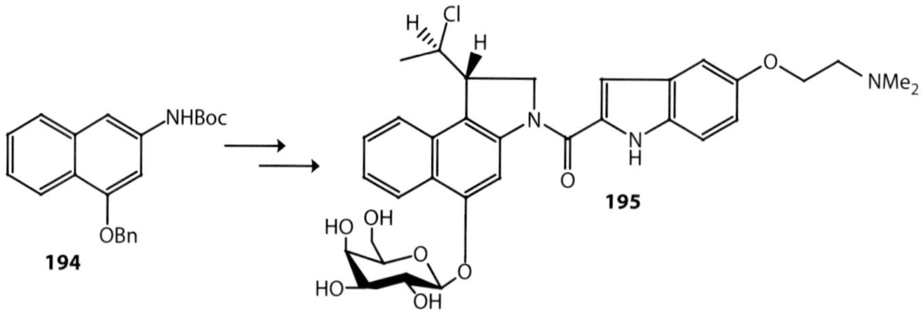

FIGURE 3.12 An aminonaphthalene derivative **194** as a precursor in the synthesis of the duocarmycin prodrug **195**.

3.3.6 C–C Bond Formation: Enantioselective Examples

Employing a series of crude enzyme lysates, containing hydroxynitrile lyase (HNL), Rutjes and coworkers [95] demonstrated the ability to perform the enantioselective synthesis of cyanohydrins under continuous flow. Up to this point, enzymatic transformations had been performed within micro reactors using either purified or immobilized enzymes; however, the authors of this work acknowledged that when utilized on a commercial platform, enzymes are typically used as either crude cell lysates or partially purified preparations. With this in mind, the authors set about demonstrating the ability to manipulate crude cell lysates within a borosilicate glass reactor (Micronit Microfluidics BV, NL), selecting the addition of HCN **199** to aldehydes as a model reaction (Scheme 3.51).

Under biphasic conditions, the authors investigated the nucleophilic addition of cyanide to a series of aldehydes employing two reactant solutions, the first an organic phase containing the aldehyde (0.23 M) and an internal standard in methyl *tert*-butyl ether and the second an aqueous phase containing potassium cyanide (0.23 M) and the crude lysate ((*S*)-HNL **200**), 10% v/v; brought to pH 5 with citric acid prior to filling the reactant syringe.

Under these conditions, the authors investigated the effect of reaction time (1 to 30 min) on the cyanohydrin formation, with reaction progress determined using offline analysis of the denatured samples by GC and enantioselectivity determined using chiral HPLC. Gratifyingly, the authors were able to manipulate the cell lysate within the micro reactor without problems, obtaining moderate to high conversions

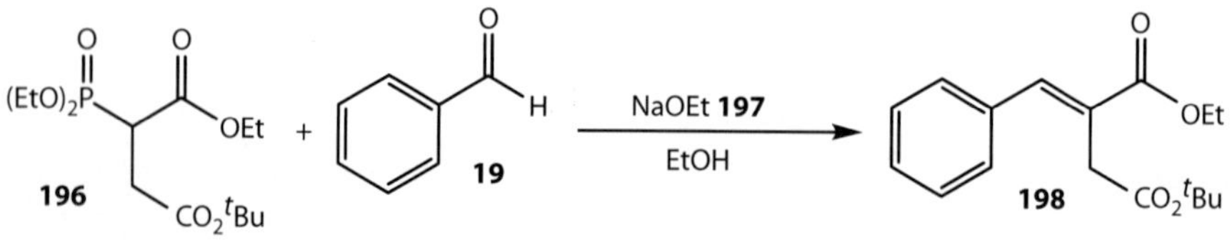

SCHEME 3.50 Illustration of the HWE-olefination used in the synthesis of a pharmaceutically useful aminonaphthalene derivative **195**.

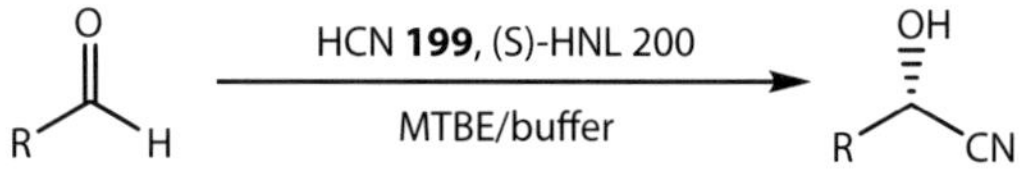

Liquid-Phase Micro Reactions

$$R-CHO \xrightarrow[\text{MTBE/buffer}]{\text{HCN 199, (S)-HNL 200}} R-CH(OH)-CN$$

SCHEME 3.51 Schematic illustrating the model reaction selected to demonstrate the use of cell lysates within micro reactor systems.

in 5 min (benzaldehyde **19** = 65% and 95% *ee*); in all cases slug flow was observed. In a second investigation, the authors screened (*S*)-HNL **200** for activity towards a series of aldehydes and as Table 3.33 illustrates, the reaction was sensitive to the substituent on the aldehyde.

In the case of piperonal, a detailed screen of reaction conditions was performed utilizing (*R*)-HNL, whereby the authors rapidly evaluated 58 reaction conditions, over a period of 4 h, using only 150 μL of the crude cell lysate.

3.3.7 C-Hetero Bond Formation: Aza-Michael Addition

Recognizing the synthetic utility of the aza-Michael addition, Löwe et al. [96] evaluated the addition of secondary amines to α,β-unsaturated carbonyl containing compounds (Scheme 3.52) in order to identify if there were any processing advantages associated with performing this reaction under continuous flow.

Conventionally, the reaction is performed using long reaction times, not due to the kinetics of the reaction, but owing to the highly exothermic nature, and reversibility, of the reaction; at >200°C thermal cleavage occurs to afford the amine and α,β-unsaturated precursors. Consequently, reactions are performed via dropwise

TABLE 3.33
Summary of the Results Obtained for the Enzymatic Formation of Cyanohydrins under Flow Conditions

Aldehyde	Conversion (%)	ee (%)
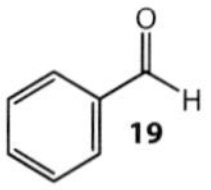 benzaldehyde (19)	64	99
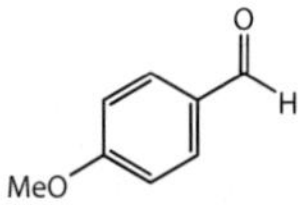 4-methoxybenzaldehyde	32	100
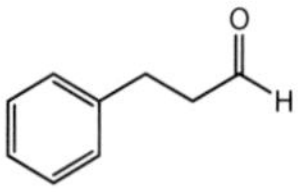 3-phenylpropanal	95	87
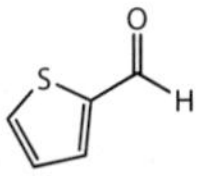 thiophene-2-carbaldehyde	30	98

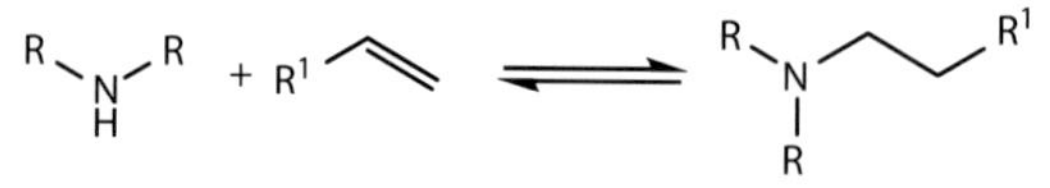

SCHEME 3.52 General schematic illustrating the aza-Michael addition investigated under continuous flow.

addition of the α,β-unsaturated compound to the amine, to facilitate efficient removal of heat, leading to lengthy reaction times.

By performing the reaction within a slit-type interdigital micromixer (Internal features = 40 μm (wide) × 300 μm (deep)) and a tube reactor (volume = 74.5 or 9.8 mL), the authors demonstrated that through the efficient dissipation of heat, the reaction could be accelerated, exposing the underlying reaction kinetics. As Table 3.34 illustrates, in the cases of three secondary amines and two α,β-unsaturated carbonyl containing compounds, while the products generated were obtained in equivalent yield to those reported within the open literature, the materials synthesized under continuous flow were obtained with space time yields ranging from 102 to 652 times faster than those prepared in batch vessels.

Pyridine Synthesis: Bagley et al. [97,98] also demonstrated the use of the Michael addition as a key step in the formation of substituted pyridines via the

TABLE 3.34
Comparison of the Results Obtained in a Flow Reactor with the Published Literature Methods for a Series of Aza-Michael Additions

Amine	α,β-Unsaturated Compound	Product	Yield (%) Batch Method	Yield (%) Micro Reactor
Dimethylamine	Acrylonitrile	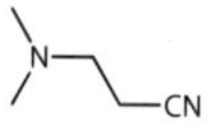	96	95.6
Diethylamine	Acrylonitrile	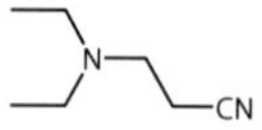	85	97.1
Piperidine	Acrylonitrile	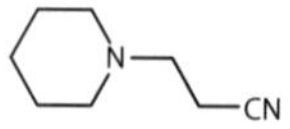	90	99.7
Dimethylamine	Ethyl acrylate	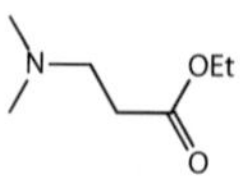	87	92.5
Diethylamine	Ethyl acrylate	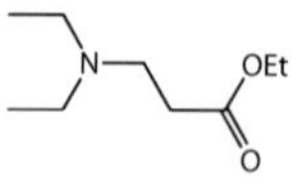	85	99.1
Piperidine	Ethyl acrylate	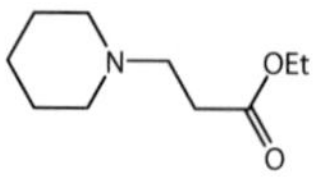	—	99.8

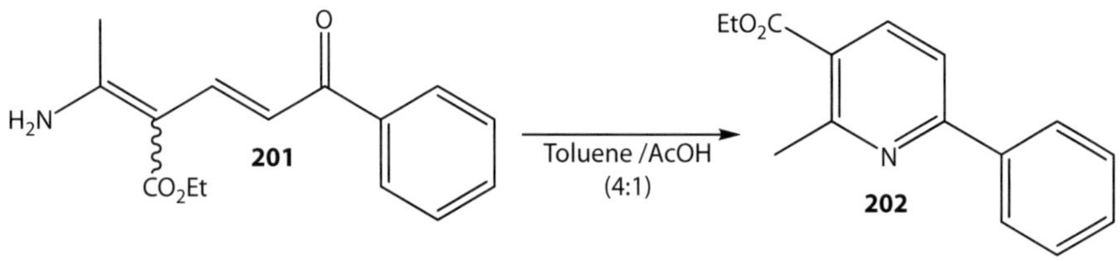

Liquid-Phase Micro Reactions

SCHEME 3.53 Illustration of the Bohlmann–Rahtz reaction performed in a microwave-assisted flow reactor.

Bohlmann–Rahtz reaction (Scheme 3.53), initially utilizing a sand-filled glass reactor and subsequently using the commercially available CEM Discover® (USA) flow reactor equipment.

Pumping a solution of aminodienone **201** (0.1 M) in toluene/AcOH (4:1) (1.5 mL min^{-1}) through the sand-filled glass reactor, while irradiating at 300 W, the authors were able to heat the reaction mixture to 100°C obtaining excellent conversion of the enamine **201** to ethyl 2-methyl-6-phenylnicotinate **202**. After quenching the reaction products with aqueous sat. NaHCO$_3$ and extracting into EtOAc, the authors isolated the desired pyridine **202** in 98% yield and excellent purity. Using this approach, the authors confirmed that the glass tube reactor afforded improved heating efficiency compared to sealed tube reactors while enabling the synthesis of large volumes of material without affecting the yield and purity of the product synthesized.

In an extension to this, the authors have more recently performed the reaction using the Syrris AFRICA® (UK) system, whereby reaction temperatures of 100°C afforded cyclodehydration of aminodienones in excellent conversion. Using the Uniqsis FlowSyn™ (UK) platform, the authors subsequently demonstrated the upscaling of the technique for the production of pyridine derivatives at a rate of 0.5 mmol min^{-1}.

3.3.8 C-Hetero Bond Formation: Alkylation of Amines

Pyridine Derivatization: In order to demonstrate the increased processing control obtained within continuous flow reactors, Kappe and coworkers [20] performed the nucleophilic substitution of alkyl halides within a stainless-steel flow reactor capable of accessing reaction temperatures of 350°C and pressures of 200 bar.

As Scheme 3.54 illustrates, the model reaction selected was the alkylation of morpholine **203** (0.04 M) with 2-chloropyridine **47** (0.04 M), which typically takes days, to afford 4-(pyridine-2-yl)morpholine **204**, when performed in batch. As discussed in previous examples, the choice of reaction solvent for flow reactions is governed by the solubility of reactants, intermediates, and products. In this case, the by-product morpholinium hydrochloride was the limiting factor and the authors selected NMP as the reaction solvent coupled with a relatively low substrate concentration. Using a reactor temperature of 270°C (70 bar) and a 10 min residence time, the authors were pleased to obtain the target compound in 82% yield.

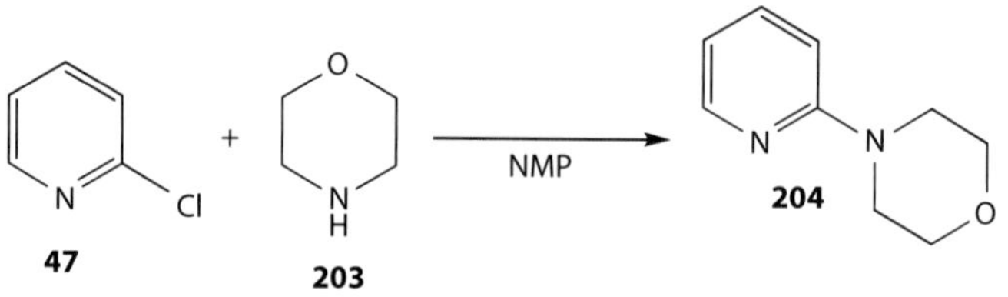

SCHEME 3.54 Illustration of the types of alkylation reactions performed under continuous flow at extreme temperatures and pressures.

3.3.9 C-Hetero Bond Formation: Synthesis of Triazoles

Championed by K. Barry Sharpless, click chemistry [99] involves the rapid construction of synthetically useful compounds via the employment of efficient and predictable synthetic steps. Owing to the ease with which complex core motifs can be generated, the protocol has been widely evaluated, with a high degree of uptake within the areas of high-throughput screening and combinatorial chemistry. Although applications of the technique have been met with moderate success, the high consumption of frequently scarce or expensive reagents is restricting the number of permutations investigated by researchers.

Target-Guided Synthesis: With this in mind, Wang et al. [100] investigated the design and fabrication of an integrated microfluidic device (PDMS) capable of rapidly combining small aliquots of reactants and addressing them into individual micro wells as a means of increasing the number of reactions that can be assessed in a given period of time. Using the synthesis of potential biligand inhibitors, via the enzyme-catalyzed reaction of azides and terminal acetylenes (Scheme 3.55), coupled with improvements in standard screening techniques, the authors initially focused on reducing the quantities of reagents required per reaction (Table 3.35) and subsequently on the time taken to screen the products generated.

Having successfully reduced the quantities of reactants employed (20- to 50-fold) and analysis time (40 min sample^{-1} to 15 s sample^{-1}), the authors subsequently demonstrated the synthetic utility of their reactor toward a series of different reaction conditions; in the presence of bovine carbonic anhydrase II (bCAII) **205**, with an without an inhibitor, in the absence of bCAII **205** and in the presence of a Cu(I)-catalyst (reference), employing the acetylenes **206a–f** and azides **207a–p** illustrated in Figure 3.13.

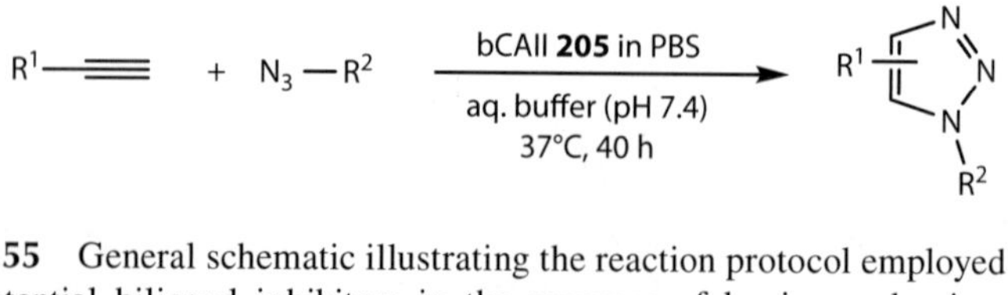

SCHEME 3.55 General schematic illustrating the reaction protocol employed for the synthesis of potential biligand inhibitors in the presence of bovine carbonic anhydrase II (bCAII) **205**.

TABLE 3.35
Comparison of Reactants Consumed Using Standard and Microfluidic Techniques

	Type of Reactor		
Parameter	96-Well Plate	1st Generation	2nd Generation
Number of reactions	96	32	1024
Enzyme (bCAII **205**) (µg)	94.00	19.00	0.36
Acetylene (nmol)	6.00	2.40	0.12
Azide (nmol)	40.00	3.60	0.12
Total reaction volume (µL)	100.00	4.00	0.40
Sample preparation time	Few min	58 s	15 s
Detection method	LC–MS	LC–MS	MS/MS
Hit identification time	40 min	40 min	15 s

Using this approach, the authors were able to rapidly identify 35 compounds as hits and 4 as modest hits, as illustrated in Figure 3.14; the use of ESI–MS MRM technology also provided the additional advantage of computerized interpretation of the results. As such, the authors predict that this type of device has the potential to increase the number and type of enzyme inhibitory molecules identified.

Scalable Click Chemistry: As a part of an investigation into the synthesis of N^1-alkylated 5-amino-1,2,3-triazole carboxamides, Storz and coworkers [101] identified a problem in the development of a generic and scalable synthetic route to this structural scaffold (Figure 3.15).

Reterosynthetic analysis of the molecule led the researchers to a [3 + 2] cycloaddition of an organic monoazide with an *in situ* generated imine; however, the hazardous nature of small molecular weight alkyl azides prohibited the use of this approach on a synthetically useful scale. As an alternative, the authors proposed the use of thio-derived monoazides as a means of increasing the stability of the azide whilst enabling facile removal of the sulfur substituent by means of a Raney-Ni **208** desulfurization (Scheme 3.56).

To evaluate this theory, the authors firstly prepared β-azidoethyl phenyl sulfide **209**, which unlike ethyl azide, was found to be a distillable, thermally stable (<155°C) colorless liquid. Although the use of azide **209** improved the handling issues associated with alkyl azides, the exothermic nature of the proposed [3 + 2] cycloaddition reaction meant that extra precautions were required for this reaction. As such, the authors investigated the synthesis of N^1-alkylated 5-amino-1,2,3-triazole carboxamides utilizing an automated reaction platform containing glass micro reactors (total reaction volume = 250 or 1000 µL).

After an initial parameter screen of inorganic (KO^{t}Bu, LiHMDS, and NaOH **11**) and organic (Et$_3$N **27**) bases, along with a series of polar reaction solvents (NMP, DMF, THF, and EtOH), the authors identified NaOH **11** and NMP as the reaction conditions targeted for further investigation due to the speed of the reaction and solubility of all reactants, products and intermediates.

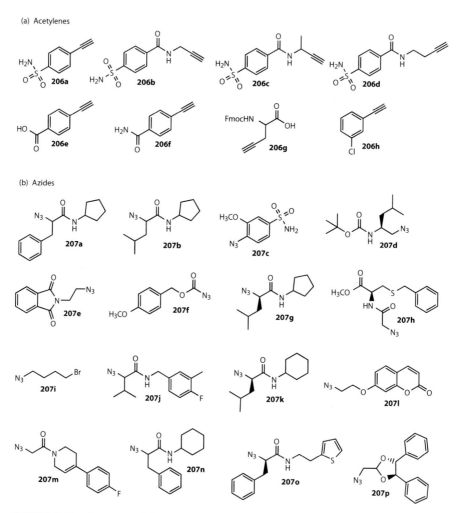

FIGURE 3.13 Summary of the reactants employed in the large-scale click chemistry screen: (a) acetylenes and (b) azides evaluated.

Using the aforementioned conditions, the effect of reaction temperature (65–95°C), stoichiometry of cyanoacetamide **210** (1.5–2.0 eq.) and NaOH **11** (1.5–2.0 eq.) was investigated using a fixed reaction time of 1 min; with reaction progress quantified using LC–MS analysis. As summarized in Table 3.36, using this approach the authors were able to rapidly identify the optimal reaction conditions for the synthesis of 5-amino-1-(2-phenylthioeth-1-yl)-1,2,3-triazole-4-carboxamide **211** as being a reaction time of 1 min, 2.0 eq. of cyanoacetamide **210** and NaOH **11**, with a reaction temperature of 65°C. With this information in hand, the flow reactor was operated continuously for a period of 40 min, collecting the reaction products in DI H_2O, after which time the reaction products were filtered and dried to afford the thio-derived compound **211** in 90% yield (4.55 g). Hydrogenation of thiazole **211** with RaNi **208**

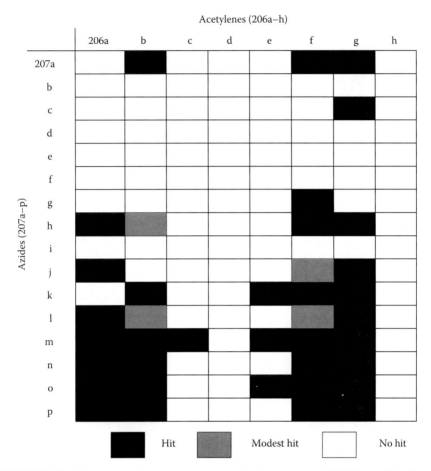

FIGURE 3.14 Illustration of the compounds identified as a hit using MRM-based identification (hit >15%, modest-hit 10–15% and no hit <5% response).

in MeOH (60 psi) for 48 to 60 h afforded complete conversion to 5-amino-1-ethyl-1,2,3-triazole-4-carboxamide **212** on 90% yield.

Using this approach, the authors were able to develop a safe, reliable reaction route to the synthesis of N^1-alkylated 5-amino-1,2,3-triazole carboxamides in higher yield and purity compared to previous reports utilizing low-molecular weight alkyl

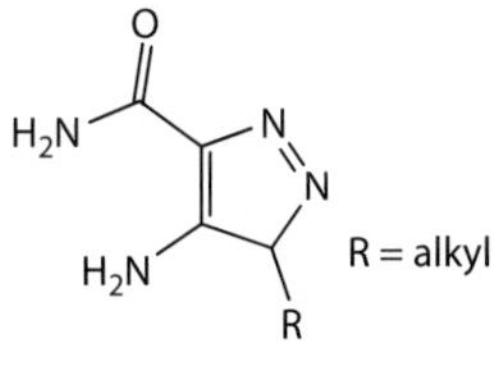

FIGURE 3.15 Illustration of the target scaffold in the development of biologically active molecules.

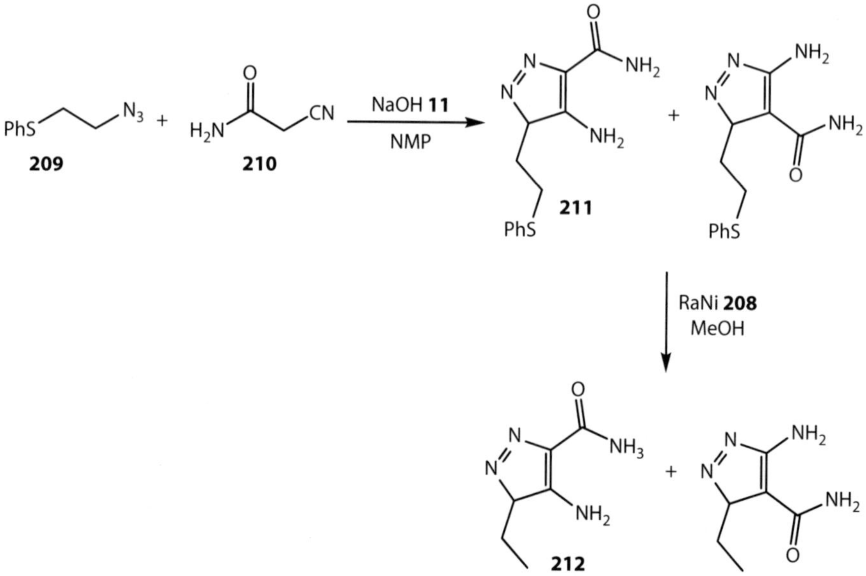

SCHEME 3.56 Illustration of the reaction protocol employed for the scalable synthesis of N^1-alkylated 5-amino-1,2,3-triazole carboxamides using thio-derived monoazides.

TABLE 3.36
Summary of the Results Obtained for the Synthesis of N^1-Alkylated 5-Amino-1,2,3-Triazole Carboxamides under Continuous Flow

Temperature (°C)	Cyanoacetamide 210	Equivalents NaOH 11	Thiazole (%)	Residual Azide 209 (%)
65	1.5	1.5	30	70
65	1.5	2.0	98	2
65	2.0	1.5	98	2
65	2.0	2.0	100	0
75	1.5	1.5	47	46
75	1.5	2.0	70	30
75	2.0	1.5	98	2
75	2.0	2.0	76	23
85	1.5	1.5	5	95
85	1.5	2.0	60	40
85	2.0	1.5	98	0
85	2.0	2.0	95	5
95	1.5	1.5	7	93
95	1.5	2.0	2	98
95	2.0	1.5	3	97
95	2.0	2.0	0	100

azides. In addition, the application of β-azidoethyl phenyl sulfide **209** also represents a scalable synthetic protocol for the rapid generation of this compound class.

Copper-Tube Reactor: In a more recent example of triazole synthesis under flow, Bogdan and Sack [102] exploited the high surface-to-volume ratio obtained within continuous flow reactors, by utilizing a copper tube reactor (dimensions = 750 μm (i.d.), Conjure, USA) which served as a catalytic surface for the click reaction. To enable rapid screening of reaction conditions with minimal reactant consumption, the authors employed segmented flow (400 μL), with reactions separated by the fluorous solvent, perfluoromethyldecalin (PFMD).

Investigating a one-pot click reaction, the authors reacted 4-ethynyltoluene (0.25 M) with sodium azide (0.5 M) and the respective alkyl halide (0.5 M), varying the reaction time, temperature, and solvent. Using this approach, the authors identified DMF as the reaction solvent, a reaction time 5 min and a reactor temperature of 150°C as being optimal. Depending on the reaction solvent selected, the rate of copper leaching was observed to vary, that is, the use of EtOH at 150°C resulted in 78 ppm Cu compared to DMF which resulted in 300 ppm leaching from the reactor; the treatment of the reaction products with a scavenger resin, QuadraPure TU, was however, found to reduce the residual copper to <5 ppm.

In summary, employing a range of alkyl halides and terminal acetylenes, the authors were able to synthesize 30 1,4-disubstituted-1,2,3-triazoles in yields ranging from 21% to 88% yield (determined by ^{1}H NMR spectroscopy) without the need for any added copper catalyst.

Under the aforementioned conditions, the authors were able to readily scale the reaction by continuously dosing single reactants into the flow reactor, using this approach they were able to synthesize triazoles at a rate of 0.58 g h^{-1}, which extrapolates to 13.8 g day^{-1}.

3.3.10 C-Hetero Bond Formation: Addition of Hydrazine to Carbonyl Compounds

The product of the condensation reaction between a hydrazine derivative and a carbonyl containing compound, such as an aldehyde or ketone, is called a hydrazone and in the case of 1,3-diketones, the hydrazone can spontaneously cyclize to afford synthetically useful heterocyclic products called pyrazoles. Due to the pharmaceutical relevance of the pyrazole core motif, Wiles et al. [103] investigated the reaction of a series of 1,3-diketones with hydrazine derivatives within a borosilicate glass micro reactor (channel dimensions = 350 μm (wide) × 52 μm (deep) × 2.5 cm (long)), using EOF as the pumping mechanism. Reactions were performed by bringing together solutions of 1,3-diketone (1.0 M) and the hydrazine derivative (1.0 M) in anhydrous THF at a T-mixer and collecting the reaction products in anhydrous THF at the reactor outlet. Employing a positive applied field at the inlets and maintaining the reactor outlet as the common ground (0 V cm^{-1}), reactants were driven through the microchannel where they met, mixed, and reacted. Analysis of the reaction products was performed offline using GC–MS and conversions calculated with respect to residual 1,3-diketone. As Table 3.37 illustrates, the authors demonstrated the reaction of an array of 1,3-diketones, obtaining quantitative conversions in all cases utilizing hydrazine monohydrate

TABLE 3.37
Summary of the Results Obtained for the Synthesis of 1,2-Azoles under Continuous Flow Conditions

[Reaction scheme: R-CO-CHR¹-CO-R² + NH₂NHR³ in THF → 1,2-azole with N—NR³, R, R², R¹ substituents]

R	R¹	R²	R³	Applied Field (V cm⁻¹)[a]	Conversion (%)
CH_3	H	CH_3	H	292, 318	100
Ph	H	CH_3	H	364, 341	100
$-(CH_2)_4-$	Ph	H	260, 303	100	
Ph	H	Ph	H	386, 364	100
Ph	CH_3	CH_3	H	292, 318	100
Ph	H	CH_3	CH_2Ph	318, 318	42 (100)[b]

[a] All reaction products were collected at 0 V cm⁻¹.
[b] Stopped flow was used to increase residence time.

213. Reduced conversions were obtained for benzyl hydrazine hydrochloride (42%); however, increasing the reaction time by employing stopped flow enabled the authors to readily increase conversion to the N-substituted azole.

Concurrently, Warrington and coworkers [104] reported the Knorr reaction in a glass pressure-driven reactor (channel dimensions = 100 μm (wide) × 25 μm (deep)), employing slugs (2.5 μL) of reactants in order to minimize the volume of material used in the reaction screen. Unlike the previous example, the authors employed 0.01 M solutions of the 1,3-diketone and 0.8 M of the hydrazine derivative. Employing MeOH as the reaction solvent and aqueous MeOH (50%) as a diluent, reaction products were analyzed using online HPLC-UV and LC–MS. Using this approach, the authors were able to introduce slugs of different diketones into the system, every 120 s, enabling the rapid construction of a compound library (7 × 3), as illustrated in Table 3.38.

3.4 ELIMINATION REACTIONS

3.4.1 Dehydration Reactions

Fukase and coworkers [105] recently demonstrated the use of an IMM micro mixer as a means of developing continuous flow methodology for the efficient dehydration of β-hydroxyketones to afford the respective unsaturated product, as illustrated in Scheme 3.57.

Upon investigating a series of reaction conditions, including reaction temperature and time, the authors developed a generic protocol whereby the β-hydroxyketone in dioxane, in this case 4-hydroxy-5-methylhexan-2-one **214** (1.0 × 10⁻² M), was mixed with p-TsOH.H₂O **156** (1.0 × 10⁻² M in dioxane) in the micro mixer, the reaction

TABLE 3.38
Illustration of the Compound Library Generated Using an Automated Glass Micro Reactor Setup; the Percentage Denotes the Conversion to Product Determined Using Online Detection

	Ph	MeO-C₆H₄	F-C₆H₄	Br-C₆H₄	O₂N-C₆H₄	PhN=N	4-Cl-2-NO₂-C₆H₃
N₂H₄	99	99	99	99	99	35	57
H₂N–NHMe	99	99	99	99	99	99	56
HO–HN–NH₂	99	99	99	99	99	85	49

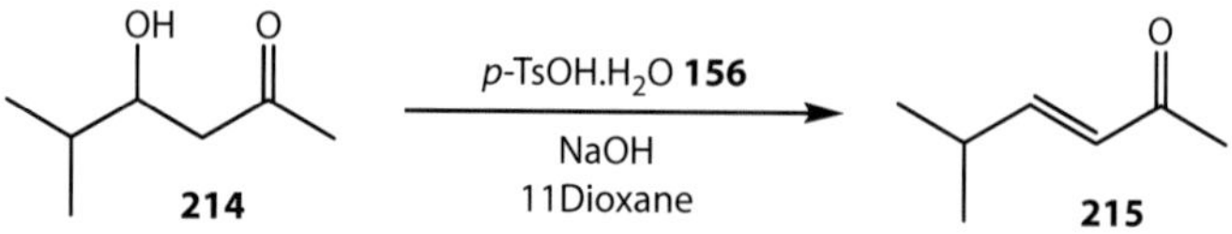

SCHEME 3.57 Schematic illustrating the model reaction used to develop an efficient method for the dehydration of β-hydroxyketones.

mixture was then heated to 110°C within a micro tube reactor prior to quenching with aq. NaOH **11** (1.0 M). Using this approach, the authors found the optimal residence time to be 47 s (200 μL min^{-1}) which afforded the dehydrated product, (*E*)-5-methylhex-3-en-2-one **215**, in a respectable yield of 71%. Compared to previous batch methodology, the continuous flow approach developed provides a facile method for the dehydration of β-hydroxyketones and as discussed in Chapter 7, the authors subsequently used the methodology for the synthesis of large volumes of the natural product Pristane.

3.4.2 Dehalogenations: Tris(Trimethylsilyl)Silane-Mediated Reductions

Using the silane reducing agent tris(trimethylsilyl)silane (TTMSS) **216**, Seeberger and coworkers [106] reported a highly efficient method for radical-based reductions, focusing on the deoxygenation and dehalogenation of an array of organic substrates.

Employing a glass micro reactor (Syrris, UK), with a reaction volume of 1 mL, the authors investigated a series of deoxygenation reactions using toluene as the reaction solvent, identified after a brief solvent screen, and 1.2 eq. of TTMSS **216** in the presence of 10 mol% of the initiator AIBN. At a reactor temperature of 130°C, the authors identified a residence time of 5 min as being optimal for this transformation, as illustrated in Table 3.39.

Initially, a series of xanthate **217** and imidazoylthiocarbamate **218** derivatives of dodecanol were investigated, affording the parent dodecane **219** in 89% and 94% respectively. The chloroxanthate derivative **220** was subsequently investigated, affording the respective chloroalkane **221** in 70% yield. The scope of the technique was subsequently probed for a series of industrially relevant small molecules, centering on the formation of deoxy sugars (91–92% yield).

Gratified by the success of the deoxygenations, the authors extended the investigation to encompass the radical-based reduction of halides. As illustrated in Table 3.39, iodo- **222**, chloro- **223**, and bromo-dodecane **224** were reduced to dodecane **219** in 85%, 67%, and 83%, respectively and phenanthrene **225** was readily synthesized from 9-bromophenanthrene **226** in 92% yield.

Tributyltin Hydride-mediated Reductions: Ryu and coworkers [107] also demonstrated a series of radical-mediated reductions, this time employing 2,2′-azobis(2,4-dimethylvaleronitrile) (V-65) as an initiator and tributyltin hydride as the mediator, finding that reaction times could be reduced to 1 min for the reduction of bromo-dodecane **224**. Subsequent investigations highlighted even better yields could be obtained if the initiator V-70 (2,2′-azobis(4-methoxy-2,4-dimethylvaleronitrile)) was utilized. As Table 3.40 illustrates, excellent yields were obtained

TABLE 3.39
Illustration of the Radical-Based Reductions Performed under Continuous Flow Using Tris(Trimethylsilyl)Silane 216

Substrate	Product	Isolated Yield (%)
n-C$_{12}$H$_{25}$OC(S)SMe **217**	n-C$_{12}$H$_{26}$ **219**	89
n-C$_{12}$H$_{25}$OC(S)OIm **218**	n-C$_{12}$H$_{26}$ **219**	94
ClC$_{10}$H$_{20}$OC(S)SMe **220**	n-C$_{10}$H$_{21}$Cl **221**	70
(imidazole thiocarbonyl sugar structure)	(deoxy sugar structure)	92
(bis-thio carbonate sugar structure)	(deoxy sugar structure)	93
n-C$_{12}$H$_{25}$I **222**	n-C$_{12}$H$_{26}$ **219**	85
n-C$_{12}$H$_{25}$Cl **223**	n-C$_{12}$H$_{26}$ **219**	67
n-C$_{12}$H$_{25}$Br **224**	n-C$_{12}$H$_{26}$ **219**	83
1-bromoadamantane	adamantane	86
(iodo sugar structure)	(deoxy sugar structure)	93
226 (bromophenanthrene)	**225** (phenanthrene)	92

with reactant concentrations ranging from 0.05 to 0.4 M and temperatures of 80–130°C, depending on the substrate employed.

See also Chapter 7 for an example of TTMSS-mediated reductions for the synthesis of ligans on a gram-scale.

Biphasic Dehalogenations Using Enzymes: Goto and coworkers [108] reported the first example of an enzymatic reaction performed in a two-phase liquid flow system, demonstrating the dehalogenation of *p*-chlorophenol **227** to hydroquinone **228**, which subsequently underwent oxidation to afford benzoquinone **229**; as illustrated in Scheme 3.58.

Employing a Pyrex glass reactor (channel dimensions = 100 µm (wide) × 25 µm (deep) × 20.0 cm (long)), containing the guide structure illustrated in Figure 3.16 and

TABLE 3.40
A Selection of Tin Hydride Radical Reductions Performed under Continuous Flow Conditions

Aryl Halide	Alkane	Yield (%)
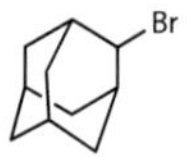 1-bromoadamantane	 adamantane	98
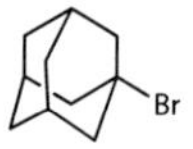 2-bromoadamantane	 adamantane	97
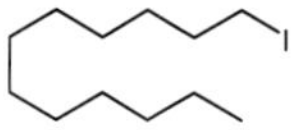 1-iodooctane	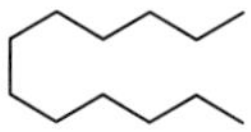 octane	94
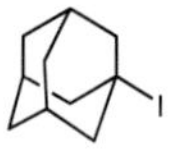 1-iodoadamantane	 adamantane	97
 2-iodoadamantane	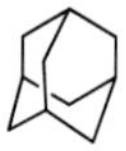 adamantane	55

prior to use, the authors derivatized the outlet channel of the organic stream with trichloro(octadecyl)silane to enable efficient separation of the two-phase flow upon exiting the reactor (Figure 3.17). After pretreating the reactor, the *Laccase*-catalyzed degradation of *p*-chlorophenol **227** was carried out using the following methodology; an *iso*-octane solution of *p*-chlorophenol **227** (1×10^{-4} M) was pumped into the reactor from the ODS-modified inlet and an aqueous succinic buffer (5.0×10^{-2} M) solution containing *Laccase* (1.86 g L^{-1}) was introduced from the second inlet. The effect of flow rate on the degradation of *p*-chlorophenol **227** was subsequently investigated over 0.75–30.0 µL min^{-1}, with the organic and aqueous phases monitored upon collection by HPLC. As expected, due to a change in interfacial contact time, the proportion of dehalogenation was found to decrease with increasing flow rate. Employing a flow rate of 1.6 µL min^{-1}, the authors were able to efficiently degrade >60% of the

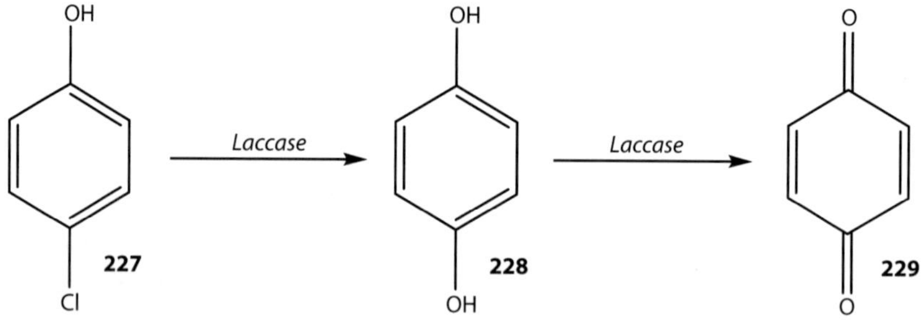

SCHEME 3.58 Illustration of the enzymatic degradation of *p*-chlorophenol **227** to benzoquinone **229** performed in a two-phase flow reactor.

Liquid-Phase Micro Reactions

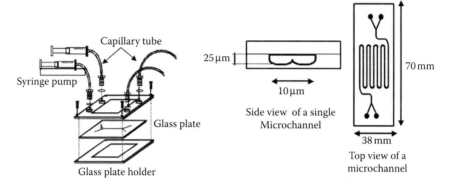

FIGURE 3.16 Schematic illustrating the microfluidic device employed for the two-phase enzymatic degradation of *p*-chlorophenol **227** to benzoquinone **229**. (Maruyama, T. et al. 2003. Enzymatic degradation of p-chlorophenol in a two-phase flow microchannel system, *Lab Chip* 3: 308–312. Reproduced by permission of the Royal Society of Chemistry.)

p-chlorophenol **227** obtaining a J_m value of 8.2×10^{-6} mol m^{-2} s^{-1}, an enhancement of 304; compared with 2.7×10^{-8} mol m^{-2}s^{-1} obtained in a batch vessel.

The observed enhancement obtained within the micro reactor was attributed to the alignment of the enzyme at the phase boundary and the reduced diffusion distance between the organic and aqueous layers.

3.5 OXIDATIONS

While oxidation reactions are a common tool to the research chemist, performing such transformations on a large scale can be problematic, due to the formation of by-products (largely attributed to over-oxidation of the target material), the generation of large exotherms, along with the use and required disposal of toxic reactants and by-products. In an attempt to address these limitations of an otherwise synthetically useful transformation, a series of research groups have investigated the performance of oxidations under continuous flow.

3.5.1 Oxidations: Inorganic Oxidants

Iron(III) Nitrate Catalysis: Having previously demonstrated the ability to dramatically enhance the rate of benzyl alcohol **35** oxidation in the presence of microwave irradiation, Jachuck et al. [109] set about designing a flow reactor capable of removing the heat effect of microwave irradiation and enabling the identification of any potential "microwave effects."

Employing a PTFE reactor (volume = 270 μL), coupled to a stainless-steel heat exchanger (volume = 6 mL), the authors evaluated the isothermal oxidation of benzyl alcohol **35** to benzaldehyde **19**, in the presence of iron(III) nitrate. Utilizing online FT–IR spectroscopy, the authors were able to rapidly evaluate the effect of reactant residence time (3.2–16.2 s) and microwave intensity (0–39 W) on the oxidation reaction. In the presence of microwave irradiation, the authors observed an increase in the formation of benzaldehyde **19** with time, obtaining 75% conversion at a residence

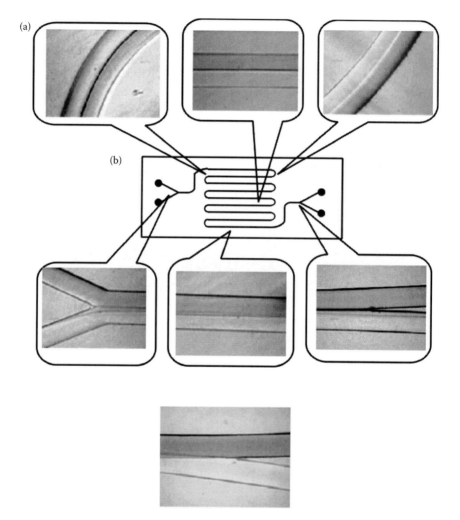

FIGURE 3.17 Illustration of the two-phase flow obtained within the glass microchannel when employing water and *iso*-octane as the solvent system (a) unmodified and (b) ODS modified channel. (Maruyama, T. et al. 2003. Enzymatic degradation of p-chlorophenol in a two-phase flow microchannel system, *Lab Chip* 3: 308–312. Reproduced by permission of the Royal Society of Chemistry.)

time of 16.2 s (38.7 W). Using this approach, the authors were able to identify that in both isothermal and adiabatic scenarios, the reaction was second order. However, by isolating the thermal microwave effects from the system, the authors were able to conclude that the reaction was accelerated by microwave irradiation.

3.5.2 Oxidations: Swern–Moffat Oxidation

Of the available oxidation methods, the Swern oxidation (Scheme 3.59) is one of the most versatile techniques, however, the need to perform these reactions at reduced

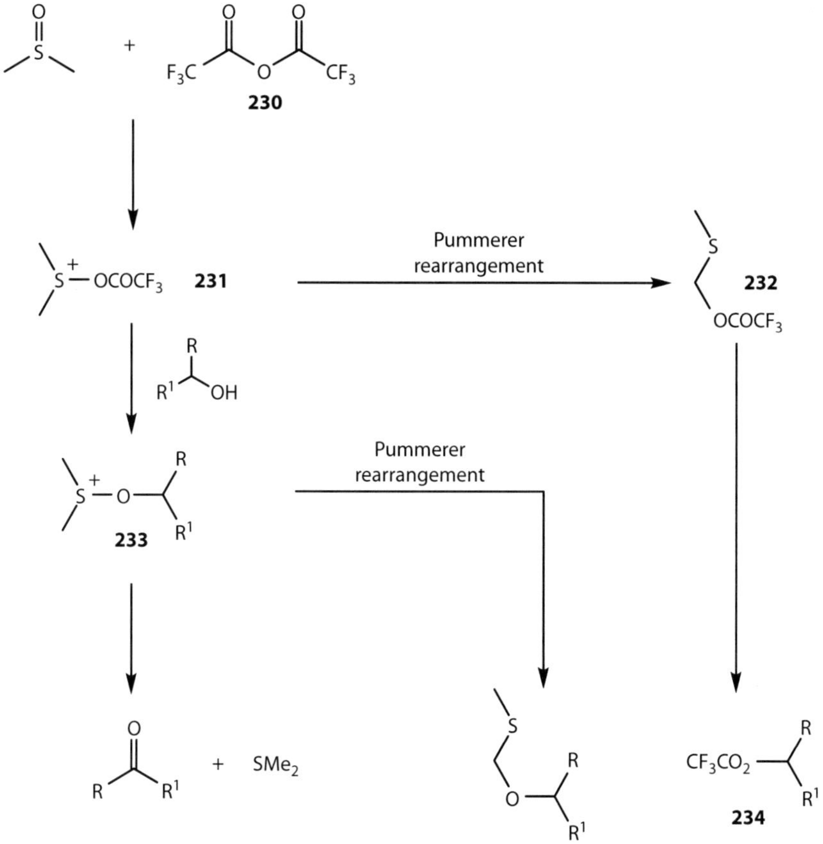

SCHEME 3.59 Illustration of the proposed mechanism for the Swern–Moffat oxidation using trifluoroacetic anhydride **230**.

temperatures ($\leq -50°C$) means that this synthetic route is rarely exploited at scale due to the costs associated with operating plant equipment at reduced temperatures. With these economic factors in mind, several researchers in both academia and industry have investigated the transformation under continuous flow as a means of increasing the practicalities associated with this versatile oxidation pathway.

As Scheme 3.59 illustrates, DMSO reacts with trifluoroacetic anhydride **230** to afford the trifluoroacetoxydimethylsulfonium salt **231**, which is known to be stable below $-30°C$. Above $-30°C$, the intermediate **231** undergoes rearrangement to afford **232**; therefore, in order to affect the desired transformation, the cation **231** must undergo reaction with the alcohol typically at temperatures of $-50°C$ or lower. In the third step, the alkoxydimethylsulfonium salt **233** is treated with an organic base to afford the respective carbonyl compound. At this stage, an additional Pummerer reaction can occur to afford methylthiomethyl ether (MTME) **234** along with the formation of the TFA ester of **232**. Owing to the number of competing reaction pathways, it is imperative that the reaction temperature is controlled in order to minimize the proportion of by-products obtained alongside the target oxidation product.

With this in mind, Yoshida and coworkers [110] investigated the Swern–Moffat oxidation under continuous flow, using a stainless-steel tubular reactor setup, with reactant delivery controlled by four syringe pumps. At the first mixer (IMM, GmbH), the authors reacted DMSO (4.0 M in DCM) with TFAA **230** (2.4 M in DCM) at a total flow rate of 2 mL min^{-1}. The resulting reaction mixer was pumped into a second mixer where it mixed and reacted with the alcohol (1.0 M in DCM) under investigation, prior to reacting with Et$_3$N **27** (1.45 M in DCM). Up to this point, the reactor was housed in a cooling bath (−20°C) and the reaction products collected in a sample tube at 30°C. Optimization of the reaction was performed by varying the residence times within each section of the reactor, with the reaction products analyzed and quantified using offline GC-FID.

As Table 3.41 illustrates, in all cases when compared with batch, elevated reactor temperatures could be employed to afford the target carbonyl compounds in higher yield and purity. The authors attribute this success to the fast transfer of the highly unstable reactive intermediates, enabling reaction prior to thermal decomposition.

Kemperman and coworkers [111] subsequently reported a continuous flow Swern oxidation of a series of alcohols, commenting on numerous processing advantages

TABLE 3.41
Summary of the Results Obtained for the Swern Oxidation of a Series of Alcohols Performed under Continuous Flow Conditions

Alcohol	Reactor	Reaction Time (s)	Temperature (°C)	Yield (%) Product	TFA Ester	MTME
1-Decanol	Flow	2.4	−20	75	8	19
	batch	0.01	0	70	6	22
		0.01	20	71	6	22
		—	−20	11	1	90
2-Octanol	Flow	2.4	−20	95	5	2
	batch	0.01	0	86	4	3
		0.01	20	89	3	2
		—	−20	20	2	75
Cyclohexanol	Flow	2.4	−20	88	6	5
	batch	0.01	0	89	7	1
		0.01	20	88	5	2
		—	−20	19	2	70
		—	−70	83	10	5
Benzyl alcohol	Flow	2.4	−20	91	—	8
	batch	0.01	0	78		14
		0.01	20	75	—	16
		—	−20	49	—	50

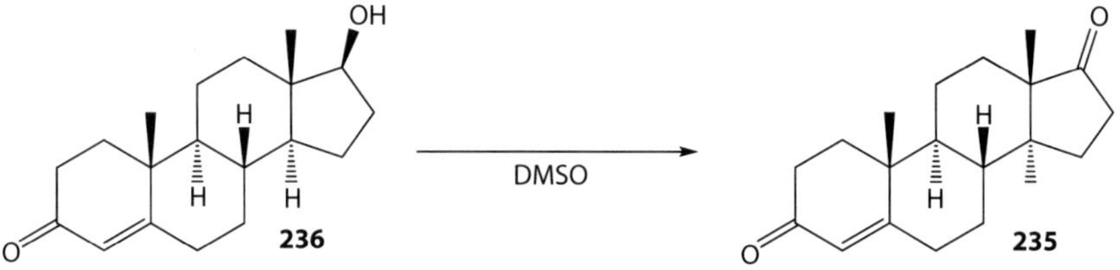

SCHEME 3.60 Illustration of the continuous flow oxidation of testosterone **236** to 4-androstene-3,17-dione **235**.

compared with batch, such as minimal accumulation of trifluoroacetoxydimethylsulfonium salt **231** and alkoxydimethylsulfonium salt **233**, along with minimal Pummerer rearrangement. The authors also found that the reaction could be successfully performed at 0–20°C rather than −70°C as conventionally applied.

Using the synthesis of 4-androstene-3,17-dione **235** as a model (Scheme 3.60), the authors were able to operate the flow reactor continuously for 1.5 h, converting testosterone **236** to afford the target compound **235** at a throughput of 64 g h^{-1}; illustrating the scalability of the developed continuous flow methodology.

In the same year, McConnell et al. [112] reported a semicontinuous process utilizing the Swern oxidation as a key step in the synthesis of heptenulose, reporting the ability to increase the reactor temperature from −60°C in batch to −10°C in the flow process.

3.5.3 Oxidations: TEMPO-Mediated Oxidations

Owing to the fact that aldehydes find widespread application as precursors for cyanohydrins, β-hydroxyketones, oximes, hydrazones, and epoxide, to name but a few, their synthetic versatility means that efficient methods are required for their large-scale preparation. With techniques such as Dess Martin periodinane, activated DMSO and IBX currently limited to a laboratory scale, Fritz-Langhals [113] investigated the development of a continuous flow method for the hydroxy-TEMPO (4-hydroxy-2,2,6,6-tetramethyl-piperidine-1-oxyl) **237** mediated oxidation of activated alcohols such as 2-hydroxyethyl-isobutyrate **238** (Scheme 3.61).

Utilizing a biphasic reaction system where the aqueous phase comprised sodium hypochlorite **238** (1.8 M, solution 1) and sodium bromide **239** (0.19 M, solution 2), and the organic phase (DCM) contained the 2-hydroxyethyl-isobutyrate **237** (0.51 M) and hydroxy-TEMPO **236** (2 mol%, solution 3), the oxidation reaction was evaluated in batch and in a titanium tubular reactor (dimensions = 3.0 mm (i.d.) × 20 m (long)) setup. When performed in batch the target aldehyde, 2-oxoethyl isobutyrate **240** was obtained in high yields (69%); however, with increasing batch size (>5 g), competing carboxylic acid **241** and subsequent ester **242** formation was detected; resulting in a diminished aldehyde **240** yield (20% on a 3 mol scale). As the competing ester **242** formation was attributed to inefficient heat removal with increasing reactor size, the authors proposed that the use of a tubular flow reactor would afford the rapid mixing, short contact times and efficient heat removal required to selectively oxidize 2-hydroxyethyl-isobutyrate **237** to 2-oxoethyl-isobuyrate **240**. Within the flow

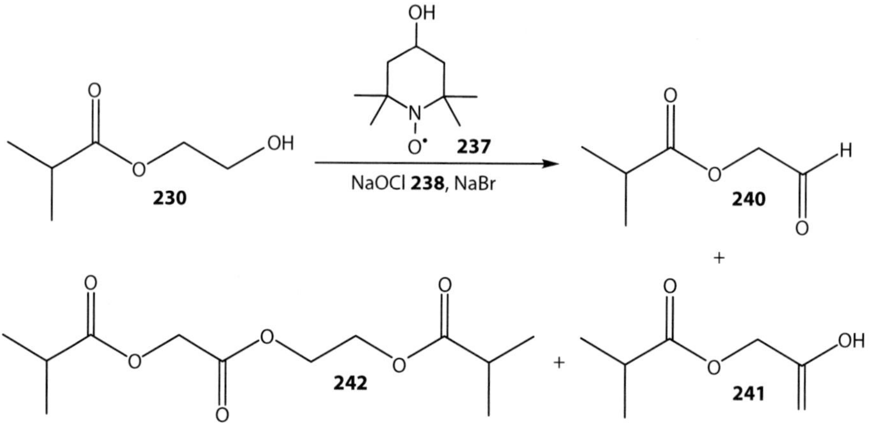

SCHEME 3.61 Illustration of the bleach (NaOCl) **239** oxidation of 2-hydroxyethyl-isobutyrate **238**, utilizing hydroxy-TEMPO **237** and the ester by-product **242** formed.

reactor, the three reactant solutions were introduced into the tubular reactor at the following flow rates; solutions 1 = 17 L h^{-1}, 2 = 2.6 L h^{-1}, and 3 = 48 L h^{-1} and the reaction products collected in a 20 L receiver flask containing aqueous sodium thiosulfate. Using this approach, the authors obtained 2-oxoethyl-isobutyrate **240** in 60% yield, demonstrating the ability to produce aliphatic aldehydes on a scale of 60 mol day^{-1} from a single tubular reactor.

3.5.4 Oxidations Using Oxone

More recently, Yamada et al. [114] reported the development of a microflow reactor suitable for performing the oxidative cyclization of alkenols using the potentially explosive reagent Oxone **243**. The authors employed carbinols as their synthetic target due to the presence of such functionality within therapeutic agents and biologically active compounds. Utilizing a heated tubular reactor (dimensions = 1 mm (i.d.) × 5 cm (long)), the authors evaluated the oxidative cyclization of (Z)-4-decen-1-ol **244** (5.0 × 10^{-2} M) in ^{i}PrOH with an aqueous solution of Oxone **243** (0.1 M); reactions were terminated using aq. Na$_2$S$_2$O$_3$ (30%) as a quench agent and analyzed offline by GC and NMR spectroscopy. After screening a series of reaction conditions, the authors identified a reaction time of 5 min and a reactor temperature of 80°C as being optimal for the synthesis of the cyclic ether *treo*-1-(2-tetrahydrofuranyl)hexan-1-ol **245**; which was obtained in 99% conversion. As Table 3.42 illustrates, the reaction protocol developed was found to be generic toward a series of alkenols, affording a range of cyclic ethers in high to excellent conversions.

3.5.5 Oxidations: Epoxidations under Flow Conditions

Owing to the synthetic versatility of chalcone epoxides as active ingredients in the pharmaceutical industry, the enantioselective oxidation of chalcones represents a commercially interesting reaction and as such formed the basis of an investigation by

TABLE 3.42
Illustration of the Reaction Protocol Employed for the Continuous Flow Oxidation of Alkenols to Afford a Series of Biologically Relevant Carbinols

Alkenol	Product	Conversion (%)
244	245	99
		90
		88
		90[a]
		70[a]

[a] Flow rate = 2 µL min^{-1} (per reactant), residence time = 10 min, and reactor temperature = 80°C.

Kee and Gavriilidis [115] into the development of a continuous flow process for the gram-scale synthesis of such compounds.

Employing an SU-8/PEEK micro reactor (Footprint = 11 cm × 8.5 cm), illustrated in Figure 3.18, the authors investigated the enantioselective oxidation of chalcones in the presence of the solution phase catalyst poly-L-leucine **246**.

To access the high system throughputs targeted, the authors found it necessary to incorporate staggered herringbone micromixers (dimensions = 200 µm (wide) × 85 µm (deep) × 4 cm (long)) into the reaction channels, thus enabling the rapid mixing of reactants, coupled with larger reaction channels (dimensions = 2000 µm (wide) × 330 µm (deep)) to enable access to the desired residence times at high flow rates.

Using the aforementioned reactor, the authors introduced two reactant solutions into the reactor from separate inlets, the first containing a solution of 1,8-diazabicyclo[5.4.0]undec-7-ene (DBU) **41** (0.88 M in THF:MeCN) and the second poly-L-leucine **246** (53.9 g L^{-1}) and peroxide **247** (0.53 M). Employing a total flow rate of

FIGURE 3.18 Photograph illustrating the SU-8/PEEK micro reactor used for the gram-scale production of (2*R*,3*S*)-epoxide **249** under continuous flow. (Reprinted with permission from Suet-Ping Kee et al. 2009. *Org. Process Research & Dev.* 13(5): 941–951. Copyright (2010) American Chemical Society.)

10 µL min^{-1}, through a reaction channel of 45 cm (length), afforded a time of 30 min, for the adsorption of the peroxy anion onto the catalysts surface. In the second half of the reactor, a solution of chalcone **248** (0.16 M in THF:MeCN) was introduced into the reactor and the solutions mixed within a second staggered herringbone mixer prior to entering the reaction channel, after 16 min (20 µL min^{-1} total flow rate) the reaction products were collected offline and quenched using a solution of sodium sulfite prior to analysis by chiral HPLC (Scheme 3.62).

Using this approach, the authors investigated the effect of reactor temperature (15–35°C) on the epoxidation reaction, finding the optimal condition to be 23.1°C which afforded (2*R*,3*S*)-epoxide **249** in 86.7% conversion and an enantioselectivity of 87.6%, finding the results compared favorably with data generated using a slit laminar flow model (Table 3.43). Further, elevated temperatures were investigated as

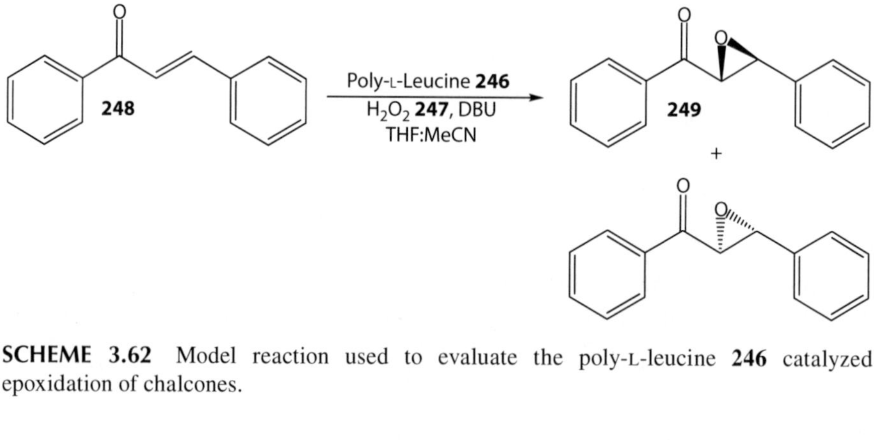

SCHEME 3.62 Model reaction used to evaluate the poly-L-leucine **246** catalyzed epoxidation of chalcones.

TABLE 3.43
Comparison of the Actual and Predicted Results Obtained for the Epoxidation of Chalcone 248

Reaction Type	Conversion (%)	ee 249 (%)
PEEK micro reactor	86.7	87.6
Slit flow reactor model	89.6	92.4
Continuous tubular reactor	88.4	88.8
Continuous tubular reactor model	88.3	92.4

a means of increasing the throughput of the reactor; however, this was found to erode enantioselectivity as the background epoxidation reaction dominated and the poly-L-leucine **246** was observed to decompose.

Under the optimal conditions, the authors were able to develop a continuous flow method for the enantioselective synthesis of chalcone epoxides affording single reactor throughputs of ~0.5 g day^{-1}.

In a second example of continuous flow epoxidation, Kraft and coworkers [116] demonstrated the oxidation of cyclohexene **250**, using H_2O_2 **247**, as part of a two-step process for the synthesis of *trans*-1,2-cyclohexanediol **251**, as depicted in Scheme 3.63. When performed using conventional methodology, the exothermic nature of these reaction steps dictates the need for slow, dropwise, addition of reactants, dilute reactant solutions, and careful temperature control; all resulting in a low system throughput.

To perform the investigation under continuous flow, the authors employed a T-mixer and PTFE tube (dimensions = 1 mm (i.d.)) reactor, with fluidic control obtained through the use of a series of syringe pumps. Due to the efficient thermal dissipation obtained in such a system, the authors were able to perform the reaction at three times the reactant concentration compared to batch.

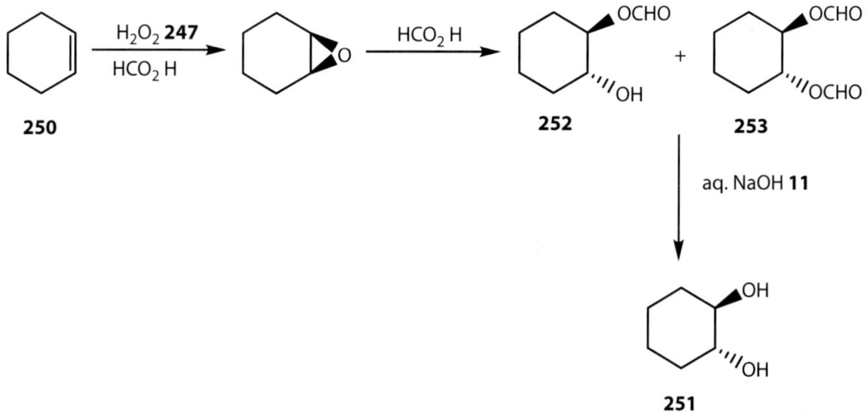

SCHEME 3.63 Illustration of the two-step synthetic route to *trans*-1,2-cyclohexanediol **251**.

To conduct step 1 of the reaction, epoxidation and ring opening, the authors employed two reactant solutions, the first a mixture of formic acid and H_2O_2 **247** (2.9 M, 29 mL h^{-1}) and the second cyclohexene **250** (9 mL h^{-1}). The reactant streams were mixed at a T-mixer and reacted in a heated coil (50°C), prior to cooling and product collection. The reaction products were subsequently analyzed by GC and the conversion quantified. Using this approach, the authors readily identified a reaction time of 1 min as optimal for step 1, compared to 120 min required in batch (for a 1 M solution of formic acid and H_2O_2 **247**). After the destruction of excess H_2O_2 **247** by treatment with sodium hydrogen sulfite, and concentration *in vacuo*, the oily residue was dissolved in aq. MeOH (10% MeOH) and circulated through a second flow reactor immersed in a heated bath (70°C) with a solution of 20% aq. NaOH **11**. Using a reaction time of 30 s, the authors were able to obtain complete saponification of the mono- **252** and diesters **253** to afford the target diol **251** in 88% yield. Compared to the literature approach, the authors found the use of a flow reactor advantageous as it not only enabled a reduced quantity of NaOH **11** to be employed, but also minimized the formation of a colored by-product frequently observed; resulting in the formation of *trans*-1,2-cyclohexanediol **251** as a colorless solid.

3.5.6 Oxidation: Deprotection of Amines

In order to access multistep synthetic processes, protecting groups are frequently employed, it is, however, important to consider the method of removal when designing a synthetic pathway as this step can often lead to loss of precious materials due to inefficiency or problems associated with product purification. With this in mind, Rutjes and coworkers [117] recently reported the use of D-optimal design to evaluate the deprotection of the *p*-methoxyphenyl-protected amine, 4-methoxy-*N*-(1-phenylethyl)aniline **254**, under continuous flow. As Scheme 3.64 illustrates, the authors selected an industrially attractive protocol which involved the use of periodic acid **255** in the presence of an equivalent of sulfuric acid **26** and water to afford the free amine, 1-phenylethylamine **256**, and 1,4-benzoquinone **229** as the by-product.

Utilizing an automated sampling platform and a glass micro reactor (Reactor volume = 7 µL), the authors investigated the effect of reaction time (0.5–4 min), reaction temperature (60–90°C), and stoichiometry (1–4) on the formation of 1-phenylethylamine **256**. To conduct the reactions, a solution of 4-methoxy-*N*-(1-phenylethyl)aniline **254** and H_2SO_4 **26** (0.2 M) in aq. MeCN (50:50) was pumped

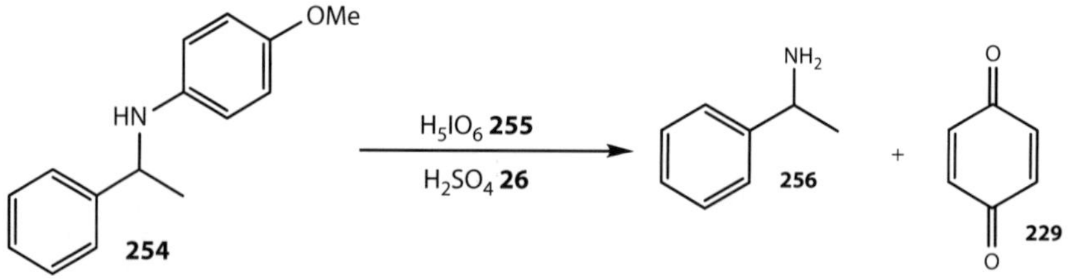

SCHEME 3.64 Schematic illustrating the reaction protocol investigated for the deprotection of 4-methoxy-*N*-(1-phenylethyl)aniline **254** using an automated micro reactor platform.

into the micro reactor, where it mixed with a solution of periodic acid **255** (0.2 M) in aq. MeCN (50:50); after a defined period of time, a solution of aq. NaOH **11** and sodium dithionite was added from a third inlet to quench the reaction. The reaction products were then analyzed offline by HPLC, using internal standardization, whereby reactant stoichiometry was found to have little effect compared to reaction time and temperature. Analysis of the data generated from the 51 reactions performed enabled the authors to identify a reaction time of 1.3 min, stoichiometry of 3.2, and a reaction temperature of 60°C as the optimal conditions, affording the deprotected 1-phenylethylamine **256** in >99% conversion.

Using this information, the authors scaled the reaction from a 7 µL reactor to a 950 µL stainless-steel reactor, where with minimal reoptimization, to prevent boiling of the reaction mixture, a throughput of 0.21 g h^{-1} **256** was obtained.

3.6 REDUCTIONS

3.6.1 Transition Metal Free Reductions

Using only catalytic amounts of lithium-*tert*-butoxide **257** in ^{i}PrOH, Sedelmeier et al. [118] reported the development of an efficient, transition metal free, method for the reduction of ketones under continuous flow.

Using the X-cube™ (Hungary) stainless-steel tubular reactor (reactor volumes = 4, 8, or 16 mL), coupled with a scavenger cartridge, containing a tosyl-functionalized resin, the authors investigated the reduction of a series of aromatic and aliphatic ketones to their respective 1° or 2° alcohols

Employing ketone concentrations of 0.3–0.4 M in ^{i}PrOH, and 10 mol% of LiO^{t}Bu **257**, a reactor temperature of 180°C (160 bar) and a residence time of 30 min, the authors obtained the target alcohols in excellent yield and purity. With reaction products passed through the scavenging cartridge requiring no additional purification after removal of the organic solvent.

As Table 3.44 illustrates, the reaction conditions were found to be versatile, enabling the efficient reduction of substituted aromatic ketones, aliphatic derivatives, and ketones in the presence of nitriles. Halogenated ketones were however observed to undergo a small degree of dehalogenation, typically 5% when using the aforementioned protocol. This methodology therefore affords the user a safer procedure compared to the use of hydride reductions or high-pressure hydrogenations performed in the presence of toxic metal catalysts.

3.6.2 Dibal-H Reductions

The availability of a diverse array of aldehydes, as synthetic precursors/raw materials, is central to many drug discovery and production processes. With this in mind, Ducry and Roberge [119] investigated the diisobutylaluminum hydride (Dibal-H) **258** promoted reduction of methyl butyrate **259** to butyraldehyde **260**, evaluating whether the use of continuous flow processing could afford a selective route to the aldehyde **260** without concomitant over-reduction to the alcohol **261** frequently observed in batch processes.

TABLE 3.44
Illustration of a Selection of the Ketones Reduced Using a Metal-Free Continuous Flow Process

$$R-CO-R^1 \xrightarrow[\text{160 bar}]{\substack{\text{LiO}t\text{Bu } \mathbf{257} \text{ (cat.)} \\ i\text{PrOH, 180°C}}} R-CH(OH)-R^1$$

Ketone	Alcohol	Yield (%)
4-methylphenyl methyl ketone	1-(4-methylphenyl)ethanol	94
phenyl cyclopropyl ketone	phenyl(cyclopropyl)methanol	92
(4-bromophenyl)(phenyl)methanone	(4-bromophenyl)(phenyl)methanol	92[a]
2-acetylpyridine	1-(pyridin-2-yl)ethanol	97
2-thienyl cyclopropyl ketone	thiophen-2-yl(cyclopropyl)methanol	88
heptan-2-one	heptan-2-ol	90

[a] 5% Dehalogenated alcohol was also detected.

Using operating temperature as the variable, the authors investigated its effect on yield and selectivity, for the model reaction illustrated in Scheme 3.65, conducting the reaction in a multi-input glass micro reactor, developed by Corning Reactor Technologies (France) and an ER-25 from Ehrfeld Mikrotechnik (Germany).

In order to obtain the desired selectivity, batch reductions are performed under cryogenic conditions, −65 to −55°C; however, using a multi-injection concept, the authors found it possible to conduct the reactions at −20°C; obtaining butyraldehyde **260** in 89% (11% n-butanol **261**), compared with only 63% **260** in batch.

In addition to the obvious advantages associated with obtaining products in higher yield and purity, the ability to perform reactions at higher temperatures is advantageous from a processing perspective as it reduces the operating costs associated with performing such reactions on a production scale.

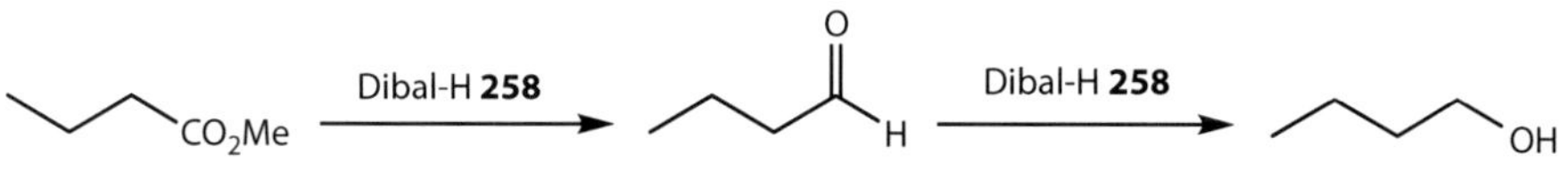

SCHEME 3.65 Schematic illustrating the synthetic route selected for the Dibal-H **258** mediated reduction of methyl butyrate **259** performed in a multi-injection micro reactor.

3.7 METAL-CATALYZED CROSS-COUPLING REACTIONS

Cross coupling reactions performed in the presence of metal catalysts, most commonly Ni or Pd based, are one of the most widely studies classes of transformation studied owing to the synthetic utility of the resulting products in pharmaceuticals, agrochemicals, dye-stuffs, and materials. With this in mind, the following section is broken down into a series of named reactions that represent common transformations in the synthetic chemists' toolbox.

3.7.1 Suzuki–Miyaura Reaction

Microwave Heating: Combining microwave and micro reaction technology, Comer and Organ [120] highlighted some of the advantages attainable when performing synthetic transformations within glass capillary reactors (dimensions = 200 µm (i.d.)). Using the Suzuki coupling reaction of 4-iodooct-4-ene **262** (0.2 M) and 4-methoxyboronic acid **263** (0.24 M), in the presence of catalytic quantities of palladium tetrakis(triphenylphosphine) **264**, the authors investigated the effect of microwave irradiation on the formation of 1-methoxy-4-(1-propylpent-1-enyl)benzene **265**; as illustrated in Scheme 3.66.

Employing 100 W microwave irradiation and a residence time of 28 min, the authors obtained quantitative conversion to the product **265**, leading them to investigate the reaction of an array of aryl halides and boronic acids, obtaining the target compounds in isolated yields ranging from 37% to 100%. In addition to reduced reaction times, the authors report a major advantage of this technique being the suppression of side reactions, leading to increased product purities and hence higher yield.

Conventional Heating: Other authors to use this mode of operation include Wilson et al. [121] who demonstrated the use of a borosilicate glass coil reactor (volume = 4 mL) for the synthesis of 4-(benzofuran-2-yl)benzaldehyde **266**, comparing

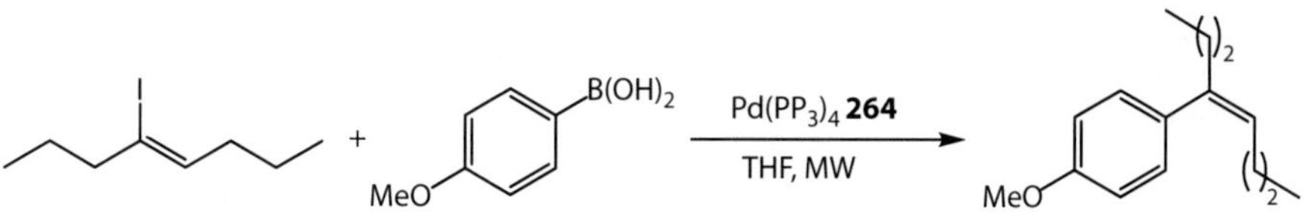

SCHEME 3.66 Illustration of the model reaction selected to illustrate the advantages associated with combining emerging technologies.

TABLE 3.45
Illustration of the Suzuki Coupling Reaction Performed Using a Variety of Techniques

[Reaction scheme: 4-bromobenzaldehyde **77** (Br-C6H4-CHO) + benzofuran-2-ylboronic acid **267** (B(OH)2) → 4-(benzofuran-2-yl)benzaldehyde **266**, using PdCl$_2$(PPh$_3$)$_2$ and Et$_3$N **27**]

Heating Method	Temperature (°C)	Flow Rate (mL min^{-1})	Reaction Time (min)	Conversion (%)[a]
None	rt	—	360	11[b]
Sealed μ-wave tube	140	—	6	86
Continuous flow μ-wave reactor	120	0.25	8	73
Continuous flow μ-wave reactor	130	0.25	8	82
Continuous flow μ-wave reactor	140	0.25	8	84[b]

[a] Determined by ^{1}H NMR spectroscopy.
[b] Isolated yield.

the results obtained with more conventional heating methods, as summarized in Table 3.45. Employing EtOH as the solvent and Et$_3$N **27** (2 eq.) as the base, the authors evaluated the effect of reactor temperature on the coupling of 4-bromobenzaldehyde **77** (1.1 eq.) with benzofuran-2-ylboronic acid **267** performed in the presence of 20 mol% PdCl$_2$(PPh$_3$)$_2$ **268**. For each flow reaction, 50 mL of the reaction mixture was cycled through the reactor, over a period of 5 h, after which time reaction products were analyzed by HPLC in order to quantify the proportion of 4-(benzofuran-2-yl)benzaldehyde **266** synthesized. As Table 3.45 illustrates, in the absence of heating the reaction proceeds slowly, affording only 11% conversion to the furan derivative **266**; in comparison, heating the flow reactor to 120°C this was increased to 73% and subsequently 84% at 140°C. Employing a simple postreaction clean-up, comprising filtration of the reaction products through a plug of silica gel, to remove any Pd residues, the authors were able to crystallize the product **266** directly from the filtrate.

Biphasic Couplings: Employing a coiled perfluoroalkoxy alkane (HP-PFA) tube reactor (dimensions = 750 μm (i.d.), volume = 3 mL), Benali et al. [122] demonstrated the coupling of fluorous spacer technology with microwave heating to enable the rapid optimization of a series of homogeneous Suzuki–Miyaura coupling reactions, while minimizing the volume of reactants employed.

Employing DMF as the reaction solvent and the inert, immiscible solvent perfluoromethyldecalin (PFMD) as the spacer, the authors investigated the effect of plug volume (200, 300, 500, 1000, and 2000 μL) on reaction reproducibility, observing

TABLE 3.46
Comparison of the Results Obtained for the Suzuki–Miyaura Coupling Reaction Performed Using Fluorous Spacer Technology

Aryl Bromide	Yield (%) Batch	Yield (%) Flow
4-ethylphenyl Br	—	100
4-formylphenyl Br	100	91
4-methoxyphenyl Br	100	96
2-methylphenyl Br	49	100
4-nitrophenyl Br	100	86

that for plug volumes larger than 300 µL, comparable conversions were obtained. With this information in hand, the authors investigated the reaction of an array of aryl halides with a series of substituted boronic acids, obtaining moderate-to-excellent yields in all cases (Table 3.46). Using this inventive approach to reaction optimization, the authors could readily increase the throughput of the reactor by simply removing the PFMD spacer solvent, providing access to the biaryl derivatives in throughputs of 0.5 g h^{-1}.

Continuous Solvent Recycling: In 2009, Theberge et al. [123] demonstrated the use of fluorous-tagged catalysts as a means of facilitating catalyst recovery and recycle (Table 3.47), details of the investigation can be found in Chapter 8.

3.7.2 HECK REACTION

The Heck reaction is a synthetically useful method for the arylation of alkenes, largely due to its tolerance to both activated and none activated alkenes, as well as a

TABLE 3.47
Summary of the Results Obtained for the Suzuki–Miyaura Couplings Performed Using a Droplet Reactor Generated in a Capillary-Based Flow Reactor

Aryl Halide	Boronic Acid	Residence Time (h)	Yield (%)[a]
HO–C₆H₄–Br	C₆H₅–B(OH)₂	3	90
HO–C₆H₄–Br	HO₂C–C₆H₄–B(OH)₂	3	77
HO₂C–C₆H₄–Br	C₆H₅–B(OH)₂	0.75	99
3-CO₂H–C₆H₄–Br	4-Me–C₆H₄–B(OH)₂	1	91
5-Br-furan-2-CO₂H	C₆H₅–B(OH)₂	8	63

[a] Determined by HPLC analysis.

wide range of functional groups. With this in mind, several research groups have investigated the organopalladium-catalyzed Heck reaction under continuous flow as a means of identifying any processing advantages associated with the technique.

Using the CYTOS Lab System (Germany), Schwalbe et al. [124] employed a sequential screening approach, to establish the effectiveness of four palladium complexes toward a series of Heck reactions, at three catalyst concentrations (0.46–0.75 mmol) and three reactor temperatures (105–125°C) (Table 3.48). To perform the reaction screening, two reactant solutions were employed, the first containing iodobenzene **269** (37.5 mmol) an acrylic nitrile (41.3 mmol) and tributylamine **270** and the second contained the palladium catalyst (0.43 mmol); both solutions were made up to 100 mL using anhydrous DMF.

Using DMF as the spacer solvent coupled with a fraction collector, the use of a plug flow reactor enabled the authors to separate the reactions performed within the continuous reactor. The reactions were each quenched offline using aq. HCl (1 M) and the reaction products evaluated with respect to the percentage coupling product formed. As Table 3.48 illustrates, using this approach, the authors readily identified PdCl₂(PPh₃)₂ **268** as the most active catalyst for the synthesis of cinnamonitrile **271**, obtaining the target compound in 76% conversion at 105°C and a residence time of 23 min. Increasing the reactor temperature to 125°C, the authors were able to dramatically reduce the proportion of catalyst **268** required, obtaining cinnamonitrile

TABLE 3.48
Summary of the Heck Reactions Performed under Flow Conditions Using the CYTOS® Lab System

269 + R → (Pd-catalyst, Bu₃N 270, DMF) → product, R = CN, R = CO_2Et

Product	Temperature (°C)	Catalyst	Amount (%)	Conversion (%)
271 (CN)	105	Pd(OAc)₂	2.0	32
	105	Pd(OAc)₂/P(ᵗBu)₃	2.0/4.0	53
	105	Pd[(PPh₃)₄]	1.2	76
	105	Pd[(PPh₃)₂Cl₂] **268**	1.2	76
	125	Pd[(PPh₃)₂Cl₂] **268**	0.6	80[a]
	125	Pd[(PPh₃)₂Cl₂] **268**	0.6	83[b]
193 (OEt)				

[a] E/Z 4.31:1.
[b] E/Z >98:1.

271 in 86% yield with an E/Z ratio of 4.3:1. Under the same reaction conditions, the authors also demonstrated the synthesis of ethyl cinnamate **193**, obtaining the target compound in 94% yield with an E/Z ratio of >98:1.

Within a PTFE tubular reactor, Wirth and coworkers [125] recently investigated the arylation of methyl acrylate **272** with iodobenzene **269** in the presence of a series of catalysts (10 mol%) and triphenylphosphine **273** (20 mol%), to form methyl cinnamate **274** as depicted in Table 3.49.

Using this approach, the authors screened the catalysts at 70°C and identified that PdCl₂ **275** and Pd(OAc)₂ **276** were the better performing catalysts in both batch and under flow, with Pd(PPh₃)₄ **264** affording the highest yield (62%) of methyl cinnamate **274** in the flow reactor.

In order to increase the reaction efficiency further, the authors subsequently investigated the use of segmented flow, selecting perfluorodecalin as it was inert to the reaction conditions under investigation. This mode of operation was evaluated as it has been shown to increase the mixing efficiency, due to internal circulation, when compared with laminar flow. Employing segmented flow conditions within a PTFE flow reactor, comprising two T-mixers and two tubular reactors, each 500 μm (i.d.) × 20 cm (long), the authors compared the effect of catalyst **264** mol% and residence time for flow reactions performed under laminar and segmented conditions. In all

TABLE 3.49
Illustration of the Results Obtained for the Homogeneous Heck Reaction Performed in a Flask and under Continuous Flow Utilizing 10 mol% of Catalyst (Unless Otherwise Stated)

269 + 272 →(10 mol% catalyst, 20 mol% PPh$_3$ 273, AcOH, DMF)→ 274

Catalyst	Batch Reaction (%)	Flow Reaction (%)
Pt(COD)Cl$_2$	8	21
CoCl$_2$	17	28
RuCl$_3$	12	45
Ni(OAc)$_2$	23	34
PdCl$_2$ 275	30	47
Pd(OAc)$_2$ 276	26	53
PdCl$_2$ 275[a]	—	19
Pd(OAc)$_2$ 276[a]	—	21
Pd(OAc)$_2$ 276[b]	29	52
Pd(PPh$_3$)$_4$ 264	—	62

[a] 1 mol% of catalyst.
[b] 5 mol% of catalyst employed.

cases, DMF was employed as the reaction solvent and the reactors were heated to 70°C via immersion in an oil bath.

As Table 3.50 illustrates, in the case of methyl cinnamate **274**, a dramatic 29% increase in yield was initially observed as a result of employing segmented flow, further increases to 76% **274** were subsequently achieved by doubling the catalyst mol% from 5 to 10, affording methyl cinnamate **274** in 72% yield. In the case of styrene **277**, while segmented flow was again observed to increase the yield of the product, only a 19% enhancement was obtained. Using this approach, the authors investigated the effect of substituents on both the arene and alkene, along with the halide (Br and I) employed, synthesizing 10 coupling products in yields ranging from 19% to 97%. Buoyed by their findings, the authors subsequently investigated the Heck reaction of arene diazonium salts and alkenes, as illustrated in Scheme 3.67 moderate-to-excellent yields were obtained.

Vinylation of Boroinic Acids: A very recent example of a continuous flow Heck reaction was reported by Lerhed and coworkers [126] who investigated the Pd-catalyzed vinylation of a series of arylboronic acids and subsequently a series of arylations.

TABLE 3.50
Comparison of the Results Obtained under Laminar Flow and Segmented Flow Conditions

Alkene	Catalyst 264 (mol%)	Residence Time (min)	Yield of Arylated Product (%)	
			Laminar Flow	Segmented Flow
Methyl acrylate 272	5	35	36	65
Methyl acrylate 272	10	45	53	76
Styrene 277	5	35	38	57
Styrene 277	10	45	43	59

Using a PTFE reactor (volume = 2 mL), the authors mixed and reacted two reactant solutions prepared in DMF, the first contained the arylboronic acid (0.5 M) and vinyl acetate **278** (5.0 M, 10 eq.) and the second contained the catalyst (Pd(OAc)$_2$) **276**, 1.0×10^{-2} M)) and ligand 1,3-bis(diphenylphosphanyl)propane (dppp) (1.1×10^{-2} M). Reaction products were collected offline and subjected to an aqueous extraction followed by purification by silica-gel chromatography. Employing a residence time of 2 min and a reactor temperature of 150°C, the authors obtained 13 target compounds in moderate-to-excellent isolated yield (42–86%), a selection of which is summarized in Table 3.51.

Ionic Liquids: Using a low-viscosity ionic liquid, 1-butyl-3-methylimidazolium bis(trifluoromethylsulfonyl)imide ([bmim]NTf$_2$), Ryu and coworkers [127] developed a continuous flow protocol for the Mizoroki-Heck reaction whereby the coupling products were readily separated from the catalyst and ionic liquid, enabling the catalyst solution to be recycled.

Employing the reaction of iodobenzene **269** and butyl acrylate **279** as a model reaction, the authors initially investigated the effect of reaction time and temperature on the formation of *trans*-butyl cinnamate **280** (Scheme 3.68), using a Pd-carbene complex **281** and an organic base.

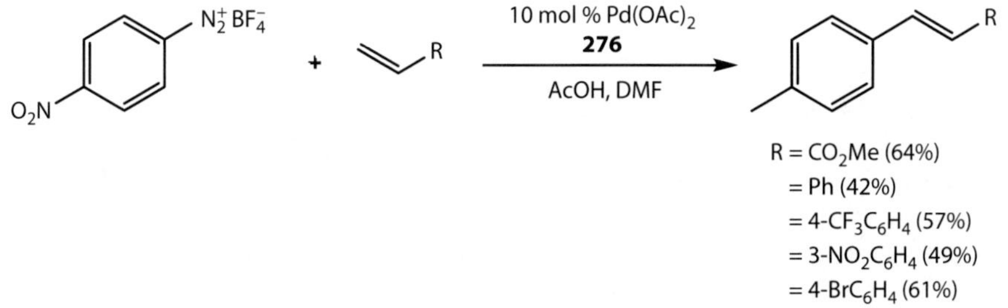

$R = CO_2Me$ (64%)
$= Ph$ (42%)
$= 4\text{-}CF_3C_6H_4$ (57%)
$= 3\text{-}NO_2C_6H_4$ (49%)
$= 4\text{-}BrC_6H_4$ (61%)

SCHEME 3.67 Illustration of the results Heck coupling of alkenes and diazonium salts, performed using segmented flow.

TABLE 3.51
A Selection of Vinyl Derivatives Synthesized Using a Continuous Flow Pd-Catalyzed Heck Reaction

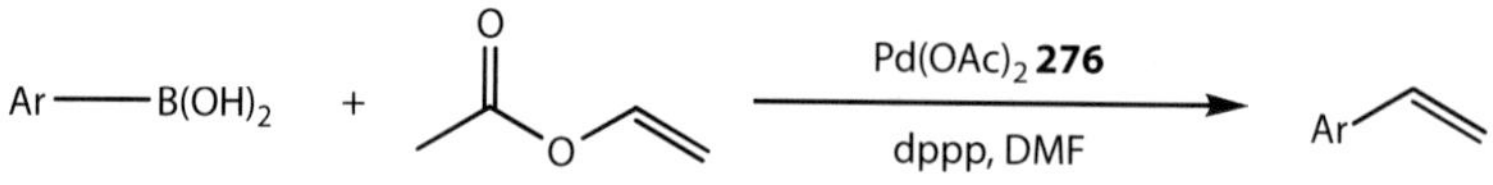

Ar—B(OH)₂ + [acetate vinyl ester **278**] →[Pd(OAc)₂ **276**, dppp, DMF]→ Ar-vinyl

Boronic Acid	Product	Yield (%)
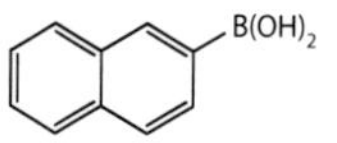 2-naphthyl-B(OH)₂	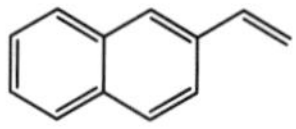 2-vinylnaphthalene	71
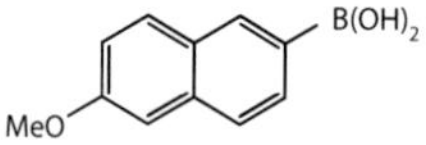 6-MeO-2-naphthyl-B(OH)₂	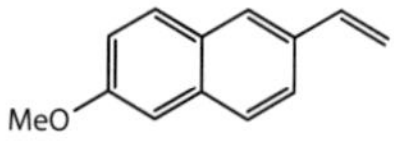 6-MeO-2-vinylnaphthalene	71
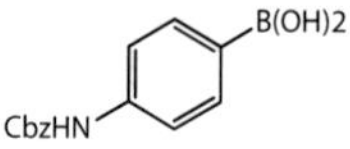 4-CbzHN-C₆H₄-B(OH)₂	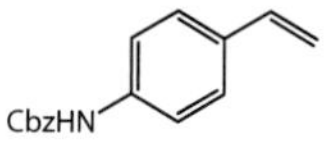 4-CbzHN-styrene	63
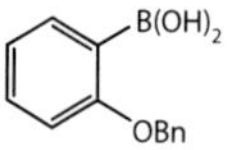 2-OBn-C₆H₄-B(OH)₂	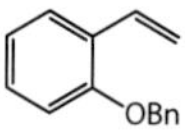 2-OBn-styrene	62
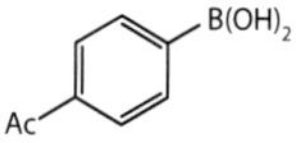 4-Ac-C₆H₄-B(OH)₂	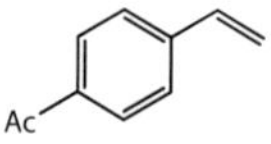 4-Ac-styrene	86

Using a CYTOS Lab System (CPC, Germany) equipped with a micromixer (channel dimensions = 100 μm (wide), volume = 2 mL) and a tubular residence time unit (volume = 15 mL), the authors reacted neat iodobenzene **269** with butyl acrylate **279** with a solution of the Pd-carbene catalyst **281** (5 mol%) in [bmim]NTf₂. Employing a reaction time of 17 min and a reactor temperature of 130°C, the authors obtained the target compound **280** in >90% yield after extraction with hexane, with the ammonium salt and catalyst **281** remaining in the ionic liquid. Initially, the ionic

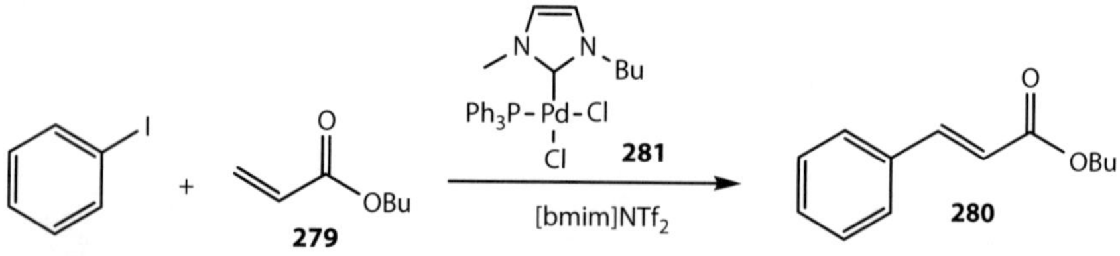

SCHEME 3.68 Illustration of the model reaction used to demonstrate the efficient recycling of an organopalladium catalyst under continuous flow.

liquid was washed with water to remove any ammonium salt formed and was subsequently employed in additional reactions. Using this approach the authors obtained comparable yields suggesting that the catalyst **281** remained active. The authors subsequently developed an automated system capable of separating the reaction products from the ionic liquid and recycling the catalyst **281** solution. Under the aforementioned conditions, the authors operated the system for a total of 11.5 h, consuming 0.7 mol of iodobenzene **269** and generating 115.3 g of *trans*-butyl cinnamate **280**; corresponding to an overall yield of 80%. During this time, the same 90 mL aliquot of catalyst **281**/ionic liquid solution was employed, demonstrating the efficient nature of the recycling technique developed.

3.7.3 Sonogashira Reaction

In addition to an extensive batch investigation into the use of ionic liquids as a reaction medium for the Sonogashira reaction, Ryu and coworkers [128] evaluated the Sonogashira coupling reaction, depicted in Scheme 3.69, using an IMM micromixer (Germany), comprising 2×15 interdigital channels (dimensions = 40 μm (wide) × 200 μm (deep)). Employing two stock solutions, the first containing iodobenzene **269** (1.21 mmol), phenyl acetylene **282** and ^{n}Bu$_2$NH and the second PdCl$_2$(PPh$_3$)$_2$ **268** (0.08 mmol) in [bmim][PF$_6$] (1.2 mL), the authors performed the reaction at a total flow rate of 200 μL min^{-1} respectively. Maintaining the device at a temperature of 110°C, via immersion in an oil bath, the authors obtained diphenylacetylene **283** in 93% yield, which compared favorably with the results obtained in batch. By conducting reactions under continuous flow, utilizing an ionic liquid as the reaction solvent, the authors demonstrated the ability to perform the PdCl$_2$(PPh$_3$)$_2$ **268**-catalyzed Sonogashira coupling reaction in the absence of a copper salt. Coupled with the solvent recycling system described previously for the Heck reaction [129], this approach provides an interesting alternative to more conventional solvent and catalyst combinations employed for the synthesis of substituted alkynes.

High temperature and pressure: Using a series of rapid mixing and heating steps, performed within a high-temperature, pressurized (25 MPa) flow reactor (HPHT-H$_2$O), Kawanami et al. [130] demonstrated the development of a highly efficient method for performing the Sonogashira couplings illustrated in Table 3.52 using water as the reaction solvent.

To perform a reaction, solutions of phenyl acetylene **282**/iodobenzene **269**, and aqueous NaOH **11** (0.2 M)/PdCl$_2$ **275** (2 mol%) were pumped at high speed into a tubular reactor where they met at a T-mixer, generating a fine dispersion of phenyl acetylene **282** (nm to μm) in the aqueous phase. The reaction mixture was then

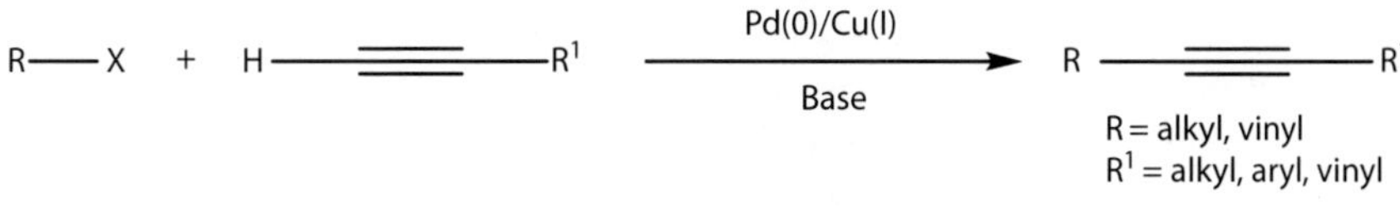

R—X + H———R^1 $\xrightarrow[\text{Base}]{\text{Pd(0)/Cu(I)}}$ R———R^1

R = alkyl, vinyl
R^1 = alkyl, aryl, vinyl

SCHEME 3.69 General reaction scheme illustrating the Sonogashira coupling reaction used for the formation of terminal and/or aryl acetylenes.

TABLE 3.52
Summary of the Results Obtained for the Sonogashira Coupling of Aryl Iodides with Phenylacetylene 282 Using PdCl$_2$ 275 as the Catalyst and Aqueous NaOH 11 as the Base, Performed for 0.1 s at 16 MPa and 250°C

Product/ R	Yield (%)	Selectivity (%)	TOF (h^{-1})
H 283	99	100	1.6×10^6
4-Me	90	100	1.6×10^6
4-OMe	91	99	1.6×10^6
4-NH$_2$	92	100	1.7×10^6
4-OH	88	98	1.6×10^6
3-CF$_3$	99	100	1.8×10^6

heated (250°C) to promote the reaction, followed by rapid cooling and phase separation, enabling facile isolation of the reaction products from the aqueous phase and the catalyst which precipitated as Pd^0. Using this approach, the authors investigated the effect of time on the formation of the coupling product, diphenylacetylene **283**, evaluating residence times ranging from 0.012 to 4.0 s; observing 1.5% conversion at 0.012 s and near quantitative formation of diphenylacetylene **283** at times >0.1 s.

Compared to conventional techniques, the combination of rapid mixing/heating and cooling enabled the authors to synthesize an array of substituted diphenylacetylenes in high yield and excellent selectivity in the absence of a copper catalyst, while utilizing an environmentally benign solvent. In addition, the turnover frequencies (TOF) obtained for the inexpensive commercially available catalyst **275** evaluated were more than 6400 times greater than those previously reported for organoamine-based palladium catalysts (2.5×10^2 h^{-1}).

Automated Sonogashira Couplings: More recently, Fukuyama and coworkers [131] reported the construction of an automated-flow micro reactor system, in collaboration with Dainippon Screen Mfg. Co. Ltd. (Japan), for the rapid optimization of reactions and the production of 10–100 g of material.

Utilizing a Sonogashira coupling reaction, illustrated in Scheme 3.70, the authors investigated the coupling of a bromothiophene **284** derivative to 4-methylphenylacetylene **285** to a matrix metalloproteinase inhibitor **286**. In an initial screening study, the authors evaluated seven reaction conditions by varying the reaction temperature (70–110°C) and reaction time (20–60 min). Using offline HPLC analysis, the authors identified a reaction time of 110°C and a reaction time of 60 min as the optimal, obtaining the thiophene derivative **284** in 88% conversion; no residual

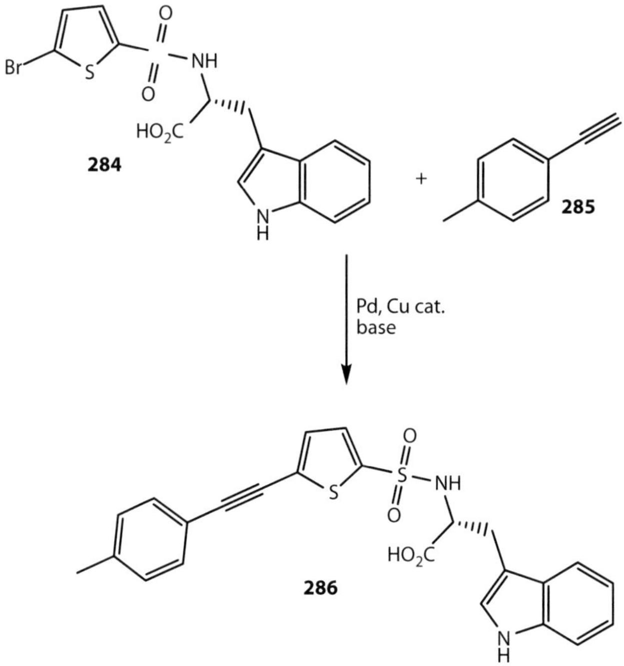

SCHEME 3.70 A Sonogashira coupling reaction performed in an automated micro reactor system.

acetylene **285** was detected owing to a small amount of homocoupling occurring. With this information in hand, the authors performed a second screen, this time investigating the effect of increased acetylene **285** stoichiometry at 110–120°C and various residence times. This time, the authors identified 120°C, 20 min and an acetylene ratio of 1.25 eq and 3 eq. of base as being optimal, affording the target compound **286** in excellent yield. At this stage, the authors operated the reactor for an 8 h period, isolating the product in 84% yield (14 g). In order to increase the productivity of the system, a larger residence time unit was fitted into the system which enabled the flow rate to be increased to 3.14 mL min^{-1} enabling the synthesis of (R)-3-(1H-indol-3-yl)-2-(5-(p-tolylethynyl)thiophene-2-sulfonamido)propanoic acid **286** at a throughput of 18.8 g h^{-1}; demonstrating the synthesis of 113 g of **286** over a 6 h period.

3.7.4 OTHER METAL-CATALYZED COUPLING REACTIONS

Stille Coupling: Using a fused-silica capillary reactor, maintained under thermal control, Weber and coworkers [132] developed a serial loading technique for the performance of multiple reactions under flow conditions and applied the technique to the screening of catalysts toward the Stille coupling reaction depicted in Table 3.53.

The methodology developed involved the injection of 0.75–1 µL of a series of catalyst solutions into a stream of iodobenzene **269** and Bu$_3$SnCH = CH$_2$ **287**. Using a switching valve, the authors were able to spatially resolve the plugs of catalyst

TABLE 3.53
Illustration of the Catalysts Evaluated and the Results Obtained Using a Serial Loading Technique within a Fused Silica Capillary Reactor

269 + Bu$_3$SnCH=CH$_2$ **287** → 277 (Pd-catalyst/ligand, THF, 50°C)

Precatalyst	Ligand	Conversion (%)
Pd$_2$dba$_3$	AsPh$_3$	49.2
Pd[(C$_6$H$_5$)$_3$P]$_4$	AsPh$_3$	38.0
PdCl$_2$(C$_6$H$_5$CN)$_2$	AsPh$_3$	43.8
PdCl$_2$(CH$_3$CN)$_2$	AsPh$_3$	50.7
PdCl$_2$[(C$_6$H$_5$)$_3$P]$_2$	AsPh$_3$	15.9
Pd(OAc)$_2$ **276**	AsPh$_3$	23.0
PdCl$_2$(CH$_3$CN)$_2$	PPh$_3$	28.5
PdCl$_2$(CH$_3$CN)$_2$	(2-furyl)$_3$P	40.0
PdCl$_2$(CH$_3$CN)$_2$	(4-FC$_6$H$_4$)$_3$P	21.1
PdCl$_2$(CH$_3$CN)$_2$	(4-ClC$_6$H$_4$)$_3$P	15.5

enabling the stacking of multiple reactions within the fused-silica capillary reactor (dimensions = 75 μm (i.d.) × 6.7 m (long)). Once all the catalysts were loaded, the flow of reactants was stopped and the reactor was heated to 50°C for 5 h, detection of the serial reactions was subsequently performed by reinstating the flow and analyzing the percentage styrene **277** formed by online GC. As Table 3.53 illustrates, the technique enables a series of catalysts to be evaluated under identical reaction conditions, affording the user the ability to rapidly screen catalyst types, catalyst concentrations, and ligands with ease. In addition, the technique was shown to afford excellent reproducibility (±0.36%) by multiple injections of individual catalyst solutions.

Buchwald–Hartwig Reaction: Using a CYTOS micromixer (CPC, Germany) coupled with a 17 mL residence time unit (Residos®), Mauger et al. [133] investigated the palladium-catalyzed aromatic amination of *p*-bromotoluene **288** with piperidine **289** (Scheme 3.71), based on the early findings of Buchwald and Hartwig.

To perform a reaction, the authors investigated the coupling of *p*-bromotoluene **288** (0.3 M) and piperidine **289** (1.0 eq.) in xylene, employing Pd(OAc)$_2$ **276** and Davephos **290** (5.45 × 10^{-3} M) as the catalyst and ligand, in the presence of the base sodium *tert*-amylate **291** (0.42 M). Employing a residence time of 7.5 min and a reactor temperature of 110°C, the authors were gratified to obtain quantitative conversion of *p*-bromotoluene **288** to *N*-(4-tolyl)-piperidine **292** in >99% selectivity. Unlike analogous batch reactions, no by-products arising from C-Br reduction were observed.

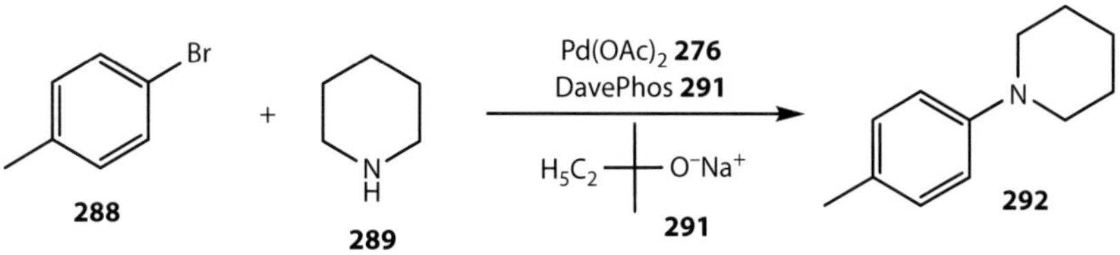

SCHEME 3.71 Schematic illustrating the palladium catalyzed aromatic amination of *p*-bromotoluene **288** performed under continuous flow conditions.

Decreasing the catalyst ratio from 18 to 9 mol%, the authors also obtained quantitative conversion to **292**, this time with an increased reaction time of 11.3 min. In order to confirm the need for a catalyst, the reaction was subsequently repeated in the absence of the catalyst and ligand, whereby a residence time of 3 days failed to produce any *N*-(4-tolyl)-piperidine **293**.

With this in mind, the authors utilized their optimum conditions to synthesize *N*-(4-tolyl)-piperidine **289** at a throughput of 150 g day^{-1}, readily increasing to 400 g day^{-1} (1.3 ton year^{-1}) when two Residos modules (volume = 47 mL) were employed.

3.8 REARRANGEMENTS

Owing to the fact that a rearrangement reaction involves the movement of a substituent from one position to another, typically along a carbon skeleton, to afford a structural isomer, these named reactions provide an atom efficient route to some synthetically complex and interesting structures. With this in mind, several research groups have investigated the performance of a wide number of rearrangement reactions under continuous flow conditions, the findings of which are discussed herein.

3.8.1 CLAISEN REARRANGEMENT

Conventional Heating: Recently, Jia and coworkers [134] exploited the excellent heat transfer obtained within microfluidic devices to conduct a series of thermally induced Claisen rearrangements within a stainless-steel tubular reactor (dimensions = 170 µm (i.d.) × 1.2 m (long)), housed within an oil bath. Using this setup, the effect of reaction temperature (200–220°C) and time (8–24 min) was evaluated for the rearrangement of 4-chlorophenyl allyl ether **293** to 2-allyl-4-chlorophenol **294**; as depicted in Scheme 3.72. At a reactor temperature of 200°C, the authors obtained 35% rearrangement in 24 min, and this was however increased to 82% **294** at 220°C; representing an increase of 68% compared to a batch reaction performed at reflux. In addition to the increase in conversion observed by HPLC, analysis of the crude reaction products by ^{1}H NMR spectroscopy confirmed the presence of only the target phenol **294** and residual starting material **293**; illustrating the cleanliness of the transformation.

Having demonstrated such an increase in reaction efficiency and cleanliness as a result of employing a continuous flow reactor, the authors investigated the rearrangement of several *para*-substituted phenyl ethers, obtaining moderate to high conversions as summarized in Table 3.54.

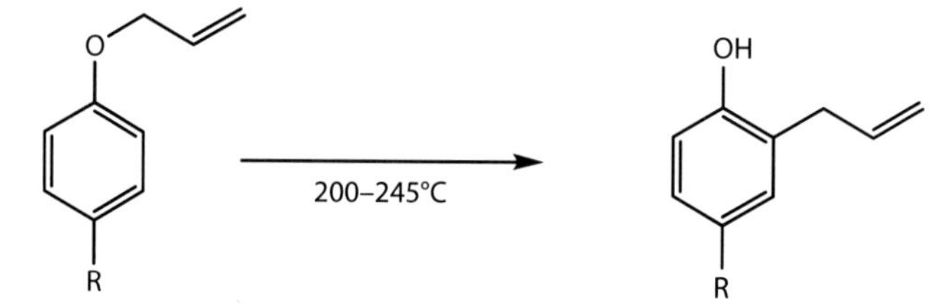

SCHEME 3.72 Schematic illustrating the Claisen rearrangements of allyl *para*-substituted phenyl ethers conducted under continuous flow.

TABLE 3.54
Summary of the Results Obtained for the Continuous Flow Rearrangement of Allyl *Para*-Substituted Phenyl Ethers

Substrate	Product	Reaction Time (min)	Temperature (°C)	Conversion (%)[a]	
				Micro Reaction	Batch Reaction
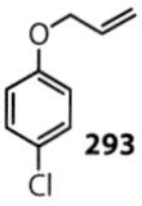 293 (Cl)	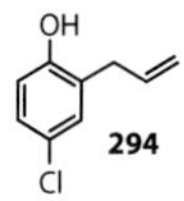 294 (Cl)	24	220	82	14
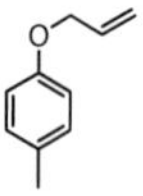 (Me)	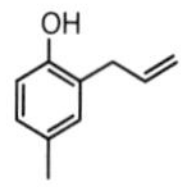 (Me)	36	200	73	2
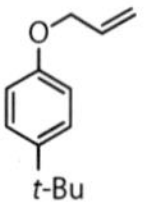 (t-Bu)	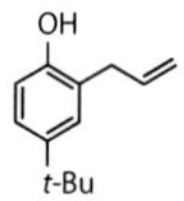 (t-Bu)	36	225	97	39
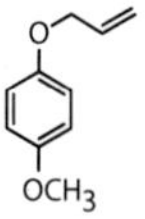 (OCH3)	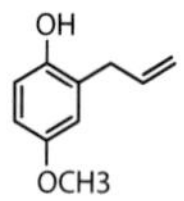 (OCH3)	24	220	Quant.[b]	32
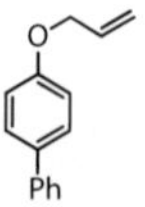 (Ph)	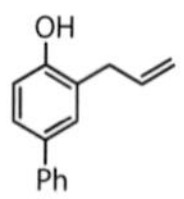 (Ph)	24	240	90[c]	37
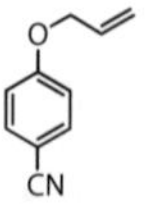 (CN)	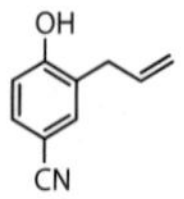 (CN)	24	245	93[c]	43

[a] Determined by HPLC analysis.
[b] No purification required.
[c] Diphenyl ether as solvent.

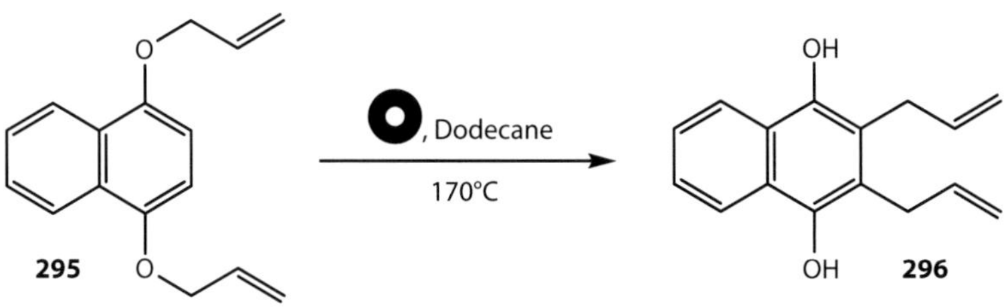

SCHEME 3.73 Claisen rearrangement performed in the presence of magnetic nanoparticles.

In addition to increased reaction yield, the use of a micro reactor for the Claisen rearrangement proved advantageous as the sealed nature of the reactor vessel enabled exclusion of oxygen from the reaction mixture, thus preventing possible oxidation of the reaction products. Furthermore, the use of relatively short reaction times reduced the risks associated with overheating and carbonization, frequently encountered when performing the rearrangements in batch.

In some instances, the reaction temperature was limited by the boiling point of the reactant under investigation; for example, 4-methylchlorophenyl allyl ether has a boiling point of 210°C; as such, the reaction temperature was not increased above 200°C. Increased reaction temperatures could, however, have been accessed by conducting the reaction under pressure, as reported within the literature by Kappe and coworkers [20], whereby temperatures of 240°C (100 bar) have been demonstrated for the rearrangement of phenyl allyl ether (95% yield).

Inductive Heating: Kirschning and coworkers [135] previously demonstrated the Claisen rearrangement of 1,4-bis(allyloxy)naphthalene **295**, to afford 2,3-diallyl-naphthalene-1,4-diol **296** (Scheme 3.73), under continuous flow however, rather than using an oil bath to heat their system, the authors employed inductive heating of magnetic nanoparticles (10–40 nm). As Figure 3.19 illustrates, this was achieved by packing the flow reactor with magnetic silica-coated nanoparticles and placing the reactor in an electromagnetic field. Utilizing a glass flow reactor (dimensions = 9 mm (i.d.) × 14 cm (long)), the authors evaluated the reactor for cyclic and continuous flow operation. Employing a solution of 1,4-bis(allyloxy)naphthalene **295** (0.4 M) in dodecane, at a flow rate of 0.5 mL min^{-1}, and a reactor temperature of 170°C, the authors obtained 2,3-diallylnaphthalene-1,4-diol **296** in 85% yield which represented an increase of 23% compared to a conventionally heated system.

Aqueous Solvent: Using a microflow reactor comprising a T-mixer and tube reactor (SUS316), housed within an electrical furnace, Kawanami and coworkers [136] investigated the effect of reactant concentration, reaction time, temperature, and pressure on the Claisen rearrangement of allyloxybenzene **297** to 2-allylphenol **298** (Table 3.55) in an aqueous solvent. For comparative purposes, the authors conducted the reaction in batch, using conventional heating and microwave irradiation, prior to evaluating a flow process. As Table 3.55 illustrates, employing a residence time of 149 s, at 5 MPa and 265°C, the authors were able to convert 98% of allyloxybenzene **297** to 2-allylphenol **298** with 98% selectivity, compared to 73% yield and 74% selectivity at 81 s. Unlike analogous batch reactions, the use of a flow reactor enabled the

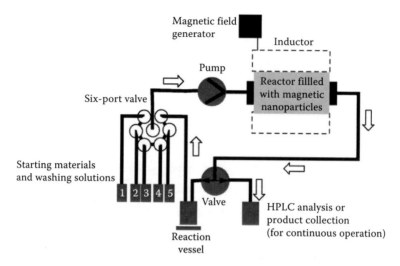

FIGURE 3.19 Schematic illustrating the experimental setup used to exploit the inductive heating of magnetic nanoparticles in an electromagnetic field.

reaction to be performed without significant side reactions such as hydrolysis, hydration, or pyrolysis dominating. As an extension to this, the Johnson–Claisen reaction between cinnamyl alcohol **299** and triethyl orthoformate **300** was also investigated, whereby the target compound, ethyl 3-phenylbut-3-enoate **301**, was obtained in 95% yield (Scheme 3.74).

TABLE 3.55
Comparison of Various Methodologies Evaluated for the Noncatalytic Claisen Rearrangement

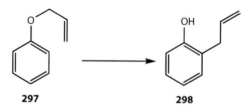

Method	Concentration (mol kg^{-1})	Temperature (°C)	Pressure (MPa)	Time	Selectivity (%)	Yield (%)
Conventional heating	7.5	220	0.1	6 h	—	85
Microwave	—	325–361	0.1	10 min	—	21
Flow (solvent free)	6.9	265	5.0	360 s	68	37
Flow (H$_2$O)	0.77	265	5.0	81 s	74	73
Flow (H$_2$O)	0.27	265	5.0	149 s	98	98

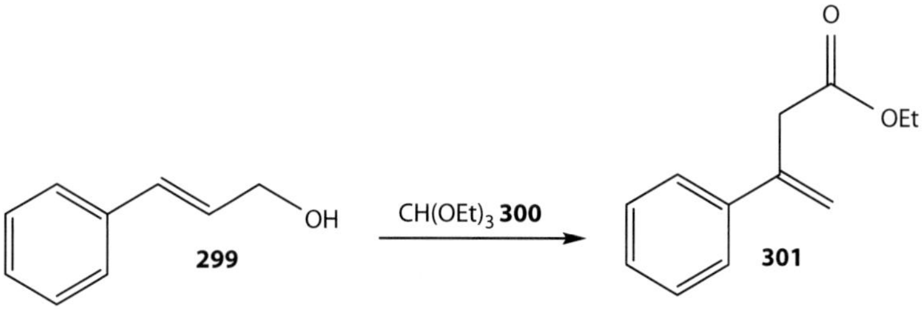

SCHEME 3.74 Illustration of the Johnson–Claisen reaction performed under high temperature and pressure in the absence of a catalyst.

3.8.2 NEWMAN–KWART REARRANGEMENT

In response to the need to generate a quantity of 4′-*tert*-butyl-2,6-dimethylbiphenyl-4-thiol **302** for an early-stage development project, Tilstan et al. [137] utilized previously reported continuous flow methodology [138,139] as a means of synthesizing the biaryl target **302**, via a Newman–Kwart rearrangement (Figure 3.20), in volumes inaccessible using the conventional batch protocol.

The continuous flow reactor employed for this transformation was an in-house fabricated system and comprised a high-pressure HPLC pump connected to a stainless-steel tube (1/18″ o.d. × 23 m long) reactor, housed within an oven. The system contained five serial back-pressure regulators to enable the effect of reaction temperature to be evaluated over the range of 250–320°C.

Prior to evaluating the Newman–Kwart rearrangement under flow conditions, 4′-*tert*-butyl-2,6-dimethylbiphenyl-4-ol **303** was converted into the respective *O*-thiocarbamate **304**; recrystallization from cyclohexane afforded the precursor **304** in high purity and 79% yield (Scheme 3.75).

As Table 3.56 illustrates, conducting the rearrangement in diglyme, at 2 mL min^{-1}, the authors obtained quantitative conversion of *O*-thiocarbamate **304** to *S*-thiocarbamate **305** at a reactor temperature of 320°C, with an accompanying 10%

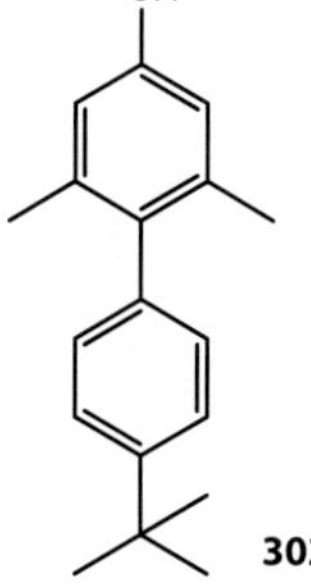

FIGURE 3.20 Illustration of the target compound, 4′-*tert*-butyl-2,6-dimethylbiphenyl-4-thiol **302**, that formed the basis of the continuous flow development project.

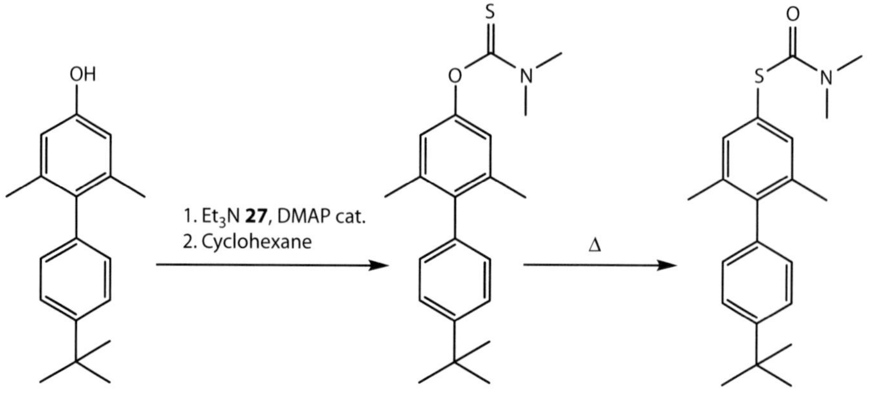

SCHEME 3.75 Reaction protocol employed for the synthesis of 4′-*tert*-butyl-2,6-dimethyl-biphenyl-4-thiol **303** under continuous flow.

unidentified by-product. Employing a lower reaction temperature (280°C) and a longer reaction time (1 mL min^{-1}), *S*-thiocarbamate **305** was prepared in 98% conversion and 97% purity on a 10 g scale. Unfortunately, isolation of the product from large quantities of diglyme was not found to be straightforward; consequently, the authors investigated the use of a lower boiling ether as solvent; dimethoxyethane (DME).

This time employing a reaction temperature of 300°C, the authors obtained 99.1% conversion of **304** to **305**, with the remaining 0.9% comprising of unreacted **304** (0.6%) and an unidentified impurity (0.3%). Through using DME as the reaction solvent, the authors found product isolation was facile, achieved by distillation of the DME followed by crystallization of **305** (upon addition of *n*-heptane). Using this approach, the target *S*-thiocarbamate **306** was obtained in 93% isolated yield,

TABLE 3.56
Summary of the Results Obtained for the Newman–Kwart Rearrangement Conducted in Diglyme

Reactor Temperature (°C)	Conversion to 305 (%)
250	50[a]
280	80[a] (98[b])
300	87
320	100 (10% impurity)

[a] Flow rate = 2 mL min^{-1}.
[b] 1 mL min^{-1}.

TABLE 3.57
Illustration of the Newman–Kwart Rearrangements Performed in a High-Temperature and High-Pressure Tubular Reactor

Product (R)	Temperature (°C)	Pressure (bar)	Yield (%)
CN	220	60	>99
OMe	300	80	>99

99.2% purity at a throughput of 1.5 kg 24 h^{-1}. In a second example, Kappe and coworkers [20] demonstrated the use of a stainless-steel reactor for the transformation (Table 3.57).

3.8.3 HOFMANN REARRANGEMENT

Owing to the fact that the Hofmann rearrangement has recently found application in key synthetic steps in the formation of natural products [140,141] and pharmaceutical agents, such as Tamiflu [142], Palmieri et al. [143] investigated the performance of this synthetically useful reaction in a commercially available fused silica tube reactor (Advion NanoTek LF™, USA), complete with 200–400 µL reactant loops.

Preliminary experiments focused on identifying the optimum reaction conditions for the reaction including reactant concentration, solvent, stoichiometry, base, reaction temperature, and reaction time; performed on a 50–100 µg scale. Upon completion of this initial screen, the authors identified DBU **41** (2 eq.), *N*-bromosuccinimide (NBS) **142**, ethanol or methanol as the solvent, 120°C reactor temperature and a residence time of 1 min (flow rate = 15 µL min^{-1}) as the optimal conditions for the transformation. As Table 3.58 illustrates, using this approach the authors were able to rearrange a series of amides, selected to highlight any substituent effects, to afford fourteen carbamates in moderate to high yield (41–80%); isolated after purification of the bulk collected reaction products through a short silica-gel column. Using this approach, the authors demonstrated a dramatic reduction in reaction time (25-fold) compared to an analogous reaction performed under optimized batch conditions.

In a second tubular flow reactor (Uniqsis FlowSyn™, UK), with a volume of 20 mL, the authors subsequently investigated the scale-up of the Hofmann rearrangement for the synthesis of methyl phenylcarbamate **306**, methyl-*p*-tolylcarbamate **307,** and methyl-*m*-tolylcarbamate **308** isolating the target compounds in 88%, 74%, and 37% yield, respectively, compared to 79%, 80%, and 61% in the Advion system.

TABLE 3.58
Summary of the Results Obtained for a Series of Hofmann Rearrangements Performed in a Heated, Continuous Flow Reactor

$$R-C(=O)-NH_2 \xrightarrow[R^1OH]{NBS\ 142,\ DBU\ 41} R-NH-C(=O)-OR^1$$

R	R¹	Product Yield (%)
Ph	Me	79 **306**
4-MeO-C₆H₄	Me	80
3-MeO-C₆H₄	Me	62
2-naphthyl	Me	71
4-MeO-C₆H₄	Et	46
Ph	Et	32
4-Me-C₆H₄	Me	80 **307**
2-Me-C₆H₄	Me	78
3-Me-C₆H₄	Me	67 **308**
2-EtO-C₆H₄	Me	74
4-Cl-C₆H₄	Me	77
3-F-C₆H₄	Me	41

TABLE 3.58 (continued)
Summary of the Results Obtained for a Series of Hofmann
Rearrangements Performed in a Heated, Continuous Flow Reactor

R	R¹	Product Yield (%)
MeO₂C-C₆H₄-	Me	55
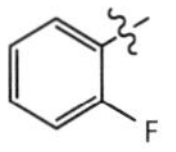 2-F-C₆H₄-	Me	57

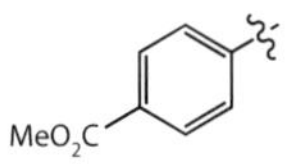

3.8.4 Fisher Indolization

In 2005, Bagley et al. [97] demonstrated the development of a simple, continuous flow microwave reactor, in which they investigated a series of microwave-assisted organic reactions; including the Fischer Indole synthesis (Scheme 3.76). Employing a glass tube reactor, packed with sand, to afford a network of microchannels, housed within a microwave cavity the authors were able to evaluate the effect of microwave irradiation on the indolization of phenyl hydrazine **309** (0.55 M) and cyclohexanone **22** (0.5 M) to afford 2,3,4,9-tetrahydro-1H-carbazole **310**. Using 150 W, the authors were able to heat the reactor, measured by an *in situ* IR probe, to 150°C affording the target compound **310** in 91% yield with a throughput of 2 g h⁻¹.

Employing a stainless-steel, coiled tube reactor (dimensions = 1000 µm (i.d.), tube volume = 4 mL or 16 mL), Kappe and coworkers [20] also investigated the effect of reactor temperature on the indolization of phenyl hydrazine **309** (0.5 M) and cyclohexanone **22** (0.5 M) using acetic acid/2-propanol (3:1) as the reaction solvent (Scheme 3.76).

Using HPLC detection, the authors identified a reactor temperature of 200°C, pressure of 75 bar and a residence time of 3 min as the optimal conditions for the indolization, obtaining the tetrahydrocarbazole **310** in 96% yield; similar to the results obtained previously by Bagley et al. [97].

Having identified the optimal conditions, the authors subsequently increased the volume of the tube reactor, from 4 to 16 mL, and demonstrated the synthesis of 25 g of tetrahydrocarbazole **310** with 1 h of continuous processing.

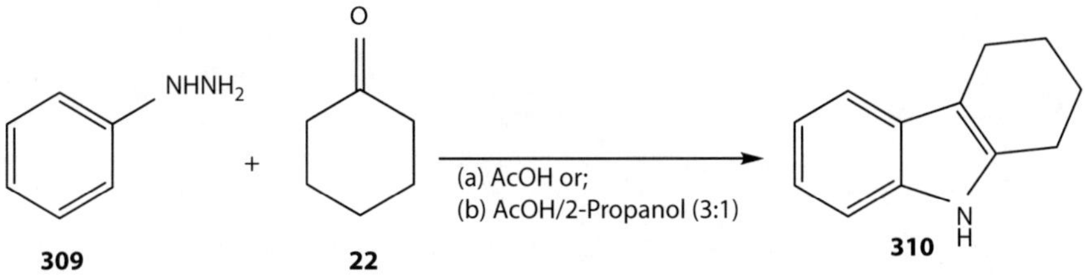

SCHEME 3.76 Schematic illustrating the Fischer indole synthesis performed in (a) a microwave flow reactor and (b) a pressurized flow reactor.

TABLE 3.59
Summary of the Results Obtained for the Fisher Indolization Performed in a Series of Glass Micro Reactors

Ketone	Product	Batch	Flow (Solution Phase)	Flow (Solid-Supported)
			Yield (%)	
(methyl ethyl ketone)	(2,3-dimethylindole)	88	96	98
(ethyl acetoacetate)	(ethyl indole-2-carboxylate)	68	88	98
(cyclopentanone)	(tetrahydrocyclopenta[b]indole)	69	68	74
(cyclohexanone)	(tetrahydrocarbazole)	76	52	56

As part of a research project into continuous flow radio syntheses, Wahab et al. [144] required a facile route to the indole core motif. With this in mind, the authors investigated a series of methods for the solution phase synthesis of substituted indoles using glass micro reactors. Employing glacial acetic acid as the reaction solvent and sulfuric acid (10%) as the catalyst, the authors identified an optimal reaction temperature of 105°C afforded the target compounds in moderate-to-excellent chromatographic yield, as Table 3.59 illustrates, with throughputs ranging from 1.9 to 2.3 mg h^{-1}.

Based on their need to perform subsequent derivatization of the indoles formed, the authors found the use of acetic acid problematic due to difficulties associated with its efficient removal. With this in mind, the authors investigated the use of an ion exchange resin (Amberlite-IR-120) to promote the indolization, enabling the use of EtOH as the reaction solvent. Employing a reactor temperature of 70°C, the authors obtained the target compounds in yields ranging from 56% to 98%, with increases in all cases observed compared to the solution phase approach. In addition to affording ease of product isolation, compared to the solution phase technique this solid-supported methodology proved advantageous, resulting in a 2.2-fold increase in reactor throughput for all compounds studied; typically 13–20 mg h^{-1}.

3.8.5 CURTIUS REARRANGEMENT

In 2007, Sahoo et al. [145] reported the development of a flow protocol for the synthesis of carbamates via the Curtius rearrangement of isocyanates, as generalized in

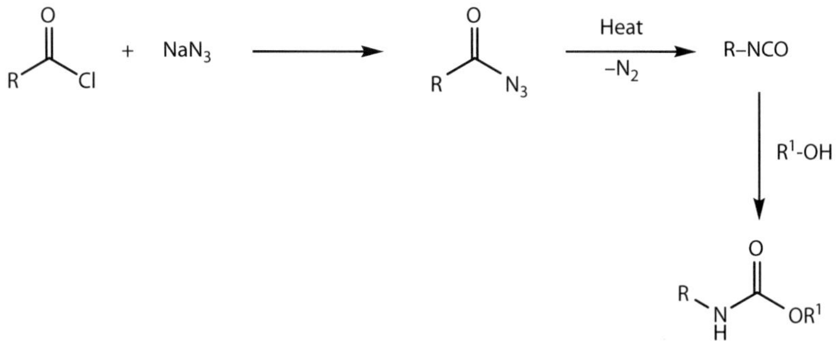

SCHEME 3.77 Schematic illustrating the general reaction protocol employed for the synthesis of carbamates via the Curtius rearrangement.

Scheme 3.77. The authors selected carbamates as a synthetic target, not only because they serve as useful building blocks, but also require the use of hazardous azides as reactive intermediates and the separation of N_2, evolved during their thermal decomposition.

Utilizing a silicon-glass micro reactor, the authors employed a phase transfer reaction for the conversion of an acyl chloride (toluene) to an acyl azide, using sodium azide **79**. Coupling the reactor outlet to a membrane separator, the authors readily removed water from the reaction mixture, prior to thermal rearrangement of the acyl azide (in the presence of a solid acid catalyst) to afford the respective isocyanate. The nitrogen evolved during this process was then removed from the reaction stream using a gas–liquid separator and the resulting liquid stream reacted with a series of alcohols to afford the respective carbamate. Using this approach, the authors were able to efficiently synthesize phenyl isocyanate at a reactor temperature of 105°C with a residence time of 60 min, affording the target compounds at a throughput of 80–120 mg day^{-1}. For a description of the separation techniques developed by the authors, refer to Chapter 8.

3.8.6 DIMROTH REARRANGEMENT

Using a microwave-assisted continuous flow reactor, comprising a 10 mL glass vial filled with glass beads (2 mm), designed to afford a device that contained a series of microchannels, Kappe and coworkers [146] investigated the Dimroth rearrangement of a series of 2-amino-6*H*-1,3-thiazines as a means of accessing otherwise difficult to prepare dihydropyrimidine-2-thiones (DHPM's). Initially performing the reactions in batch, using sealed tube reactors, the authors identified that unsubstituted thiazines (R^1 = H) more readily rearranged in toluene at concentrations of 0.1 M (210°C), with *N*-substituted thiazines transformed using NMP as the solvent at 200°C.

Performing reactions under continuous flow, the authors found that they were able to generate the target DHPM's in 88% yield, by pumping solutions of thiazine (0.17 M) through the preheated flow cell (200°C) at a flow rate of 0.33 mL min^{-1}. The ability to transfer batch reactions to flow therefore provided the authors with a

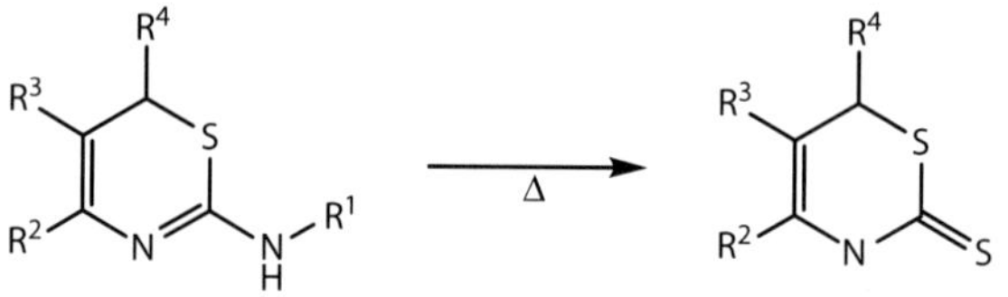

SCHEME 3.78 Generalized schematic illustrating the Dimroth rearrangement as a means of preparing dihydrpyriminine-2-thiones.

scalable technique (0.34 mmol h^{-1}) for the synthesis of such structurally interesting compounds (Scheme 3.78).

3.9 MULTISTEP/MULTICOMPONENT LIQUID–PHASE REACTIONS

3.9.1 Multicomponent Synthesis of Heterocycles

Building on their experience in the successful implementation of microwave-assisted continuous flow organic synthesis (MACOS), Bremner and Organ [147] demonstrated numerous advantages associated with application of this technique toward a series of multicomponent reactions, for the preparation of medicinally relevant heterocyclic compounds.

In the first instance, the authors investigated the synthesis of tetrahydropyrazolo[3, 4-*b*]quinolin-5(6*H*)-ones, achieved via the reaction of equimolar quantities of dimedone **311**, 5-amino-3-methyl-1*H*-pyrazole **312** with a series of substituted benzaldehydes, as illustrated in Table 3.60.

Employing an in-house fabricated reactor, comprising a stainless-steel mixer, connected to a glass capillary reactor (dimensions = 1180 μm (i.d.)), housed within a single-mode microwave chamber, the authors evaluated the effect of residence time and temperature on the formation of the target quinolines. Reactants were introduced into the reactor from three separate inlets, as solutions in DMF (5.0 M) at a total flow rate of 60 μL min^{-1}, in the absence of microwave irradiation only a trace amount of product was observed, increasing to >91% conversion at 170 W.

As Table 3.60 illustrates, in the presence of microwave irradiation, a series of quinoline derivatives were obtained in moderate-to-excellent yield, after removal of the reaction solvent, purification by silica gel chromatography and recrystallization from EtOH. Due to the high solvating capacity of DMF, the reactions were able to be performed at high concentrations affording throughputs in the range of 6 mmol h^{-1}.

In second example, the authors evaluated the use of their MACOS approach as a tool in the preparation of a series of aminofurans. As Table 3.61 illustrates, the technique was found to be tolerant to a diverse array of functionalities on the substituted benzaldehyde, with no efforts made to vary the acetylene derivative, dimethylacetylene dicarboxylate (DMAD) **313**, or isocyanide, cyclohexylisocyanide **314**. Again, moderately high concentrations of reactants were employed, using DMF as the reaction solvent and a stoichiometry of 1:1.2:1.2 aldehyde:acetylene:isocyanide, with reactions performed at a total flow rate of 60 μL min^{-1} and 180 W. The reaction products were again subjected to offline purification using silica gel chromatography

TABLE 3.60
Summary of the Results Obtained for the Synthesis of Tetrahydropyrazolo[3,4-B]Quinolin-5(6H)-Ones Performed Using MACOS

R	Conversion (%)[a]	Isolated Yield (%)[b]
N(CH$_3$)$_2$	95	94
CN	100	55
CO$_2$Me	100	88
Br	100	80
OH	94	94
OMe	91	71

[a] Conversion determined by ^{1}H NMR spectroscopy.
[b] Isolated yield determined after purification by silica gel chromatography.

and recrystallization from EtOH, affording the tetrasubstituted furans in moderate-to-high isolated yield.

Using this approach, the authors found it quicker to optimize reactions conditions when compared with the iterative approach of batch reactions; the authors also comment on the techniques potential for the production of chemicals via the use of capillary bundles [148].

3.9.2 MULTISTEP SYNTHESIS OF 1,2,4-OXADIAZOLES

In 2008, Cosford and coworkers [149] employed two glass micro reactors and a capillary reactor in series to perform the multistep synthesis of 1,2,4-oxadiazoles, as depicted in Scheme 3.79. The core motif was selected for investigation as it is found in a series of biologically active molecules, such as the S1P1 agonist **315** and mGlu5 receptor antagonist **316** illustrated in Figure 3.21. In a typical batch protocol, the arylnitrile **317** is reacted with hydroxylamine hydrochloride **82** to afford the aldoxime intermediate **318** which subsequently cyclizes with an acyl chloride **319** to afford the respective 1,2,4-oxadiazole **320** in low to moderate yield. While the aldoxime **322** is formed readily, the cyclization is often problematic, requiring the use of high reaction temperatures (sealed tube) and long reaction times.

With this in mind, the authors embarked upon the development of a continuous flow protocol as a means of gaining rapid access to the 1,2,4-oxadiazoles **320** in

TABLE 3.61
Summary of the Results Obtained for the Synthesis of Tetrasubstituted Furans Using MACOS

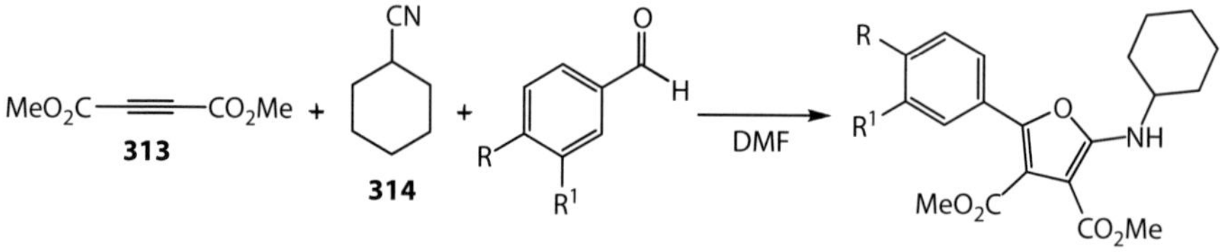

$MeO_2C{\equiv}CO_2Me$ (**313**) + cyclohexyl isocyanide (**314**) + ArCHO → tetrasubstituted furan (DMF)

R	R^1	Conversion (%)[a]
NO_2	H	83 (79)[b]
H	NO_2	70
CF_3	H	76 (76)[b]
CO_2Me	H	71
F	H	57
Cl	H	55
OMe	H	30

[a] Conversion determined by ^1H NMR spectroscopy.
[b] Isolated yield determined after purification by silica gel chromatography.

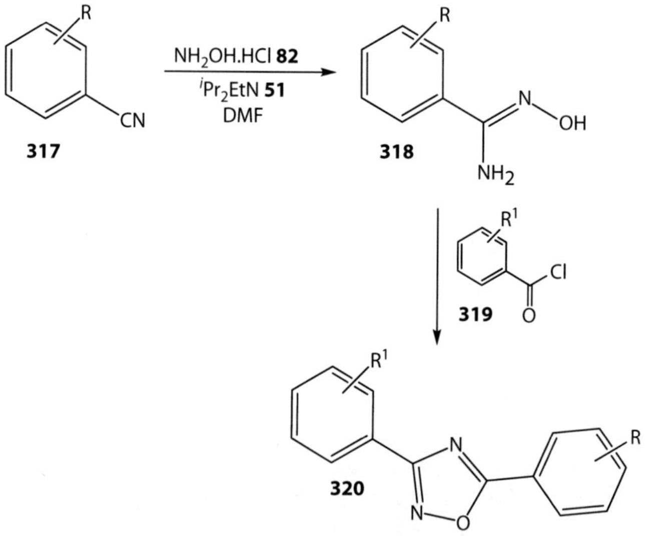

SCHEME 3.79 General reaction scheme illustrating the reaction protocol employed for the continuous flow synthesis of 1,2,4-oxadiazoles.

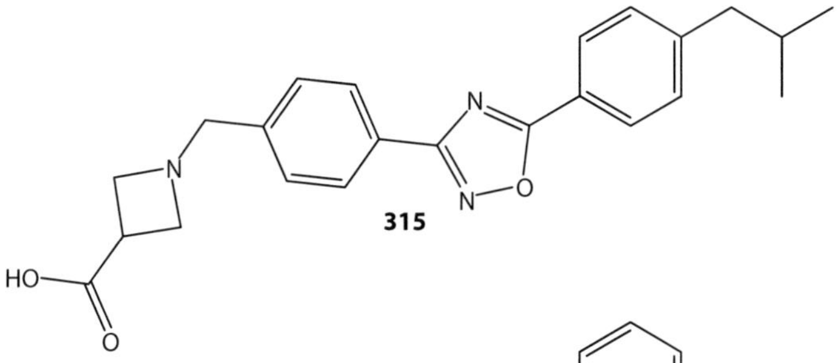

FIGURE 3.21 Illustration of biologically active 1,2,4-oxadiazoles.

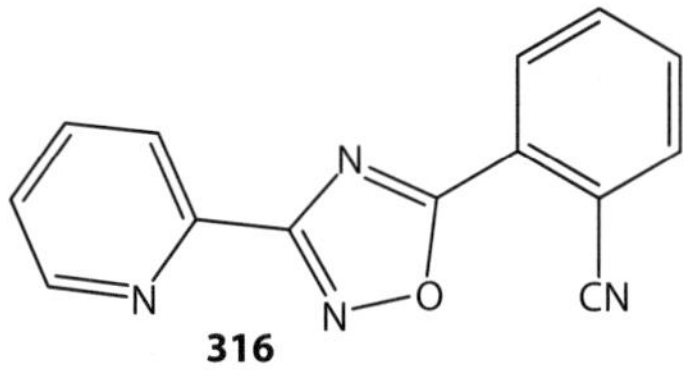

higher yield and purity. Employing DMF as the reaction solvent and a glass micro reactor (1000 µL), the authors initially investigated the formation of the aldoxime **318** by reacting the arylnitrile **317** (0.5 M, 65 µL min^{-1}) with H$_2$NOH.HCl **82** (0.4 M, 95 µL min^{-1}) in the presence of Hunig's base **51** (1.2 M) at 150°C. Analysis of the reaction products using LC-MS confirmed quantitative conversion of the arylnitrile **317** to the aldoxime **318** was achieved when employing a reaction time of 6 min. In a separate reactor, the authors investigated the cyclization of the arylnitrile **317** with an aryl chloride **319** (1.0 M), finding that the reaction proceeded to completion with a reaction time of 10 min and a reactor temperature of 200°C (7.5–8.5 bar).

Coupling the two reaction steps together was however found to be problematic, and upon combining the output stream of the first reactor with a second reactor, in which the cyclization was performed, the reaction was found to be unsuccessful. Upon investigating the reaction further, it was found to be necessary to cool the aldoxime **318** (0°C) prior to addition of the aryl chloride **319** which was reacted with the aldoxime **318** at room temperature for 2 min prior to heating of the reaction mixture to 200°C. Using this approach, the authors were able to perform the multistep synthesis illustrated in Scheme 3.79 affording the bis-substituted 1,2,4-oxadiazoles **3Me** in isolated yield ranging from 40% to 63% (Table 3.62).

3.9.3 Continuous Flow Synthesis of Ibuprofen

As part of their research into the development of new methodologies for the rapid and efficient synthesis of important small molecules, McQuade and coworkers [150] recently disclosed the results of their investigation into the continuous flow synthesis of Ibuprofen **321**. As Scheme 3.80 illustrates, the reaction pathway selected involved a Friedel-Crafts acylation, followed by a 1,2-aryl migration and an ester hydrolysis to afford the target compound **321** over three reaction steps. When conducted in batch, it was not found to be possible to perform the reactions in a single flask due to

TABLE 3.62
A Selection of the Results Obtained for the Multistep Synthesis of 1,2,4-Oxadiazoles under Continuous Flow

Arylnitrile	Aryl Chloride	1,2,4-Oxadiazole	Yield (%)[a]
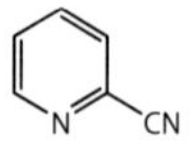 2-pyridyl-CN	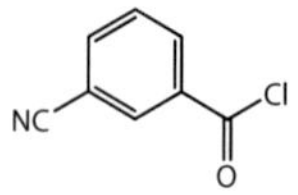 3-NC-C6H4-COCl	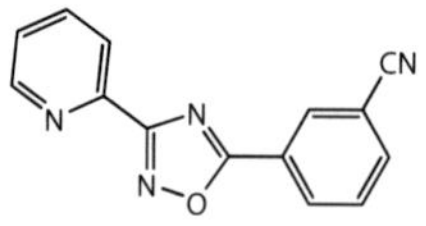 3-(2-pyridyl)-5-(3-cyanophenyl)-1,2,4-oxadiazole	45
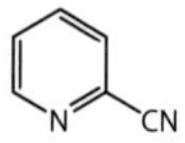 2-pyridyl-CN	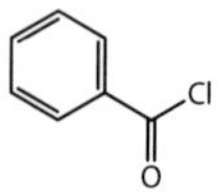 C6H5-COCl	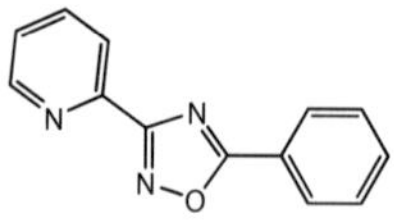 3-(2-pyridyl)-5-phenyl-1,2,4-oxadiazole	45
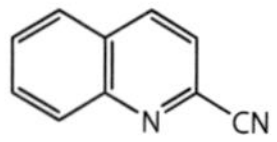 2-quinolyl-CN	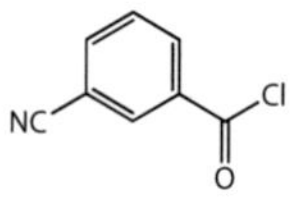 3-NC-C6H4-COCl	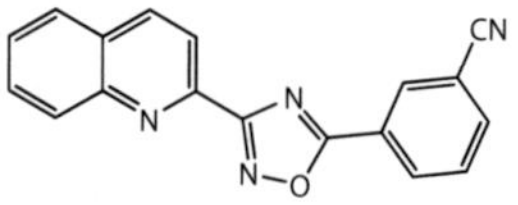 3-(2-quinolyl)-5-(3-cyanophenyl)-1,2,4-oxadiazole	63
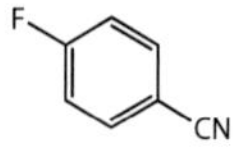 4-F-C6H4-CN	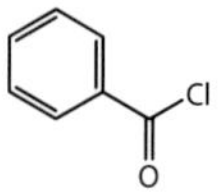 C6H5-COCl	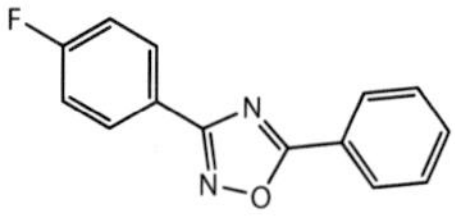 3-(4-fluorophenyl)-5-phenyl-1,2,4-oxadiazole	63
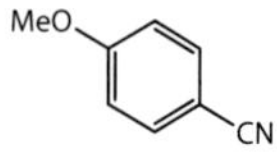 4-MeO-C6H4-CN	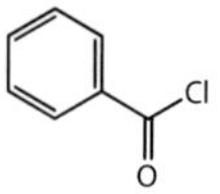 C6H5-COCl	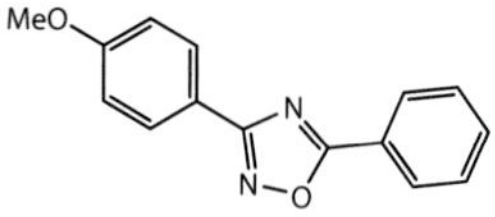 3-(4-methoxyphenyl)-5-phenyl-1,2,4-oxadiazole	40

[a] Isolated yield after purification using preparative HPLC.

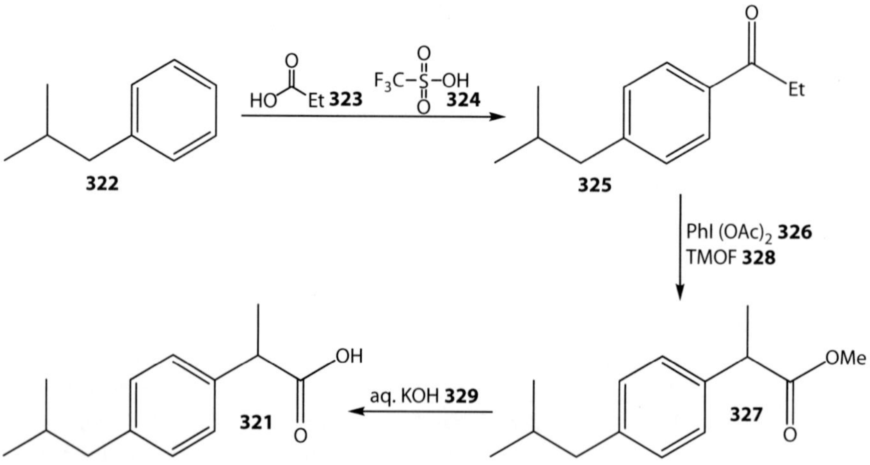

SCHEME 3.80 Schematic illustrating the reaction pathway investigated for the multistep synthesis of Ibuprofen **321** in a PFA tubular reactor.

the exothermic nature of the addition of the acidic reaction mixture to a base in order to perform the saponification step. It was however envisaged that performing the reaction within a tubular micro reactor, comprising PFA tubing (750 μm i.d.) and ETFE interconnects, control over reaction exotherms could be obtained thus, affording a facile and safe method of producing the nonsteroidal anti-inflammatory drug Ibuprofen **321**; without the intensive energy requirements of a conventional batch process.

The first step of the reaction sequence involved the Friedel–Crafts acylation of *iso*-butylbenzene **322** (4.3 M) with propionic acid **323** (4.3 M), in the presence of the catalyst triflic acid **324** (11.3 M). Employing a reaction time of 5 min (reactor volume = 220.9 μL, total flow rate = 43.8 μL min^{-1}) and reactor temperatures ranging from 50 to 150°C, the authors obtained 1-(4-*iso*-butylphenyl)ethanone **325** in conversions ranging from 15 to 91% as determined by offline GC analysis using an internal standard.

Having optimized the acylation step, the authors went on to investigate the iodobenzene diacetate (PhI(OAc)$_2$) **326**-mediated 1,2-aryl migration to afford methyl-2-(4-*iso*-butylphenyl)propanoate **327**, finding that 1 equivalent of PhI(OAc)$_2$ **326** and 4 equivalents of trimethylorthoformate **328** (0.5 and 2.0 M, respectively) afforded the target ester **327** in 70% yield, with a reactor temperature of 50°C and a residence time of 2 min (reactor volume = 353.4 μL, total flow rate = 175.3 μL min^{-1}). During the optimization of this step, the authors found it necessary to cool the reaction products of the acylation step to 0°C prior to performing the 1,2-aryl migration to prevent off-gassing; this was achieved via cooling of the T-connector in an ice bath.

The final reaction step involved the saponification of the methyl ester **327** with potassium hydroxide **329** (0.5 M in MeOH:H$_2$O 4:1) and was performed at 65°C with a residence time of 3 min (reactor volume = 1325.4 μL, total flow rate = 435.3 μL min^{-1}); affording the crude target compound **321** in a throughput of 9 mg h^{-1}. The reaction products were collected in a 100 mL round-bottomed flask and DI H$_2$O (25 mL) added, prior to removal of the reaction solvent under reduced pressure. The resulting aqueous phase was extracted using diethyl ether to afford Ibuprofen **321** as a pale orange solid in 68% yield and 96% purity (determined by GC/GC-MS). The crude material **321** was then treated with activated carbon to afford Ibuprofen **321** as an off-white solid in 51% yield and 99% purity.

3.9.4 Cation-Mediated Sialylation Reactions

Using an IMM-micromixer (Germany) and stainless-steel tubular reactor, Tanaka and Fukase [151] reported the efficient, large-scale preparation of bioactive natural products, focusing on the synthesis of asparagine-linked oligosaccharides.

Utilizing highly reactive sialyl donors, such as the C-5 cyclic imides **330** and **331**, illustrated in Scheme 3.81, the authors previously illustrated an efficient α-sialylation methodology for the preparation of disaccharides. Although successful in batch on a small scale (50 mg, 92% yield **332**), when increasing the reaction size to 100 mg, the authors observed a dramatic reduction in yield (60% **332**) and accompanying by-product (glycal) formation. Based on these findings, the authors evaluated the reaction

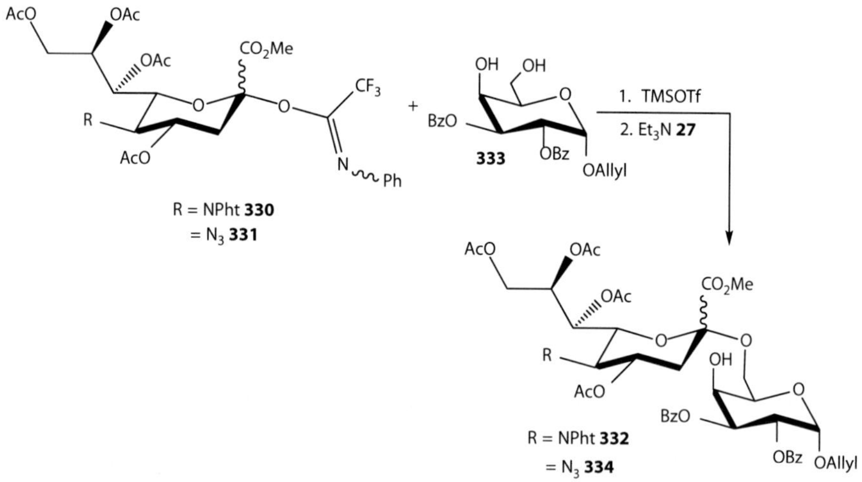

SCHEME 3.81 Illustration of the reaction protocol employed for the α(2–6)-sialylation performed under continuous flow.

of *N*-phthalimide **330** under continuous flow as a means of developing a scale-independent route to these disaccharides.

To achieve this, the authors employed an IMM micromixer (channel width = 40 μm) coupled to a tubular reactor (dimensions = 1.0 mm (i.d.) × 1.0 m (long)) in which a propionitrile solution of donor **333** and acceptor **336** were mixed with TMSOTf **337**, in DCM, at −78°C. After a residence time of 47 s, the reaction mixture was quenched with Et$_3$N **27** and the reaction products analyzed to determine the yield of disaccharide **335** formed.

As Table 3.63 illustrates, the concentration of donor (**333** and **334**), acceptor (**336**), along with TMSOTf **337** stoichiometry had a dramatic effect on the reaction yield. Optimal conditions of donor (0.2 M), acceptor (0.1 M), and TMSOTf **337** (0.15 M) were found to afford the target α-sialoside **335/338** in >99% yield and high

TABLE 3.63
Summary of the Optimization Process Used for the Continuous Flow Synthesis of α-Sialosides

Donor Concentration 330 (M)	Acceptor Concentration 333 (M)	TMSOTf Concentration 337 (M)	Yield of α-Sialoside (%)	α:β
0.15	0.1	0.08	332 14	α only
0.15	0.1	0.15	332 88	α only
0.2	0.1	0.15	334 > 99	α only
0.2	0.1	0.15	334 > 99	20:2

α-selectivity; affording access for the first time to these compounds efficiently on a 5 to 10 g scale.

3.9.5 Oligosaccharide Synthesis

Building on their experience of disaccharide synthesis under continuous flow [152], Seeberger and coworkers [153] investigated the synthesis of oligosaccharides, such as homotetramer **335**, using the iterative steps outlined in Scheme 3.82. Initially, the authors focused on optimizing the reaction conditions required for the glycosylation step and subsequent Fmoc-deprotection, this was achieved employing a silicon-glass microreactor, with an internal volume of 78.3 μL capable of mixing three reactant streams and performing *in situ* reaction quenching.

In order to optimize the protocol for the glycosylation step, the authors introduced a solution of the nucleophile into the reactor from inlet 1, glycosyl phosphate **336** (2.0 eq.) from inlet 2 and the activator, TMSOTf **337** (2.0 eq.) from inlet 3. Employing a deprotective quench, consisting of base (25% in DMF) and TBAF **4**, the effect of reactant residence time (10 s to 10 min) and temperature (0–20°C) was investigated. The aforementioned reaction screen identified a residence time of 30 s and a reactor temperature of 20°C as being optimum for the synthesis of monoglycoside **338** in 99% yield, representing a significant improvement compared to conventional methodology whereby reaction conditions of 30 min, at −78 to −40°C were employed.

After fluorous solid-phase extraction (FSPE) and treatment with silica gel, the monoglycoside **338** was reacted with glycosyl phosphate **336** to afford the respective disaccharide **339** in 97% yield, with a residence time of 20 s (20°C). Until now,

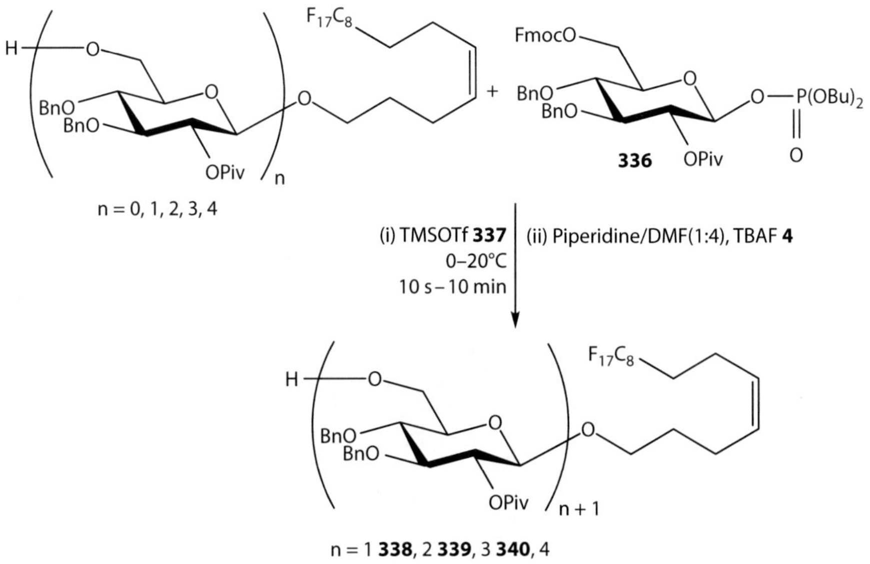

SCHEME 3.82 Illustration of the iterative process used to synthesize oligosaccharides under continuous flow.

residual starting material was detected within the reaction product; as such the authors increased the proportion of glycosyl phosphate **336** and TBAF **4** which afforded the trisaccharide **340** in 90% yield (60 s) and in the final step, the tetrasaccharide was obtained in 95% yield.

3.9.6 Synthesis of Indole Alkaloids Using Metal-Coated Capillary Reactors

Employing a metal-coated capillary flow reactor (dimensions = 1180 μm (i.d.)), Organ and coworkers [154] investigated the use of a two-step aryl amination/cross-coupling reaction sequence as a means of preparing a series of indole alkaloids, as illustrated in Scheme 3.83.

Employing a metal-coated microcapillary and a premixed reaction mixture containing the alkene (1.2 eq.), 2-bromoaniline **341** (1 eq.), Pd-PEPPSI-IPr **342** (2.5 mol%) and sodium *tert*-butoxide **343** (3.0 eq.) in toluene, the authors initially investigated the effect of microwave irradiation on the formation of the respective indole; determined by ^{1}H NMR spectroscopy. Using this approach, the authors concluded that a Pd-coating was more effective than Au (48%), affording quantitative conversion of 2-bromoaniline **341** to the target indole.

In order to identify the role of the Pd coating, that is, catalytic or thermal effect, the authors performed the reaction in the absence of the catalyst **342**, whereby no reaction was observed and with the catalyst in the absence of the coating, again no product was formed. Based on these observations, the authors concluded that the reaction was catalyzed in the presence of Pd-PEPPSI-IPr **342** and the metal coating served to couple the microwave irradiation and enhance heating of the reaction mixture within the capillary reactor.

With this information in hand, the authors investigated the scope of the optimized reaction conditions, finding them to be applicable to a range of coupling partners, as depicted in Scheme 3.84, leading the authors to synthesize 21 substituted indoles, in high-to-excellent isolated yield; after purification by column chromatography.

3.9.7 Iododeamination under Flow

As previously discussed, the generation or use of diazonium intermediates in industrial-scale synthetic processes requires extensive evaluation and assessment prior to implementation; as such, the development of safe and efficient alternatives have been

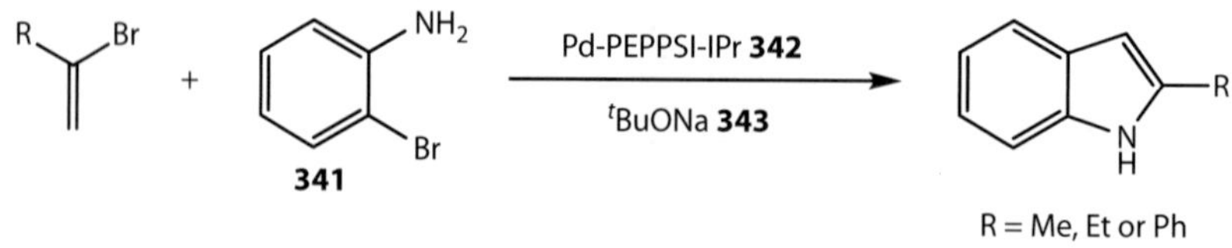

SCHEME 3.83 General scheme illustrating the synthesis of indole alkaloids performed using a metal-coated capillary reactor.

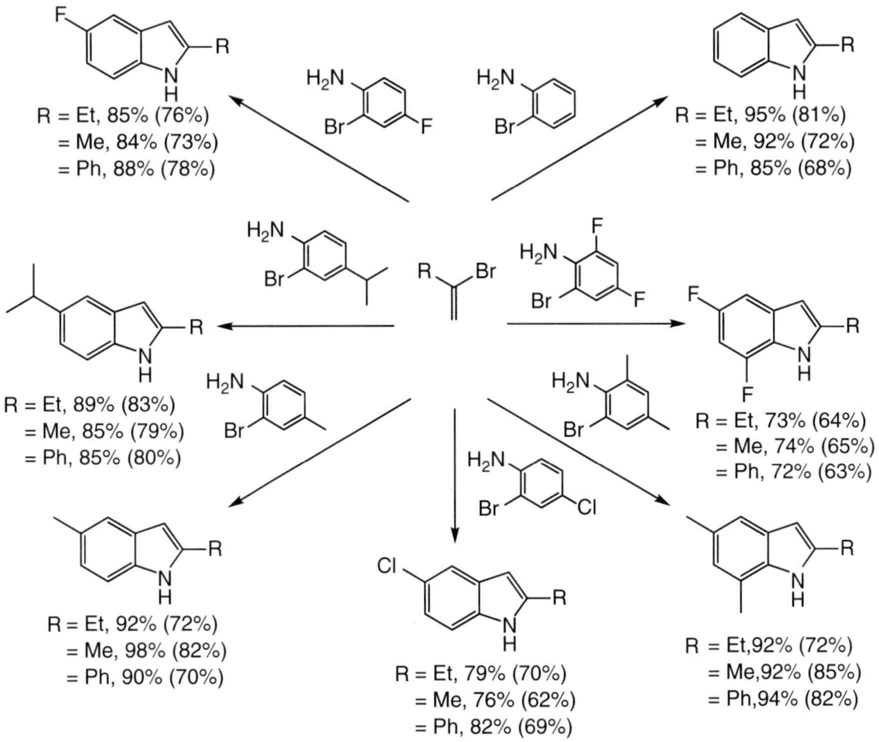

SCHEME 3.84 Schematic illustrating the array of indole alkaloids synthesized under continuous flow conditions (results in parentheses represent the isolated yield).

sought. In the past, continuous flow methodology has been shown to afford many processing advantages in the preparation and handling of diazonium intermediates and with this in mind, Malet-Sanz et al. [155] developed a continuous flow procedure for the iododeamination of aromatic and heteroaromatic amines.

As Scheme 3.85 illustrates, initial investigations were conducted using the iododeamination of 2-amino-5-bromobenzonitrile **343** as a model reaction. Based on the literature precedent, where reactions were heated to 35°C for 10 min followed

SCHEME 3.85 Illustration of the model reaction used for the development of a continuous flow iododeamination protocol.

by 1 h at room temperature, a flow protocol was developed utilizing a Vapourtec R series (UK) flow chemistry system. From two separate inlets, a solution of the amine **343** (0.15 M) in MeCN was mixed with a solution of iodine **344** (0.15 M) and *tert*-butylnitrite **345** (0.23 M, 1.5 eq.) in MeCN, the resulting reaction mixture was heated to 35°C for 12.5 min, then cooled to room temperature for 50 min, and collected in a solution of aqueous $Na_2S_2O_3$. Prior to extraction, the aqueous solution was concentrated *in vacuo* to remove the MeCN and the residue extracted into ethyl acetate. The organic layer was concentrated *in vacuo* and the organic residue purified by column chromatography to afford 5-bromo-2-iodobenzonitrile **346** in 50% yield. Based on these initial findings, the authors investigated the effect of increased reaction temperature, coupled with a reduced reaction time, finding that the reaction yield could be increased to 66% **346** at 60°C with a residence time of 25 min (10 mL reactor), comparing favorably with the 26% 5-bromo-2-iodobenzonitrile **346** obtained in batch.

The flow protocol was subsequently used to evaluate the substituent effect on the reaction and as Table 3.64 illustrates, substrates containing electron-withdrawing groups in the *para*- and *meta*-position reacted well, steric hindrance in the *ortho*-position led to a reduction in yield however the absence of electron-withdrawing groups was found to result in a low yield of the target compounds. In all cases, however, the proportion of by-products formed was reduced as a result of employing a flow protocol and is attributed to increased control of reaction parameters such as temperature and reaction time; preventing the formation of iodination or ArH by-products.

TABLE 3.64
A Selection of the Results Obtained for the Continuous Flow Iododeamination Performed under Continuous Flow

Amine	Product	Yield (%)[a]
NC—C₆H₄—NH₂ (para)	NC—C₆H₄—I **350**	91
NC—C₆H₄—NH₂ (meta)	NC—C₆H₄—I (meta)	81
nPr-substituted aniline (with methyl)	nPr-substituted iodoarene	51
EtO₂C-pyrazole with H₂N, N-Ph	EtO₂C-pyrazole with I, N-Ph	5

[a] All yields reported are isolated.

In a final comparison of the flow methodology with batch, 4-iodobenzonitrile **347** protocol was scaled from 1.50 to 42.00 mmol using a 40 mL flow reactor. Using this approach, the authors were able to prepare 8.8 g of 4-iodobenzonitrile **347** in 91% isolated yield in 7 h compared to 82% in batch.

3.9.8 Radical Additions under Flow

Ryu and coworkers [156,157] demonstrated the advantages associated with micro reaction technology for a series of intermolecular reactions, focusing on thermally induced radical addition (Scheme 3.86) of alkenes to alkoxyamines removing the need for toxic trialkyl tin mediators.

As Table 3.65 illustrates radical generation occurs by C–O bond homolysis, this is then followed by addition of the C-radical to an olefin, upon which the adduct radical is trapped by the aminoxyl radical (nitroxide) to form the corresponding alkoxyamine product. During a detailed study, the authors found that the structure of the alkoxyamine had a dramatic influence on the reaction outcome, concluding that alkoxyamine **348** reacted most efficiently to afford the respective adduct.

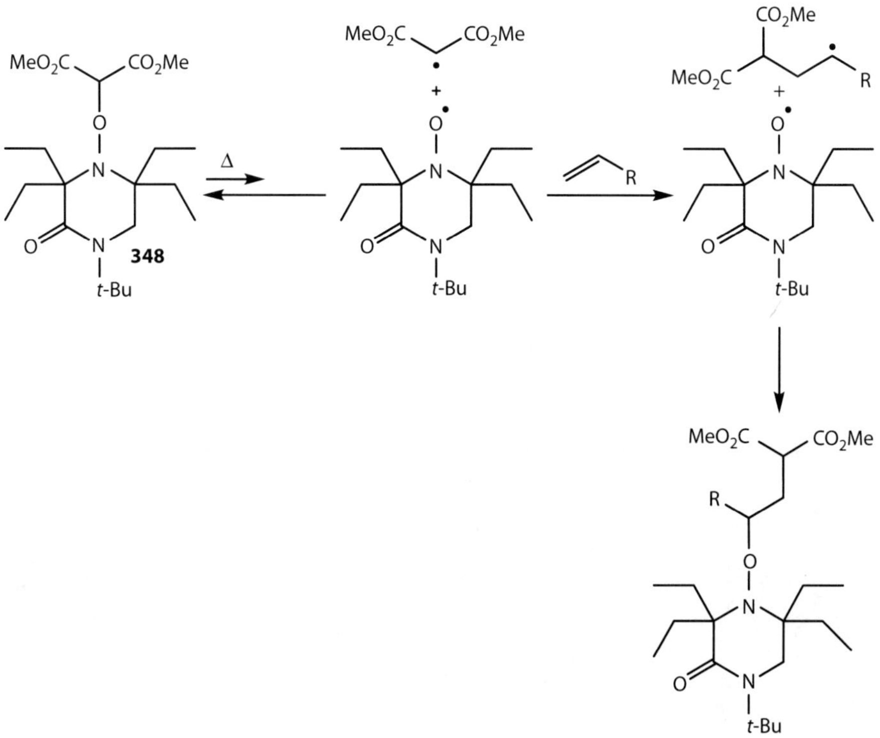

SCHEME 3.86 Schematic illustrating the carboaminoylations of olefins via a novel alkoxyamine **348**.

TABLE 3.65
Comparison of Batch and Flow Radical Carboaminoxylations

Alkene[a]	Time (min)	Batch Yield (%)	Flow Reactor (%)
1-Octene	10	45	73
1-Octene	10	72	95
4-Phenyl-1-butene	5	37	65
4-Phenyl-1-butene	10	55	86
β-Pinene	5	60	77
2-Methyl-1-nonene	5	60	81
Butyl vinyl ether	5	65	93
Vinylphthalimide	5	58	68

[a] 2 eq.
[b] 5 eq. wrt the alkoxyamine **348**.

With this information in hand, the authors evaluated the effect of performing such radical additions under continuous flow in order to identify any advantages associated with this mode of operation. To perform such reactions, the authors employed a stainless-steel tubular reactor (dimensions = 1 mm (i.d.) × 50 cm (long)) heated to 125°C with a single reactant inlet and outlet. Employing a reactant solution of alkoxyamine **348** and alkene (0.07 M) in DMSO, the authors investigated the radical generation under identical reaction times to those used in batch. As Table 3.65 illustrates, significantly higher yields were obtained as a result of performing reactions in a tubular reactor compared to reactions performed under analogous conditions in batch.

The authors attributed this observation to the high thermal efficiency obtained in their microflow system compared to a stirred batch reactor. Furthermore, the ability to rapidly cool the reaction products obtained within their flow reactor reduced the formation of by-products previously attributed to thermal decomposition when conducted in batch. These findings led the authors to reexamine TEMPO-malonate addition chemistry which previously proved problematic due to the need for high reactor temperatures to afford effective addition to alkenes, again observing higher product yields under flow conditions (at 180°C).

3.10 SUMMARY

We hope that from the diverse array of synthetic examples provided, you have been able to obtain an understanding of the advantages associated with the use of micro reaction technology, not only from an investigative standpoint but also from a production perspective. For additional examples of industrially interesting flow synthesis using liquid-phase steps, please refer to Chapter 7.

REFERENCES

1. Hartman, R. L. and Jensen, K. J. 2009. Microchemical systems for continuous-flow synthesis, *Lab Chip* 9: 2495–2507.
2. Jähnisch, K., Hessel, V., Löwe, H., and Maerns, M. 2004. Chemistry in microstructured reactors, *Angew. Chem. Int. Ed.* 43: 406–446.
3. Cukalovic, A., Monbaliu, J. M. R., and Stevens, C. V. 2010. Microreactors technology as an efficient tool for multicomponent reactions, *Top. Heterocycl. Chem.* 23: 161–198.
4. Wiles, C. and Watts, C. 2007. Recent advances in micro reaction technology, *Chem. Commun.* 443–467.
5. Wiles, C. and Watts, P. 2009. Continuous-flow organic synthesis: A tool for the modern medicinal chemist, *Future Med. Chem.* 1(9): 1593–1612.
6. Hessel, V., Hardt, S., and Löwe, H. 2004. *Chemical Micro Process Engineering: Fundamentals, Modelling and Reactions*, Germany: Wiley-VCH.
7. Yoshida, J. 2008. *Flash Chemistry: Fast Organic Synthesis in Microsystems*, UK: Wiley.
8. Hessel, V., Renken, A., Schouten, J. C., and Yoshida, J. (Eds) 2009. *Micro Process Engineering, A Comprehensive Handbook Volume 2: Devices, Reactions and Applications*, Germany: Wiley-VCH.
9. Schouten, J. C. (Ed.) 2010. *Micro Systems and Devices for (Bio)chemical Processes*, The Netherlands: Academic Press.
10. Wirth, T. (Ed.) 2008. *Microreactors in Organic Synthesis and Catalysis*, Germany: Wiley VCH.
11. Wiles, C., Watts, P., Haswell, S. J., and Pombo-Villar, E. 2002. The regioselective preparation of 1,3-diketones with a micro reactor, *Chem. Commun.* 1034–1035.
12. Wiles, C., Watts, P., Haswell, S. J., and Pombo-Villar, E. 2002. The regioselective preparation of 1,3-diketones, *Tetrahedron Lett.* 43: 2945–2948.
13. Ueno, M., Hisamoto, H., Kitamori, T., and Kobayashi, S. 2003. Phase-transfer alkylation reactions using microreactors, *Chem. Commun.* 936–937.
14. Nagaki, A., Takizawa, E., and Yoshida, J. -I. 2009. Generations and reactions of N-(t-butylsulfonyl)aziridinyllithiums using microreactors, *Chem. Lett.* 38(11): 1060–1061.
15. Odedra, A. and Seeberger, P. H. 2009. 5-(Pyrrolidin-2-yl)tetrazole-catalyzed aldol and Mannich reactions: Acceleration and lower catalyst loading in a continuous flow reactor, *Angew. Chem. Int. Ed.* 48: 2699–2702.
16. Pipus, G., Plazl, I., and Koloini, T. 2000. Esterification of benzoic acid in a microwave tubular reactor, *Chem. Eng. J.* 76: 239–245.
17. Wiles, C., Watts, P., Haswell, S. J., and Pombo-Villar, E. 2003. Solution phase synthesis of esters within a micro reactor, *Tetrahedron* 59: 10173–10179.
18. Lu, S., Watts, P., Chin, F. T., Hong, J., Musachio, J. L., Briard, E., and Pike, V. W. 2004. Synthesis of ^{11}C- and ^{18}F-labeled carboxylic esters within a hydrodynamically-driven micro reactor, *Lab Chip* 4(1): 523–525.
19. Sato, M., Matsushima, K., Kawanami, H., and Ikuhsima, Y. 2007. A highly selective, high-speed, and hydrolysis free O-acylation in subcritical water in the absence of a catalyst, *Angew. Chem. Int. Ed.* 46: 6284–6288.

20. Razzaq, T., Glasnov, T. N., and Kappe, C. O. 2009. Continuous-flow microreactor chemistry under high-temperature/pressure conditions, *Eur. J. Org. Chem.* 1321–1325.
21. Hallmark, B., Mackley, M. R., and Gadala-Maria, F. 2005. Hollow microcapillary arrays in thin plastic films, *Adv. Eng. Mater.* 7(6): 545–547.
22. Hornung, C. H., Mackley, M. R., Baxendale, I. R., and Ley, S. V. 2007. A microcapillary flow disk reactor for organic synthesis, *Org. Proc. Res. Dev.* 11: 399–405.
23. Belder, D., Ludwig, M., Wang, L., and Reetz, M. T. 2006. Enantioselective catalysis and analysis on a chip, *Angew. Chem. Int. Ed.* 45: 2463–2466.
24. Hamper, B. C. and Tesfu, E. 2007. Direct uncatalyzed amination of 2-chloropyridien using a flow reactor, *Synlett* 14: 2257–2261.
25. Wilson, N. S., Sarko, C. R., and Roth, G. P. 2004. Development and applications of a practical continuous flow microwave cell, *Org. Proc. Res. Dev.* 8: 535–538.
26. Löwe, H., Axinte, R. D., Breuch, D., Hofmann, and Changhong, L. 2010. Passive and intrinsic cooling of highly exothermic reactions-synthesis of ionic liquids in the lab and pilot scale, *11th International Conference on Microreaction Technology*, Kyoto, Japan, 16–17.
27. Böwing, A. G., Jess, A., and Wasserscheid, P. 2005. Kinetik und reasktions-technik der synthese ionischer flüssigkeiten, *Chem. Ing. Technik* 77(9): 1430–1439.
28. Renken, A., Hessel, V., Löb, P., Miszczuk, R., Uerdingen, M., and Kiwi-Minsker, L. 2007. Ionic liquid synthesis in a microstructured reactor for process intensification, *Chem. Eng. J.* 46: 840–845.
29. Wiles, C., Watts, P., Haswell, S. J., and Pombo-Villar, E. 2004. Stereoselective alkylation of an Evans auxillary derivative within a pressure-driven micro reactor, *Lab Chip* 4: 171–173.
30. Schwalbe, T., Autze, V., and Wille, G. 2002. Chemical synthesis in microreactors, *Chimia*, 56(11): 636–646.
31. Chevalier, B., Lavric, E. D., Cerato-Noyerie, C., Horn, C. R., and Woehl, P. 2008. Microreactors for industrial multi-phase applications, *Chem. Today* 26(2): 38–42.
32. Hooper, J. and Watts, P. 2007. Expedient synthesis of deuterium-labeled amides with micro reactors, *J. Labelled Compd. Radiopharm.* 50: 189–196.
33. Sato, M., Matsushima, K., Kawanami, H., Chatterjee, M., Yokoyama, T., Ikushima, Y., and Suzuki, T. M. 2009. Highly efficient chemoselective N-acylation with water microreaction system in the absence of catalyst, *Lab Chip* 9, 2877–2880.
34. Lui, S., Chang, C., Paul, B. K., and Remcho, V. T., 2008. Convergent synthesis of polyamide dendrimer using a continuous flow microreactor, *Chem. Eng. J.* 135S: S333–S337.
35. Singh, B. K., Stevens, C. V., Acke, D. R. J., Parmar, V. S., and van der Eyken, E. V. 2009. Copper-mediated N- and O-arylations with arylboronic acids in a continuous flow microreactor: A new avenue for efficient scalability, *Tetrahedron Lett.* 50: 15–18.
36. Kopach, M. E., Murray, M. M., Braden, T. M., Kobierski, M. E., and Williams, O. L. 2009. Improved synthesis of 1-(Azidomethyl)-3,5-bis-trifluoromethyl)benzene: Development of batch and microflow azide processes, *Org. Proc. Res. Dev.* 13: 152–160.
37. Brandt, J. C. and Wirth, T. 2009. Controlling hazardous chemicals in microreactors: Synthesis with iodine azide, *B. J. Org. Chem.* 5(30).
38. Nieuwland, P. J., Hanssen, B., Koch, K., Janssen, P., Delville, M., and Rutjes, F. P. J. T. 2010. Azide synthesis in microreactors: From optimization to lab-scale production, *11th International Conference on Microreaction Technology*, Kyoto, Japan, 208–209.
39. Kopach, M. E., Murray, M. M., Braden, T. M., Kobierski, M. E., and Williams, O. L. 2009. Improved synthesis of 1-(Azidomethyl)-3,5-(trifluoromethyl)benzene: Development of batch and microflow azide processes, *Org. Proc. Res. Dev.* 13: 152–160.
40. Riva, E., Gagliardi, S., Mazzoni, C., Passarella, D., Rencurosi, A., Vigo, D., and Martinelli, M. 2009. Efficient continuous flow synthesis of hydroxamic acids and suberoylanilide hydroxamic acid preparation, *J. Org. Chem.* 74: 3540–3543.

41. Bedore, M. W., Zabrenko, N., Jensen, K. F., and Jamison, T. F. 2010. Aminolysis of epoxides in a microreactor system: A continuous flow approach to β-amino alcohols, *Org. Proc. Res. Dev.* 14: 432–440.
42. Gustafsson, T., Gilmour, R., and Seeberger, P. H. 2008. Fluorination in microreactors, *Chem. Commun.* 3022–3024.
43. Baumann, M., Baxendale, I. R., Martin, L. J., and Ley, S. V. 2009. Development of fluorination methods using continuous-flow microreactors, *Tetrahedron* 65: 6611–6625.
44. Schwalbe, T., Autze, V., Hohmann, M., and Stirner, W. 2004. Novel innovation systems for a cellular approach to continuous process chemistry from discovery to market, *Org. Proc. Res. Dev.* 8: 440–454.
45. Schwalbe, T., Kursawe, A., and Sommer, J. 2005. Application report on operating cellular process chemistry plants in fine chemical and contract manufacturing industries, *Chem. Eng. Technol.* 28(4): 408–419.
46. Goto, S., Velder, J., El Sheikh, S., Sakamoto, Y., Mitani, M., Elmas, S., Adler, A., et al. 2008. Butyllithium-mediated coupling of aryl bromides with ketones under *in-situ*-quench (ISQ) conditions: An efficient one-step protocol applicable to microreactor technology, *Synlett* 9: 1361–1365.
47. Yoshida, J., Nagaki, A., and Yamada, T. 2008. Flash chemistry: Fast chemical synthesis by using microreactors, *Chem. Eur. J.* 7450–7459.
48. Usutani, H., Tomida, Y., Nagaki, A., Okamoto, H., Nokami, T., and Yoshida, J. 2007. Generation and reactions of *o*-bromophenyllithium without benzyne formation using a microreactor, *J. Am. Chem. Soc.* 129: 3046–3047.
49. Nagaki, A., Kim, H., and Yoshida, J. 2008. Aryllithium compounds bearing alkoxycarbonyl groups: generation and reactions using a microflow system, *Angew. Chem. Int. Ed.* 47: 7833–7836.
50. Kim, H., Nagaki, A., and Yoshida, J. 2010. Aryllithium compounds bearing alkoxycarbonyl groups. Generation and reactions using a microflow system, *11th International Conference on Microreaction Technology*, Kyoto, Japan, 40–41.
51. Nagaki, A., Kim, H., Takabayashi, N., Tomida, Y., Usutani, H., and Yoshida, J. 2010. Generation and reactions of unstable aryllithiums using integrated microflow systems, *11th International Conference on Microreaction Technology*, Kyoto, Japan, 168–169.
52. Tomida, Y., Nagaki, A., and Yoshida, J. 2010. Carbolithiation of conjugated enzymes with aryllithiums in microflow systems, *11th International Conference on Microreaction Technology*, Kyoto, Japan, 20–21.
53. Nagaki A., Kim, H., Usutani, H., Matsuo, C., and Yoshida, J. 2010. Generation and reaction of cyano-substituted aryllithium compounds using microreactors, *Org. Biomol. Chem.* 8: 1212–1217.
54. Nagaki, A., Takizawa, E., and Yoshida, J. 2009. Oxiranyl anion methodology using microflow systems, *J. Am. Chem. Soc.* 131(5), 1654–1655.
55. Nagaki, A., Kim, H., and Yoshida, J. 2009. Nitro-substituted aryl lithium compounds in microreactor synthesis: Switch between kinetic and thermodynamic control, *Angew. Chem. Int. Ed.* 48: 8063–8065.
56. Riva, E., Gagliardi, S., Martinelli, M., Passeralla, D., Vigo, D., and Rencurosi, A., 2010. Reaction of Grignard reagents with carbonyl compounds under continuous flow conditions, *Tetrahedron* 66: 3242–3247.
57. Hu, R., Lei, M., Xiong, H., Mu, X., Wang, Y., and Yin, X. 2008. Highly selective acylation of ferrocene using microfluidic chip reactor, *Tetrahedron Lett.* 49: 387–389.
58. Suga, S., Nagaki, A., and Yoshida, J. 2003. Highly selective Friedel–Crafts monoalkyation using micromixing, *Chem. Commun.* 354–355.
59. Doku, G. N., Haswell, S. J., McCreedy, T., and Greenway, G. M. 2001. Electric field-induced mobilisation of multiphase solution systems based on the nitration of benzene in a micro reactor, *Analyst* 126: 14–20.

60. Kulkarni, A. A., Kalyani, V. S., Joshi, R. A., and Joshi, R. R. 2009. Continuous flow nitration of benzaldehyde, *Org. Proc. Res. Dev.* 13: 999–1002.
61. Liu, L., Lu, C., and Zhang, X. 2007. Regioselective nitration of toluene with nitric acid in the presence of sodium dihydrogen phosphate catalysts, *Fine Chem.* 24: 1139–1141.
62. Burns, J. R. and Ramshaw, C. 2002. A microreactor for the nitration of benzene and toluene, *Chem. Eng. Commun.* 189: 1611–1628.
63. Ferstl, W., Loebbecke, S., Antes, J., Krause, H., Haeberl, M., Schmalz, D., Muntermann, H. et al. 2004. Development of an automated microreaction system with integrated sensorics for process screening and production, *Chem. Eng. J.* 101: 431–438.
64. Panke, G., Schwalbe, T., Stirner, W., Taghavi-Moghadam, S., and Wille, G. 2003. A practical approach of continuous processing to high energetic nitration reactions in microreactors, *Synthesis* 2827–2830.
65. Shen, J., Zhao, Y., Chen, G., and Yuan, Q. 2009. Investigation of nitration processes of *iso*-octanol with mixed acid in a microreactor, *Chin. J. Chem. Eng.* 17(3): 412–418.
66. Braune, S., Pöchlauer, P., Reintjens, R., Steinhofer, S., Winter, M., Lobet, O., Guidat, R., Woehl, P., and Guermeur, C. 2008. Selective nitration in a microreactor for pharmaceutical production under cGMP conditions, *Chem. Today* 26(5): 1–4.
67. Thayer, A. M. 2009. Handle with care, *Chem. Eng. News* 87(11): 17–19.
68. Benali, O., Deal, M., Farrant, E., Tapolczay, D., and Wheeler, R. 2008. Continuous flow microwave-assisted reaction optimization and scale-up using fluorous spacer technology, *Org. Proc. Res. Dev.* 12: 1007–1011.
69. Löb, P., Löwe, H., and Hessel, V. 2004. Fluorinations, chlorinations, and brominations of organic compounds in micro reactors, *J. Fluor. Chem.* 125: 1677–1694.
70. Wahab, B. 2010. Advances in organic synthesis: Expedient radiosynthesis of substituted indoles for pharmaceutical lead development within microreactors, Thesis, The University of Hull.
71. Löb, P., Hessel, V., Klefenz, H., Löwe, H., and Mazanek, K. 2005. Bromination of thiophene in micro reactors, *Lett. Org. Chem.* 2(8): 767–779.
72. Salimi-Moosavi, H., Tang, T., and Harrison, D.J. 1997. Electroosmotic pumping of organic solvents and reagents in microfabricated reactor chips, *J. Am. Chem. Soc.* 119, 8716–8717.
73. Wootton, R. C. R., Fortt, R., and de Mello, A. J. 2002. On-chip generation and reaction of unstable intermediates—monolithic nanoreactors for diazonium chemistry: Azo dyes, *Lab Chip* 2: 5–7.
74. Günther, P. M., Möller, F., Henkel, T., Köhler, J. M., and Groß, G. A. 2005. Formation of monomeric and novolak azo dyes in nanofluid segments by use of a double injector chip reactor, *Chem. Eng. Technol.* 28(4): 520–527.
75. Wille, Ch., Gabski, H. –P., Haller, Th., Kim, H., Unverdorben, L., and Winter, R. 2004. Synthesis of pigments in a three-stage microreactor pilot plant—An experimental technical report, *Chem. Eng. J.* 101: 179–185.
76. Pennemann, H., Forster, S., Kinkel, J., Hessel, V., Löwe, H., and Wu, L. 2005. Improvement of dye properties of the azo pigment yellow 12 using a micromixer-based process, *Org. Proc. Res. Dev.* 9: 188–192.
77. Fortt, R., Wootton, R. C. R., and de Mello, A. J. 2003. Continuous-flow generation of anhydrous diazonium species: Monolithic microfluidic reactors for the chemistry of unstable intermediates, *Org. Proc. Res. Dev.* 7(5): 762–768.
78. Asano, Y., Miyamoto, T., Togashi, S., and Endo, Y. 2010. Optimization for Sandmeyer reaction including an unstable diazonium salt using microreactors, *11th International Conference on Microreaction Technology*, Kyoto, Japan, 178–179.

79. Löbbecke, S., Ferstl, W, Panic, S., and Türke, T. 2005. Concepts for modularization and automation of microreaction technology. *Chem. Eng. Technol.* 28(4): 484–493.
80. Tanaka, K. and Fukase, K. 2009. Renaissance of traditional organic reactions under microfluidic conditions: A new paradigm for natural products synthesis. *Org. Proc. Res. Dev.* 13: 983–990.
81. Tanaka, K., Motomatsu, S., Koyama, K., and Fukase, K. 2008. Efficient aldol condensation in aqueous biphasic system under microfluidic conditions. *Tetrahedron Lett.* 49: 2010–2012.
82. Mu, X. J., Yin, X. F., and Wang, Y. G. 2005. The Claisen–Schmidt reaction carried out in microfluidic chips, *Synlett* 2005: 3163–3165.
83. Mikami, K., Yamanaka, M., Islam, M. N., Kudo, K., Seino, N., and Shinoda, M. 2003. Fluorous nanoflow system for the mukaiyama aldol reaction catalyzed by the lowest concentration of lanthanide complex with bis(perfluorooctanesulfonyl)amide ponytail, *Tetrahedron*, 59: 10593–10597.
84. Wiles, C., Watts, P., Haswell, S. J., and Pombo-Villar, E. 2001. The aldol reaction of silyl enol ethers within a micro reactor, *Lab Chip* 1: 100–101.
85. Fernandez-Suarez, M., Wong, S. Y. F., and Warrington, B. H. 2002. Synthesis of a three-member array of cycloadducts in a glass microchip under pressure driven flow, *Lab Chip* 2: 170–174.
86. Bula, W. P., Verboom, W., Reinhoudt, D. N., and Gardeniers, H. J. G. E. 2007. Multichannel quench-flow microreactor chip for parallel reaction monitoring, *Lab Chip* 7: 1717–1722.
87. Wiles, C., Watts, P., Haswell, S. J., and Pombo-Villar, E. 2002. 1,4-addition of enolates to α,β-unsaturated ketones within a micro reactor, *Lab Chip* 2: 62–64.
88. Miyake, N. and Kitazume, T. 2003. Microreactors for the synthesis of fluorinated materials, *J. Fluorine Chem.* 122: 243–246.
89. Snyder, D. A., Noti, C., Seeberger, P. H., Schael, F., Bieber, T., and Ehrfeld, W. 2005. Modular microsystems for homogeneously and heterogeneously catalyzed chemical synthesis, *Helv. Chim. Acta* 88: 1–9.
90. Hallmark, B., Mackley, M. R., and Gadala-Maria, F. 2005. Hollow microcapillary arrays in thin plastic films, *Adv. Eng. Mater.* 7(6): 545–547.
91. Damm, M., Glasnov, T. N., and Kappe, C. O. 2010. Translating high-temperature microwave chemistry to scalable continuous flow processes, *Org. Proc. Res. Dev.* 14: 215–224.
92. Miyake, N. and Kitazume, T. 2003. Microreactors for the synthesis of fluorinated materials, *J. Fluorine Chem.* 122: 243–246.
93. Snyder, D. A., Noti, C., Seeberger, P. H., Schael, F., Bieber, T., and Ehrfeld, W. 2005. Modular microsystems for homogeneously and heterogeneously catalyzed chemical synthesis, *Helv. Chim. Acta* 88: 1–9.
94. Tietze, L. F. and Liu, D. 2008. Continuous-flow microreactor multi-step synthesis of an aminonaphthalene derivative as starting material for the preparation of novel anticancer agents, *Arkivoc* vii: 193–210.
95. Koch, K., van den Berg, R. J. F., Nieuwland, P. J., Wijtmans, R., Schoemaker, H. E., van Hest, J. C. M., and Rutjes, F. P. J. T. 2007. Enzymatic enantioselective C–C bond formation in microreactors, *Biotechnol. Bioeng.* 99(4): 1028–1033.
96. Löwe, H., Hessel, V., Löb, P. and Hubbard, S. 2006. Addition of secondary amines to α,β-unsaturated carbonyl compound and nitriles by using microstructured reactors, *Org. Proc. Res. Dev.* 10: 1144–1152.
97. Bagley, M. C., Jenkins, R. L., Caterina Lubinu, M., Mason, C., and Wood, R. 2005. A simple continuous flow microwave reactor, *J. Org. Chem.* 70: 7003–7006.
98. Bagley, M. C., Fusillo, V., Jenkins, R. L., Lubinu, M. C., and Mason, C. 2010. Continuous flow processing from microreactors to mesoscale: The Bohlmann–Rhatz cyclodehydration reaction, *Org. Biomol. Chem.* 8: 2245–2251.

99. Tron, G. C., Pirali, T., Billington, R. A., Canonico, P. L., Sorba, G., and Genazzani, A. A. 2008. Click chemistry reactions in medicinal chemistry: Application of the 1,3-dipolar cycloaddition between azides and alkynes, *Med. Res. Rev.* 28(2): 278–308.
100. Wang, Y., Lin, W.Y., and Liu, K. 2009. An integrated microfluidic device for large-scale *in situ* click chemistry screening, *Lab Chip* 9: 2281–2285.
101. Tinder, R., Farr, R., Heid, R. Zhao, R. Rarig, R. S., and Storz, T. 2009. A convenient and stable synthon for ethyl azide and its evaluation in a [3 + 2]-cycloaddition reaction under continuous flow, *Org. Proc. Res. Dev.* 13: 1401–1406.
102. Bogdan, A. R. and Sack, N. W. 2009. The use of copper flow reactor technology for the continuous synthesis of 1,4-disubstituted 1,2,3-triazoles, *Adv. Synt. Catal.* 351: 849–854.
103. Wiles, C., Watts, P., Haswell, S. J., and Pombo-Villar, E. 2004. The application of microreaction technology for the synthesis of 1,2-azoles, *Org. Proc. Res. Dev.* 8: 28–32.
104. Garcia-Egido, E., Spikmans, V., Wong, S. Y. F., and Warrington, B. H. 2003. Synthesis and analysis of combinatorial libraries performed in an automated micro reactor system, *Lab Chip* 3: 73–76.
105. Tanaka, K., Motomatsu, S., Koyana, K., Tanaka, S., Fukase, K. 2007. Large-scale synthesis of immunoactivating natural product, pristane, by continuous flow microfluidic dehydration as the key step, *Org. Lett.* 9: 299–302.
106. Odedra, A., Geyer, K., Gustafsson, T., Gilmour, R., and Seeberger, P. H. 2008. Safe, facile radical-based reduction and hydrosilylation reactions in a microreactor using tris(trimethylsilyl)silane, *Chemical Communications.* 3025–3027.
107. Fukuyama, T., Kobayashi, M., Rahman, M. T., Kamata, N., and Ryu, I. 2008. Spurring radical reactions of organic halides with tin hydride and TTMSS using microreactors, *Org. Lett.* 10(4): 533–536.
108. Maruyama, T., Uchida, J., Ohkawa, T., Futami, T., Katayama, K., Nishizawa, K., Sotowa, K. Kubota, F., Kamiya, N., and Goto, M. 2003. Enzymatic degradation of *p*-chlorophenol in a two-phase flow microchannel system, *Lab Chip* 3: 308–312.
109. Jachuck, R. J. J., Selvaraj, D. K., and Varma, R. S. 2006. Process intensification: Oxidation of benzyl alcohol using a continuous isothermal reactor under microwave irradiation, *Green Chem.* 8: 29–33.
110. Kawaguchi, T., Miuata, H., Ataka, K., Mae, K., and Yoshida, J. 2005. Room temperature Swern oxidations by using a microscale flow system, *Angew. Chem. Int. Ed.* 44: 2413–2416.
111. Van der Linden, J. J. M., Hilberink, P. W., Kronenburg, C. M. P., and Kemperman, G. J. 2008. Investigation of the Moffat–Swern oxidation in a continuous microreactor system, *Org. Proc. Res. Dev.* 12: 911–920.
112. McConnell, J. R., Hitt, J. E., Daugs, E. D., and Rey, T. A. 2008. The Swern oxidation: Development of a high-temperature semicontinuous process, *Org. Proc. Res. Dev.* 12: 940–945.
113. Fritz-Langhals, E. 2005. Technical production of aldehydes by continuous bleach oxidation of alcohols catalyzed by 4-hydroxy-TEMPO, 9(5): 577–582.
114. Yamada, Y. M. A., Torii, K., and Uozumi, Y. 2009. Oxidative cyclization of alkenols with oxone in a miniflow reactor, *Beilstein J. Org. Chem.* 5: 18.
115. Kee, S. P. and Gavriilidis, A. 2009. Design and performance of a microstructured peek reactor for continuous poly-L-leucine catalyzed chalcone epoxidation, *Org. Proc. Res. Dev.* 13: 941–951.
116. Hartung, A., Keane, M. A., and Kraft, A. 2007. Advantages of synthesizing *trans*-1,2-cyclohexanediol in a continuous flow microreactor over a standard glass apparatus, *J. Org. Chem.* 72: 10235–10238.
117. Koch, K., van Weerdenburg, B. J. A., Verkade, J. M. M. Nieuwland, P. J., Rutjes, F. P. J. T., and van Hest, J. C. M. 2009. Optimizing the deprotection of the amine protecting

p-methoxyphenyl group in an automated microreactor platform, *Org. Proc. Res. Dev.* 13: 1003–1006.
118. Sedelmeier, J., Ley, S. V., and Baxendale, I. R. 2009. An efficient and transition metal free protocol for the transfer hydrogenation of ketones as a continuous flow process, *Green Chem.* 11: 683–685.
119. Ducry, L. and Roberge, D. M. 2008. Dibal-H reduction of methyl butyrate into butyraldehyde using microreactors, *Org. Proc. Res. Dev.* 12: 163–167.
120. Comer, E. and Organ, M. G. 2005. A microreactor for microwave-assisted capillary (continuous flow) organic synthesis (MACOS), *J. Am. Chem. Soc.* 127(22): 8160–8167.
121. Wilson, N. S., Sarko, C. R., and Roth, G. P. 2004. Development and applications of a practical continuous flow microwave cell, *Org. Proc. Res. Dev.* 8: 535–538.
122. Benali, O., Deal, M., Farrant, E., Tapolczay, D., and Wheeler, R. 2008. Continuous flow microwave-assisted reaction optimization and scale-up using fluorous spacer technology, *Org. Proc. Res. Dev.* 12: 1007–1011.
123. Theberge, A. B., Whyte, G., Frenzel, M., Fidalgo, L. M., Wootton, R. C. R., and Huck, W. T. S. 2009. Suzuki–Miyaura coupling reactions in aqueous microdroplets with catalytically active fluorous interfaces, *Chem. Commun.* 6225–6227.
124. Schwalbe, T., Autze, V., Hohmann, M., and Stirner, W. 2004. Novel innovation system for a cellular approach to continuous process chemistry from discovery to market, *Org. Proc. Res. Dev.* 8: 440–454.
125. Ahmed-Omer, B., Barrow, D. A., and Wirth, T. 2009. Heck reactions using segmented flow conditions, *Tetrahedron Lett.* 50: 3352–3355.
126. Odell, L. R., Lindh, J., Gustafsson, T., and Lerhed, M. 2010. Continuous flow palladium(ii)-catalyzed oxidative heck reactions with arylboronic acids, *Eur. J. Org. Chem.* 2270–2274.
127. Liu, S., Fukuyama, T., Sato, M., and Ryu, I. 2004. Continuous microflow synthesis of butyl cinnamate by a Mizoroki–Heck reaction using a low-viscosity ionic liquid as the recycling reaction medium, *Org. Proc. Res. Dev.* 8: 477–481.
128. Fukuyama, T., Shinmen, M., Nishitani, S., Sato, M., and Ryu, I. 2002. A copper-free Sonogashira coupling reaction in ionic liquids and its application to a microflow system for efficient catalyst recycling, *Org. Lett.* 4: 1691–1694.
129. Liu, S., Fukuyama, T., Sato, M., and Ryu, I. 2004. Continuous microflow synthesis of butyl cinnamate by a Mizoroki–Heck reaction using a low-viscosity ionic liquid as the recycling reaction medium, *Org. Proc. Res. Dev.* 8: 477–481.
130. Kawanami, H., Matsushima, K., Sato, M., and Ikushima, Y. 2007. Rapid and highly selective copper-free Sonogashira coupling in high-pressure, high-temperature water in a microfluidic system, *Angew. Chem. Int. Ed.* 46: 5129–5132.
131. Sugimoto, A., Fukuyama, T., Rahman, M., and Ryu, I. 2009. An automated-flow microreactor system for quick optimization and production: Application of 10 and 10-gram order productions of a matrix metalloproteinase inhibitor using a Sonogashira coupling reaction, *Tetrahedron Lett.* 50: 6364–6367.
132. Shi, G., Hong, F., Liang, Q., Fang, H., Nelson, S., and Weber, S. G. 2006. Capillary-based serial-loading parallel microreactor for catalyst screening, *Anal. Chem.* 78: 1972–1979.
133. Mauger, C., Buisine, O., Caravieihes, S., and Mignani, G., 2005. Successful application of microstructured continuous reactor in the palladium catalyzed aromatic amination, *J. Organomet. Chem.* 690: 3627–3629.
134. Kong, L., Lin, Q., Lu, X., Yang, Y., Jia, Y., and Zhou, Y. 2009. Efficient Claisen rearrangement of allyl *para*-substituted phenyl ethers using microreactors, *Green Chem.* 11: 1108–1111.

135. Ceylan, S., Friese, C., Lammel, C., Mazac, K., and Kirschning, A. 2008. Inductive heating for organic synthesis by using functionalized magnetic nanoparticles inside microreactors, *Angew. Chem. Int. Ed.* 47: 8950–8953.
136. Sato, M., Kawanami, H., Chatterjee, M., Otabe, N., Tuji, T., Ikushima, Y., Yokoyama, T., and Suzuki, T. M. Highly selective non-catalytic Claisen re-arrangement in a high pressure and high temperature water microreaction system, *11th International Conference on Microreaction Technology*, Kyoto, Japan, 50–51.
137. Tilstan, U., Defrance, T. and Giard, T. 2009. The Newman–Kwart rearrangement revisited: Continuous process under supercritical conditions, *Org. Proc. Res. Dev.* 13: 321–323.
138. Zhang, X., Stefanick, S. and Villani, F. 2004. Application of microreactor technology in process development, *Org. Proc. Res. Dev.* 8: 455–460.
139. Lin. S., Moon. B., Porter, K. T., Rossman, C. A., Zennie, T., and Wemple, J. 2000. A continuous procedure for preparation of *para* functionalized aromatic thiols using Newman–Kwart chemistry, *Org. Prep. Proceed. Int.* 32: 547–555.
140. Greshock, T. J. and Funk, R. L. 2006. An approach to the total synthesis of welwistatin, *Org. Lett.* 8: 2643–2645.
141. Poullennec, K. G. and Romo, D. 2003. Enantioselective total synthesis of (+)-dibromophakellstatin, *J. Am. Chem. Soc.* 125: 6344–6345.
142. Satoh, N., Akiba, T., Yokoshima, S. and Fukuyama, T. 2007. A practical synthesis of (−)-oseltamivir, *Angew. Chem. Int. Ed.* 46: 5734–5736.
143. Palmieri, A., Ley, S. V., Hammond, K., Polyzos, A. and Baxendale, I. R. 2009. A microfluidics flow chemistry platform for organic synthesis: The Hofmann rearrangement, *Tetrahedron Lett.* 50: 3287–3289.
144. Wahab, B., Ellames, G., Passey, S., and Watts, P. 2010. Synthesis of substituted indoles using continuous flow micro reactors, *Tetrahedron* 66: 3861–3865.
145. Sahoo, H. R., Kralj, J. G., and Jensen, K. F. 2007. Multi-step continuous-flow microchemical synthesis involving multiple reactions and separations, *Angew. Chem. Int. Ed.* 46: 5704–5708.
146. Glasnov, T. N., Vugts, D. J., Konigstein, M. M., Desai, B., Fabian, W. M. F., Orru, R. V. A., and Kappe, C. O. 2005. Microwave-assisted dimroth re-arrangement of thiazines to dihydropyrimidinethiones: Synthetic and mechanistic aspects, *QSAR Comb. Sci.* 25(5–6): 509–518.
147. Bremner, W. S. and Organ, M. G. 2007. Multicomponent reactions to form heterocycles by microwave-assisted continuous flow organic synthesis, *J. Comb. Chem.* 9: 14–16.
148. Comer, E., and Organ, M. G. 2005. A microcapillary system for simultaneous, parallel microwave-assisted synthesis, *Chem. Eur. J.* 11(24): 7223–7227.
149. Grant, D., Dahl, R., and Cosford, N. D. P. 2008. Rapid multistep synthesis of 1,2,4-oxadiazoles in a single continuous microreactor sequence, *J. Org. Chem.* 73: 7219–7223.
150. Bogdan, A. R., Poe, S. L., Kubis, D. C., Broadwater, S. J., and McQuade, D. T. 2009. The continuous flow synthesis of ibuprofen, *Angew. Chem. Int. Ed.* 48(45): 8547–8550.
151. Tanaka, K. and Fukase, K. 2009. Renaissance of traditional organic reactions under microfluidic conditions: A new paradigm for natural products. *Synthesis* 13: 983–990.
152. Ratner, D. M., Murphy, E. R., Jhunjhunwala, M., Snyder, D. A., Jensen, K. F., and Seeberger, P. H. 2005. Microreactor-based reaction optimization in organic chemistry-glycosylation as a challenge, *Chem. Commun.* 578–580.
153. Carrel, F. R., Geyer, K., Codée, J. D. C., and Seeberger, P. H. 2007. Oligosaccharide synthesis in microreactors, *Org. Lett.* 9(12): 2285–2288.
154. Shore, G., Morin, S., Mallik, D., and Organ. M. G. 2008. Pd PEPPSI-IPr-mediated reactions in metal-coated capillaries under MACOS: The synthesis of indoles by sequential aryl amination/Heck couplings, *Chem. Eur. J.* 14: 1351–1356.

155. Malet-Sanz, L., Madrzak, J., Holvey, R. S., and Underwood, T. 2009. A safe and reliable procedure for the iododeamination of aromatic and heteroaromatic amines in a continuous flow reactor, *Tetrahedron Lett.* 50: 7263–7267.
156. Wienhöfer, I. C., Studer, A., Rahman, M. T., Fukuyama, T. and Ryu, I. 2009. Microflow radical carboaminoxylations with a novel alkoxyamine, *Org. Lett.* 11(11): 2457–2460.
157. Fukuyama, T., Rahman, M. T., Ryu, I, Wienhöfer, I. C., and Struder, A. 2010. Microflow radical carboaminoxylations with alkoxyamines, *11th International Conference on Microreaction Technology*, Kyoto, Japan, 198–199.
158. Suet-Ping Kee et al. 2009. Design and performance of a microstructured PEEK reactor for continuous poly-lleucine-catalysed chalcone epoxidation. *Org. Process Research and Dev.* 13(5): 941–951.
159. Tanaka et al. 2009. Renaissance of traditional organic reactions under microfluidic conditions, a new paradigm for natural product synthesis, *Org. Proc. Res. Dev.* 13(5): 983–990.

4 Multi-Phase Micro Reactions

As observed with liquid-phase micro reactions, a diverse array of reactors has been reported for the performance of heterogeneous flow reactions, ranging from in-house fabricated devices to commercially available equipment. The aim of this chapter is to describe a selection of chemical examples, which serve to illustrate the types of reactions accessible through the incorporation of chemical/biochemical catalysts, reagents, and scavengers within such devices.

Throughout this chapter different methods are described for the incorporation of solid materials into microflow reactors, these range from packed beds [1], monoliths [2,3], and wall coatings [4,5] to *in situ* fabricated membranes [6], all of which have applications to which they are best suited. In addition to the material provided herein, for topical discussions on the subject of multiphase micro reactions, see reviews by Kirschning et al. [7], Kiwi-Minsker and coworkers [8], Miyazaki and Maeda [9], Westermann and Melin [10], Weinberg and coworkers [11] to mention a few. For examples of gas–liquid–solid reactions, in particular, examples of hydrogenation reactions refer to Chapter 2.

4.1 NUCLEOPHILIC SUBSTITUTION

4.1.1 C–O Bond-Forming Reactions: Esterifications

In order to reduce the proportion of acidic waste generated during production-scale syntheses and increase productivity, several authors have investigated the performance of heterogeneously catalyzed esterifications under continuous flow conditions.

Packed-Bed Reactor: Employing a miniaturized, stainless steel packed-bed reactor, Kulkarni et al. [12] evaluated the use of the macroreticular resin Amberlyst-15 for the esterification of acetic acid **1** with butanol; investigating the effect of reaction time (50–3000 s) and temperature (20–80°C) on the formation of butyl acetate. Using this approach, the authors observed 50% conversion at 155 s (80°C), increasing to 70% conversion at 1238 s; compared to a homogeneous reaction where 5% butyl acetate was obtained at a residence time of 301 s.

Utilizing this strategy, the authors were able to obtain results that were consistent with the literature and proposed that the fabrication of larger reaction plates would enable throughput to be increased in order to produce sufficient material for production purposes.

Biocatalytic Esterification: Also using a packed-bed reactor, this time a tubular glass device (dimensions = 1.65 mm (i.d.) × 3.0 cm (long)), Woodcock et al. [13]

reported the biocatalytic esterification of a series of aliphatic carboxylic acids using Novozyme-435 **2**. After screening a series of solvents (MeCN, hexane, toluene, DCM), the authors identified hexane as the most promising solvent for use under flow conditions. Employing the carboxylic acid and alcohol in a 1:1 ratio (0.2 M) and a total flow rate of 1 µL min^{-1}, the authors investigated the effect of alkyl chain length on both the acid and alcohol components. As Table 4.1 illustrates under the aforementioned optimized conditions, the target alkyl esters could be obtained in high to excellent conversion; as determined by offline GC-FID analysis. With the exception of butyl laurate and butyl hexanoate, whereby reductions in conversion were observed over a 2 h time frame (~35% decrease), all other flow reactions were observed to be stable over extended periods of operation.

Monolith Reactor: Using a sulfonic acid derived inorganic monolith, housed within a heat shrink PTFE tube, Coq and coworkers [14] recently investigated the transesterification of triacetine **3** with MeOH, Scheme 4.1, comparing the results obtained with batch and packed-bed reactors. Employing a substrate solution containing triacetine **3** (0.70 M) in MeOH, the authors investigated the reactor at a flow rate of 0.5 mL min^{-1}. Using GC-FID analysis of the reaction products, the authors quantified the proportion of transesterification, obtaining 79% conversion to methyl acetate **4**. Continuously operating the system under the aforementioned conditions affords a productivity of 82×10^{-5} mol min^{-1} g^{-1} representing a 32.8 times enhancement compared to batch and 2.9 times increase compared to a packed-bed reactor.

Large-Scale Preparation of Vitamins: Owing to the industrial importance of vitamins as a bulk product, Karge and coworkers [15] demonstrated the development of lipase-catalyzed transformations as a key step in the large-scale preparation of

TABLE 4.1
Summary of the Results Obtained for the Lipase Catalyzed Synthesis of Alkyl Esters

R^1	R^2	Conversion (%)
C$_5$H$_{11}$	Me	92
C$_7$H$_{15}$	Me	92
C$_{11}$H$_{23}$	Me	91
C$_5$H$_{11}$	Et	80
C$_7$H$_{15}$	Et	92
C$_{11}$H$_{23}$	Et	95
C$_5$H$_{11}$	Bu	100
C$_7$H$_{15}$	Bu	99
C$_{11}$H$_{23}$	Bu	90

Reaction conditions = 0.2 M in hexane (1:1 acid:alcohol ratio), total flow rate = 1 µL min^{-1}.

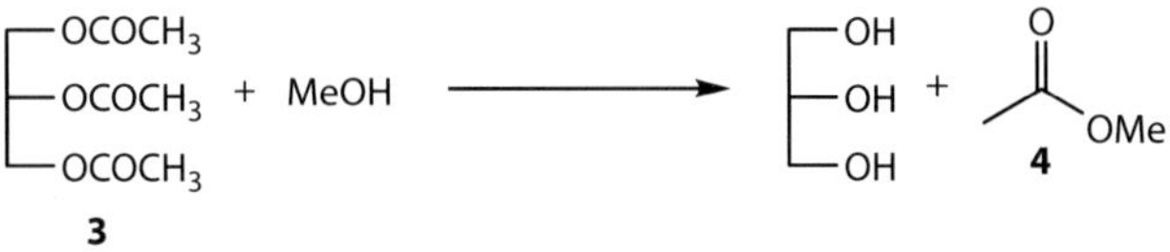

```
    ┌─OCOCH₃                      ┌─OH      O
    ├─OCOCH₃  + MeOH    ──────►   ├─OH  +   ║
    └─OCOCH₃                      └─OH      ─OMe
         3                                   4
```

SCHEME 4.1 Illustration of the model transesterification reaction used to evaluate SO₃H-derived inorganic monoliths under flow conditions.

Vitamin A, based on the drivers at Roche Vitamins, which include a desire to lower running/investment costs and to gain competitive advantage through technology.

With this in mind, the authors investigated the use of a continuous flow reactor as a means of accessing (E)-retinyl acetate **5** from intermediate **6** (Scheme 4.2), without the formation of mono- and diacetylated products that usually accompany the target product formation. Using commercially available Chirazyme L2-C2 **7**, the authors investigated the use of vinyl acetate **8** as the acyl donor within a fixed-bed reactor. Employing a reactor temperature of 50°C and a total flow rate of 1 mL min⁻¹, the authors developed a continuous flow technique for the synthesis of (E)-retinyl acetate **5** in quantitative conversion. In order to obtain a process suitable for long-term use, the authors implemented an EDTA precolumn which served to purify reactant streams thus, preventing denaturation of the biocatalyst **7**. Under the aforementioned conditions, the authors operated the reactor for 100 days with no loss of productivity. In order to access greater quantities of material, the authors subsequently scaled the reactor to a mini-plant and at a substrate concentration of 30 wt.% and a throughput of 10 g **6** min⁻¹ the authors were able to obtain 1.6 kg day⁻¹ of monoester **9**. To reduce the costs of the process further, the by-product acetaldehyde and residual vinyl acetate **8** were separated from the product by distillation and recycled.

Pleased with the synthetic strategy developed, the authors also reported application of the technique toward the synthesis of Vitamin E ester **10**, using the selective hydrolysis of diacetate **11** to afford the monoacetate precursor **12** (Scheme 4.3).

Ester Hydrolyses: Using six-channel micro tubular reactors (dimensions = 0.2 mm (i.d.) × 30 cm (long)) containing an enzyme functionalized mesoporous silica thin film, Endo and coworkers [16] reported the lipase catalyzed hydrolysis of esters.

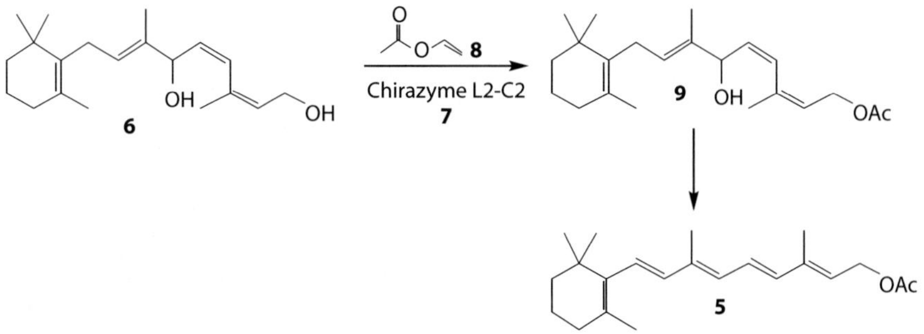

SCHEME 4.2 Schematic illustrating the biocatalytic transformation of intermediate **9** to Vitamin A ester **5**.

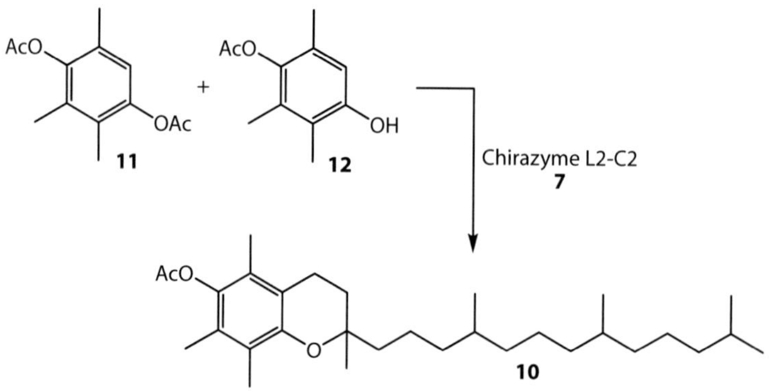

SCHEME 4.3 Lipase-catalyzed ester hydrolysis used in the synthesis of Vitamin E ester **10**.

Employing the hydrolysis of 4-nitrophenyl acetate **13** to 4-nitrophenol **14** as a model reaction, Scheme 4.4, the authors investigated the effect of precursor concentration (12.5–100.0 mg mL^{-1} DMSO) in DMSO/MES buffer and flow rate (0.3–12 µL min^{-1}) at 25°C. Under the aforementioned conditions, the authors were able to determine the rate constant of the product **14** formation in Michaelis–Menten kinetics whereby $k_2 = 3.65 \times 10^{-2}$ mol s^{-1} kg^{-1} lipase, comparing favorably with those determined under batch conditions.

Owing to the reversible nature of esterifications, the ability to control reaction time and readily separate the reaction mixture from the acid or biocatalyst means that higher quality products can be obtained from flow processes.

4.1.2 C–N Bond-Forming Reactions: Azidations

In Chapter 3, a series of techniques are presented for the synthesis of azides, citing their high reactivity and thermal instability as reasons for developing continuous processing tools. In an extension to this, Ley and coworkers [17] described the development of an azide monolith (2.00 mmol g^{-1}) within a flow reactor (dimensions = 15 mm (i.d.) × 10 cm (long)) in which a series of acyl chlorides (1.0 M) were converted into their respective acyl azide with a reaction time of 13 min. Using this approach, the authors were able to isolate a series of acyl azides in high yield and purity by simply removing the reaction solvent (MeCN) *in vacuo*. Compared to previous solution

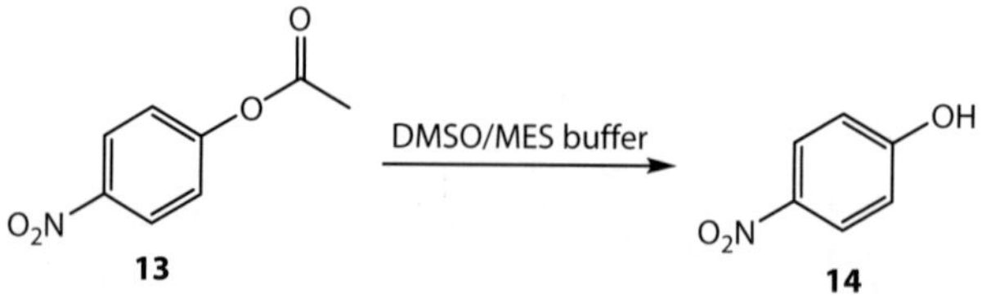

SCHEME 4.4 Illustration of the model reaction used to determine reaction kinetics under micro flow conditions.

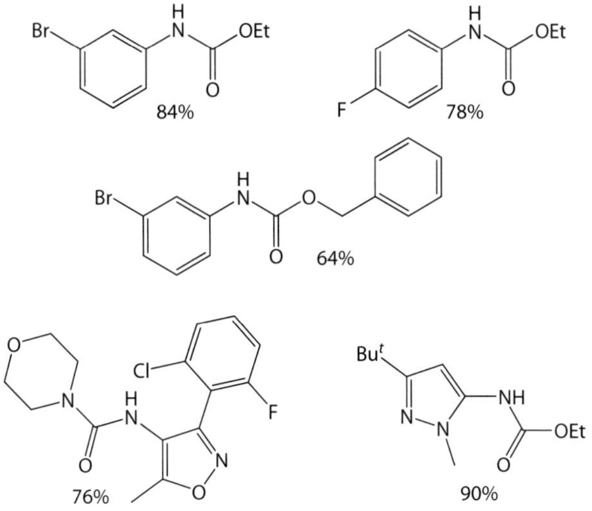

FIGURE 4.1 Illustration of a selection of products obtained via the *in situ* azidation of acyl chlorides and their subsequent Curtius rearrangement and isocyanate trapping with nucleophiles.

phase examples, this approach is advantageous as the resulting products are water free and as such can directly be rearranged to the respective isocyanate. With this in mind, the authors demonstrated the conversion of a series of acyl chlorides into isocyanates and their subsequent reaction with a series of nucleophiles, as illustrated in Figure 4.1.

4.2 ELECTROPHILIC SUBSTITUTION

Unlike liquid–liquid reactions, due to the noncatalytic nature of reactants employed in electrophilic substitution reactions, few transformations of this type have been reported using solid-supported reagents.

4.2.1 BROMINATIONS

As part of an investigation into the development of a continuous flow protocol for the synthesis of Casein Kinase I inhibitors (Figure 4.2), Venturoni et al. [18] investigated the use of polymer-supported pyridine hydrobromide **15** as a means of preparing α-bromoketones, which upon reaction with 3-amino-6-chloropyridazine **16** subsequently furnished the respective imidazopyridazine (Scheme 4.5).

Employing a glass column reactor, containing polymer-supported hydrobromide **15** (5.0 g, 2.0 mmol g^{-1}), the authors investigated the effect of ketone substitution (0.20 M in MeOH) and residence time on the resulting product distribution. Conducting the reaction at a flow rate of 0.25 mL min^{-1}, equivalent to a reaction time of 13 min, the authors obtained a mixture of mono- and dibrominated products;

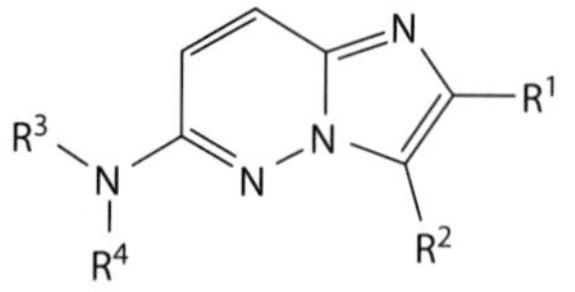

FIGURE 4.2 Schematic illustrating the imidazo[1,2-*b*]pyridazine core motif identified in a series of Casein Kinase I inhibitors.

however, reducing the residence time to 5 min the target α-bromoketones were obtained in quantitative yield upon removal of the reaction solvent. See Section 4.9.6 for a description of the multistep synthesis of Caesin Kinase I inhibitors.

4.2.2 Phosgene Synthesis

Micro reaction technology is widely reported as a means of not only improving process safety, but also providing the opportunity to produce chemicals at the site of use. One such example of this was reported by Jensen and coworkers [19] and focused on the development of a silicon packed-bed reactor for the synthesis of phosgene ($COCl_2$) **17** owing to its widespread use in the synthesis of pharmaceuticals and pesticides (Scheme 4.6).

The reactor under investigation consisted of a silicon etched channel (dimensions = 625 μm (wide) × 300 μm (deep) × 2.0 cm (long)), capped with a Pyrex cover plate. The catalyst was loaded under vacuum and the packed-bed filled with activated carbon (particle size = 53–73 μm) and reactions performed by pumping over a stream of carbon monoxide and chlorine gas in a ratio of 2:1 at 4.5 std. cm³ m⁻¹. To optimize the reaction conditions, the effect of temperature on the process was

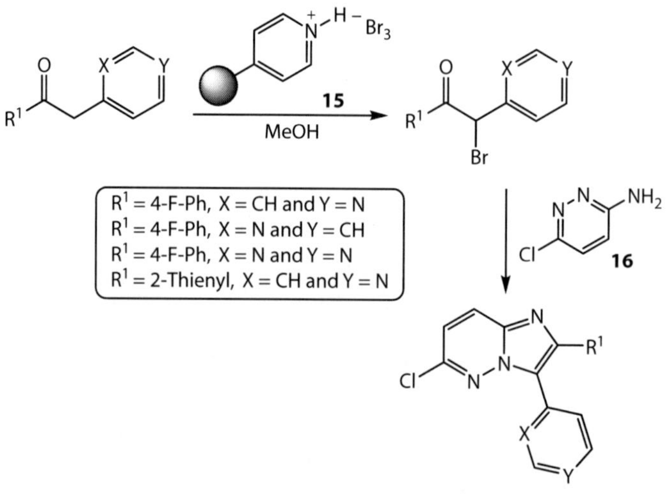

SCHEME 4.5 Schematic illustrating the reaction protocol used for the α-bromination of ketones using polymer-supported pyridine hydrobromide **15**.

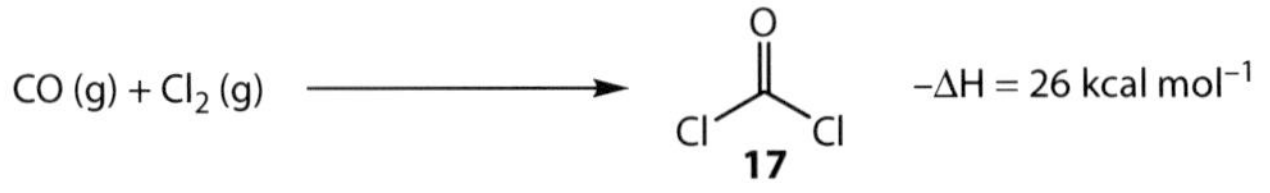

$$CO(g) + Cl_2(g) \longrightarrow \underset{\mathbf{17}}{Cl-\overset{O}{\underset{\|}{C}}-Cl} \quad -\Delta H = 26 \text{ kcal mol}^{-1}$$

SCHEME 4.6 Illustration of the protocol employed for the synthesis of phosgene **17**.

investigated by incrementally increasing the reactor temperature from 25°C to 220°C, with reaction products assessed by MS for chlorine conversion; over the range of 150–210°C, the authors observed increasing Cl_2 conversion from 60% to 100%. Temperatures above 250°C were not investigated due to the potential of etching the device, which is undesirable for long-term, safe operation of the units. Under the optimal conditions reported the productivity of a single reactor was calculated to be 3.5 kg year^{-1} **17**, increasing the flow rate of the reactants to 8 std. cm^3 min^{-1}, enabled the authors to increase productivity to 9.3 kg year^{-1}.

Using this approach, the authors report that the increased heat and mass transfer obtained afford greater safety control due to the suppression of thermal gradients when compared to standard macroscale reactors. With this in mind, employing a 10-channel reactor, with 1:1 CO:Cl_2 at a rate of 8.0 std. cm^3 min^{-1}, production rates of 100 kg year^{-1} would be attainable and therefore, operating these units in parallel, at site synthesis of phosgene **17** will be possible.

4.3 NUCLEOPHILIC ADDITION

Of the wide array of synthetically useful nucleophilic addition reactions available to the chemist, those researchers evaluating reactions under heterogeneous flow conditions have focused on base-catalyzed transformations, with particular attention paid to the Knoevenagel condensation, Henry reaction and Michael addition. More recently, the reaction scope has been extended to include the metal-catalyzed Diels–Alder reaction and enzyme-promoted benzoin condensations.

4.3.1 C–C Bond-Forming Reactions: Knoevenagel Condensation

Packed-Bed Reactors: In 2004, Haswell and coworkers [20] reported the development of a capillary flow reactor (dimensions = 500 μm (i.d.) × 3.0 cm (long)) in which they incorporated a series of solid-supported organic bases (catalyst = 5 mg). Using EOF as the pumping mechanism, the authors investigated the effect of flow rate and supported base on the condensation reaction between ethylcyanoacetate **18** and benzaldehyde **19** to afford 2-cyano-3-phenyl acrylic acid ethyl ester **20** (Scheme 4.7). Initially investigating the catalytic effect of silica-supported piperazine **21**, the authors employed an applied field which resulted in pumping of a premixed solution of benzaldehyde **19** and ethylacetoacetate **18** (1.0 M respectively) in MeCN through the packed-bed and into a collection vessel containing MeCN. The reaction products were periodically collected and analyzed by GC whereby the percentage product **20** formed was quantified.

Using this approach, the authors were able to tune the residence time within the packed bed to afford complete conversion of the starting materials **18** and **19** into the

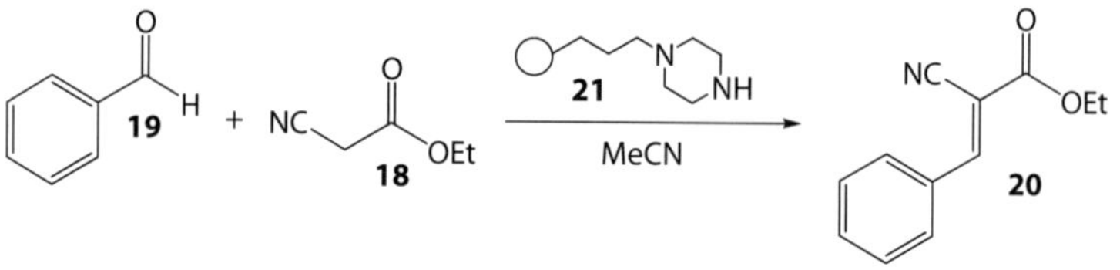

SCHEME 4.7 Illustration of the model reaction used to investigate the performance of the Knoevenagel condensation reaction using the pumping mechanism.

ethyl ester derivative **20**, evaporation of the reaction solvent then afforded the target compound **20** as an analytically pure material. Substituting benzaldehyde **19** for a series of aromatic aldehydes, the authors were able to rapidly generate a small compound library, employing analogous reaction conditions whereby isolated yields ranged from 98.9% to 99.9%.

In a second example, the authors investigated the use of wide bore capillary reactors (dimensions = 3 mm (i.d.)) as a means of increasing system throughput, this time employing 100 mg of catalyst material **21** [21]. Again employing EOF as the pumping mechanism, the authors obtained the target compound in excellent yield and purity (99.9% and 99.7%, respectively), this time affording a throughput of 0.75 g h^{-1} compared with 8.43×10^{-3} g h^{-1} obtained previously.

McQuade and coworkers [22] subsequently reported the use of polymer-supported tertiary bases packed within FEP tubing (dimensions = 1.6 mm (i.d.)) for the same transformation at 60°C. Using a 1,5,7-triazabicyclo[4.4.0]undec-3-ene derived polymer support the authors obtained 2-cyano-3-phenyl acrylic acid ethyl ester **20** in 93% conversion with a residence time of 5 min. Under the aforementioned conditions, the authors were able to produce the condensation product at a throughput of 200 mg h^{-1}.

Monolithic Reactors: More recently, Coq and coworkers [14] demonstrated the use of amine functionalized inorganic monoliths, housed within a heat shrinkable polymer tube, for activity toward the Knoevenagel condensation reaction between benzaldehyde **19** and ethylcyanoacetate **19** (Scheme 4.7). Using DMSO as the reaction solvent, the authors pumped a premixed solution of ethylcyanoacetate **18** (0.68 M) and benzaldehyde **19** (0.80 M) through the monolith at a flow rate of 0.5 mL min^{-1}. Under the aforementioned conditions, the authors obtained 80% conversion to the target compound **20** equating to a productivity of 178 [mol min^{-1} g^{-1}] × 10^{-5} and demonstrating 7.4 times increase compared to batch and 1.64 times increase compared to a packed-bed reactor; supporting findings from previous work utilizing polymeric monoliths [23].

Wall-Coated Micro Reactors: Again employing the Knoevenagel condensation reaction between benzaldehyde **19** and malononitrile as a model, Verboom and coworkers [24] investigated the catalytic activity of a nanostructure-based polymer brush as a wall-coated catalyst within silicon micro channels.

Using a process of surface functionalization via surface-initiated polymerization, the authors were able to construct 1,5,7-triazabicyclo[4.4.0]dec-5-ene (TBD) derived polymer monolayer (thickness = 150 nm) within the micro channel. The resulting micro channels (dimensions = 100 μm (wide) × 100 μm (deep) × 103 cm (long)) were then evaluated and compared to a series of nonfunctionalized polymeric brushes.

Employing malononitrile in a 16- to 48-fold excess (1.2×10^{-3} M) with respect to benzaldehyde **19** (7.5×10^{-5} M), the authors investigated the effect of reaction time on the formation of 2-benzylidene malononitrile at 65°C; quantified using in-line UV detection. Using this approach, the authors were able to determine that the reaction behaved as pseudo first order (1.3×10^{-2} s^{-1}) with the respective rate constants obtained proportional to the concentration of malononitrile. In addition, the authors observed no degradation in catalytic activity of the polymer brushes after being used 25 times; repetition of the above investigation 30 days later also confirmed stability of the polymeric coating.

In addition, Yeung and coworkers [25] demonstrated the condensation of benzaldehyde **19** and ethylacetoacetate in the presence of a Cs-exchanged zeolite X catalyst coated on the walls of a series of SS-316L micro channels (dimensions = 300 μm (wide) × 600 μm (deep) × 2.5 cm (long)). Using this approach, the authors were able to obtain the target product, 2-acetyl-3-phenylacrylic acid ethyl ester, in 60% conversion, and 78% selectivity.

4.3.2 C–C BOND-FORMING REACTIONS: MICHAEL ADDITIONS

In an early example of enantioselective synthesis under flow, Hodge and coworkers [26] reported the performance of a Michael addition using a polymer-supported 3° amine catalyst **22** for the reaction between methyl-1-oxoindan-2-carboxylate **23** (0.5 M) and methyl vinyl ketone **24** (0.53 M) to afford adduct **25** (Scheme 4.8).

Performing reactions at 20–50°C, the authors investigated the effect of flowing reactant mixtures through the packed bed at a flow rate of 117–350 μL min^{-1}, which corresponded to a residence time of 6 h. Employing Amberlyst-21 as the catalyst, the

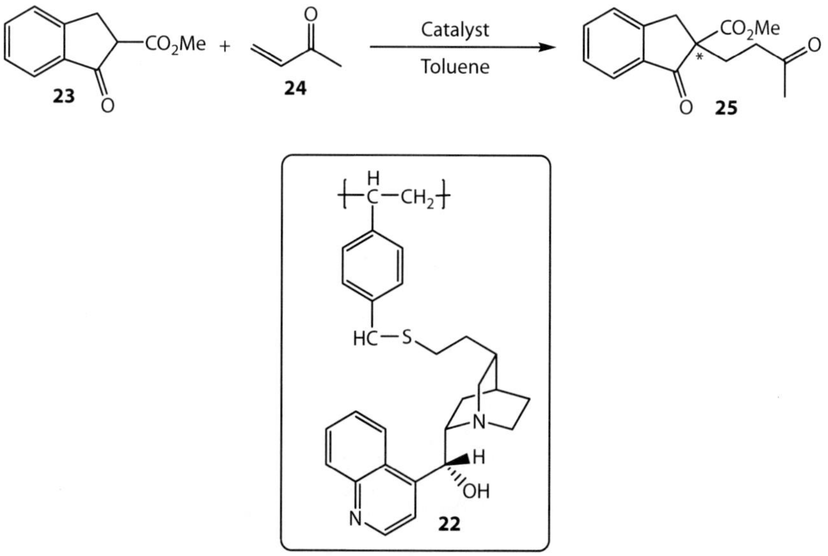

SCHEME 4.8 Illustration of the model Michael addition used to demonstrate the performance of enantioselective addition reactions under continuous flow conditions.

authors obtained yields ranging from 14% to 99% **25** with no enantioselectivity. Replacing the catalyst with a polymer-supported cinchonidine **22**, the authors obtained 92% to 96% **25** with 51–52% *ee* forming the (*S*)-enantiomer in excess. Unlike conventional batch processes, *ee*'s were not affected by an increase in reactor temperature (50°C), which enabled the authors to increase productivity of their reactor while obtaining purities equivalent to those obtained in batch.

4.3.3 C–C Bond-Forming Reactions: Henry Reaction

Using a combination of commercially available meso reactors (Uniqsis, UK) and packed-bed reactors, Ley and coworkers [27] demonstrated the development of a continuous flow protocol for the synthesis of α-ketoesters, based on a desire to increase the ease of synthesizing such compounds with high structural diversity.

In the first instance, the authors investigated the continuous flow synthesis of nitro-olefins via the base-catalyzed Henry reaction, as Scheme 4.9 illustrates. Using a tubular reactor, solutions of nitro alkane (0.1 M) and ethyl glyoxylate **26** (0.1 M) in toluene were mixed and pumped through a packed-bed containing MP-carbonate polymer **27** (0.36 g, 2.8 mmol g^{-1}) at a flow rate of 100 µL min^{-1}. The product stream was collected, dried over Na$_2$SO$_4$ and concentrated *in vacuo* prior to dissolution in toluene (25 mL). In a second flow reaction, the nitroalkanol was reacted with trifluoroacetic anhydride **28** (1.5 eq.) and Et$_3$N **29** (2.5 eq.) at room temperature with a residence time of 3.3 h. The reaction product was subjected to an offline aqueous extraction and column chromatography to afford the target nitroolefin.

In a third step, a solution of the nitroolefin (0.15 M) in MeCN was pumped through a packed-bed containing QuadraPure–BZA **30** (1.09 g, 5.5 mmol g^{-1}) at a flow rate of 200 µL min^{-1} and captured by the solid-supported reagent. The packed-bed was subsequently heated to 65°C and an aliquot of 1,1,3,3-tetramethylguanidine **31** (0.3 M) in MeCN passed through at a flow rate of 100 µL min^{-1}. This was followed by a wash solution of MeCN (20 mL) in order to remove any impurities prior to hydrolyzing the

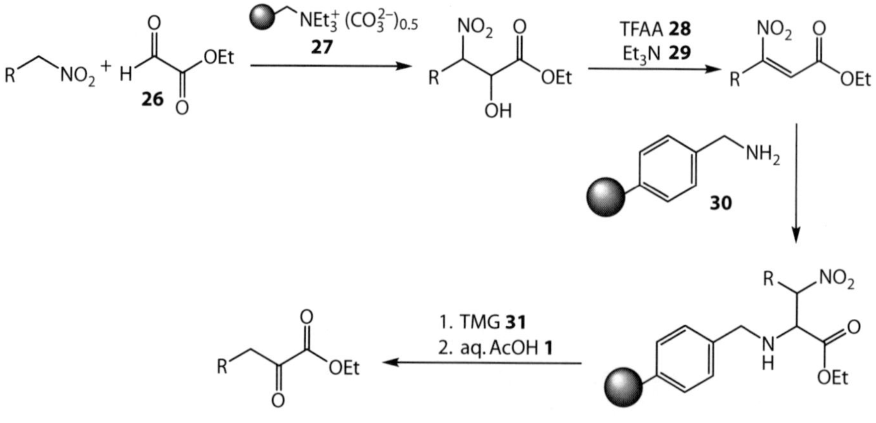

SCHEME 4.9 Schematic illustrating the steps involved in the synthesis of nitroolefins and α-ketoesters under continuous flow.

immobilized enamino acid ester with aqueous acetic acid **1** (Scheme 4.8). The eluent was collected, along with the column washings (EtOAc, 40 mL) and subjected to an offline aqueous extraction which afforded the target α-ketoester in high purity.

By utilizing the principle of catch, react, and release, the authors were able to synthesize ten α-ketoesters in moderate isolated yields and excellent purities (≥97% by NMR spectroscopy), as summarized in Table 4.2.

TABLE 4.2
Illustration of the α-Ketoesters Synthesized Using a Catch-React and Release Strategy under Flow Conditions

α-Ketoester	Yield (%)
	36
	45
	44
	31
	44
	42
	43
	45
	41
	48

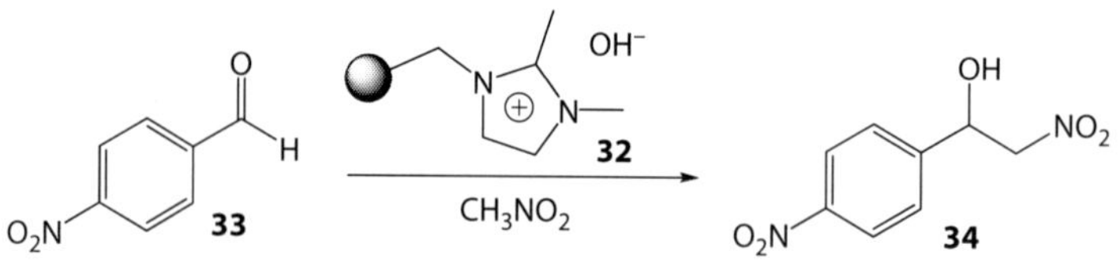

SCHEME 4.10 Illustration of the base-SILLP **32**, evaluated for activity toward the Henry reaction performed under continuous flow conditions.

Using a polymer-supported ionic liquid-like phase (SILLP) **32**, Garcia-Verdugo and coworkers [28] demonstrated the continuous flow Henry reaction illustrated in Scheme 4.10. To perform a reaction, the authors dissolved 4-nitrobenzaldehyde **33** (1.66 mmol) in nitromethane (25 eq.) and the solution recirculated through a packed-bed reactor at room temperature and upon completion, the reactor was purged with MeOH. Using this approach, the authors investigated the effect of anion on the supported reagent and polymer type, finding that monolithic supports reduced the system back-pressure and the hydroxyl anion afforded the most stable catalyst. With this in mind, the authors were able to develop a system capable of synthesizing 2-nitro-1-(4-nitrophenyl)ethanol **34** in 95% yields with no variation in product quality after 47 bed volumes had been reacted. In addition, using a continuous flow technique the authors could readily isolate any unreacted nitromethane and recycle it within the system.

4.3.4 C–C Bond-Forming Reactions: Diels–Alder

Depositing thin Pd films (<10 µm) on the inside of capillary tube reactors (dimensions = 1180 µm (i.d.) × 12 cm (long)) and employing microwave irradiation, Shore and Organ [29] demonstrated the ability to dramatically enhance the rate of reaction for a series of Diels–Alder reactions. From the data collected and the investigation conducted, the results of which are summarized in Table 4.3, the authors concluded that the Pd film had a dual role, firstly that of a heat source and secondly as a catalyst. When conducted in an oil bath in the absence of a Pd-film (205°C), the authors obtained only 54% of the target (1R,4S)-dimethyl 7-oxabicyclo[2.2.1]hepta-2,5-diene-2,3-carboxylate **35** and was subsequently increased to 72% upon employing a Pd-coated capillary. Employing microwave irradiation as an alternative heating technique, the authors obtained 10% conversion to **35** in the absence of Pd, 47% conversion when the reactor was coated on the outside and 90% **35** when the Pd-coating was in contact with the reactant solution; confirming the presence of a catalytic effect. With this in mind, the authors investigated a series of cycloadditions, obtaining all target compounds in >90% conversion.

4.3.5 C–C Bond-Forming Reactions: Benzoin Condensation

Building on their experience in the field of polymer-assisted solution phase synthesis (PASS) under continuous flow conditions, Kirschning and coworkers [30] demonstrated the use of immobilized His$_6$-tagged proteins, Figure 4.3, as highly potent biocatalysts toward the benzoin reaction (Scheme 4.11).

TABLE 4.3
Summary of the Reaction Conditions Investigated to Identify the Source of Rate Enhancement for the Continuous Flow Diels-Alder Reaction

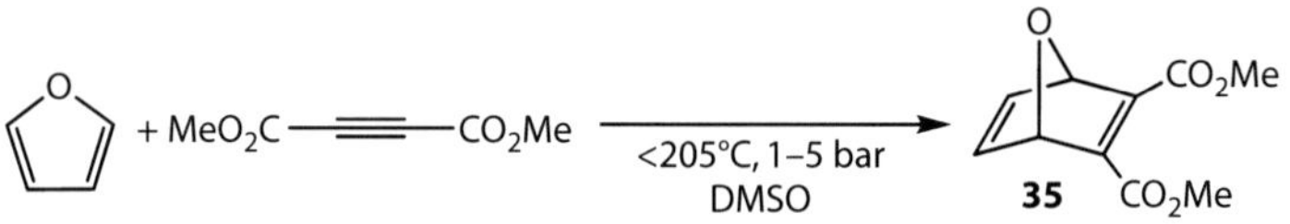

Capillary Treatment	Temperature (°C)	Heat Source	Conversion (%)[a]
None	205	Oil bath	54
Pd-coated, inside	205	Oil bath	72
None	115	MW	10
Pd-coated, outside	205	MW	47
Pd-coated, inside	205	MW	90

[a] Conversion determined by ^{1}H NMR spectroscopy.

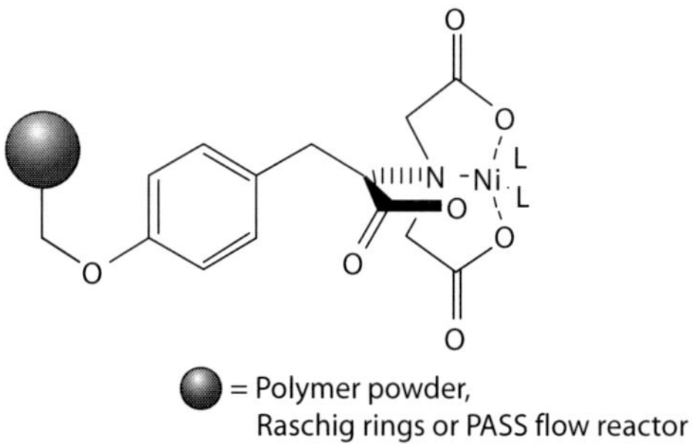

FIGURE 4.3 Schematic representation of the tyrosine-based support used for the immobilization of His$_6$-tagged proteins for use in PASS flow reactors.

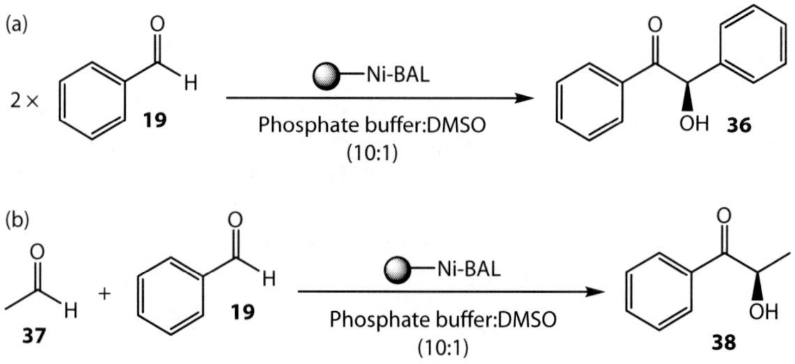

SCHEME 4.11 Illustration of the (a) benzoin and (b) cross-benzoin condensation reactions performed using immobilized benzaldehyde lyase (BAL).

To evaluate the activity of the resulting biocatalyst, the authors initially investigated the benzoin reaction of benzaldehyde **19** (5.5×10^{-2} M) in phosphate buffer (pH 7, containing 10% DMSO), a reaction temperature of 37°C. Recirculating the reaction mixture through the packed-bed reactor the authors were able to obtain reaction times of 90 min (1 mL min^{-1}) affording 99.5% (*R*)-benzoin **36** as quantified by offline GC analysis. Subsequent investigations into the use of more concentrated substrate solutions (0.2–1.0 M) were met with recovery of 94% benzaldehyde **19**, meaning that longer reaction times were necessary to access higher conversions (typically 9 h).

As an extension to this investigation, the authors evaluated the cross-benzoin reaction between acetaldehyde **37** (2.3×10^{-1} M) and benzaldehyde **19** (4.6×10^{-2} M), to synthesize (*R*)-2-hydroxy-1-phenylpropan-1-one **38**, which was obtained in 92% isolated yield with 8% (*R*)-benzoin **36** formed as a minor by-product.

4.3.6 C–C Bond-Forming Reactions: Trifluoromethylation under Continuous Flow

Owing to the metabolic stability of organofluorine compounds, the efficient incorporation of fluorine into molecules is a synthetically useful procedure. Based on previous findings within their research group, Baumann et al. [31] therefore, demonstrated the use of a trimethylammonium fluoride monolith (6.5 g, 0.75 mmol g^{-1}) within a glass reactor (10 mm (i.d.) × 10 cm (long)) for the continuous flow trifluoromethylation of a series of aldehydes in the presence of Ruppert reagent (TMS–CF$_3$) **39**. To perform a reaction, the authors brought together at a T-mixer solutions of TMS–CF$_3$ **39** (0.3 M) and the aldehyde (0.25 M) in THF, prior to pumping through the monolithic reactor, containing the fluoride monolith, at a total flow rate of 300 µL min^{-1}. Upon exiting the monolithic reactor, the reaction mixture entered a heated (40°C) residence time unit (volume = 20 mL) which afforded a reaction time of 2 h. The crude reaction products were subsequently purified using a series of in-line scavengers, PS-benzaldehyde (1.2 mmol g^{-1}), QuadraPure–SA (3.0 mmol g^{-1}) and PS–TsNHNH$_2$ (2.84 mmol g^{-1}) contained within glass packed-bed reactors (10 mm (i.d.) × 10 cm (long)) and the resulting reaction product analyzed by ^{1}H and ^{19}F NMR spectroscopy.

Using this approach, the authors were able to obtain a series of trifluoromethylated aldehydes in high to excellent yield, as summarized in Table 4.4.

In addition to demonstrating wide substrate compatibility for this transformation, the authors also evaluated the oxidation of 2,2,2-trifluoro-1-(5-nitrobenzo[*d*][1,3]dioxol-6-yl)ethanol **40**, using manganese dioxide **41**, illustrating the ability to access trifluormethylated ketones, such as 2,2,2-trifluoro-1-(6-nitrobenzo[*d*][1,3]dioxol-5-yl)ethanone **42** in 93% yield (Scheme 4.12) (see Chapter 3 for other examples of oxidation reactions performed under flow).

4.3.7 C–C Bond Formation: Aldol Reaction

Owing to the industrial importance of 2-methylpentenal **43**, a compound widely employed in the fragrance, flavors, and cosmetic industries, Poliakoff and coworkers

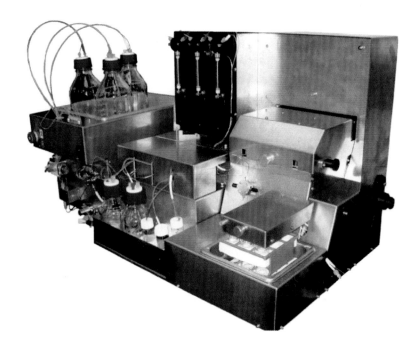

FIGURE 1.14 The Propel system from Accendo Corporation, USA. (Reproduced by permission of Accendo Corporation, 2010, Belmont, CA, USA.)

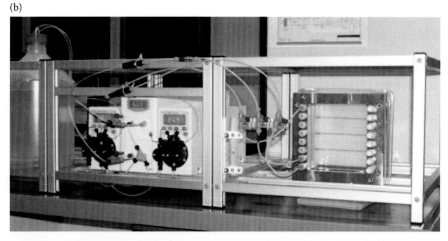

FIGURE 1.16 Glass-based flow reactor platforms: (a) Labtrix and (b) Plantrix from Chemtrix BV, The Netherlands. (Reproduced by permission of Chemtrix BV, 2010, Geleen, The Netherlands.)

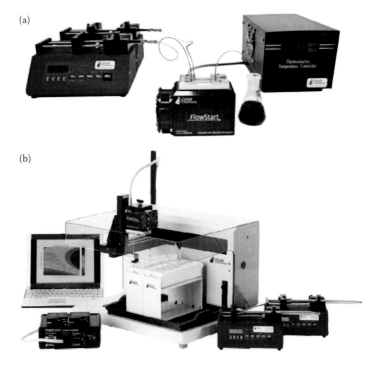

FIGURE 1.17 Glass-based flow reactor platforms: (a) FlowStart and (b) FlowScreen, from FutureChemistry BV, The Netherlands. (Reproduced by permission of FutureChemistry BV, 2010, Njimegen, The Netherlands.)

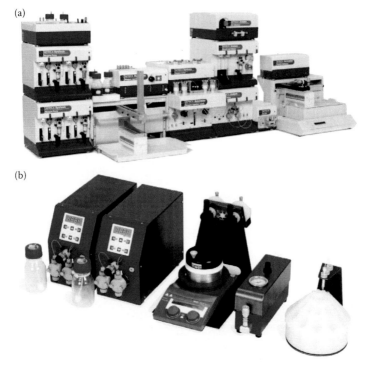

FIGURE 1.18 Glass-based flow reactor platforms: (a) AFRICA and (b) FRX system, from Syrris Ltd., UK. (Reproduced by permission of Syrris Ltd., 2010, Royston, UK.)

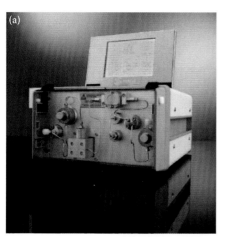
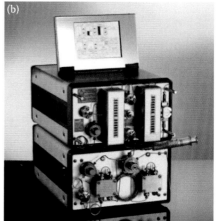
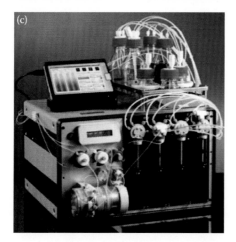

FIGURE 1.19 Tubular reactor systems: (a) H-Cube, (b) X-Cube, and (c) O-Cube from ThalesNano Inc., Hungary. (Reproduced by permission of ThalesNano Inc., 2010, Zahony, Hungary.)

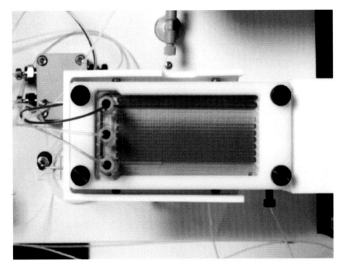
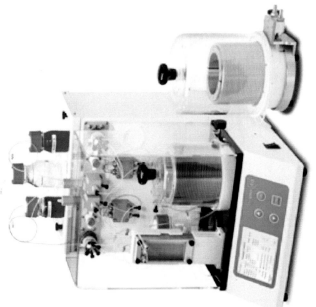

FIGURE 1.20 FlowSyn, a tubular and mesoreactor system from Uniqsis, UK. (Reproduced by permission of Uniqsis Ltd., 2010, Shepreth, UK.)

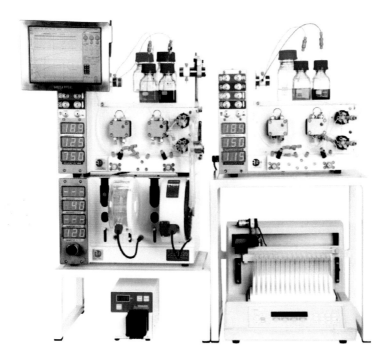

FIGURE 1.21 Vapourtec R-series, a flexible tubular flow reactor system from Vapourtec Ltd., UK. (Reproduced by permission of Vapourtec Ltd., 2010, Bury St Edmonds, UK.)

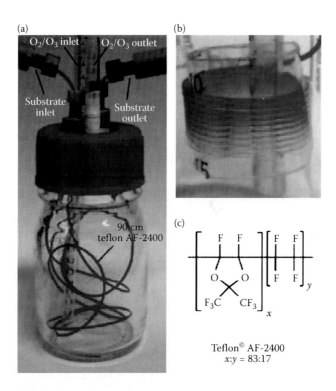

FIGURE 2.3 Photograph of the experimental set-up employed for the continuous flow ozonolysis of alkenes; (a,b) reactor filled with Sudan Red to enable visualization of the process and (c) porous tube structure. (Reproduced with permission from O'Brien, M., Baxendale, I. R., and Ley, S. V. 2010. *Org. Lett.* 12(7): 1596–1598. Copyright (2010) American Chemical Society.)

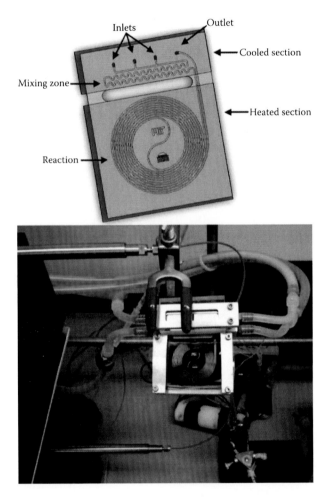

FIGURE 3.5 Illustration of the silicon micro reactor used to evaluate the continuous flow synthesis of β-amino alcohols which utilized water cooling to maintain low temperatures at the chuck. (Reproduced with permission from Bedore, M. et al. 2010. *Org. Proc. Res. Dev.* 14: 432–440. Copyright (2010) American Chemical Society.)

TABLE 4.4
Summary of the Results Obtained for the Trifluoromethylation of Aldehydes under Continuous Flow

Aldehyde	Product	Yield (%)
2-nitrobenzaldehyde	1-(2-nitrophenyl)-2,2,2-trifluoroethanol	84
5-(1-methyl-5-(trifluoromethyl)-1H-pyrazol-3-yl)thiophene-2-carbaldehyde	corresponding trifluoromethyl carbinol	88
pyridine-3-carbaldehyde	2,2,2-trifluoro-1-(pyridin-3-yl)ethanol	86
2-bromobenzaldehyde	1-(2-bromophenyl)-2,2,2-trifluoroethanol	80
3,5-dimethylisoxazole-4-carbaldehyde	corresponding trifluoromethyl carbinol	88
6-nitrobenzo[d][1,3]dioxole-5-carbaldehyde	corresponding trifluoromethyl carbinol	76
1-methyl-1H-indole-2-carbaldehyde	corresponding trifluoromethyl carbinol	79
benzofuran-2-carbaldehyde	corresponding trifluoromethyl carbinol	84
5-chloro-3-methyl-1-phenyl-1H-pyrazole-4-carbaldehyde	corresponding trifluoromethyl carbinol	87
3-phenylbutanal	1,1,1-trifluoro-4-phenylpentan-2-ol	83

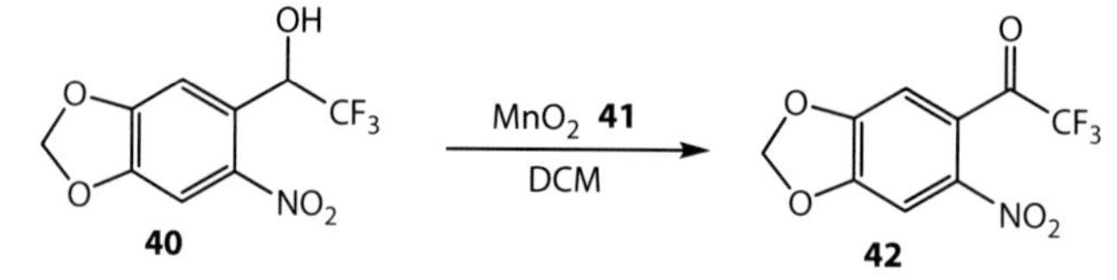

SCHEME 4.12 Illustration of the facile transformation of a trifluoroalcohol **40** to a trifluoroketone **42** using MnO$_2$ **41**.

[32] investigated the use of supercritical carbon dioxide (scCO$_2$) as a reaction solvent as a means of developing a more environmentally attractive route to the compound **43** (Scheme 4.13). Under conventional conditions, the enal **43** is synthesized via the NaOH **44** catalyzed aldol reaction whereby 99% conversion of the raw material is consumed and the target product **43** is formed in 86% selectivity. In addition to the choice of an environmentally benign solvent, the authors aimed to further increase the cleanliness of the reaction by employing solid-supported acids and bases as catalysts packed into a 316 SS tube reactor (dimensions = 12 mm (o.d.) × 10 cm (long)).

Employing an aldehyde **45** flow rate of 100 µL min^{-1}, a CO$_2$ flow rate of 1.0 mL min^{-1} (at 10 MPa), the effect of temperature and catalyst on the reaction selectivity was investigated readily identifying Amberlyst-15 **46** as a viable catalyst for the selective synthesis of 2-methylpentenal **43**. Conducting further optimization reactions, the authors confirmed that conducting reactions in the absence of CO$_2$ resulted in insignificant enal **43** formation (<130°C), compared with 48% conversion and 92% selectivity (90°C) in the presence of CO$_2$. Reducing the flow rates to 75 µL min^{-1} and 0.75 mL min^{-1} respectively, resulted in a slight increase in conversion (51%) and an increase in selectivity toward **43** (95%). Further, increases in reactor temperature did however lead to reductions in selectivity. With this information in hand, the authors looked at increasing the packed-bed length with conversions increased to 80% with a doubling of the reactor length (>99% selectivity). Gratified by these findings, the authors evaluated the use of a mixed packed-bed, containing

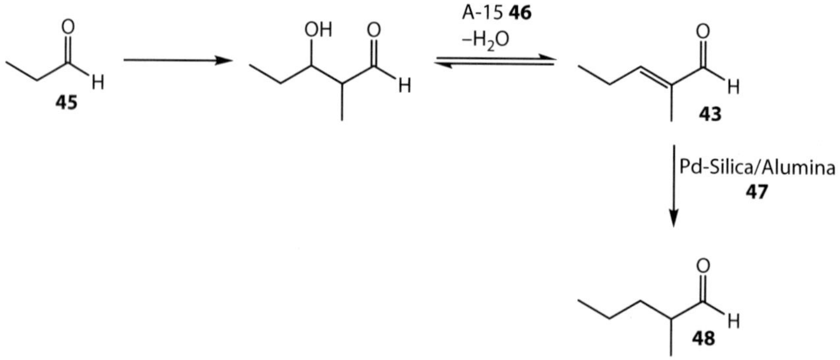

SCHEME 4.13 Illustration of the acid-catalyzed synthesis of 2-methylpentenal **43** and subsequent hydrogenation to afford 2-methylvaleraldehyde **48**.

Amberlyst-15 **46** and Pd on silica/alumina **47**, successfully performing the *in situ* hydrogenation of **43** to 2-methylvaleraldehyde **48**.

4.3.8 C–N Bond Formation: Cycloaddition Reactions

The advent of Click chemistry [33] and the identification of its synthetic utility have led many researchers to use this powerful synthetic tool in the synthesis of an array of biologically active intermediates and products. More recent contributions to the field have included the development of solid-supported catalysts **49**, by Girard et al. [34], to facilitate the ease of catalyst recovery, and product isolation and more recently the incorporation of such materials into continuous flow reactor modules by Ley and coworkers [35] (Scheme 4.14).

Coupling the immobilized copper (I) iodide reagent **49** with two immobilized scavengers, QuadraPure-TU (thiourea) **50**, and a phosphane resin **51** retained within separate glass packed-bed reactors (6.6 mm (i.d.) × 150 mm (long)), the authors investigated the synthesis of 1,4-disubstituted-1,2,3-triazoles by pumping a premixed solution of organic azide (0.15–0.20 M) and terminal alkynes (0.1 M), in DCM, through the reactors at a flow rate of 30 µL min^{-1}. The crude reaction products were then concentrated using a solvent evaporator and evaluated by LC–MS, ^{1}H and ^{13}C NMR spectroscopy. A selection of the 14 compounds synthesized is presented in Table 4.5 and illustrate the high yielding nature of this synthetic protocol, applicable for the gram-scale synthesis of such compounds.

As observed with all forms of compound library generation, the incorporation of structural diversity is limited to those substrates that can be readily purchased or synthesized. With this in mind, the authors subsequently reported the *in situ* preparation of alkynes, from commercially available aldehydes, as a means of increasing the type of alkynes available for incorporation into the 1,4-disubstituted-1,2,3-triazoles [36].

To achieve this, the authors employed the Bestmann–Ohira reagent **52** (0.10 M), in the presence of KO^{t}Bu **53** (0.12 M) and the aldehyde (≤0.15 M), at 100°C in a PTFE tubular flow reactor (residence time = 30 min). Upon exiting the flow reactor, the crude alkyne was pumped through a series of scavenger columns, selected in order to remove any unreacted starting materials and acidic residues from the alkyne (Scheme 4.15). Using this approach, the terminal alkynes were obtained in moderate yield (65–82%) and excellent purity; as determined by ^{1}H NMR spectroscopy.

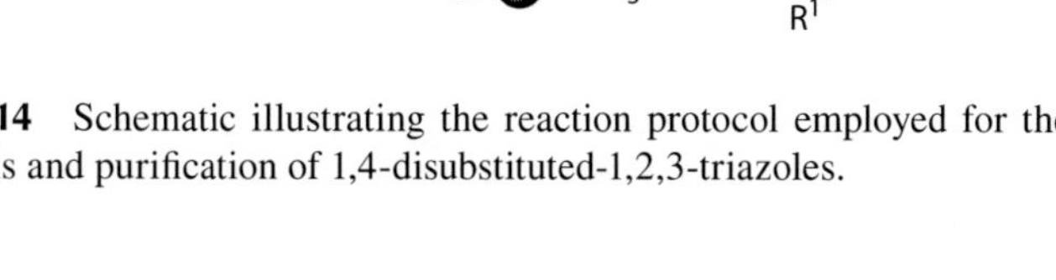

SCHEME 4.14 Schematic illustrating the reaction protocol employed for the continuous flow synthesis and purification of 1,4-disubstituted-1,2,3-triazoles.

TABLE 4.5
A Selection of the 1,4-Disubstituted-1,2,3-Triazoles Synthesized under Continuous Flow Using Girard's Recyclable Catalyst 49

Triazole	Yield (%)
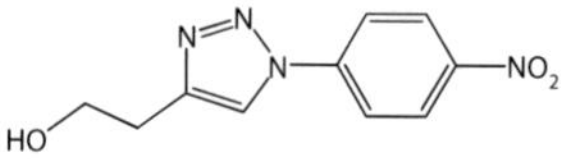 1-(4-nitrophenyl)-4-(2-hydroxyethyl)-1,2,3-triazole	93
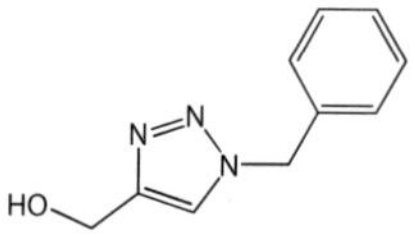 1-benzyl-4-(hydroxymethyl)-1,2,3-triazole	85
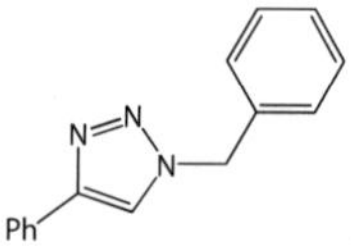 1-benzyl-4-phenyl-1,2,3-triazole	85
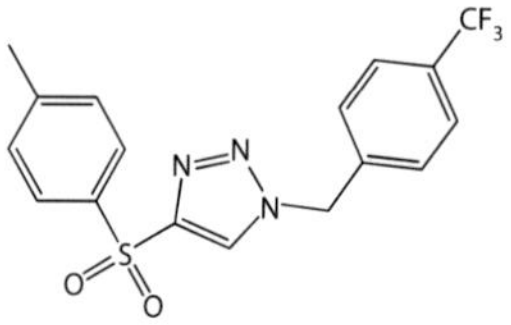 1-(4-trifluoromethylbenzyl)-4-(4-methylphenylsulfonyl)-1,2,3-triazole	91
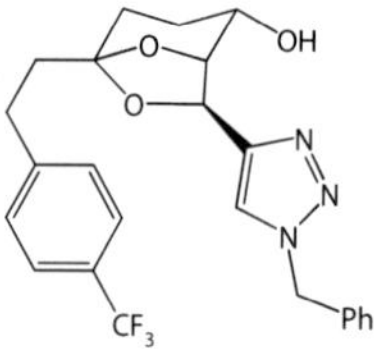 sugar-derived triazole with CF$_3$ and Ph substituents	88

As an extension to this, the authors coupled the acetylene flow reactor to the packed-bed reaction set-up previous described to demonstrate the synthesis of 1,4-disubstituted-1,2,3-triazoles from commercially available aldehydes. Using this approach, the authors were able to generate highly reactive, unstable or hazardous intermediates with ease, enabling the generation of diverse compound libraries (Table 4.6).

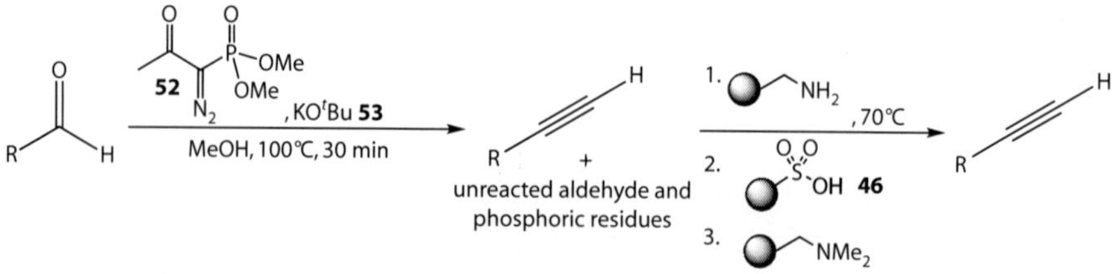

SCHEME 4.15 Illustration of the reaction protocol employed for the continuous flow synthesis of substituted alkynes from commercially available aldehydes.

TABLE 4.6
Summary of the Results Obtained for the Continuous Flow Synthesis of 1,4-Disubstituted-1,2,3-Triazoles from Substituted Aldehydes

R	R¹	Yield (%)
4-(MeO₂C)C₆H₄–	benzyl	62 (95)[a]
2-naphthyl	benzyl	65 (95)
4-biphenyl	4-F-benzyl	55 (>95)[b]

[a] Product purity.
[b] Aldehyde generated via the *in situ* oxidation of the respective 1° alcohol; therefore, a three-step process.

Mechanistic Investigations: More recently, Kappe and coworkers [37] investigated the copper-catalyzed azide–alkyne cycloaddition reaction using copper in charcoal (Cu/C) **54**, along with a series of other heterogeneous copper metal sources, under continuous flow as a means of gaining mechanistic insight into the cycloaddition reaction.

Using the cycloaddition of benzyl azide **55** and phenylacetylene **56**, to afford 1-benzyl-4-phenyl-1*H*-1,2,3-triazole **57** as a model reaction, Scheme 4.16, the authors investigated the effect of reaction time and temperature on the Cu/C **54** catalyzed reaction within a packed-bed reactor. With initial results appearing promising, the authors obtained the triazole **57** in >99% isolated yield at a residence time of 3 min and a reactor temperature of 170°C (X-Cube™, ThalesNano Inc., Hungary). The authors subsequently investigated the reaction at higher flow rates and although no change in conversion was obtained, stripping of the Cu from the support was observed, via a proposed route of desorption, during the catalytic process; as previously observed for the Mizoroki–Heck reaction [38].

Using ICP-MS analysis for quantification purposes, the authors subsequently investigated the effect of reactant composition on the percent Cu leached at a fixed temperature (170°C), pressure (20 bar) and flow rate (1.5 mL min⁻¹); highlighting increased Cu release in the presence of the benzyl azide **55** (59 µg Cu) organic base Et₃N **29** (38.6 µg Cu) and 1,2,3-triazole **57** (71 µg Cu). With the reaction mixture

SCHEME 4.16 Illustration of the model azide–alkyne cycloaddition employed in the mechanistic evaluation of 1,2,3-triazole **57** synthesis under continuous flow conditions.

resulting in the leaching of 600 µg g^{-1} of product **57**, exceeding the proportion of Cu residues permitted in pharmaceutical compounds, 15 mg kg^{-1}, the authors employed an in-line scavenger cartridge containing QuadraPure™ TU 50 resin or activated charcoal; enabling a reduction in Cu from 600 mg kg^{-1} to <1 and 6 mg kg^{-1} respectively. Due to the presence of leaching, the authors subsequently investigated the lifetime of the catalyst within the reactor, processing 1 L of reaction mixture (0.25 M azide **55** and 0.27 M alkyne **56**) prepared in acetone. Samples were taken each hour over a period of 11 h and assessed via HPLC-UV (conversion), ICP-MS (Cu contamination) and isolated yield. During the initial 4 h, the authors obtained the target compound **57** in >99% yield and 98% purity (32 g); however, after this time a significant drop in conversion was obtained and the isolated product contained significant quantities of unreacted starting materials. ICP-MS analysis of the catalytic material from within the packed-bed confirmed >60% leaching had occurred, with a decrease from 8.7 to 3.3% Cu. This investigation therefore, enabled the authors to confirm that the reaction involves a zerovalent Cu, connected to a surface layer of Cu_2O, and a homogeneous mechanism is in operation. Although this protocol is not suitable for large-scale preparation, from the volumes of materials produced it can be seen that the technique is useful for the preparation of 1,2,3-triazoles in 10′s of g's.

4.3.9 C–O Bond-Forming Reactions: Acetalizations

With EOF-based pumping techniques viewed as being limited in application due to their intolerance to acidic reactant mixtures, Haswell and coworkers [39] investigated the use of a series of solid-supported acid catalysts within capillary based (dimensions = 500 µm (i.d.) × 3.0 cm (long)) as a means of increasing the reaction scope associated with the pumping technique.

Using the acetalization of a series of aldehydes (1.0 M) using trimethylorthoformate **58** (2.5 M) in MeCN, the authors investigated the effect of reaction time on the degree of aldehyde protection obtained. Performing all reactions at room temperature, the authors identified flow rates ranging from 0.5 to 2.0 µL min^{-1} to be optimal for the transformation, as illustrated in Table 4.7. Again the reaction products were readily isolated upon evaporation of the reaction solvent, affording the target dimethyl acetals in analytical purity; as determined by analysis using ^{1}H NMR spectroscopy, MS and elemental analysis. In addition to the protection of aldehydes, the authors were also able to demonstrate the acid-catalyzed deprotection of acetals

TABLE 4.7
Summary of the Results Obtained for the Heterogeneously Catalyzed Acetalization under EOF Conditions

Aldehyde	Flow Rate (µL min^{-1})	Yield (%)
Benzaldehyde **19**	1.75	99.8
4-Bromobenzaldehyde	1.00	99.9
4-Chlorobenzaldehyde	1.60	99.8
4-Cyanobenzaldehyde	2.00	99.6
2-Naphthaldehyde	1.40	99.8
Methyl-4-formylbenzoate	0.60	99.9
3,5-Dimethoxybenzaldehyde	0.50	98.8

when using MeOH as the reaction solvent, a feature exploited for the multistep synthesis of α,β-unsaturated compounds in a subsequent investigation (see Section 4.9 for details).

4.3.10 C–S Bond-Forming Reactions: Thioacetalizations

In an extension to their previous investigations, Haswell and coworkers [40] evaluated the chemoselectivity associated with the thioacetalization of 4-acetylbenzaldehyde **59** under continuous flow conditions. Identifying the need for significantly different reaction times for the protection of aldehydes (~65 µL min^{-1}) and ketones (~40 µL min^{-1}) as their respective 1,3-dithiolane or 1,3-dithiane, the authors proposed that the accurate reaction control attainable within flow reactors would enable them to efficiently and chemoselectivity protect the aldehydic functionality of 4-acetylbenzaldehyde **59** without the need for protection of the ketone moiety (Scheme 4.17). Employing a premixed solution of 4-acetylbenzaldehyde **59** (1.0 M) and 1,3-propanedithiol **60** (1.0 M) in MeCN, the authors employed a flow rate of 65.2 µL min^{-1} enabling the protection of the aldehyde moiety **61** in quantitative conversion and selectivity; compared to batch where a significant proportion of dithioacetalization **62** was observed.

4.4 ELIMINATION REACTIONS

4.4.1 Dehydration Reactions

Using a heated device, Wilson and McCreedy [41] demonstrated an early example of catalyst incorporation into glass/PDMS micro reactors demonstrating the *in situ* deposition of the super acid catalyst sulfated zirconia **63**. Prior to investigation, the

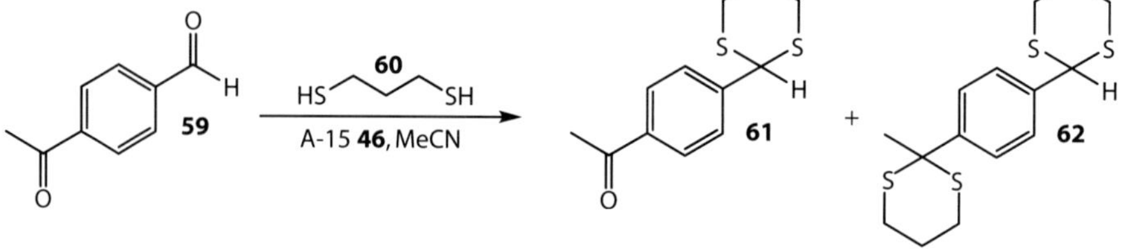

SCHEME 4.17 Schematic illustrating the reaction products obtained in batch when protecting 4-acetylbenzaldehyde **59** as the 1,3-dithiane **61**.

alcohols were degassed with N_2 and reactions performed by pumping the materials through the reactor at 3 µL min^{-1} (155°C), with reaction products collected offline and analyzed by GC-FID to determine the percentage dehydration that had occurred. Using this approach, the authors were able to dehydrate hexan-1-ol **64** in 85–95% conversion to hex-1-ene **65** (Scheme 4.18) and EtOH to ethene with some cracking.

Renken and coworkers [42] subsequently demonstrated the use of a micro channel stacked plate reactor for the dehydration of 2-propanol to propene **66** using γ-alumina deposited within the micro channels (Scheme 4.19). Each plate contained 34 quadrangular channels (dimensions = 300 µm (wide) × 300 µm (deep) × 2 cm (long)) and were constructed from photo-etched stainless steel, affording results in agreement with theoretical data.

4.4.2 Dehydration Reactions

See Chapter 5 for a discussion of heterogeneously catalyzed dehalogenations.

4.5 OXIDATION REACTIONS

With the selective oxidation of primary alcohols difficult to achieve, due to further oxidation of the target aldehyde furnishing the often undesirable carboxylic acid, several research groups have investigated the reaction under continuous flow with the aim of increasing reaction selectivity through the ability to accurately control reaction time.

Noncatalytic Oxidations: In 2006, Wiles and coworkers [43] exploited the high levels of reaction control attainable within micro reactors to quantitatively oxidize an array of primary alcohols to the aldehyde or carboxylic acid, with the product obtained dependent solely on the reaction time employed.

Utilizing a borosilicate glass reactor (3 mm (o.d.) × 5 cm (long)), packed with silica-supported Jones' reagent **67** (0.15 g, 0.15 mmol), the authors initially investigated

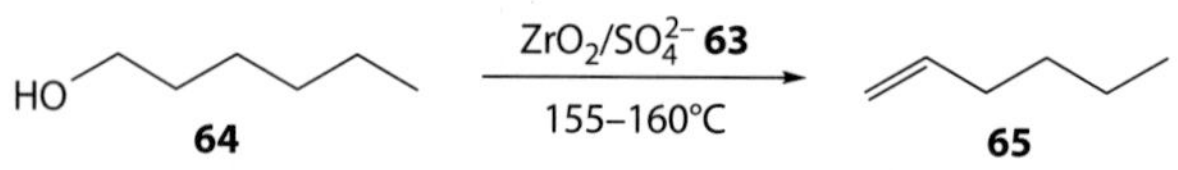

SCHEME 4.18 Illustration of a model dehydration reaction performed using sulfated zirconia within a packed-bed reactor.

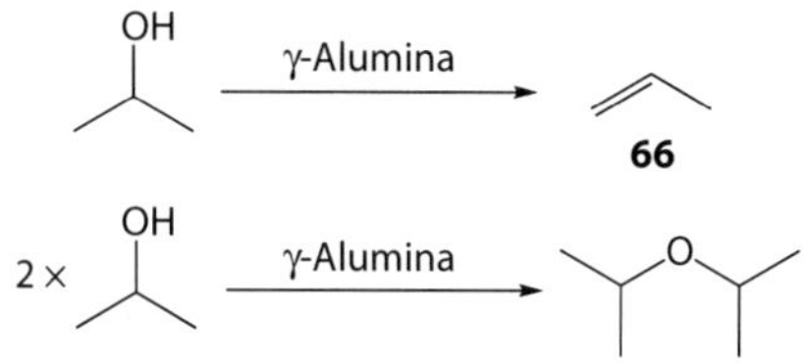

SCHEME 4.19 Model reaction used to demonstrate the catalytic activity of a γ-alumina wall-coated micro channel reactor.

the effect of reactant residence time on the product distribution obtained for the oxidation of benzyl alcohol **68**.

Owing to the stoichiometric nature of the oxidant employed, dilute reactant solutions (1×10^{-2} M in DCM) were used in order to prevent rapid deactivation of the solid-supported material during the optimization process. To evaluate a particular reaction condition, the authors collected 15 samples from the reactor outlet and quantified the product distribution using offline GC–MS analysis. Using this approach the authors evaluated a range of flow rates, between 50 and 1000 μL min^{-1}, observing at low flow rates over-oxidation to benzoic acid **69** was obtained and at high flow rates, benzaldehyde **19** was obtained. Further investigations enabled the authors to identify 650 μL min^{-1} (9.7 s) as the optimal condition for the chemoselective synthesis of benzaldehyde **19,** and 50 μL min^{-1} (126 s) for the over-oxidation product, benzoic acid **69**; as expected, at 300 μL min^{-1} (21 s) a mixture of reaction products **19** and **69** was obtained.

Having demonstrated the ability to chemoselectively oxidize 1° alcohols, the authors subsequently operated the flow reactor under the optimum conditions for 15 min, after which time the reaction products were collected, concentrated *in vacuo* and dissolved in CDCl$_3$ prior to analysis by ^{1}H NMR spectroscopy and ICP–MS. Using this approach, the authors were able to unequivocally confirm the selectivity of the process and quantify any chromium residues within the reaction products (<6.9×10^{-5} wt.% Cr); which were found to be significantly lower than those from batch prepared compounds.

To demonstrate the generality of the high degree of reaction control obtained, the authors evaluated the oxidation of a further 14 1° alcohols using the previously optimized reaction conditions. As Table 4.8 illustrates, excellent selectivities were obtained in all cases and importantly, no functional group incompatibilities were observed. Although this example serves to illustrate the reaction control attainable in such systems, a limitation of the technique is the stoichiometric nature of the oxidizing agent employed (turning from orange to green upon exhaustion); consequently for this technique to be harnessed as a production tool, a regeneratable oxidant would be required.

4.5.1 Catalytic Oxidations

As previously discussed, a limitation of the solid-supported oxidants reported within the literature has been the need for stoichiometric quantities, as such when coupled

TABLE 4.8
Summary of the Results Obtained Illustrating the Selective Oxidation of 1° Alcohols under Continuous Flow

Alcohol	Flow Rate (µL min^{-1})	Product Distribution (%)[a]	
		Aldehyde	Carboxylic Acid
Benzyl alcohol **68**	650	100 (99.1)[b] **19**	0
	50	0	100 (99.6) **69**
3,5-Dimethoxybenzyl alcohol	650	100 (99.5)	0
	50	0	100 (99.3)
4-Bromobenzyl alcohol	650	100 (99.0)	0
	50	0	100 (98.3)
4-Chlorobenzyl alcohol	650	100 (99.3)	0
	50	0	100 (99.4)
4-Cyanobenzyl alcohol	650	100 (98.5)	0
	50	0	100 (99.0)
Methyl-4-formylbenzyl alcohol	650	100 (99.2)	0
	50	0	100 (95.6)
4-Methylbenzyl alcohol	650	100 (99.2)	0
	50	0	100 (95.6)
4-Benzyloxybenzyl alcohol	650	100 (99.5)	0
	50	0	100 (99.8)
4-Aminobenzyl alcohol	650	100 (100)	0
	50	0	100 (99.8)
4-Dimethylaminobenzyl alcohol	650	100 (99.3)	0
	50	0	100 (99.6)
Biphen-4-yl methanol	650	100 (99.7)	0
	50	0	100 (99.5)
(5-Nitrothiophen-2-yl)-methanol	650	100 (99.8)	0
	50	0	100 (99.7)
2-Benzyloxybenzyl alcohol	650	100 (99.7)	0
	50	0	100 (99.9)
2-Naphthalen-2-yl methanol	650	100 (99.9)	0
	50	0	100 (99.9)
4-Acetylbenzylalcohol	650	100 (99.8) **59**	0
	50	0	100 (99.8)

[a] $n = 15$.
[b] Values within parentheses represent the isolated yield (%).

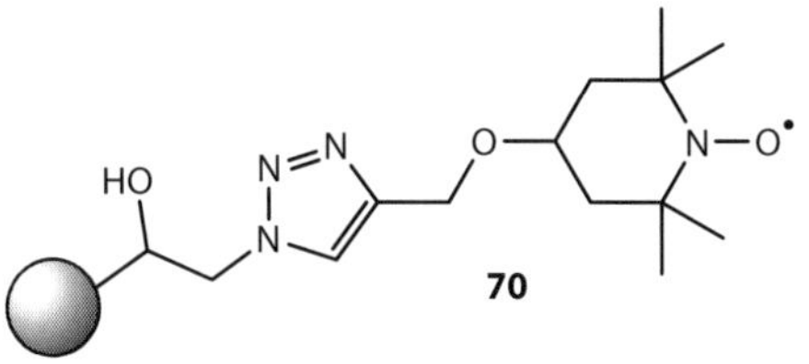

Multi-Phase Micro Reactions

[structure of compound 70 — polymer-supported TEMPO]

70

FIGURE 4.4 Illustration of the polymer-supported TEMPO **70** oxidizing agent evaluated under continuous flow.

with continuous flow reactor methodology the packed-beds developed have a limited lifetime before reagent regeneration or replacement is required.

In a recent example by Bogdan and McQuade [44] the authors developed a solid-supported TEMPO derivative **70** (Figure 4.4), based on a modified Amberzyme Oxirane (AO) resin, and evaluated the oxidants efficiency within a fluoroelastomeric packed-bed reactor (1.6 mm (i.d.) × 60 cm (long)) (Figure 4.5). Using a homogeneous cooxidant, sodium hypochlorite (NaOCl) **71**, the authors were able to demonstrate the catalytic oxidation of primary and secondary alcohols affording the target aldehydes in high yields and purities; with no leaching observed (Table 4.9).

Using the reactor set-up illustrated in Figure 4.5, a typical experimental procedure involved pumping a solution of alcohol (0.2 M) and cyclooctane (0.02 M) in

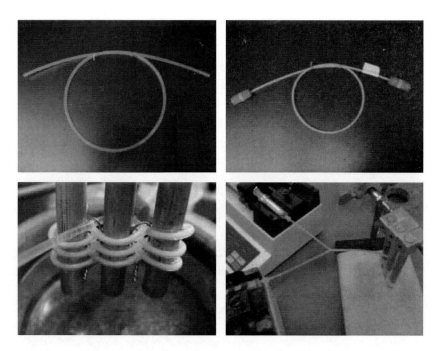

FIGURE 4.5 Photographs illustrating the reactor set-up utilized for the continuous flow oxidation of alcohols to aldehydes. (Reproduced from Bogdan, A. and McQuade, D. T. 2009. *B. J. Org. Chem.* 5(17): 1–7. With permission of Beilstein Institute.)

TABLE 4.9
Summary of the Results Obtained for the Continuous Flow Oxidation of Alcohols Using Polymer-Supported TEMPO 70 as an Oxidizing Agent

Alcohol	Product	Residence Time (min)	Conversion (%)[a]	Yield (%)
PhCH₂OH (**68**)	PhCHO (**19**)	4.8	>99	95
4-Cl-C₆H₄CH₂OH	4-Cl-C₆H₄CHO	4.8	>99	93
Cyclohexyl-CH₂OH (**68**)	Cyclohexyl-CHO	4.8	89	86
Cyclohexanol	Cyclohexanone	4.8	95	84
PhCH(OH)CH₃	PhC(O)CH₃	4.8	>99	95
PhCH(OH)CH₂CH₃	PhC(O)CH₂CH₃	4.8	89	85

[a] Determined by GC analysis using cyclooctane as an internal standard.
[b] Alcohol (0.2 M), NaOCl **71** (1.5 eq.).
[c] Alcohol (0.1 M), NaOCl **71** (3.0 eq.).

DCM into the reactor from one inlet and a solution of aqueous NaOCl **71** (0.25 M, pH 9.1) and KBr (0.5 M) from a second inlet. The reactants met at a Y-interconnect prior to passing through the packed bed containing AO-TEMPO **70** (0.30 g) submerged in an ice bath. The reaction products were collected in a screw cap vial (4 mL) and upon completion of the reaction the organic fraction was separated from the aqueous phase and analyzed by GC. Using this approach, the conversion of alcohol into aldehyde was determined, using cyclooctane as an internal standard, and the reactor washed with DI H_2O, in order to prepare the system for the next flow reaction.

As Table 4.9 illustrates, moderate to excellent isolated yields were obtained in all cases, with secondary alcohols requiring the use of 3.0 equivalents of NaOCl **71** compared with 1.5 equivalents **71** for primary alcohols. In addition to successfully demonstrating the use of a catalytic oxidant **70** within a flow reactor, the authors found the use of a fluoroelastomeric reactor advantageous compared the metal based reactors previously demonstrated for this transformation as no wall oxidation was observed [45].

$scCO_2$ for the Aerobic Oxidation of Alcohols: Using palladium nanoparticles stabilized on PEG-modified silica as catalysts, Leitner and coworkers [46] reported the aerobic oxidation of a series of alcohols using $scCO_2$ as the reaction solvent. Employing a stainless steel tubular reactor (dimensions = 7.5 mm (i.d.) × 50 cm (long)) packed with a mixture of quartz filler and Pd_{561}-PEG-modified silica, the authors investigated the effect of reaction time on the oxidation of simple alcohols at 80°C. With an average reaction time of 1.2 h for a single pass, the authors collected the reaction products in cold traps (at −35°C) and periodically sampled the contents by GC and ^1H NMR spectroscopy. As Table 4.10 illustrates, under the aforementioned conditions, a series of synthetically useful aldehydes were synthesized in moderate to high conversion (46.4–98.9%) and high selectivity (>98.0%) demonstrating the synthetic utility of this technique.

4.5.2 EPOXIDATIONS

The synthetic utility of epoxides along with their industrial application in resins, foams, and adhesives has led to several authors investigating their synthesis in micro

TABLE 4.10
Summary of the Results Obtained for the Pd Nanoparticles Catalyzed Oxidation of Alcohols Using $scCO_2$ as Solvent

Substrate	Product	t/h	Conversion (%)	Selectivity (%)	TON
⌒⌒OH	⌒⌒=O	4	98.9	98.0	45
Ph⌒⌒OH	Ph⌒⌒=O	5	96.8	98.5	47
Ph-CH(OH)-CH₃	Ph-C(=O)-CH₃	12	58.8	99.5	29
Ph-CH₂-OH	Ph-CHO	16	96.5	98.8	45
cyclooctanol	cyclooctanone	18	46.4	98.2	22

fabricated reactors. The following section details a selection of those techniques reported with the aim of demonstrating the diverse array of methods available.

An early example of continuous flow epoxidations was reported in 2002 by Gavriilidis and coworkers [47] and involved the use of a silicon micro reactor whereby the micro channel (dimensions = 1000 μm or 500 μm (wide) × 250 μm (deep)) walls were coated with a titanium silicate-1 (TS-1) catalytic coating (Thickness = 6 μm) via a method of selective seeding and thermally activated in air prior to applying the channel cover plate. In order to demonstrate the catalytic activity of the zeolite reactor, the authors investigated the epoxidation of 1-pentene **72** to afford 1,2-epoxypentene **73** as illustrated in Scheme 4.20.

To perform a reaction, the authors prepared a premixed stock solution containing 1-pentene **72** (0.9 M), H_2O_2 **74** (0.18 M), MTBE (0.2 M) as the internal standard and MeOH as the reaction solvent. Under pressure-driven flow, the authors investigated the effect of flow rate (30–120 μL h^{-1}) on the alkene **72** epoxidation, analyzing the reaction products by GC-FID. Comparing the two channel dimensions, the authors identified over 50% enhancement in epoxide **73** formation when employing 500 μm wide channels compared with 1000 μm wide channels, an observation which is attributed to an increased surface-to-volume ratio. The authors also report deactivation of the catalytic reactor after 100 h and comment that further investigations are required to identify the source of fouling/deactivation.

Chemo-enzymatic: Using the commercially available oxidant, H_2O_2 **74** (100 volumes) or the stable urea:H_2O_2 complex (UHP) **75**, and Novozyme-435 **2**, an immobilized form of *Candida antarctica* lipase B, Wiles and coworkers [48] investigated the chemo-enzymatic oxidation of a series of alkenes in high yield and purity. Employing the epoxidation of 1-methylcyclohexene **76** as a model reaction, the authors investigated the *in situ* biocatalytic hydrolysis of EtOAc to acetic acid **1** and subsequent perhydrolysis to afford peracetic acid **77**, this was then followed by the peroxy acid epoxidation of the alkene **76** (Scheme 4.21). To conduct this sequence of events under continuous flow, the authors prepared a premixed stock solution containing the alkene **76** (0.1 M) and UHP **75** (2 eq.) in EtOAc. The stock solution was then pumped through a glass packed-bed reactor (dimensions = 3.0 mm (i.d.) × 3.6 cm (long)), containing 0.10 g of Novozyme-435 **2** and the effect of residence time (0.5–5.2 min) and reactor temperature (27–70°C) evaluated. The reaction products were analyzed by periodic sampling of the reactor effluent and product conversion quantified using GC-FID analysis. Using this approach the reaction was optimized and the authors identified a residence time of 2.6 min and a reactor temperature of 70°C afforded the target epoxide **78** in quantitative conversion. To ensure the system developed was stable for long-term use, the reactor was operated continuously for 24 h with frequent sampling confirming no loss of enzyme activity. For full characterization

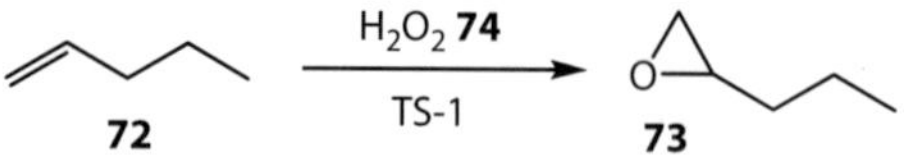

SCHEME 4.20 Illustration of the model reaction used to investigate a zeolite-based micro reactor.

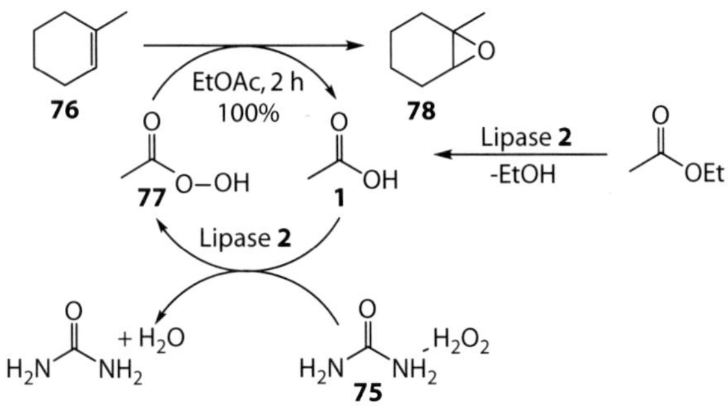

SCHEME 4.21 Schematic illustrating the chemo-enzymatic epoxidation of 1-methylcyclohexene **76** to 1-methylcyclohexene oxide **78**.

purposes, the reactor was operated for 24 h and the reaction products concentrated *in vacuo* and subjected to an aqueous extraction, prior to determination of the isolated yield, throughput and analysis by ^{1}H NMR spectroscopy. With this information in hand, the generality of the technique was investigated using a series of aliphatic and aromatic alkenes. As Table 4.11 illustrates, the reaction conditions were found to

TABLE 4.11
A Selection of the Epoxides Generated Using a Chemo-Enzymatic Approach under Continuous Flow Conditions

Alkene	Temperature (°C)	Residence Time (min)[a]	Conversion (%)	Yield (%)[b]
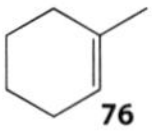 **76**	70	2.6 (2)	100.0	99.1 (6.7)[c]
	70	2.6	57.2	—
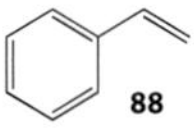 **88**	70	5.2 (33)	100.0	99.2 (3.6)
	70	2.6	32.1	—
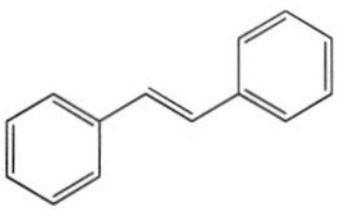	70	5.2	100.0	99.1 (5.9)
	70	2.6	31.9	—
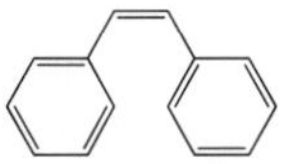	70	5.2	100.0	99.5 (5.9)
	70	2.6 (40)	100.0	97.6 (5.9)

Values within parentheses represent
[a] The reaction time required under batch conditions (h).
[b] The isolated yield obtained after continuous operation of the reactor for 24 h.
[c] The throughput (mg h^{-1}) using the optimized conditions.

be suitable for the preparation of a diverse array of epoxides in isolated yields ranging from 97.6% to 99.2%. While the technique reported offers a simple method for the preparation of epoxides, for the synthesis of chiral epoxides it would be necessary to employ a chemzyme [49].

4.6 METAL-CATALYZED CROSS-COUPLING REACTIONS

As can be seen from Chapter 3, C–C bond-forming reactions constitute the most widely studied organic transformation within continuous flow reactors, which is testament to their synthetic utility at both a research and production level. As such, the following section details an array of techniques evaluated for the incorporation of metal-based catalysts into continuous flow reactors with the aim being to increase catalytic efficiency and reduce the need for postreaction purifications.

4.6.1 Suzuki–Miyaura Reaction

Packed-Bed Reactors: In an extension to their earlier work (Chapter 3), whereby Kirschning and coworkers [50] reported the use of silica-coated magnetic nanoparticles as a means of introducing heat into micro reaction channels, the authors investigated the surface functionalization of magnetic particles as a means of also incorporating a catalytic surface within the flow reactor.

As Scheme 4.22 illustrates, the authors employed the reductive precipitation of ammonium-bound tetrachloropalladate salts to afford nanoparticles functionalized with Pd^0 **79**, demonstrating their application in a series of cross-coupling reactions such as the Suzuki–Miyaura reaction and the Heck reaction.

As illustrated in Table 4.12, conducting the reactions on a 1.0 mmol scale and a reaction temperature ranging from 100°C to 120°C, depending on the substrate employed, the authors obtained moderate to excellent yields as a result of recirculating the reaction mixture, at a flow rate of 2 mL min^{-1}, through the packed bed for 1 h. Although this approach combined a novel method of heating with solid-supported catalysis, the authors noted Pd leaching (34 ppm for the Suzuki–Miyaura reaction and 100 ppm for the Heck reaction), as such improvements in catalyst stability would be required before this approach could be adopted as a potential production tool.

Supercritical CO_2 as a Reaction Solvent in Continuous Flow Systems: In addition to the use of conventional solvent systems and reaction conditions within flow reactors, Leeke et al. [51] demonstrated the use of supercritical carbon dioxide

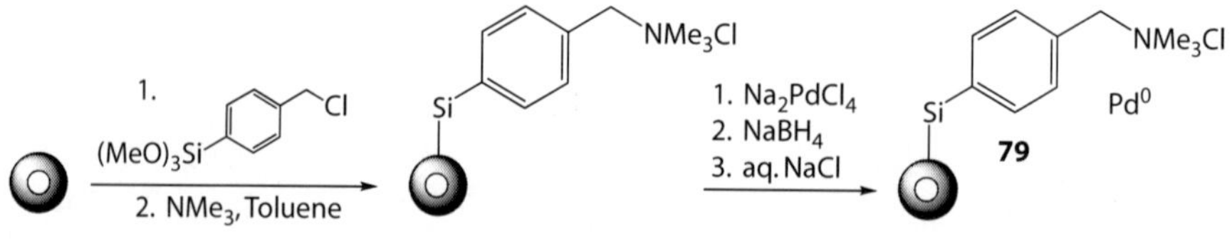

SCHEME 4.22 Schematic illustrating the synthetic protocol employed for the surface functionalization of magnetic nanoparticles.

TABLE 4.12
Summary of the Results Obtained for the Cross-Coupling Reactions Conducted in the Presence of Pd⁰ Functionalized Magnetic Nanoparticles 79

Halide	Boronic Acid/Alkene	Conditions	Yield (%)
4-bromoacetophenone (Br, COMe)	PhB(OH)₂	A	77
3-bromo-4-cyanobenzene (NC, Br)	PhB(OH)₂	A	83
4-iodoacetophenone (I, COMe)	styrene **88**	B	76
4-iodoanisole (I, OMe)	styrene **88**	B	84
2-iodothiophene	styrene **88**	B	63

A: 1.5 eq. of phenyl boronic acid, 1.0 eq. aryl halide, 2.4 eq. CsF, 2.8 mol% in DMF/H₂O, at 100°C; B: 1.0 eq. aryl halide, 3.0 eq. styrene, 3 eq. n-Bu₃N, 2.8 mol% in DMF, at 120°C.

($scCO_2$) as a reaction solvent for the Suzuki–Miyaura reaction conducted under continuous flow.

Compared to organic solvent systems, the use of $scCO_2$ as a reaction solvent is widely viewed as a green, cheap, nontoxic solvent which offers many processing advantages such as ease of product isolation; peripheral equipment however, remains specialized. With this in mind, the authors investigated the Pd-catalyzed Suzuki–Miyaura reactions within a small bore continuous flow reactor as it represented an industrially relevant reaction. Following their previous work [52] into the use of PdEnCat™ **80** under continuous flow, Leeke and coworkers evaluated the synthesis of 4-phenyltoluene **81** using commercially available precursors 4-tolylboronic acid **82** and iodobenzene **83** (Scheme 4.23).

Using a high-pressure packed-bed reactor (48.5 cm (long) × 25.4 mm (i.d.)), containing 123 g of PdEnCat™ **80** (0.4 mmol g⁻¹ Pd), and a liquid piston pump to deliver the reagents, the authors investigated the effect of concentration (0.05–0.72 M), temperature (80–100°C) and pressure (102–250 bar) as a means of optimizing the process.

Under the aforementioned conditions, the authors found the ratio of $scCO_2$:MeOH to be critical, with the highest conversions (81% **81**) obtained with a ratio of 10:1; furthermore, the most successful reactions were performed at 166 bar and 100°C attributed to the formation of a monophasic mixture. In this instance, MeOH had a dual role, firstly enabling the concentration of reactants to be tuned and secondly

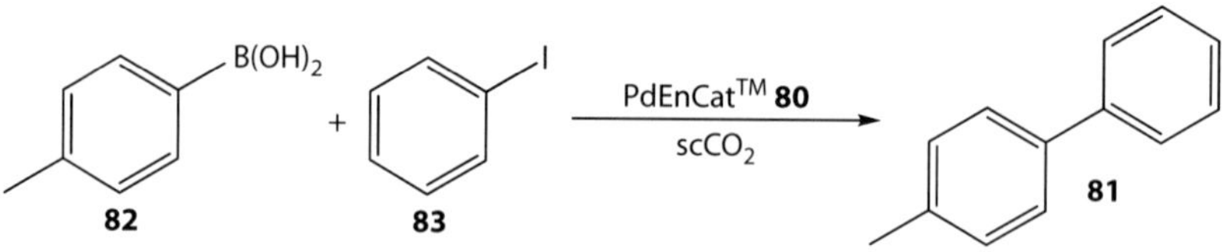

SCHEME 4.23 Model reaction used to demonstrate the processing advantages associated with the use of scCO$_2$ as a reaction solvent within continuous flow systems.

preventing the precipitation of ammonium salts and reaction by-products from the scCO$_2$. Having identified the optimum conditions, the authors analyzed the reaction products obtained by ICP-MS where Pd content was found to be <0.8 ppm; affording 4-phenyltoluene **81** at throughput of 3.6 g h^{-1}. In addition to decreasing the proportion of metal within the reaction products, the use of scCO$_2$ compared to organic solvents, reduced the CO$_2$ emissions by up to 20 kg kg^{-1} **81** as the solvent does not need to be incinerated and can be scrubbed prior to recycling.

Membrane Reactors: In addition to packed-beds, monoliths and wall-coatings, porous membranes offer a method for the incorporation of catalytic surfaces into microfabricated reactors. Using this approach, Uozumi et al. [53] reported the deposition of a poly(acrylamide)-triarylphosphane-palladium membrane (PA-TAP-Pd) (1.3 µm (wide) × 0.37 mm g^{-1} Pd) within a glass micro channel (dimensions = 100 µm (wide) × 40 µm (deep) × 1.4 cm (long)). The reactor was subsequently used to perform the Suzuki–Miyaura reaction by introducing two reactant solutions into the heated reactor (50°C), the first containing the aryl iodide (6.3 × 10^{-3} M) in EtOAc/2-PrOH (1:2.5) and the second the aryl boronic acid (9.4 × 10^{-3} M) in aqueous Na$_2$CO$_3$ (1.8 × 10^{-2} M). Employing a flow rate of 2.5 µL min^{-1}, which equates to a residence time of 4 s the authors collected the biphasic reaction products at the reactor outlet and analyzed the organic fraction by GC and ^{1}H NMR spectroscopy. Using this approach, a series of substrates were investigated and as Table 4.13 illustrates, the respective coupling products were obtained in yields ranging from 88% to 99%. Reactions were repeated in the absence of a membrane whereby no coupling products were obtained; confirming that the membrane had a high catalytic activity.

See Chapter 3 for a discussion of Suzuki–Miyaura cross-coupling reactions performed using homogeneous catalysts under conventional and microwave heating.

4.6.2 Heck Coupling Reactions

Raschig Ring Reactors: Using poly(vinylpyridine) glass composite materials shaped as Raschig rings, Kirschning and coworkers [54] investigated the coordinative immobilization of an oxime-based palladacycle as a means of developing a heterogeneous catalyst for incorporation into PASS-flow reactors. Employing the Mizoroki–Heck reactions illustrated in Table 4.14 the authors investigated the effect of substituents on the aryl halide and alkene at 120°C; operating in a single pass mode, incomplete conversions were obtained. Consequently, a fixed flow rate of 2 mL min^{-1} was employed and the reaction mixture cycled through the reactor for the reported reaction time; using this approach, excellent yields were obtained for all precursors

TABLE 4.13
A Selection of the Suzuki-Miyaura Coupling Reactions Performed in a Membrane Reactor

R_1	R_2	Yield (%)
H 83	4-MeO	99
H 83	3-Me	96
H 83	2-Me	99
3-EtCO$_2$	4-Me	99
3-Cl	4-MeO	88
4-CF$_3$	4-MeO	99

investigated. In addition, the authors also demonstrated the system toward a series of Suzuki–Miyaura cross coupling reactions, obtaining the target compounds in excellent yields.

It is important to note that leaching of Pd was observed and the authors found it necessary to incorporate thio-derived Raschig rings into the system to sequester any

TABLE 4.14
A Selection of the Results Obtained for the Mizoroki-Heck Reaction Performed under Continuous Flow

R^1	R^2	Residence Time (h)	Yield (%)[a]
4-Ac	CO$_2$Cy	2	99
4-Ac	CO$_2$t-Bu	4	99
4-Ac	CN	4	97 (5:1)[b]
4-Me	Ph	24	97
4-OMe	Ph	24	92
2-NH$_2$	Ph	8	99

[a] Unless otherwise stated the E-isomer was obtained exclusively.
[b] The data in parentheses represents the E/Z ratio. Catalyst loading = 3.57×10^{-5} mmol Pd mg^{-1}.

free Pd. While the reaction products were free of any trace metal contaminants, this does mean that the heterogeneous catalyst will have a limited lifetime, something that should be considered when applying catalysts under continuous flow conditions; especially when looking to apply the technique to production cases.

Monolithic Reactors: In addition to the use of packed-bed reactors, authors have also reported the use of polymer monoliths as a means of preparing flow reactors with low back-pressures. A recent example of this was reported by Jones et al. [55], who focused on the development of monolithic substrates with backbones analogous to Merrifield and Wang resins, onto which precatalysts of the type $PdCl_2(L^2)$ (L^2 = 4-(4-benzyloxyphenyl)-2-methylthiomethylpyridine) were loaded. Using the Mizoroki–Heck and Suzuki–Miyaura reactions, the authors were able to confirm that the monolithic catalysts exhibited the same trends in reactivity when varying the aryl halide. As Table 4.15 illustrates, for the Heck reaction the authors investigated the reaction of a series of phenyl halides (I = **83**, Br = **84**) with *n*-butyl acrylate **85** to afford butyl-3-phenylacrylate **86** within fused-silica capillary flow reactors (dimensions = 250 µm (i.d.)). Employing a premixed solution of the acrylate, aryl halide and Et_3N **29** in DMA and a flow rate of 0.1 mL min^{-1}, the authors investigated the coupling reaction over a period of 24 h at 120°C. Upon completion of the reaction, the reaction products were collected and analyzed by GC-FID and compared to synthetic standards. Analysis of the reaction products by ICP-MS also confirmed <0.05% leaching of Pd over a 24 h period.

Packed-Bed Reactors: Under microwave irradiation and conventional heating, Kappe and coworkers [56] compared the use of heterogeneous catalysis under batch

TABLE 4.15

Summary of the Results Obtained for the Heck Reaction Performed Using a Monolithic Catalyst under Continuous Flow Conditions

X	Yield (%)
I (**83**)	98
Br (**85**)	45

Residence time = 39 min, aryl halide (0.1 M), alkene (0.15 M), Et_3N **29** (0.2 M) in DMA at 120°C.

and flow conditions for the Mizoroki–Heck reaction. Employing the commercially available X-cube™ (ThalesNano Inc., Hungary) for flow studies, the authors investigated the coupling of aryl iodide **83** with butyl acrylate **85** (1.5 eq.) in the presence of Et$_3$N **29** (1.5 eq.) using Pd/C **87** (0.31 g) and MeCN as the reaction solvent. Investigating a range of flow rates and temperatures, the authors identified 0.5 mL min^{-1} and a reactor temperature of 150°C as optimal, with faster flows and lower temperatures returning unreacted starting materials. In comparison to batch experiments, the authors observed greater by-product formation under flow, attributing the findings to the use of pressure, potentially resulting in separation of reactants within the system-reducing the local concentration of one or more of the reactants. In addition, the authors observed the formation of Pd deposits within the product solutions collected, confirming leaching of the Pd. Although comparable batch reactions were performed without problem, the authors concluded that Pd–C was not a suitable catalyst for the Mizoroki–Heck reaction under continuous flow conditions due to the involvement of a homogeneous Pd species.

Wall-Coated Reactors: Employing Pd-PEPPSI-IPr in metal-coated capillary reactors (dimensions = 1180 µm (i.d.)), Organ and coworkers [57] investigated the synthesis of indoles via a sequential aryl amination and Heck coupling reaction performed in the presence of microwave irradiation. Using this approach, the authors found it possible to obtain the target indoles in high yield and purity when employing a Pd coating compared with low-to-moderate yields in the presence of an Ag coating. Under optimal conditions, Organ and coworkers were able to use their Pd-coated flow reactor for the synthesis of wide array of indole alkaloids, obtaining the target compounds in high yield and purity (Scheme 4.24); for a detailed discussion of the reactor and process see Chapter 3, Section 3.9.6.

Building on the array of Heck reactions performed using heterogeneous catalysts, Plucinski and coworkers [58] demonstrated the coupling of Heck and hydrogenation reactions in a compact, continuous flow reactor. Employing Pd/C **87** as both the Heck and hydrogenation catalyst, the authors investigated the coupling of iodobenzene **83** and styrene **88** to afford 1,2-diphenylethene **89** (Scheme 4.25), subsequent hydrogenation afforded the target 1,2-diphenylethane **90**.

In comparison with a batch reactor, the flow reactor was found to be more efficient, employing significantly reduced reaction times (hour to 7 min), no ligand and lower operating temperatures and pressures. Although the authors observed leaching of Pd/C **87**, as previously illustrated by other researchers, employing reverse flow the authors found that the catalytic activity of the system could be maintained.

4.6.3 OTHER METAL-CATALYZED COUPLING REACTIONS

In addition to the relatively large number of Suzuki and Heck couplings performed under continuous flow conditions, other metal-catalyzed coupling reactions have received less attention (see Chapter 3 for examples of liquid phase reactions); however, those investigated have observed the same advantages including increased catalyst turnover and product purity.

Kumada-Coupling Reactions: With this in mind, Styring and coworkers [59] expanded their investigation of coupling reactions to incorporate the Kumada

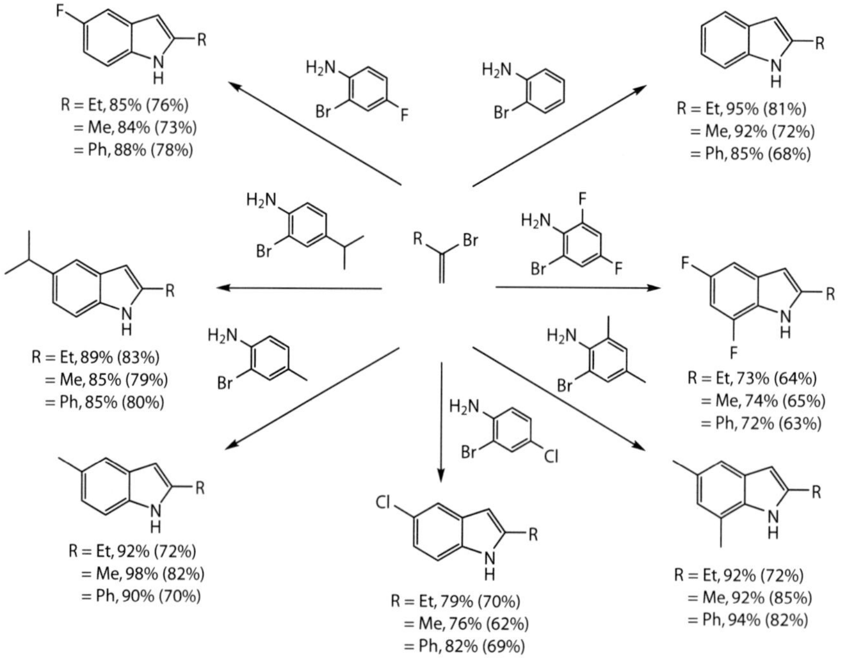

SCHEME 4.24 Summary of the indole alkaloids synthesized using a metal-coated capillary reactor (results in parentheses represent the isolated yield).

coupling, focusing on the reaction between 4-bromoanisole **91** and a phenyl magnesium chloride **92**, to afford 4-methoxybiphenyl **93**, as depicted in Scheme 4.26. Employing a packed-bed reactor (dimensions = 3 mm (i.d.) × 2.5 cm (long)), containing a silica-supported nickel catalyst **94**, the authors investigated the effect of flow rate on the conversion of a 0.5 M stock solution (in THF) into 4-methoxybiphenyl **93**

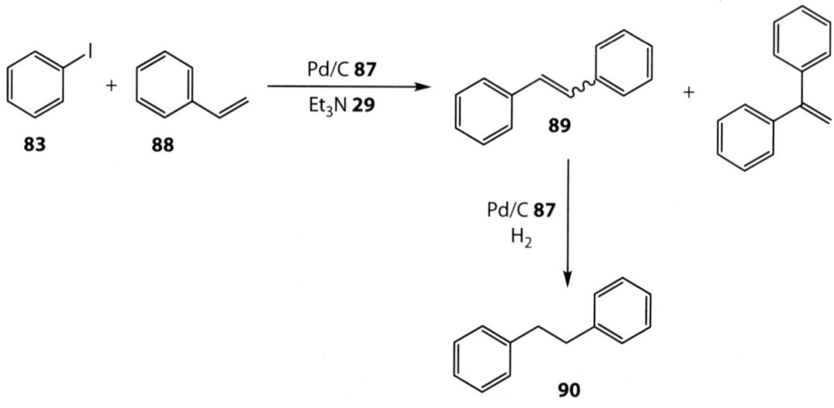

SCHEME 4.25 Illustration of the combined Heck and hydrogenation steps used for the continuous synthesis of 1,2-diphenylethane **90**.

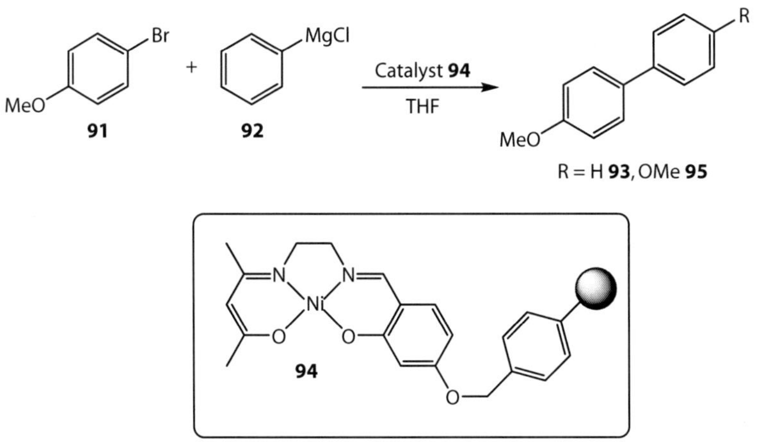

Multi-Phase Micro Reactions

SCHEME 4.26 Illustration of the Kumada coupling reaction investigated under continuous flow conditions.

at room temperature. As Table 4.16 illustrates, in addition to the target compound **93**, by-products 4,4-dimethoxybiphenyl **95** and anisole were also detected by GC analysis, with product distribution barely affected by reactant residence time in the presence of the catalyst.

Observing the reaction over an extended period of time, the authors were able to confirm catalyst stability for 3 h, however, due to the build-up of magnesium salts performance was found to degrade upon further use; studies are now underway to investigate the implementation of wash steps between reactions to remove such materials. Importantly, analysis of the reaction products by ICP–MS confirmed minimal Ni leaching was observed compared to analogous reactions performed under batch conditions.

Sonogashira Cross Coupling: Employing polymer monoliths as solid supports for a series of Pd catalysts, Smith and coworkers [60] reported their findings on the heterogeneously catalyzed Sonogashira coupling reaction illustrated in Scheme 4.27.

TABLE 4.16
Summary of the Effect of Reaction Time on the Kumada Coupling Reaction Performed under Continuous Flow Conditions

			By-products	
Flow Rate (μL min^{-1})	Product (%)	Starting Material (%)	Anisole (%)	4,4'-Dimethoxybiphenyl (%)
6	64	5	23	8
13	64	6	20	10
20	59	11	23	7
33	54	17	23	6

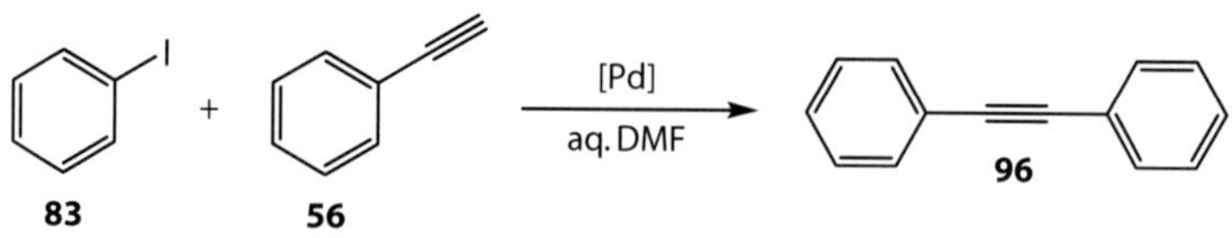

SCHEME 4.27 Model reaction used to evaluate the Sonogashira coupling reaction using a Pd-monolith.

Forming the polymer monoliths within capillaries (dimensions = 250 μm (i.d.)) using thermal initiation, the authors were able to produce porous structures capable of withstanding pressures up to 500 psi. Once formed, the support was derivatized with a series of ligands (1,10-phenanthroline (phen) and N-heterocyclic carbene (NHC)) and the capillaries evaluated for the Sonogashira coupling between iodobenzene **83** and phenylacetylene **56** in aqueous DMF. Employing a reactor temperature of 80°C, the authors analyzed the reaction products collected over a period of 45 min by GC-FID and were gratified to find 96% diphenylacetylene **96** for the phen catalyst and 95% **96** for the NHC catalyst. Operating the reactor over a period of 8 days, the authors confirmed minimal Pd leaching (~4%), illustrating the potential of the technique for continuous Sonogashira coupling reactions.

Allyl Bromide Addition: Using palladium on alumina **97** (pellets = 3.0 × 3.5 mm), Fukuyama et al. [61] subsequently demonstrated the reaction of allyl bromide **98** with phenylacetylene **56** in a stainless-steel packed-bed reactor (dimensions = 4.0 mm (i.d.) × 50.0 cm (long)) investigating the effect of reaction time and temperature. Under solvent-free conditions, the authors pumped a solution of allyl bromide **98** (10 mL, 120 mmol) and phenylacetylene **56** (10 mmol) through the preheated reactor (40°C) at a flow rate of 2.2 mL min^{-1}. Using a reaction time of 2 h, the authors obtained 87% conversion and a mixture of **99** and **100** in a ratio of 79:21 (Scheme 4.28) comparing favorably with batch results whereby 57% conversion was obtained (75:25 **99**:**100**).

Allylic Arylation: As an alternative to packed-bed or wall-coated reactors, Uozumi and coworkers [62] exploited the laminar flow conditions attainable within microfluidic devices to install a series of catalytic membranes within glass micro channel reactors (dimensions = 100 μm (wide) × 40 μm (deep) × 4 cm (long)). Using two immiscible reactant streams, the authors were able to construct porous membranes at the fluid interface, evaluating three different catalytic surfaces, all Pd complexes with various polymeric structures, (1) poly(acrylamide)-triaryl phosphane (PA-TAP-Pd), (2) poly(4-vinylpyridine) and (3) Poly[4,4'-bipyridyl-*co*-1,4-bis(bromomethyl)benzene]. Using these devices, the authors investigated the activity of the membranes toward the allylic arylation illustrated in Scheme 4.29.

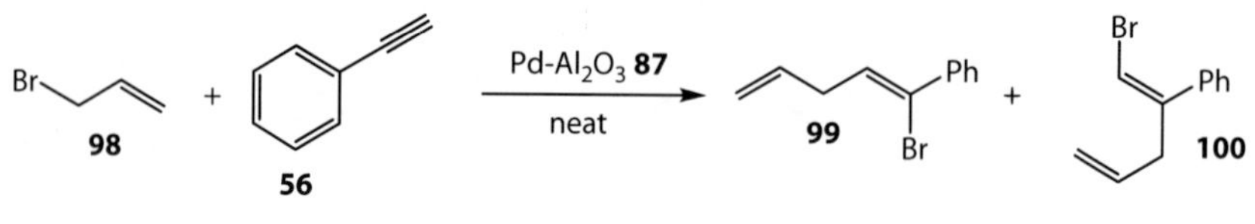

SCHEME 4.28 Illustration of the solvent-free coupling reaction promoted by Pd-Al$_2$O$_3$ **87**.

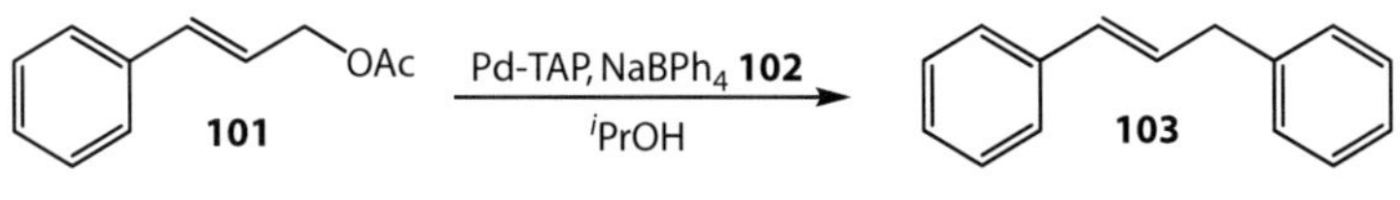

SCHEME 4.29 Illustration of the model reaction selected to evaluate a series of Pd-membranes deposited longitudinally within glass microchannels.

To perform a reaction, the authors introduced two reactant solutions into the membrane divided reactor, the first contained cinnamyl acetate **101** (5.8×10^{-3} M in ^{i}PrOH) and the second sodium tetraphenylborate **102** (5.8×10^{-2} M ^{i}PrOH). Employing a flow rate of 3 µL min^{-1} and a reactor temperature of 50°C, the authors compared the three membrane types for their activity toward the synthesis of 1,3-diphenyl-1-propene **103**, with quantification performed by offline GC analysis. In all cases, the authors obtained the target compound **103** in >99% selectivity however, the conversions ranged from 56% to 99% depending on the membrane employed. From this example, the authors concluded that the PA-TAP-Pd membrane was the most efficient and set about investigating the synthetic scope of the technique.

Using the previous reaction conditions, the authors found that the PA-TAP-Pd membrane afforded all allylic arylation products in excellent selectivity, with the yield most notably affected by the structure of the allylic ester (Table 4.17).

Ullmann-Type Cross Coupling Reactions: Using a wall-coated capillary reactor, comprising of a thin film of Pd nanoparticles (500 nm) embedded in a polycyanoethylphenyl methylsiloxane coating, Trapp and coworkers [63] investigated the biarylation of aryl halides (Scheme 4.30) under continuous flow conditions.

Installing the reactor, which comprised of a 10 m Pd-coated capillary column (dimensions = 250 µm (i.d.)), preceded by a 1 m preseparation column and followed by a 25 m separation column, into a gas chromatograph, the authors investigated the catalytic activity of the catalytic reactor by injecting slugs of reactants onto the column (200°C). Under these conditions, multiple reactions could be performed in series as the reactants were spatially and temporally resolved, obtaining online analysis of the reaction products. Injecting a library of 10 aryl iodides onto the column, the authors used FID detection for quantification of the products and MS detection for characterization of the products. Under the aforementioned conditions, the authors obtained the respective biaryls in yields ranging from 5.9% to 15.3%.

Using this approach, the authors were able to perform the Ullmann-type cross coupling reactions using a ligand free, thin film Pd catalyst, and perform on-column separation of the reaction products prior to online analysis by FID or MS. The authors comment that in addition to the investigation reported, the technique shows great promise for other cross-coupling reactions, where due to potential involatility of products an analogous LC technique may be required. Clearly, this method of operation is not suited to the production of materials; however, the technique shows great promise as a screening tool enabling facile comparison of substituent effects on C–C bond-forming reactions.

Summary: While this section illustrates the diversity of C–C bond-forming reactions successfully investigated within liquid–solid micro reactors, it is important to consider the type of catalyst employed, particularly when employing metal based

TABLE 4.17
Summary of the Reactions Performed Using an *In Situ* Deposited Pa-Tap-Pd Membrane and a Reaction Time of 1 s

Allylic Ester	Arylboron Reagent	Membrane	Selectivity (%)	Yield (%)
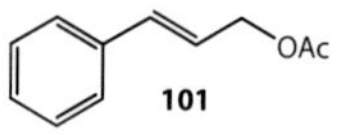 Ph-CH=CH-CH2-OAc 101	NaBPh$_4$ 102	1	>99	99
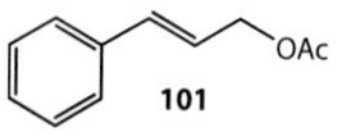 Ph-CH=CH-CH2-OAc 101	NaBPh$_4$ 102	2	>99	77
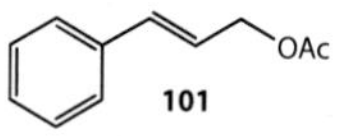 Ph-CH=CH-CH2-OAc 101	NaBPh$_4$ 102	3	>99	56
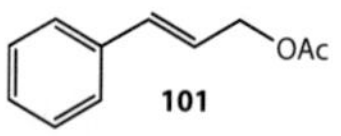 Ph-CH=CH-CH2-OAc 101	PhB(OH)$_2$-Na$_2$CO$_3$	1	>99	43
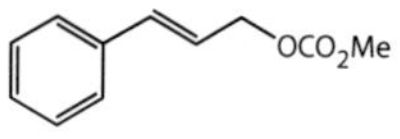 Ph-CH=CH-CH2-OCO$_2$Me	NaBPh$_4$ 102	1	>99	99
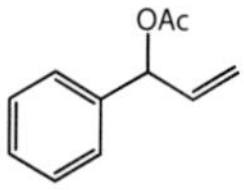 Ph-CH(OAc)-CH=CH2	NaBPh$_4$ 102	1	>99	99
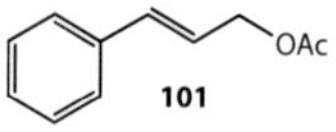 Ph-CH=CH-CH2-OAc 101	NaB(4-F-C$_6$H$_4$)$_4$	1	>99	94
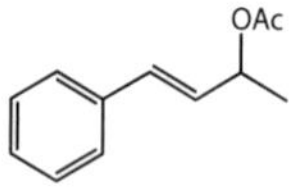 Ph-CH(OAc)-CH=CH-	NaBPh$_4$ 102	1	>99	33
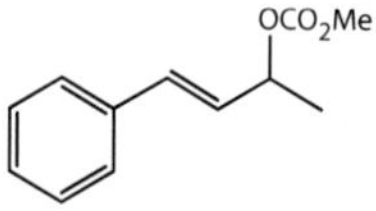 Ph-CH(OCO$_2$Me)-CH=CH-	NaBPh$_4$ 102	1	>99	57

catalysts, to ensure that any intermediates formed are immobilized for the duration of the reaction. In the case of catalysts where the metal is released from the surface and acts as a homogeneous catalyst, prior to readsorption, this can result in leaching of the metal and degradation of the catalytic material; obviously on a production-scale this is not practical or cost effective.

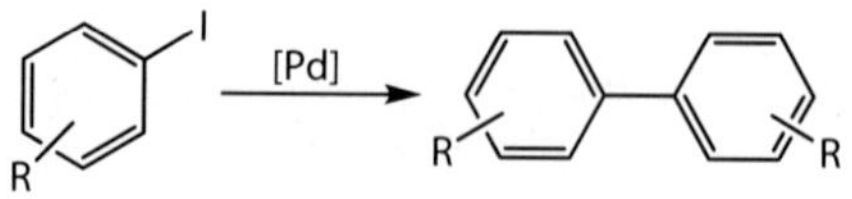

SCHEME 4.30 Illustration of the Ullmann-type C-C coupling reactions performed using a wall-coated capillary flow reactor.

4.7 REARRANGEMENTS

As discussed in Chapter 3, rearrangement reactions are of great synthetic interest due to their atom economy. With this in mind, Brasholz et al. [64] recently demonstrated the use of packed-bed flow reactors for the synthesis of 6,5,5-spiropiperidines, followed by their rearrangement to afford the respective 6,6,5-configured spiropiperidines within a microwave flow reactor (Scheme 4.31). The target compounds were selected as building blocks for the histrionicotoxin family of alkaloids and as such a facile and efficient method was sought for their preparation.

Employing a Vapourtec R2 + /R4 (UK) flow reactor system, the authors initially investigated the synthesis of the 6,5,5-configured spiropiperidine **104** via the base-catalyzed **105** conversion of ketone **106** into the oxime **107** using hydroxylamine hydrochloride **108** (7.0 eq.). This was followed by heating of the reaction mixture to 50°C within a tubular reactor. At this stage, the authors identified problems with the presence of excess **108** and found it necessary to place an amine scavenger cartridge into the system to remove the excess hydroxylamine **108**. Pumping the reaction mixture through a cartridge of QuadraPure™ AK acetoacetate **109** removed any residual amine **108**, enabling the reaction mixture to be heated to 150°C within a microwave reactor (volume = 5 mL) affording the 6,6,5-configured product **104**. Collection of the reaction products followed and analysis by ^{1}H NMR spectroscopy was used to

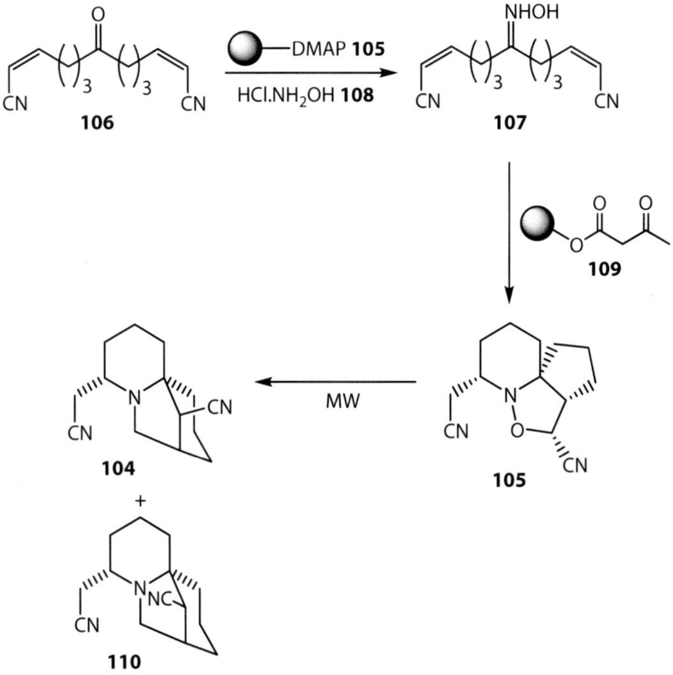

SCHEME 4.31 Illustration of the flow reaction sequence used in the synthesis of a 6,6,5-configured spiropiperidine **104**.

determine the ratio of products formed. Using this approach, the authors were able to obtain the target compound **104** in 62% yield, with 3% of the 6,5,5-intermediate **110** and 35% tricycle **105**. Further studies are currently underway in order to determine the stage at which epimerization occurs with a view to improving the synthetic process further.

4.8 ENANTIOSELECTIVE REACTIONS

4.8.1 CHEMICALLY PROMOTED REACTIONS

In an early example of enantioselective synthesis performed under continuous flow Salvadori and coworkers [65] demonstrated the enantioselective glyoxylate-ene reaction utilizing an insoluble polymer-bound bis(oxazoline) ligand (IPB-box) as a means of increasing the efficient use of the ligand and reducing the costs associated with its use for medium to large-scale preparative purposes.

Using the ene reaction of α-methylstyrene **111** with ethyl glyoxylate **26** to afford the ene product **112** as a model reaction, the authors employed stainless-steel tubular reactor (dimensions = 4.6 mm (i.d.) × 25.0 cm (long)), into which the immobilized box-Cu(OTf)$_2$ **113** catalyst was packed (0.19 mmol Cu g^{-1}). Prior to use, the packed-bed was washed with anhydrous THF, followed by DCM, in order to ensure any free copper triflate was removed. To perform a reaction, the reactor was cooled to 0°C and a premixed solution of α-methylstyrene **111** (0.32 M) and ethyl glyoxylate **26** (0.83 M) in DCM was pumped through the reactor at flow rates ranging from 15 to 25 µL min^{-1}. The reaction products were subsequently analyzed by GC to determine the conversion of α-methylstyrene **111** into the ene product **112** and chiral HPLC analysis to determine the enantioselectivity of the process (Scheme 4.32). Using this approach, the authors were able to obtain the target compound **112** in 78% yield and 88% ee comparing favorably with analogous batch reactions. Unlike batch protocols however, the use of a packed-bed reactor afforded the authors, the ability to recycle the catalysts **113** as the material underwent no physical degradation; as such the authors demonstrated five individual reactions with comparable data obtained throughout and no erosion of enantioselectivity reported.

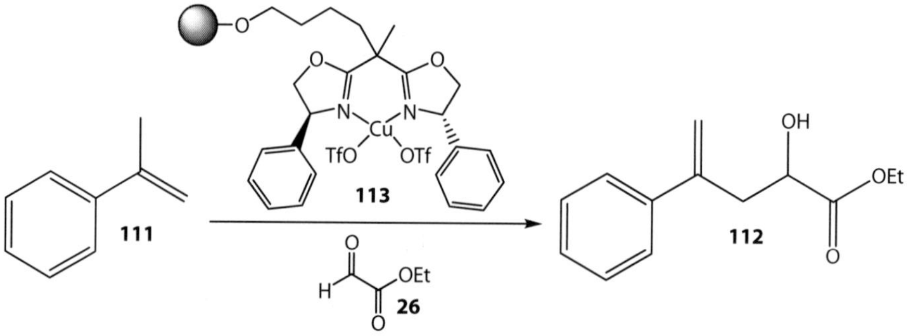

SCHEME 4.32 Schematic illustrating the model reaction used to evaluate the enantioselective glyoxylate-ene reaction under continuous flow conditions.

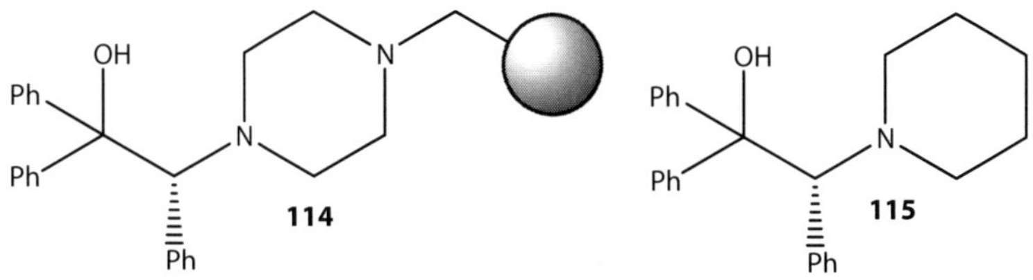

Multi-Phase Micro Reactions

FIGURE 4.6 Illustration of (R)-1,1,2-triphenyl-2-(piperidine-1-yl)ethanol **115** and its polymer-supported analog **114**.

Organic Catalyst: Pericàs et al. [66] subsequently reported the selective synthesis of 1-arylpropanols via the ethylation of substituted aromatic aldehydes in the presence of a polymer-supported β-amino alcohol **114** (Figure 4.6).

Utilizing a jacketed low-pressure chromatography column (1.0 cm (i.d.) × 7.0 cm (long)), a packed-bed containing the swollen solid-supported β-amino alcohol **114** (1.5 g, 0.98 mmol) was prepared to afford a single pass reactor. To perform a reaction, as outlined in Scheme 4.33, solutions of aldehyde (0.9 M) and diethylzinc **122** (1.1 M) in toluene were pumped into the reactor using two peristaltic pumps and the reagents mixed at a T-mixer prior to entering the packed bed. Under the aforementioned conditions, the authors evaluated the effect of reactant flow rate and reactor temperature on the synthesis of a series of (S)-arylpropanols, quantifying the selectivity and ee's offline by chiral GC analysis.

Using the synthesis of (S)-1-phenylpropanol as a model, the authors identified the optimal conditions to be a reactor temperature of 10°C and a total flow rate of 240 µL min^{-1} (equating to a residence time of 9.8 min within the packed-bed **114**) affording the target compound in 98% conversion and 93% ee. Under continuous flow, the authors obtained a production rate of 4.40 mmol h^{-1} g^{-1} **114** enabling the synthesis of 2.61 g of (S)-1-phenylpropanol in 3 h.

To explore the generality of the technique the authors subsequently investigated the ethylation of a series of substituted aromatic aldehydes using the optimized conditions identified previously and compared to batch, the use of a flow reactor afforded an increase in conversion while maintaining product selectivity and enhancing ee (Table 4.18).

In addition, the used of highly activated aldehydes, such as 4-cyanobenzaldehyde **117**, enabled production rates of 13.0 mmol h^{-1} g^{-1} **114** to be obtained with an 86-fold reduction in reaction time compared to batch.

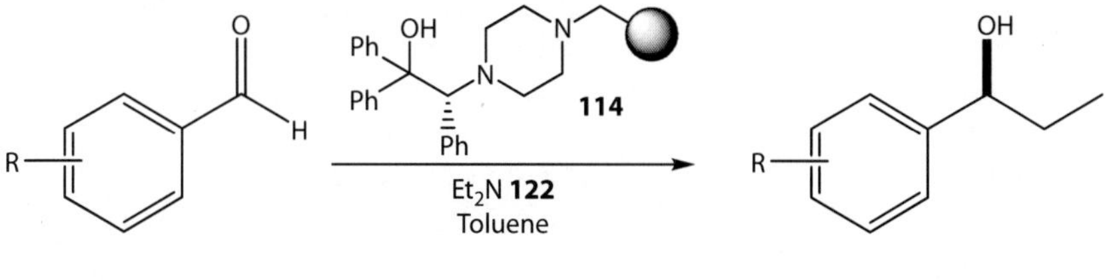

SCHEME 4.33 Schematic illustrating the general reaction protocol employed for the enantioselective synthesis of 1-aryl-propanols under continuous flow.

TABLE 4.18
Comparison of the Results Obtained in Batch and under Flow Conditions in the Presence of a Solid-Supported β-Amino Alcohol 114

Substrate	Method	Reaction Time (min)	Temperature (°C)	Conversion (%)	Selectivity (%)	ee (%)
19	Batch	240	10	99	>99	93
	Flow	9.8	10	99	98	93
123	Batch	240	10	99	99	91
	Flow	9.8	10	95	>99	87
124	Batch	240	10	87	98	92
	Flow	9.8	10	93	>99	93
(F₃C-substrate)	Batch	240	20	71	76	78
	Flow	9.8	20	95	86	82
117	Batch	240	10	>99	>99	89
	Flow	2.8	10	>99	>99	87

In an extension to this, Pericàs and coworkers [67] more recently demonstrated the enantioselective arylation of aldehydes, using a solid-supported β-amino alcohol **114** (Figure 4.6), as a means of preparing compounds bearing the diarylmethanol scaffold found in pharmaceutical agents such as (R)-neobenodine **118**, (R)-orphenadrine **119** and (S)-carbinoxamine **110** (Figure 4.7).

As a means of reducing the costs associated with the use of diarylzinc on a production-scale, the authors investigated the use of triarylboroxins as an alternative aryl source, via the *in situ* preparation of ArZnEt as depicted in Scheme 4.34.

To perform such reactions, the authors prepared a packed-bed reactor comprising of a glass column (1.0 cm (i.d.) × 7.0 cm (long)) containing the solid-supported β-amino alcohol **114** (1.1 g, 0.47 mmol g^{-1}), which was swollen with anhydrous toluene (0.24 mL min^{-1}) prior to use. A solution of arylating agent **120** ((PhBO)$_3$ **121** (12.1 mmol) and Et$_2$Zn **122** (50.4 mmol)) in toluene (33 mL) was then pumped through the packed bed (0.12 mL min^{-1}), forming an amido alcohol-Zn complex (1 h) prior to the addition of a solution of aldehyde (20.2 mmol) in toluene (33 mL) from a second pump (0.12 mL min^{-1}). The reaction products were eluted from the column and collected in a quench solution of aqueous NH$_4$Cl prior to analysis by GC, to quantify the conversion, and HPLC, to determine the *ee*. Once optimized, the

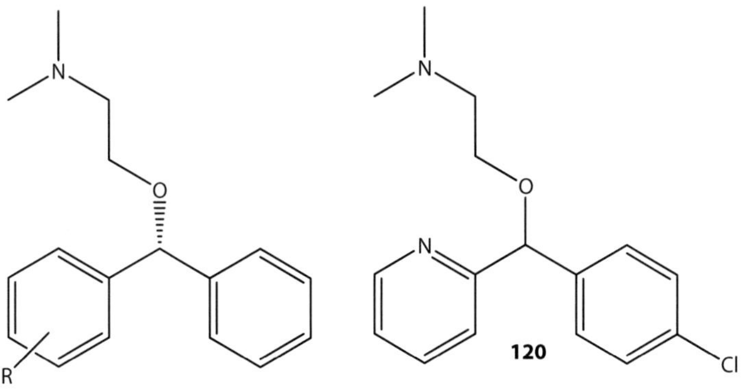

R = 4–CH₃ **118**, R = 2–CH₃ **119**

FIGURE 4.7 Illustration of a series of pharmaceutical agents containing the diarylmethanol scaffold.

reactions were performed for 4 h and the reaction products subjected to an aqueous extraction and purification by flash chromatography; affording the target diarylmethanol in yields ranging from 67% to 80% and *ee*'s of 86–93%. The results obtained are summarized in Table 4.19 which illustrate that the best results were obtained for the carbinols of 2-fluorobenzaldehyde **123**, 4-chlorobenzaldehyde **124** and 2-naphthaldehyde; obtaining stable, reproducible conversions and *ee*'s. With the exception of α-methylcinnamaldehyde, the system could be run for several hours without a significant decrease in conversion to the enantioenriched carbinols.

The synthesis of (*R*)-(4-methoxyphenyl)(phenyl)methanol **125** was also used to illustrate the formation of alternative ArZnEt species, demonstrating the scope of this technique for the multigram preparation of synthetically interesting carbinols.

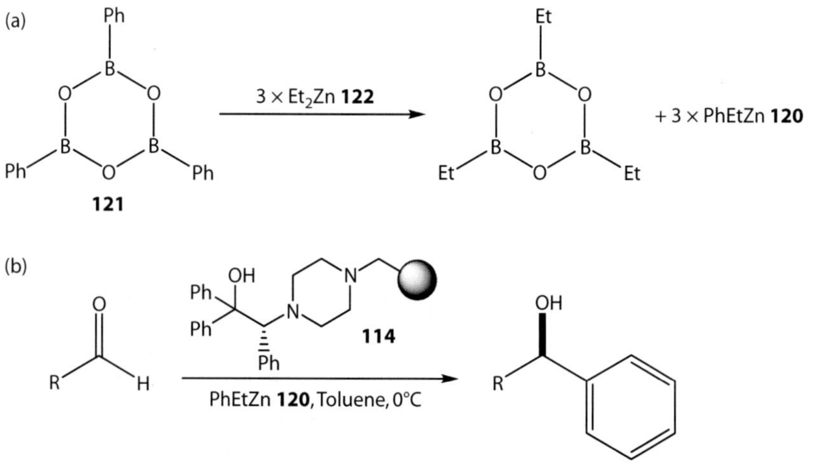

SCHEME 4.34 Schematic illustrating (a) the preparation and (b) the use of PhZnEt **120** in the synthesis of carbinols.

TABLE 4.19
Summary of the Diaryl Compounds Synthesized under Continuous Flow Utilizing Catalyst 114[a]

Product	Time (h)	Conversion (%)[b]	ee (%)[c]
(4-methylphenyl)(phenyl)methanol	1	99	83
	2	98	82
	3	98	81
	4	98	79
	Overall	80 (76)[d]	81 (93)[d]
(2-fluorophenyl)(phenyl)methanol	1	99	58
	2	99	56
	3	99	53
	Overall	91	55
(4-chlorophenyl)(phenyl)methanol	1	99	72
	2	99	70
	3	99	67
	4	99	65
	Overall	78 (67)[d]	68 (86)[d]
naphthalen-2-yl(phenyl)methanol	1	>99	81
	2	>99	74
	3	>99	70
	4	>99	69
	Overall	93	70
1,3-diphenylallyl alcohol	1	99	—
	2	>99	—
	3	82	—
	Overall	82	66
(4-methoxyphenyl)(phenyl)methanol	1	99	57
	2	>99	60
	3	>99	62
	4	97	67
	Overall	83	63

[a] All reactions performed with 1.1 g of resin **114**, 0.24 mL min^{-1} and 0.55 M maximum concentration of (PhBO)$_3$ **121**.
[b] Conversions were determined by GC, using tridecane as an internal standard.
[c] ee Determined by HPLC with a chiral column.
[d] After a single recrystallization.
[e] Determined by ^{1}H NMR.

4.8.2 Enzymatic Enantioselective Micro Reactions

Along with a selection of chemically catalyzed reactions, immobilized enzyme reactors have been investigated as a means of generating enantiopure materials utilizing continuous flow reactors. The immobilization of biocatalysts is particularly important when considering scaling a synthetic process due to the costs associated with sourcing and recovering biocatalytic material. In addition, immobilization is also advantageous as it can stabilize the biocatalyst enabling the development of more robust synthetic methodology. For topical reviews on the subject of immobilized biocatalysts within micro flow reactors, refer to Goodall and coworkers [68], Miyazaki and Maeda [69], and Fernandes [70].

Wall-Coated Reactors: Using wall coating methodology, Lin and coworkers [71] reported the development of a protocol for the enantioselective hydrolysis of esters (Scheme 4.35) as a means of efficiently recycling a Lipase enzyme. To achieve this, the authors employed a borosilicate glass reactor, containing a wet-etched serpentine micro channel (dimensions = 200 μm (wide) × 25 μm (deep) × 41 cm (long)), onto which the lipase (*Burkholderia cepacia* (BCL)) was covalently bound using a process of silanization, glutaraldehyde derivatization, and immobilization. Using a micro-BCA protein assay the authors quantified the amount of enzyme immobilized onto the micro channel surface to be 14 μg.

To perform a reaction, the authors pumped a dilute solution of (*rac*)-1-phenylethylacetate **126** (1 × 10^{-3} M) in phosphate buffer (pH 7.4) through the wall-coated reactor at a range of flow rates in order to evaluate the effect of residence time on the formation of (*R*)-1-phenylethanol **127**; quantification was achieved by offline chiral HPLC analysis. Using this approach, the authors observed increasing conversion with residence time, identifying 30 min as the optimum (21% **127**). Operation of the micro reactor under the aforementioned conditions over a period of 9 days illustrated stable conversion of the racemate **126** into (*R*)-1-phenylethanol **127** over extended periods of operation; the reactor was also shown to maintain enzyme activity after a period of 3 months (Table 4.20). With this in mind, the authors proposed that such a system may find application as a tool for substrate and enzyme screening.

Although the evaluation of wall-coated reactors has demonstrated the ability to catalyze processes in the presence of catalytic surfaces, the surface-to-volume ratio, although significantly higher than batch reactors, is still not sufficient to afford high space time yields. With this in mind, authors have investigated the use of functionalized monoliths within micro fabricated reactors as a means of increasing the catalytic surface area without observing high backpressures commonly encountered within packed-bed systems.

Monolithic Reactors: Ngamsom et al. [72] more recently reported the fabrication of a borosilicate glass reactor configured for the high-throughput screening of enzymes toward substrates, employing the biocatalytic conversion of *N*-benzoyl-l-phenylalanine

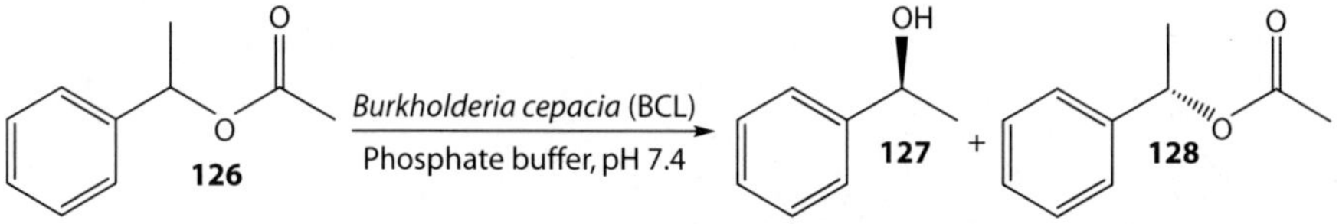

SCHEME 4.35 Illustration of the model reaction selected to demonstrate the enantioselective hydrolysis of (*rac*)-1-phenylethylacetate **126** to afford (*R*)-1-phenylethanol **127** and (*S*)-α-methylbenzyl acetate **128**.

TABLE 4.20
Comparison of Enzyme Efficiency in Batch (Free) and in a Micro Reactor (Immobilized)

Reactor	Enzyme Dosage (μg)	Ester Dosage (μg)	Yield (%)	ee (%)[a]
Micro	14	33	20	95
Batch	1400	3284	18	91

[a] Determined by offline HPLC analysis using a chiral analytical column.

129 into l-phenylalanine 130 as a model reaction (Scheme 4.36). Using a process of photoinitiation, poly(glycidylmethacrylate-*co*-ethylenedimethacrylate) monomers were formed within a series of wet-etched micro channels (300 μm (wide) × 100 μm (deep) × 1.5 cm (long)), affording a glycidyl-derived surface onto which the l-aminoacylase 131, derived from *T. litoralis*, was covalently bound (1% v/v in Tris-HCl buffer, 1 μL min⁻¹ for 3 h); with any free enzyme 131 removed with Tris-HCl buffer (pH 8.0). Once functionalized, the activity of the immobilized enzyme 131 was evaluated by pumping a solution of *N*-benzoyl-l-phenylalanine 129 (1 × 10⁻³ M) in Tris-HCl buffer (pH 8.0) through the monolith, at a range of flow rates (1–8 μL min⁻¹). The reaction products were then analyzed using two methods, a modified Cd/Ninhydrin method and by analytical HPLC, in order to determine the conversion of *N*-benzoyl-l-phenylalanine 129 into l-phenylalanine 130 and benzoic acid 69. Using these conditions, the authors confirmed enzyme activity and identified flow rates of <4 μL min⁻¹ as optimum for quantitative deacylation of *N*-benzoyl-l-phenylalanine 129. Having confirmed enzyme activity, the authors subsequently investigated the enzymes 131 stereospecificity achieved via the introduction of a racemic substrate, *N*-benzoyl-d,l-phenylalanine (1 × 10⁻³ M), into the micro reactor. Analysis of the reaction products illustrated 100% cleavage of the l-form 129, leaving *N*-benzoyl-d-phenylalanine unreacted; demonstrating retention of the catalysts 131 stereospecificity upon immobilization. The authors then performed a series of screening reactions in order to compare the results obtained within the micro reactor to those published for the free enzyme 131, prior to investigating a series of previously unreported substrates (Table 4.21).

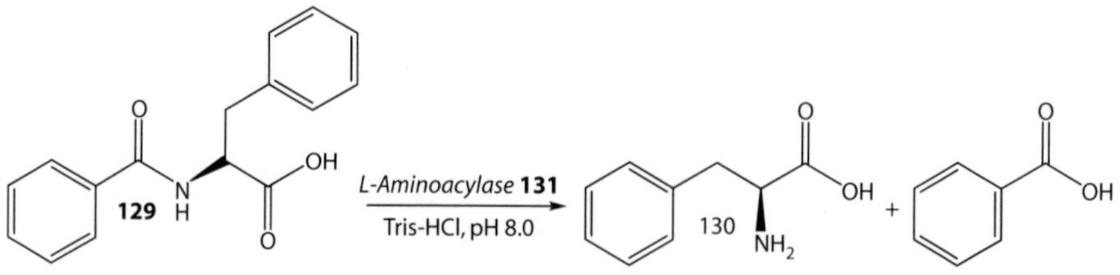

SCHEME 4.36 Model reaction selected to demonstrate the successful fabrication of a biocatalytic micro reactor.

TABLE 4.21
Substrate Screen Performed Using a Biocatalytic Micro Reactor

	Conversion (%)	
Substrate	1 μL min^{-1}	2 μL min^{-1}
N-Benzoyl-d,l-Phe	50.0[a]	—
N-Benzoyl-l-Phe **129**	100.0	100.0
N-Chloroacetyl-l-Phe	100.0	90.1
N-Acetyl-l-Phe	100.0	66.1
N-t-Boc-l-Phe	40.9	—
N-Cbz-l-Phe	30.9	—
N-Benzoyl-l-Thr	68.3	—
N-Benzoyl-l-Leu	52.2	—
N-Acetyl-d,l-Ser	—	25.6[a]
N-Benzoyl-l-Arg	43.6	—
N-Acetyl-d,l-Leu	—	20.5[a]
N-Acetyl-l-Met	38.7	—
N-Acetyl-l-Tyr	33.3	—
N-Acetyl-l-Trp	7.0	—
N-Acetyl-l-Lys	0.0	—

[a] Substrate conversions out of a maximum 50%.

In all cases, the amides were investigated as 1×10^{-3} M solutions in Tris-HCl buffer (pH 8.0) using a total flow rate of 2 μL min^{-1} and a reactor temperature of 25°C. Using this approach, it can be seen that the enzyme retained its specificity toward benzoyl- and chloroacetyl-protecting groups compared to acetyl- or other N-protecting groups, as demonstrated using the phenylalanine residue (benzoyl>chloroacetyl>acetyl>Boc>Cbz). In addition, the authors investigated the enzymes **131** activity toward a series of N-protected amino acids, finding the biocatalyst more selective toward Phe > >Ser > Leu > Met > Tyr > Trp.

Packed-bed Reactors: Csajagi et al. [73] recently demonstrated the enantioselective acylation of a series of racemic alcohols using a stainless steel packed-bed reactor containing commercially available solid-supported biocatalysts. As Scheme 4.37 illustrates, initial investigations focused on the acylation of (*rac*)-phenyl-1-ethanol

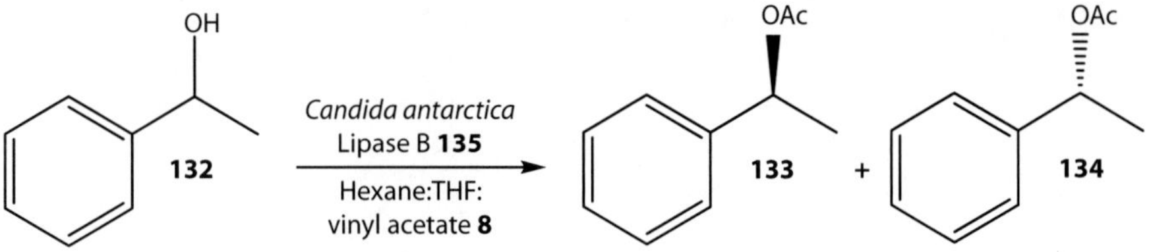

SCHEME 4.37 Schematic illustrating the model reaction used to demonstrate the enantioselective acylation of alcohols under continuous flow.

132 to afford (R)-phenylacetate **133** and (S)-phenyl-1-ethanol **134** using *Candida antarctica* Lipase B (CaLB) on acrylic beads **135** as the biocatalyst and vinyl acetate **8** as the acylating agent. Employing a packed-bed reactor, containing 0.40 g of CaLB **135**, the authors pumped a solution of (*rac*)-phenyl-1-ethanol **132** (5–50 mg mL^{-1}) in hexane:THF:vinyl acetate **8** (2:1:1) through the reactor (0.1 to 1.0 mL min^{-1}) and a reactor temperature of 25°C. Under the aforementioned conditions, the authors noted that the reactor reached steady state after 30 min, after which the reaction products were sampled and analyzed chromatographically to determine the conversion and enantioselectivity of the process.

Optimal conditions were found to be a feedstock concentration of 10 mg mL^{-1} **132**, affording the (R)-phenylacetate **133** in 50% conversion and 99.2% *ee*; analogous results were obtained in batch however, a reaction time of 24 h was required. Pleased with their findings, the authors compared several other parameters of key interest when performing kinetic resolutions, enantiomer selectivity (*E*), enantiomeric excess of both product (ee_p) and residual substrate (ee_s), the specific reaction rate (*r*); which indicates the amount of product min^{-1} g^{-1} of enzyme.

As Table 4.22 illustrates, the authors investigated the activity of eight enzymes toward the acylation of (*rac*)-phenyl-1-ethanol **132** comparing the results obtained in their packed-bed reactor with those from conventional batch reactions.

Using this approach, it can clearly be seen that the productivity for a given enzyme is higher when used within a packed-bed reactor than a stirred batch reactor, an

TABLE 4.22
Summary of the Results Obtained for the Kinetic Resolution of (*Rac*)-Phenyl-1-Ethanol 132 Catalyzed via a Series of Lipases

Enzyme	Reactor	Conversion (%)	ee_s 134 (%)	ee_p 133 (%)	E	r (μmol min^{-1} g^{-1})
CaLB **135**	Flow	50	98.8	99.2	>>200	10.2
	Batch	51	99.6	98.3	>>200	5.7
Pseudomonas cepacia, IM	Flow	50	96.8	97.4	>200	10.2
	Batch	57	99.5	80.6	53	7.5
Lipozyme™ *Mucor miehei*	Flow	19	31.1	99.1	>200	4.0
	Batch	42	78.3	98.7	>200	4.1
Lipozyme™ TL TM	Flow	18	28.9	98.9	>100	3.7
	Batch	31	55.4	97.7	>100	2.5
Amano lipase AK	Flow	52	99.8	94.6	>100	10.6
	Batch	75	99.0	47.0	13	17.3
Amano lipase PS	Flow	24	39.6	99.6	>>200	5.0
	Batch	44	84.4	99.4	>>200	4.4
Candida rugosa	Flow	15	21.7	53.2	4	3.1
	Batch	16	22.4	53.7	4	1.1
Porcine pancreas	Flow	6	13.6	97.2	81	1.2
	Batch	16	32.1	98.5	>100	1.0

observation that has been attributed a lower voidage within packed-beds (34%) compared to stirred reactors (90%). The table also illustrates that in the most part no significant improvements in E or ee were obtained; however in the cases of *Pseudomonas cepacia* and Amano Lipase AK dramatic increases in selectivity were observed. In addition to the kinetic resolution of (*rac*)-phenyl-1-ethanol **132**, the authors also investigated the resolution of (*rac*)-cyclohexylethanol and (*rac*)-phenyl-propan-2-ol due to their commercial applicability and academic interest; confirming the procedures tolerance toward other 2° alcohols.

Having demonstrated the ability to kinetically resolve a series of 2° alcohols with increased productivity within packed-bed reactors, the researchers subsequently investigated the use of this technology on a preparative scale (20 mL). As Table 4.23 illustrates, excellent yields and enantioselectivities were obtained, presenting a synthetically viable approach to the kinetic resolution of racemic alcohols. Interestingly, the authors also reported the use of lyophilized enzymes within their packed-bed system demonstrating the ability to employ free enzymes in nonaqueous flow systems.

TABLE 4.23
Summary of the Results Obtained for the Kinetic Resolution of 2° Alcohols Performed on a Preparative Scale

Compound	Yield (%)[a]	ee (%)	$[\alpha]_D^{25}$[b]	E
132 (OAc)	40	98.5	−62.8	
134 (OH)	48	99.1	+125.3	>>>200
(OAc, cyclohexyl)	26	77.4	+2.0	
(OH, cyclohexyl)	41	99.0	+7.1	>200
(OAc, phenylpropyl)	34	56.4	+4.9	
(OH, phenylpropyl)	41	85.1	−23.3	22

[a] Products isolated from the reactor output stream.
[b] Specific rotations (c 1.0, $CHCl_3$).

4.9 MULTISTEP/MULTICOMPONENT REACTIONS

Demonstrating the synthetic versatility of continuous flow reactors, numerous authors have reported the development of flow protocols combining liquid-phase and heterogeneous catalysts to achieve multiple reaction steps in a single process.

4.9.1 INDEPENDENT MULTISTEP FLOW REACTIONS

The ability to spatially resolve solid-supported catalysts within micro flow reactors has the potential to enable novel synthetic transformations to be performed. In order to evaluate this concept, Haswell and coworkers [74] initially investigated the use of a solid-supported acid, followed by a solid-supported base in order to perform an acid-catalyzed acetal hydrolysis followed by a Knoevenagel condensation reaction using the *in situ* generated aldehyde (Scheme 4.38). Using this approach, the authors were able to tune the residence time within the reactor to suit both the hydrolysis reaction and the condensation step, affording quantitative conversion of the acetal into the respective α,β-unsaturated product. To evaluate the scope of this technique, the authors evaluated the effect of acetal functionality and the activated methylene, along with various acid and base combinations, obtaining the target compounds in excellent isolated yield (>99.2%) and purity (99.9%).

4.9.2 INTEGRATED MULTISTEP SEQUENCES

In addition to examples illustrating the use of catalysts, reagents and scavengers in series, authors have also reported the fabrication of integrated reactors whereby solution phase reaction steps are followed by heterogeneously catalyzed steps.

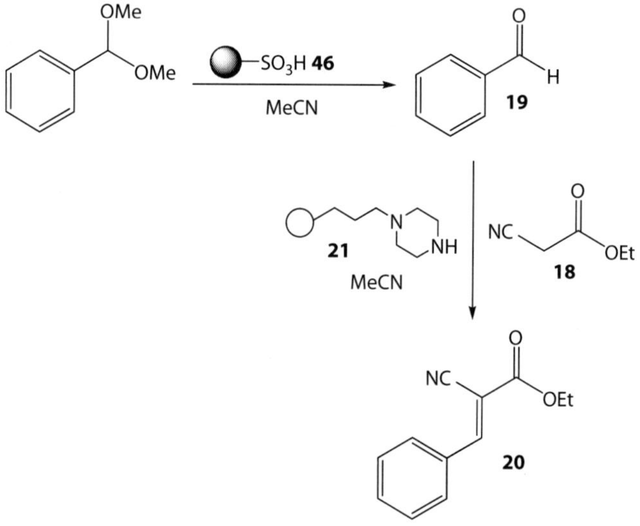

SCHEME 4.38 Illustrating of the multicatalyst approach employed using spatially resolved catalysts in a single capillary reactor.

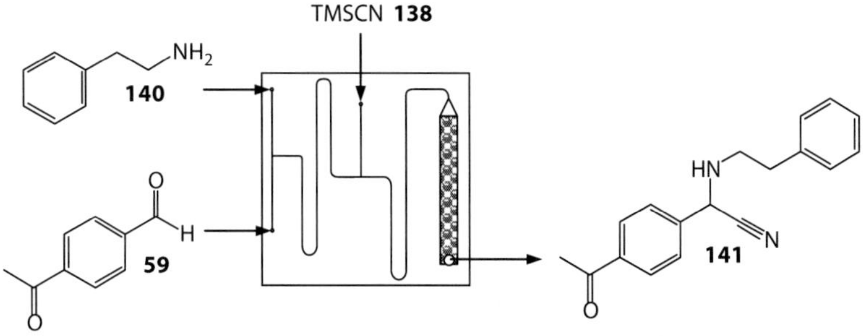

Multi-Phase Micro Reactions 273

FIGURE 4.8 Schematic illustrating the reaction manifold used for the multicomponent Strecker reaction controlled using pressure-driven flow.

One example of this was reported by Wiles and Watts [75,76] whereby an in-house fabricated borosilicate glass reactor was used for the multicomponent Strecker reaction which comprised of an initial solution phase imine formation and was followed by a Lewis Acid catalyzed addition of cyanide, to the *in situ* prepared imine **136**, to afford the respective α-aminonitrile (Figure 4.8 and Scheme 4.39).

Initially using polymer-supported ethylenediaminetetraacetic acid ruthenium (III) chloride (PS-RuCl$_3$) **137** as the catalyst (0.01 g, 0.26 mmol g^{-1}), reactions were conducted by pumping a solution of aldehyde (0.4 M in MeCN) and amine (0.4 M in MeCN) into a central micro reaction channel (dimensions = 150 μm (wide) × 50 μm (deep) × 5.6 cm (long)), from separate inlets, where they reacted to afford the intermediate aldimine. A solution of TMSCN **138** (0.2 M, 1 eq. in MeCN) was added from a third inlet where it mixed with the imine prior to entering the packed-bed where the nucleophilic addition of cyanide occurred to afford the target α-aminonitrile. The reaction products were concentrated *in vacuo* and the crude solid obtained analyzed by ^{1}H NMR spectroscopy to determine the conversion of aldehyde into

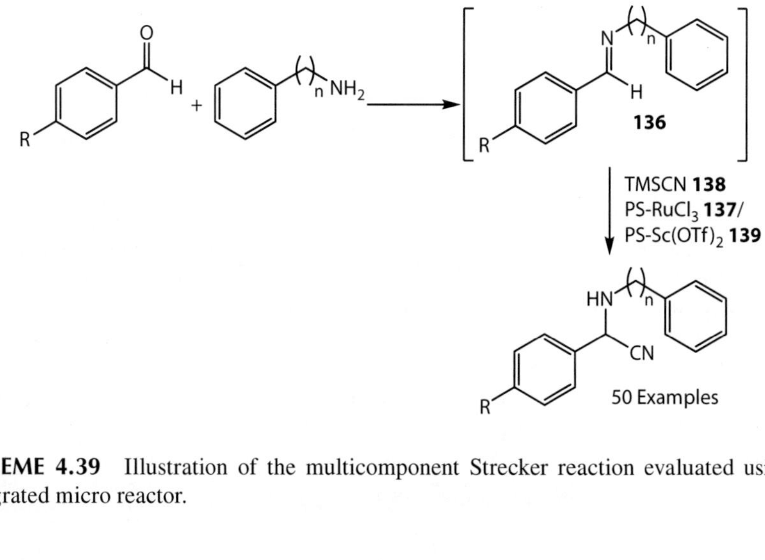

SCHEME 4.39 Illustration of the multicomponent Strecker reaction evaluated using an integrated micro reactor.

product. In addition to PS-RuCl$_3$ **137**, the authors also investigated the use of the catalyst polymer-supported scandium triflate (PS-Sc(OTf)$_2$) **139**, whereby an increase in productivity was obtained; a selection of results are presented in Table 4.24.

Compared to a multicomponent batch reaction, the stepwise flow methodology enabled the formation of α-aminonitriles in high yield and purity as no competing cyanohydrin formation occurred. Furthermore, by controlling the reaction times employed the technique could be tuned to enable the chemoselective reaction of aldehydes in the presence of ketonic functionalities. This was demonstrated using the reaction of 4-acetylbenzaldehyde **59** and 2-phenylethylamine **140** (Figure 4.8); whereby 2-(4-acetylphenyl)-2-(phenethylamino)acetonitrile **141** was obtained in 99.8% yield, as the sole reaction product.

4.9.3 Reagents and Scavengers in Series

Owing to their pharmaceutically interesting properties, Ley and coworkers [77] investigated the synthesis of 4,5-disubstituted oxazoles using a multipurpose meso fluidic reactor (Syrris, UK). Introducing two reactant solutions into the glass device, from independent inlets, the authors mixed the solutions of acyl chloride (1×10^{-3} M in MeCN) and ethylisocyanoacetate **142** (1×10^{-3} M in MeCN) at a T-mixer prior to heating (by placing the reactor on a modified hotplate), the reactants were then pumped through a packed-bed reactor containing PS-BEMP (2-*tert*-butylimino-2-diethylamino-1,3-dimethylperhydro-1,3,2-diazaphosphorane on polystyrene) **143** where the intermediate addition product cyclized to afford the target 4,5-disubstituted oxazole. The reaction products were then pumped through a scavenger cartridge, containing QuadraPure BZA **30** in order to enhance the purity of the final product. Using this approach, the authors identified a reaction time of 20–30 min as optimal depending on the substrate employed and as Table 4.25 illustrates the target compounds were obtained in high to excellent yield. In addition to varying the aromatic substituent, the authors also evaluated the effect use of aliphatic and heterocyclic acyl chlorides, again obtaining the respective 4,5-disubstituted isoxazole in excellent isolated yield (88–99%). Other examples of this mode of operation include the synthesis of (±)-Oxomaritidine [78] and Grossamide [79], details of which can be found in Chapter 7.

4.9.4 Combined Chemical and Biochemical Catalysis

In a novel example, Spain and coworkers [80] demonstrated the coupling of a metal and biochemical catalyst into a flow reactor to affect the continuous flow synthesis of aminophenols from nitroaromatic compounds. As Scheme 4.40 illustrates, the combination of zinc powder **144** and an immobilized mutase biocatalyst **145** enabled the authors to generate the hydroxylamine derivative **146** from nitrobenzene **147** and react it without isolation to afford the respective 2-aminophenol **148**. Employing an aqueous solution of nitrobenzene **147** (1.0 mM) and a total flow rate of 250 μL min^{-1}, the authors obtained 2-nitrophenol **148** in 89% with small amounts of residual nitrobenzene **147** and hydroxylaminobenzene **146**. Monitoring the reactor effluent for a period of 5 h, the authors confirmed the system stability, observing no

TABLE 4.24
Comparison of PS-RuCl₃ 137 and PS-Sc(OTf)₂ 139 as Catalysts for the Strecker Reaction under Continuous Flow Conditions

Product	PS-Catalyst	Throughput (mg h^{-1})
(4-Br-C₆H₄)CH(NHPh)(CN)	137	17.2
	139	34.4
(4-Br-C₆H₄)CH(NHCH₂Ph)(CN)	137	18.1
	139	36.1
(4-Br-C₆H₄)CH(NHCH₂CH₂Ph)(CN)	137	18.9
	139	38.0
(4-Br-C₆H₄)CH(NHCH₂CH₂CH₂Ph)(CN)	137	19.7
	139	39.5
(4-Br-C₆H₄)CH(N-pyrrolidinyl)(CN)	137	32.0
	139	69.6

TABLE 4.25
A Selection of 4,5-Disubstituted Oxazoles Synthesized Using a Combination of Solid-Supported Base and Scavenger Columns under Continuous Flow

R	Product	Yield (%)
4-Br		88
4-NO$_2$		83
4-F		94
2-CF$_3$		98
2-CN		83
3,4-OMe		83

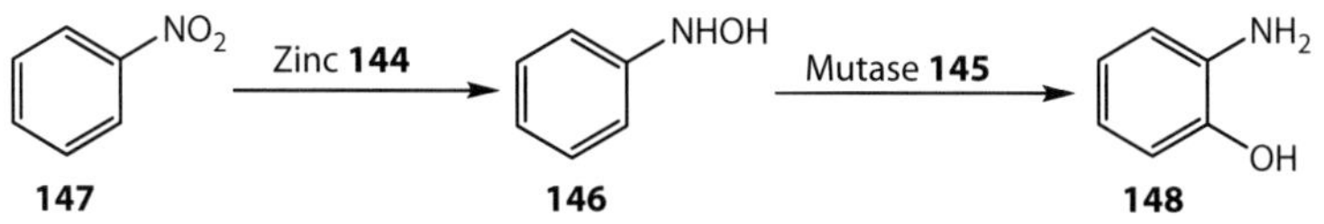

Multi-Phase Micro Reactions

SCHEME 4.40 Schematic illustrating the two catalyst protocol employed for the synthesis of 2-aminophenol **148**.

reduction in conversion efficiency. With this information in hand, the authors subsequently investigated the synthesis of N-[2-(4-amino-3-hydroxyphenyl)-2-hydroxy-1-hydroxymethylethyl]-2,2-dichloroacetamide **149** from chloramphenicol **150**. Operating the reactor under the aforementioned conditions the authors obtained the target compound **149** in quantitative conversion at a production rate of 0.24 mg h^{-1} mg^{-1} total protein (Scheme 4.41).

4.9.5 "Catch and Release" Strategies under Continuous Flow

Employing a principle often referred to as "catch and release," Ley and coworkers [81] described the development of a general flow process for the multistep assembly of peptides and its application to the synthesis of Boc and Cbz N-protected amides. Using a commercially available micro/mesofluidic pumping system (Syrris, UK), the authors devised a serial reaction process which enabled the coupling of carboxylic acids and amines under flow conditions. Reactions were performed by placing reactant solutions into a series of sample loops, connected to a flow stream which passed through a series of packed-bed reactors, containing polymer-supported reactants and scavengers, affording the target peptide as a solution at the reactor outlet.

An example of this is illustrated in Scheme 4.42 whereby a solution of an N-protected carboxylic acid **151**, a phosphonium coupling reagent **152** and diisopropylethylamine **153** is passed through a packed-bed containing polymer-supported 1-hydroxybenzotriazole **154**, sequestering the carboxylic acid **151** on the solid-support as the activated ester. Residual reagents are then washed from the column using DMF, prior to connecting a polymer-supported DMAP (PS-DMAP) **105** column and a solid-supported sulfonic acid (PS-SO_3H) **46** column in-line. In the second reaction step, a protected HCl salt **155** is passed through the PS-DMAP **105**, liberating the amine which subsequently couples with the active ester to afford the target dipeptide **156**. The reaction mixture then passes through a third and final column **46** where any unreacted amine **155** is removed. The target dipeptide **156** can then be isolated from

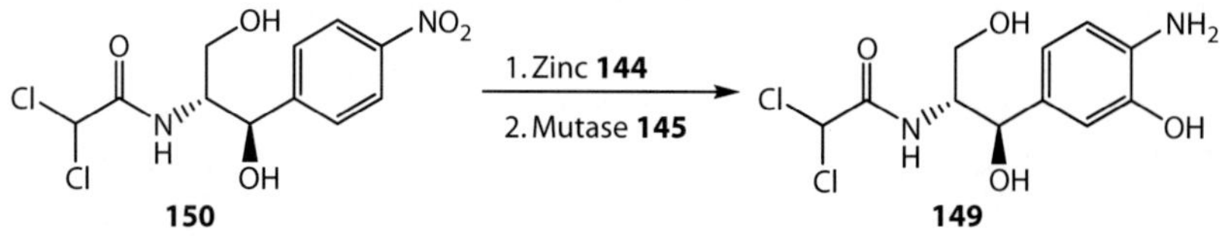

SCHEME 4.41 Illustration of the zinc/mutase cascade reaction of the antibiotic chloramphenicol **150** to the aminophenol derivative **149**.

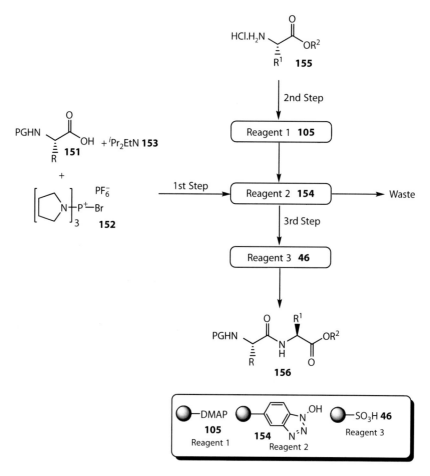

SCHEME 4.42 Schematic illustrating the "catch and release" strategy employed for the continuous flow synthesis of dipeptides.

the reaction stream via evaporation of the solvent (DMF) and analyzed by ^{1}H NMR spectroscopy and LC–MS in order to evaluate product purity (typically >95.5).

Using this approach, the authors investigated the coupling of a series of peptides obtaining products with yields ranging from 61% to 83%, as summarized in Table 4.26, observing that due to the short contact times within the reactors racemization was negligible.

Wild et al. [82] subsequently developed a solid-supported crown ether reagent suitable for the noncovalent protection of amine functionalities as a means of simplifying protection/reaction/deprotection strategies. To demonstrate the synthetic utility of this methodology, the authors evaluated the reagent **157** under "catch and release" conditions to perform the acylation of tyramine **158**. For comparison purposes, the authors initially conducted the reaction in the absence of an *N*-protecting group, therefore obtaining a complex reaction mixture containing the desired

TABLE 4.26
Summary of the Dipeptides Synthesized Using a "Catch and Release" Flow Protocol

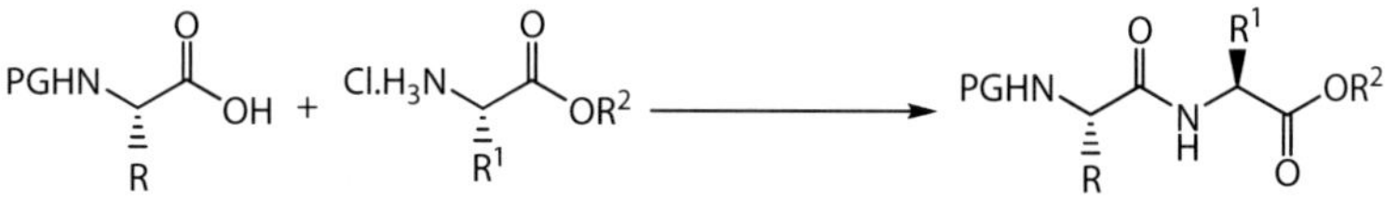

PG	R	R^1	R^2	Yield (%)
Boc	Me	CH$_2$Ph	Et	80
Boc	Me	H	Et	81
Boc	Me	CHMe$_2$	Me	83
Boc	Me	[-CH$_2$-]$_3$	Me	66
Cbz	CH$_2$Ph	CHMe$_3$	Me	79
Cbz	CH$_2$Ph	H	Et	76
Cbz	Me	CH$_2$Ph	Et	75
Cbz	Me	H	Et	78
Cbz	Me	[-CH$_2$-]$_3$	Me	61

[a] Isolated yield.
[b] Purities measured as >95% by ^{1}H NMR and LC–MS.

tyramine acetate **159** (23%), tyramine *N*-acetate **160** (12%), tyramine diacetate **161** (20%) and residual starting material (45%) (Scheme 4.43).

In comparison, using the "catch and release" strategy illustrated in Scheme 4.44, the authors were able to noncovalently protect the trifluoroacetic acid salt of tyramine **162** using the immobilized crown ether (step (a)), the material was subsequently *O*-acetylated with acetic anhydride **163** in the presence of an organic base **29** (step (c)). In the final reaction step the product was simultaneously deprotected and the crown ether regenerated using a solution of *N*,*N*,*N*′*N*′-tetramethylethylenediamine

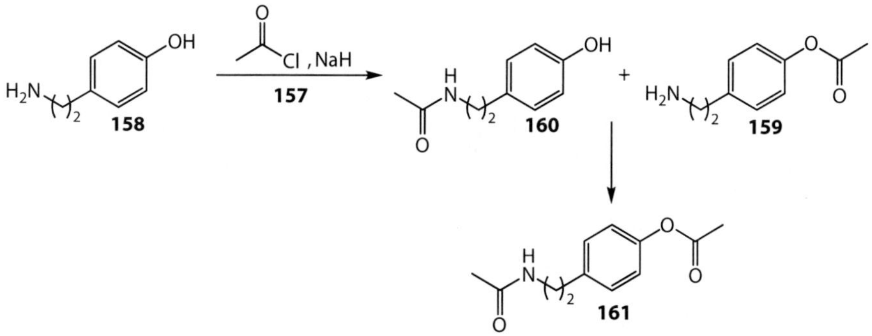

SCHEME 4.43 Schematic illustrating the array of reaction products attainable when acetylating tyramine **158** in the absence of a protecting group.

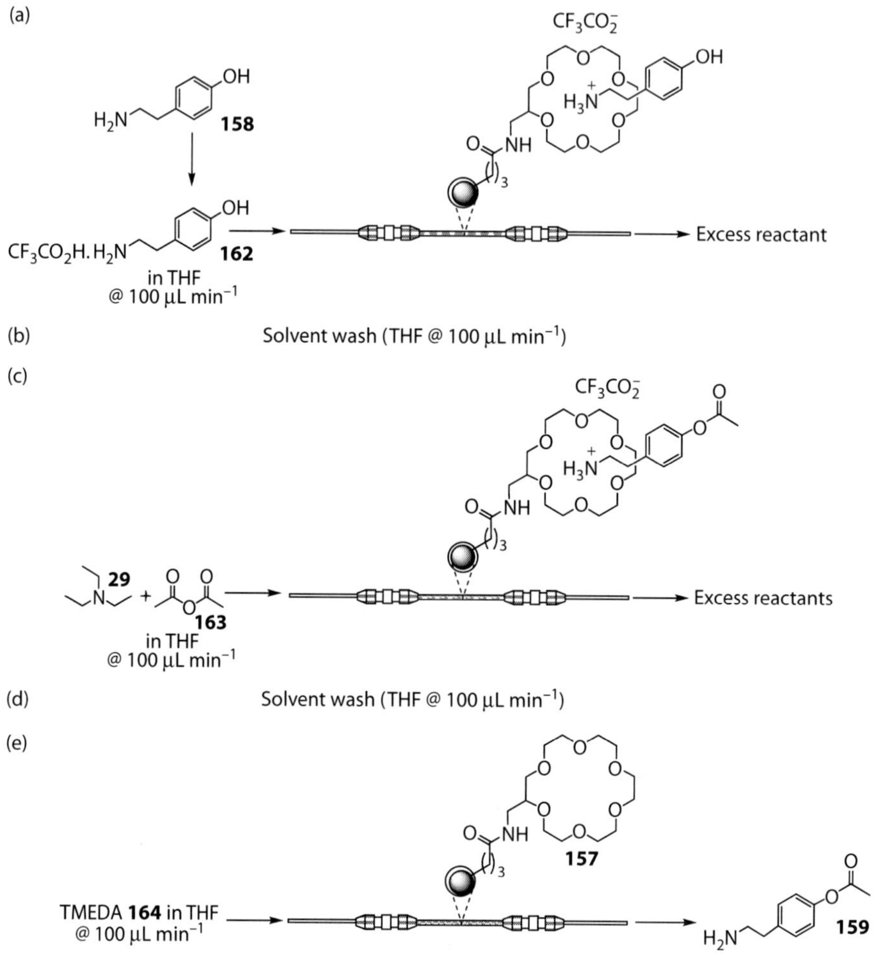

SCHEME 4.44 Schematic illustrating the reaction protocol employed for the continuous flow acetylation of tyramine **158**, using an immobilized 18-crown-6 ether derivative **157**.

164 (step (e)), affording the target compound tyramine acetate **159** in quantitative yield (2.4×10^{-2} mmol reaction^{-1}) and excellent selectivity. The generality of the protecting group strategy was subsequently evaluated, finding that HCl, TFA, and *p*-TSA salts could be readily complexed and the technique was general for a range of bifunctional compounds. Due to the various stages of the reaction methodology, it is clear that for this technique to find widespread application, automation of the process would be required.

4.9.6 Casein Kinase I Inhibitor Synthesis

During the development of continuous flow processes, the researcher must overcome technical challenges, such an example was recently reported by Ley and coworkers

[18] where owing to the thermal instability of lithiated bases such as n-BuLi **165**, a practical solution was required in order to enable reagent storage at −78°C and constant dispensation of n-BuLi **165** over several hours; this was achieved by using a dual sample loop arrangement where one was used while the other filled in preparation for use (26 min cycle). With this technique in hand, the authors investigated a series of organometallic deprotonation and substitution reactions used in the synthesis of a family of Casein Kinase Inhibitors (Figure 4.2).

To achieve this, n-BuLi **165** (0.4 M) in hexanes and the respective picoline derivative (0.3 M) in THF were premixed, at −78°C, prior to the addition of the aryl derivative (0.2 M) in THF. Upon mixing (5 mL coil), the reactants were warmed to room temperature (15 mL coil) and the reaction products collected and quenched in an aqueous solution of sat. NH_4Cl over a period of 11 h' processing 50.0 mmol of material. The reaction products subsequently underwent organic extraction, into EtOAc, with removal of the reaction solvent affording the target product as a keto–enol mixture (Scheme 4.45, Flow reaction 1).

In the second step, the authors investigated the α-bromination of the ketones from step 1, under the conditions described previously (Scheme 4.5), a packed-bed containing polymer-supported pyridine hydrobromide perbromide **15** was used to afford the target α-bromoketones in quantitative yield and purity; upon evaporation of the reaction solvent (MeOH).

In a third step, the α-bromoketone was dissolved in DMF and reacted with 3-amino-6-chloropyridazine **16** (4 equivalents) at 120°C, for 20 min, to afford the respective imidazo[1,2-*b*]pyridazine in isolated yields ranging from 52% to 82% (Scheme 4.45, Flow reaction 3).

In the final step (Scheme 4.45, Flow reaction 4), the authors investigated the introduction of structural diversity by the displacement of chlorine with a series of amines. After a short optimization, the authors identified the optimal conditions to be EtOH as the reaction solvent, 2 equivalents of amine, a residence time of 1.6 h and a reactor temperature of 177°C. The reaction products were treated with a scavenger resin to remove any residual amine and recrystallized to afford the target compound in high yield and purity.

Using the reaction conditions described above, the authors were able to generate a library containing 20 diverse analogs of a casein kinase I inhibitor and as summarized in Table 4.27, the compounds were obtained in good-to-moderate yields and excellent purity.

4.10 SUMMARY

From the selected examples described herein, it can be seen that the incorporation of solid materials, be it reagents, catalysts or scavengers, increases the number and type of synthetic processes that can be performed using continuous flow methodology. As observed with liquid-phase reactions, heterogeneous systems can be heated using conventional or microwave heating affording both flexibility in the mode of operation and tailoring of the system to suit the application of the flow process under development. In addition to increases in product purity which leads to a reduction in the need for post reaction processing, the incorporation of heterogeneous

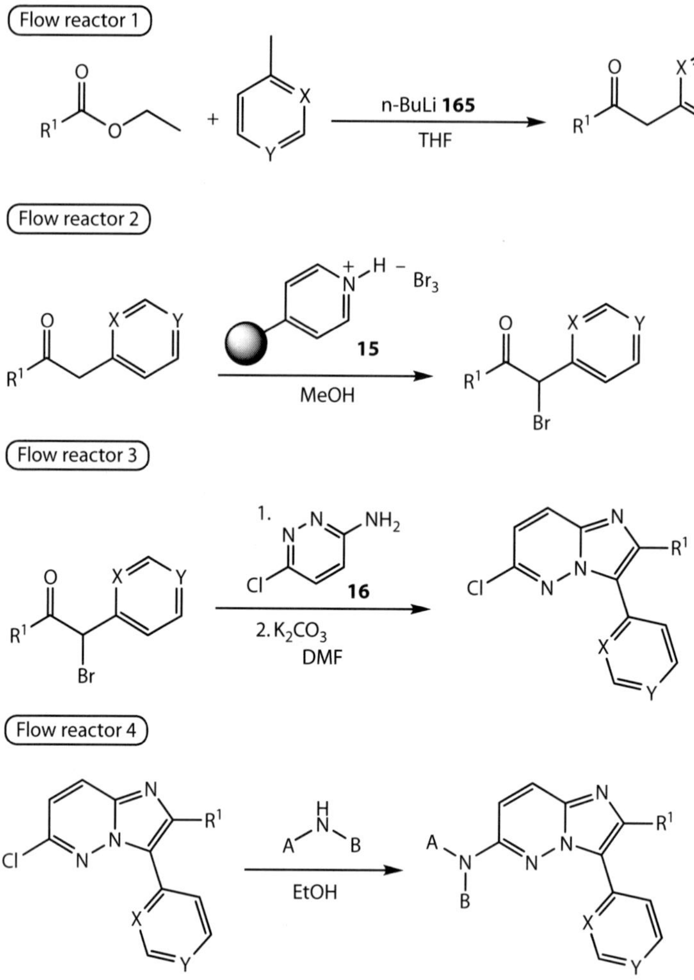

SCHEME 4.45 Summary of the multiple multiphase reaction steps performed under continuous flow for the synthesis of Casein Kinase I inhibitors.

material into flow reactor systems is also advantageous as it enables materials to be employed without degradation, compared to stirred tank reactors, making recycle and product separation facile. When employing reagents and or scavengers, the materials loading must be considered in order to determine the working time of the process. In the absence of *in situ* regeneration techniques, efficient methods for material removal/replacement or the use of disposable cartridges must be considered. With all of this in mind, it can be seen that the use of multiphase reaction systems which employ packed-bed, monolithic, membrane or wall-coated technologies have a lot to offer the research chemist in the quest to develop novel and synthetically useful techniques.

TABLE 4.27
A Selection of the Casein Kinase I Inhibitors Synthesized Using Multiphase, Multistep Continuous Flow Methodology

R^1	X	Y	Amine	Yield (%)
4-F-Ph	CH	N	Piperazine	70
4-F-Ph	CH	N	N-Methyl piperazine	61
4-F-Ph	CH	N	1-Methyl-1,4-diazepane	25
4-F-Ph	CH	N	Morpholine	63
4-F-Ph	CH	N	Piperidine	63
4-F-Ph	N	N	Piperidine	52
2-Thienyl	CH	N	Morpholine	88
2-Thienyl	CH	N	Piperidine	90
2-Thienyl	CH	N	1-Methyl-1,4-diazepane	51
2-Thienyl	CH	N	Piperazine	54
2-Thienyl	CH	N	N-Methylpiperazine	77

REFERENCES

1. Huang, J., Weinstein, J., and Besser, R. S. 2009. Particle loading in a catalyst-trap microreactor: Experiment vs. simulation, *Chem. Eng. J.* 155: 388–395.
2. Kawakami, K., Sera, Y., Sakai, S., Ono, T., and Ijima, H. 2005. Development and characterization of a silica monolith immobilized enzyme micro-bioreactor, *Ind. Chem. Res.* 44: 236–240.
3. Logan, T. C., Clark, D. S., Stachowiak, T. B., Svec, F., and Fréchet, J. M. J. 2007. Photopatterning enzymes on polymer monoliths in microfluidic devices for steady-state kinetic analysis and spatially separated multienzyme reactions, *Anal. Chem.* 79: 6592–6598.
4. Shore, G., Tsimerman, M., and Organ, M. G. 2009. Gold-film-catalyzed benzannulation by microwave-assisted, continuous flow organic synthesis (MACOS), *Beilstein J. Org. Chem.* 5(35).
5. Thomsen, M. S. and Nidetzky, B. 2009. Coated-wall microreactor for continuous biocatalytic transformations using immobilized enzymes, *Biotechnol. J.* 4: 98–107.
6. Kenis, P. A., Ismagilov, R. F., Takayama, S., and Whitesides, G. M. 2000. Fabrication inside microchannels using fluid flow, *Acc. Chem. Res.* 33: 841–847.
7. Kirschning, A., Solodenko, W., and Mennecke, K. 2006. Combining enabling techniques in organic synthesis: Continuous flow processes with heterogenized catalysts, *Chem. Eur. J.* 12: 5972–5990.
8. Kashid, M. V. and Kiwi-Minsker, L. 2009. Microstructured reactors for multiphase reactions: State of the art, *Ind. Eng. Chem. Res.* 48: 6465–6485.
9. Miyazaki, M. and Maeda, H. 2009. Microchannel enzyme reactors and their applications for processing, *Trends in Biotech.* 24: 463–470.

10. Westermann, T. and Melin, T. 2009. Flow-through catalytic membrane reactors-principles and applications, *Chem. Eng. Proc.* 48: 17–28.
11. Turner, H. W., Volpe, A. F., and Weinberg, W. H. 2009. High-throughput heterogeneous catalyst research, *Surface Sci.* 603: 1763–1769.
12. Kulkarni, A. A., Zeyer, K., Jacobs, T., and Kienle, A. 2007. Miniaturized systems for homogeneously and heterogeneously catalyzed liquid-phase esterification reaction, *Ind. Eng. Chem. Res.* 46: 5271–5277.
13. Woodcock, L. L., Wiles, C., Greenway, G. M., Watts, P., Wells, A., and Eyley, S. 2008. Enzymatic synthesis of a series of alkyl esters using novozyme 435 in a packed-bed, miniaturized, continuous flow reactor, *Biocatal. Biotransform.* 26: 501–507.
14. El Kadib, A., Chimenton, R., Sachse, A., Fajula, F., Galarneau, and Coq, B. 2009. Functionalized inorganic monolithic microreactors for high productivity in fine chemicals catalytic synthesis, *Angew. Chem. Int. Ed.* 48: 4969–4972.
15. Bonrath, W., Karge, R., and Netscher, T. 2002. Lipase-catalyzed transformations as keysteps in the large-scale preparation of Vitamins, *J. Mol. Catal. B: Enzymatic* 19–20: 67–72.
16. Kataoka, S., Endo, A., Oyama, M., and Ohmori, T. 2009. Enzymatic reactions inside a microreactor with a mesoporous silica catalyst support layer, *Appl. Catal. A: General* 359: 108–112.
17. Baumann, M., Baxendale, I. R., Ley, S. V., Nikbin, N., and Smith, C. S. 2008. Azide monoliths as convenient flow reactors for efficient Curtius rearrangement reactions, *Org. Biomol. Chem.* 6: 1587–1593.
18. Venturoni, F., Nikbin, N., Ley, S. V., and Baxendale, I. R. 2010. The application of flow microreactors to the preparation of a family of Casein Kinase I inhibitors, *Org. Biomol. Chem.* 8: 1798–1806.
19. Ajmera, S. K., Losey, M. W., Jensen, K. F., and Schmidt, M. A., 2001. Microfabricated packed-bed reactor for phosgene synthesis, *AIChE J.* 47: 1639–1647.
20. Wiles, C., Watts, P., and Haswell, S. J., 2004. An investigation into the use of silica-supported bases within EOF-based flow reactors, *Tetrahedron* 60: 8421–8427.
21. Wiles, C., Watts, P., and Haswell, S. J. 2007. The use of electroosmotic flow as a pumping mechanism for semi-preparative scale continuous flow synthesis, *Chem. Commun.* 966–968.
22. Bogdan, A. R., Mason, B. P., Sylvester, K. T., and McQuade, D. T. 2007. Improving solid-supported catalyst productivity by using simplified packed-bed microreactors, *Angew. Chem. Int. Ed.* 46: 1698–1701.
23. Nikbin, N. and Watts, P. 2004. Solid-supported continuous flow synthesis in microreactors using electroosmotic flow, *Org. Proc. Res. Dev.* 8: 942–944.
24. Costantini, F., Bula, W. P., Salvio, R., Huskens, J., Gardeniers, H. J. G. E., Reinhoudt, D. N., and Verboom, W. 2009. Nanostructure based on polymer brushes for efficient heterogeneous catalysis in microreactors, *J. Am. Chem. Soc.* 131: 1650–1651.
25. Lau, W., Yeung, K. L., and Martin-Aranda, R. 2008. Knoevenagel condensation reaction between benzaldehyde and ethylacetoacetate in microreactor and membrane microreactor, *Micro. Meso. Mat.* 115: 156–163.
26. Bonfils, F., Cazaux, I., HHodge, P., and Caze, C. 2006. Michael reactions carried out using a bench-top flow system, *Org. Biomol. Chem.* 4: 493–497.
27. Palmieri, A., Ley, S. V., Polyzos, A., Ladlow, M., and I. R. Baxendale, 2009. Continuous flow based catch and release protocol for the synthesis of α-ketoesters, *B. J. Org. Chem.* 5(23).
28. Burguete, M. I., Erythropel, H., Garcia-Verdugo, E., Luis, S. V., and Sans, V. 2008. Base supported ionic liquid-like phases as catalysts for the batch and continuous-flow Henry reaction, *Green Chem.* 10: 401–407.

29. Shore, G. and Organ, M. G. 2008. Diels–Alder cycloadditions by microwave-assisted continuous flow organic synthesis (MACOS): The role of metal films in the flow tube, *Chem. Commun.* 838–840.
30. Dräger, G., Kiss, C., Kunz, U., and Kirschning, A. 2007. Enzyme-purification and catalytic transformations in a microstructured PASSflow reactor using a new tyrosine-based Ni-NTA linker system attached to a polyvinylpyrrolidinone-based matrix, *Org. Biomol. Chem.* 5: 3657–3664.
31. Baumann, M., Baxendale, I. R., Martin, L. J., and Ley, S. V. 2009. Development of fluorination methods using continuous-flow microreactors, *Tetrahedron* 65: 6611–6625.
32. Stevens, J. G., Bourne, R. A., and Poliakoff, M. 2009. The continuous self aldol condensation of propionaldehyde in supercritical carbon dioxide: A highly selective catalytic route to 2-methylpentenal, *Green Chem.* 11: 409–416.
33. Kolb, H. C., Finn, M. G., and Sharpless, K. B. 2001. Click chemistry: Diverse chemical function from a few good reactions, *Angew. Chem. Int. Ed.* 40: 2004–2021.
34. Girard, C., Önen, E., Aufort, M., Beauvière, S., Samson, E., and Herscovici, J. 2006. Reuseable polymer-supported catalyst for the [3 + 2] Huisgen cycloaddition in automation protocols, *Org. Lett.* 8: 1689–1692.
35. Smith, C. D., Baxendale, I. R., Lanners, S., Hayward, J. J., Smith, S. C., and Ley, S. V. 2007. [3 + 2]Cycloaddition of acetylenes with azides to give 1,4-disubstituted 1,2,3-triazoles in a modular flow reactor, *Org. Biomol. Chem.* 5: 1559–1561.
36. Baxendale, I. R., Ley, S. V., Mansfield, A. C., and Smith, C. D. 2009. Multistep synthesis using modular flow reactors: Bestmann–Ohira reagent for the formation of alkynes and triazoles, *Angew. Chem. Int. Ed.* 48: 4017–4021.
37. Fuchs, M., Goessler, W., Pilger, C., and Kappe, C. O. 2010. Mechanistic insights into copper(I)-catalyzed azide–alkyne cycloadditions using continuous flow conditions, *Adv. Synth. Catal.* 352: 323–328.
38. Glasnov, T. N., Findenig, S., and C. O. Kappe, 2009. Homogeneous catalysis with $Pd(OAc)_2$ aryl bromide, *Chem. Eur. J.* 15: 1001–1010.
39. Wiles, C., Watts, P., and Haswell, S. J. 2005. Acid-catalysed synthesis and deprotection of dimethyl acetals in a miniaturized electroosmotic flow reactor, *Tetrahedron Lett.* 61: 5209–5217.
40. Wiles, C., Watts, P., and Haswell, S. J. 2007. An efficient, continuous flow technique for the chemoselective synthesis of thioacetals, *Tetrahedron Lett.* 48: 7362–7365.
41. Wilson, N. G. and McCreedy, T. 2000. On-chip Catalysis using a lithographically fabricated glass microreactor—The dehydration of alcohols using sulfated zirconia, *Chem. Commun.* 733–734.
42. Rouge, A., Spoetzl, B., Gebauer, K., Schenk, R., and Renken, A. 2001. Microchannel reactors for fast periodic operation: The catalytic dehydration of isopropanol, *Chem. Eng. Sci.* 56: 1419–1427.
43. Wiles, C., Watts, P., and Haswell, S. J. 2006. A clean and selective oxidation of aromatic alcohols using silica-supported Jones' reagent in a pressure-driven flow reactor, *Tetrahedron Lett.* 47: 5261–5264.
44. Bogdan, A. and McQuade, D. T. 2009. A biphasic oxidation of alcohols to aldehydes and ketones using a simplified packed-bed microreactor, *B. J. Org. Chem.* 5(17): 1–7.
45. Fritz-Langhals, E. 2005. Technical production of aldehydes by continuous bleach oxidation of alcohols catalyzed by 4-hydroxy-TEMPO, *Org. Proc. Res. Dev.* 9(5): 577–582.
46. Hou, Z., Theyssen, N., and Leitner, W. 2007. Palladium nanoparticles stabilized on PEG-modified silica as catalysts for the aerobic oxidation in supercritical carbon dioxide, *Green Chem.* 9: 127–132.
47. Wan, Y. S. S., Chau, J. L. H., Gavriilidis, A., and Yeung, K. L. 2002. TS-1 zeolite microengineered reactors for 1-Pentene epoxidation, *Chem. Commun.* 878–879.

48. Wiles, C., Hammond, M. J., and Watts, P. 2009. The development and evaluation of a continuous flow process for the lipase-mediated oxidation of alkenes, *Beilstein J. Org. Chem.* 5(27).
49. Beigi, M., Haag, R., and Liese, A. 2008. Continuous application of polyglycerol-supported Salen in a membrane reactor: Asymmetric epoxidation of 6-Cyano-2,2-dimethylchromene, *Adv. Synth. Catal.* 350: 919–925.
50. Ceylan, S., Friese, C., Lammel, C., Mazac, K., and Kirschning, A. 2008. Inductive heating for organic synthesis by using functionalized magnetic nanoparticles inside microreactors. *Angew. Chem. Int. Ed.*, 47: 8950–8953.
51. Leeke, G. A., Santos, R. C. D., Al-Duri, B., Seville, J. P. K., Smith, C. J., Lee, C. K. Y., Holmes, A. B., and McConvey, I. F. 2007. Continuous-flow Suzuki–Miyaura reaction in supercritical carbon dioxide, *Org. Proc. Res. Dev.* 11: 144–148.
52. Lee, C. K. Y., Holmes, A. B., Ley, S. V., McConvey, I. F., Al-Duri, B., Leeke, G. A., Santos, R. C. D., and Seville, J. P. K. 2005. Efficient batch and continuous flow Suzuki cross-coupling reactions under mild conditions, catalysed by polyurea-encapsulated palladium (II) acetate and tetra-*n*-butylammonium salts, *Chem. Commun.* 2175–2177.
53. Uozumi, Y., Yamada, Y. M. A., Beppu, T., Fukuyama, N., Ueno, M., and Kitamori, T. 2006. Instantaneous carbon–carbon bond formation using a microchannel reactor with a catalytic membrane, *J. Am. Chem. Soc.* 128: 15994–15995.
54. Mennecke, K., Solodenko, W., and Kirschning, A. 2008. Carbon–carbon cross-coupling reactions under continuous flow conditions using poly(vinylpyridine) doped with palladium, *Synthesis* 10: 1589–1599.
55. Jones, R. C., Canty, A. J., Deverell, J. A., Gardiner, M. G., Guijt, R. M., Rodemann, T., Smith, J. A., and Tolhurst, V. 2009. Supported palladium catalysis using a heteroleptic 2-methylthiomethylpyridine-N,S-donor motif for Mizoroki–Heck and Suzuki–Miyaura coupling, including continuous organic monolith in capillary microscale flow-through mode, *Tetrahedron* 65: 7474–7481.
56. Gasnov, T. N., Findenig, S., and Kappe, C. O. 2009. Heterogeneous versus homogeneous palladium catalysts for ligandless Mizoroki–Heck reactions: A comparison of batch/microwave and continuous-flow processing, *Chem. Eur. J.* 15: 1001–1010.
57. Shore, G., Morin, S., Mallik, D., and Organ, M. G. 2008. Pd PEPPSI-IPr-mediated reactions in metal-coated capillaries under MACOS: The synthesis of indoles by sequential aryl amination/Heck coupling, *Chem. Eur. J.* 14: 1351–1356.
58. Fan, X., Manchon, M. G., Wilson, K., Tennison, S., Kozynchenko, A., Lapkin, A. A., and Plucinski, P. K. 2009. Coupling of Heck and hydrogenation reactions in a continuous compact reactor, *J. Catal.* 267: 114–120.
59. Phan, N. T. S., Brown, D. H., and Styring, P. 2004. A facile method for catalyst immobilisation on silica: Nickel-catalysed Kumada reactions in mini-continuous flow and batch reactors, *Green Chem.* 6: 526–532.
60. Gömann, A., Deverell, J. A., Munting, K. F., Jones, R. C., Rodemann, T., Canty, A. J., Smith, J. A., and Guijt, R. M. 2009. Palladium-mediated organic synthesis using porous polymer monolith formed *in-situ* as a continuous catalyst support structure for application in microfluidic devices, *Tetrahedron*, 65: 1450–1454.
61. Fukuyama, T., Kippo, T., Ryu, I., and Sagae, T. 2009. Addition of allyl bromide to phenylacetylene catalyzed by palladium on alumina and its application to a continuous flow synthesis, *Res. Chem. Intermed.* 35: 1053–1057.
62. Yamada, Y. M. A., Watanabe, T., Torii, K., and Uozumi, Y. 2009. Catalytic membrane-installed microchannel reactors for one-second allylic arylation, *Chem. Commun.* 5594–5596.
63. Weber, S. K., Bremer, S., and Trapp, O. 2010. Integration of reaction and separation in a micro-capillary column reactor—palladium nanoparticle catalyzed C–C bond forming reactions, *Chem. Eng. Sci.* 65: 2410–2416.

64. Brasholz, M., Johnson, B. A., Macdonald, J. M., Polyzos, A., Tsanaktsidis, J., Saurbern, S., Holmes, A. B., and Ryan, J. H. 2010. Flow synthesis of tricyclic spiropiperidines as building blocks for the histrionicotoxin family of alkaloids, *Tetrahedron* 68(33): 6445–6449.
65. Mandoli, A., Orlandi, S., Pini, D., and Salvadori, P. 2004. Insoluble polystyrene-bound bis(oxazolin): Batch and continuous-flow heterogeneous enantioselective glyoxylate-ene reaction, *Tetrahedron Asymm.* 15: 3233–3244.
66. Pericàs, M. A., Herrerías, C. I., and Solà, L. 2008. Fast and enantioselective production of 1-Aryl-1-propanols through a single pass, continuous flow process, *Adv. Synth. Catal.* 350: 927–932.
67. Rolland, J., Cambeiro, X. C., Rodriguez-Escrich, C., and Pericàs, M. A. 2009. Continuous flow enantioselective arylation of aldehydes with ArZnEt using triarylboroxins as the ultimate source of aryl groups, *Beilstein J. Org. Chem.* 5(56).
68. Urban, P. L., Goodall, D. M., and Bruce, N. C. 2006. Enzymatic microreactors in chemical analysis and kinetic studies, *Biotech. Adv.* 24: 42–57.
69. Miyazaki, M. and Maeda, H. 2006. Microchannel enzyme reactors and their applications for processing, *Trends Biotechnol.* 24: 463–470.
70. Fernandes, P. 2010. Miniaturization in biocatalysis, *Int. J. Mol. Sci.* 11: 858–879.
71. Gao, Y., Zhong, R., Qin, J., and Lin, B. 2009. An immobilized lipase microfluidic reactor for enantioselective hydrolysis of esters, *Chem. Lett.* 38: 262–263.
72. Ngamsom, B., Hickey, A. M., Greenway, G. M., Littlechild, J. A., Watts, P., and Wiles, C. 2009. Development of a high throughput screening tool for biotransformations utilizing a thermophilic l-aminoacylase enzyme, *J. Mol. Catal. B: Enzym.* 63: 81–86.
73. Csajagi, C., Szatzker, G., Toke, R. R., Urge, L., Darvas, F., and Poppe, L. 2008. Enantiomer selective acylation of racemic alcohols by lipases in continuous-flow bioreactors, *Tetrahedron Asymm.* 19: 237–246.
74. Wiles, C., Watts, P., and Haswell, S. J. 2007. The use of solid-supported reagents for the multistep synthesis of analytically pure α,β-unsaturated compounds in miniaturized flow reactors, *LabChip* 7: 322–330.
75. Wiles, C. and Watts, P. 2008. An integrated microreactor for the multicomponent synthesis of α-aminonitriles, *Org. Proc. Res. Dev.* 12: 1001–1006.
76. Wiles, C. and Watts, P. 2008. Evaluation of the heterogeneously catalyzed Strecker reaction conducted under continuous flow, *Eur. J. Org. Chem.* 5597–5613.
77. Baumann, M., Baxendale, I. R., Ley, S. V., Smith, C. S., and Tranmer, G. K. 2006. Fully automated continuous flow synthesis of 4,5-disubstituted oxazoles, *Org. Lett.* 8: 5231–5234.
78. Baxendale, I. R., Deeley, J., Griffiths-Jones, C. M., Ley, S. V., Saaby, S., and Tranmer, G. K. 2006. A flow process for the multistep synthesis of the alkaloid natural product oxomaritidine: A new paradigm for molecular assembly, *Chem. Commun.* 2566–2568.
79. Baxendale, I. R., Griffiths-Jones, C. M., Ley, S. V., and Tranmer, G. K. 2006. Preparation of the neolignan natural product grossamide by a continuous flow process, *Synlett* 3: 427–430.
80. Luckarift, H. R., Nadeau, L. J., and Spain, J. C. 2005. Continuous synthesis of aminophenols from nitroaromatic compounds by combination of metal and biocatalyst, *Chem. Commun.* 383–384.
81. Baxendale, I. R., Ley, S. V., Smith, C. D., and Tranmer, G. K. 2006. A flow reactor process for the synthesis of peptides utilizing immobilized reagents, scavengers and catch and release protocols, *Chem. Commun.* 4835–4837.
82. Wild, G. P., Wiles, C., Watts, P., and Haswell, S. J. 2009. The use of immobilized crown ethers as *in-situ* protecting groups in organic synthesis and their application under continuous flow, *Tetrahedron* 65: 1618–1629.

5 Electrochemical and Photochemical Applications of Micro Reaction Technology

5.1 ELECTROCHEMICAL SYNTHESIS UNDER CONTINUOUS FLOW

Although electroorganic synthesis has been shown over the decades to be a powerful synthetic tool for the atom efficient formation of highly reactive intermediates, the techniques and equipment required are still widely viewed as specialized; as such the methodology is not readily employed within conventional organic laboratories. While electroorganic chemistry forms a clean method for driving reactions, by the addition and removal of electrons from precursors and intermediates, demonstrating great synthetic utility for common transformations such as oxidations and reductions, along with more complex reactions such as homo- and heterocouplings, problems associated with scaling the approach have limited the techniques application.

To address the main drawbacks of electroorganic techniques which include inhomogeneity of the electric field and energy losses due to Joule heating between electrodes, several research groups have embarked upon programs of research which have focused on the development of continuous flow reactors suitable for performing electroorganic syntheses, enabling exploitation of this clean and efficient synthetic tool at both a research and production level [1–3].

Integration of Electrodes: Many different techniques have been employed for the introduction of electrodes into microfabricated reactors, ranging from the construction of clamps/holders containing plate electrodes [4,5] to the imprinting of microband electrodes within polymeric and ceramic devices [6]. A recent example of developments in the field was communicated by Tsujimoto and coworkers [7] who reported the use of a microelectrolytic reactor (MER), containing a grooved electrode, demonstrating accelerated mass transfer and affording a uniform concentration distribution within the reactor. As Table 5.1 illustrates, compared to plate electrodes, the presence of grooved electrodes within the reactor increases the electrode surface area and increases the maximum product yield (Y_B). When compared using computational fluid dynamic simulations, the authors concluded that accelerating mass

TABLE 5.1
Comparison of Grooved and Plate Electrodes in Microfabricated Electrochemical Reactors

Electrode Type	Specific Electrode Surface Area (mm^{-1})	Maximum Product Yield (Y_B)	Channel Length (mm)
Plate	10.0	0.35	9.7
Grooved	21.2	0.47	7.0

transfer is the key to obtaining high product yields within MERs as overreaction can be prevented by reducing the reaction times required. With this in mind, the following section provides a series of examples illustrating the practical advantages associated with the use of microfabricated electrochemical reactors.

5.1.1 Electrochemical Oxidations

As can be seen from this section, oxidations represent one of the most widely studied electrochemical transformations that utilize continuous flow microstructured devices, with techniques developed presenting an opportunity to revolutionize the way the modern synthetic chemist approaches these often hazardous and unselective reactions.

Cation-Flow: In order to increase the efficiency and synthetic utility of electrochemical syntheses, Yoshida and coworkers [8] developed an electrochemical flow reactor capable of generating highly reactive intermediates, at reduced temperatures, suitable for reaction with a range of nucleophiles. Using this approach, the authors were able to continuously generate, and manipulate, conventionally unstable carbocations, providing researchers with a facile approach to the formation of C–C bonds. To illustrate proof of concept, the authors selected the use of carbamates as precursors, based on their availability and established use in conventional cation generation.

Fabricating a flow reactor comprising of a two-compartment cell—divided by a membrane (polytetrafluoroethylene (PTFE))—one containing a carbon felt anode and the second a platinum wire cathode, the authors were able to generate reactive cation intermediates while simultaneously separating and diverting any hydrogen gas generated to waste.

As Figure 5.1 illustrates, a typical reaction involved the introduction of a solution of methyl pyrrolidinecarboxylate **1** (5.0×10^{-2} M) and supporting electrolyte (Bu_4NBF_4 **2** in DCM (0.3 M)) into the anodic chamber and a solution of Bu_4NBF_4 **2** and trifluoromethanesulfonic (TfOH) acid **3** into the cathodic chamber. Using low-temperature electrolysis (−72°C, 14 mA), the cationic intermediate **4** was initially generated and transferred to a reaction vessel containing a nucleophile, which upon reaction afforded the target coupling product (at −28°C).

Using this approach, the authors investigated the reaction of **4** with a series of carbon nucleophiles, such as silanes and enol ethers affording, as Table 5.2 illustrates, the respective coupling products in moderately high conversions and selectivities.

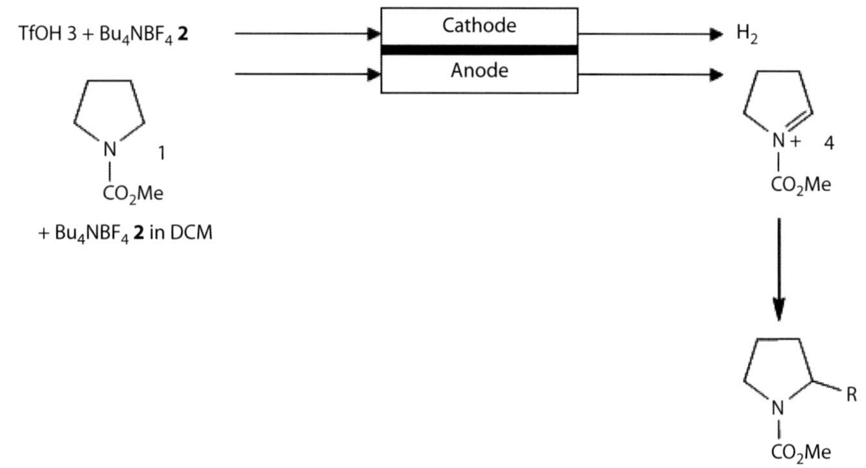

Electrochemical and Photochemical Applications

FIGURE 5.1 Schematic illustrating the flow reactor utilized for the generation of reactive intermediates; termed cation flow.

TABLE 5.2
Illustration of the Coupling Products Generated Using Cation Flow Methodology

Substrate	Nucleophile (%)	Conversion (%)	Selectivity
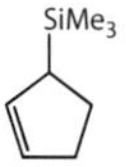 cyclopentenyl-SiMe$_3$	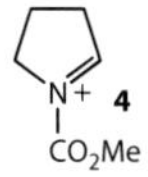 pyrrolidine-N(CO$_2$Me)$^+$ **4**	69	91
	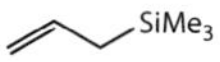 allyl-SiMe$_3$ **5**	69	100
	cyclohexenyl-SiMe$_3$	67	99
	OAc-substituted allyl	64	66
	methylallyl-SiMe$_3$	61	72
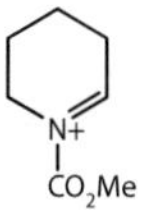	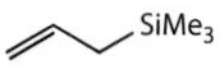 methylallyl-SiMe$_3$ **5**	60	93
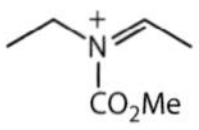 piperidinyl-N(CO$_2$Me)$^+$	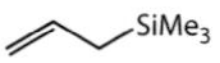 allyl-SiMe$_3$ **5**	55	98
acyclic-N(CO$_2$Me)$^+$			
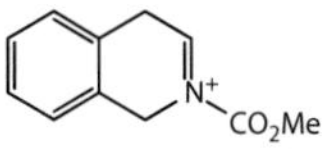 isoquinolinyl-N$^+$(CO$_2$Me)	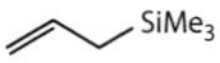 allyl-SiMe$_3$ **5**	49	67

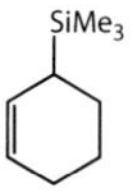

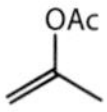

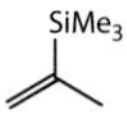

In a later development, Atobe and coworkers [9] demonstrated the construction of a microflow reactor in which both the anodic oxidation and nucleophilic reaction of the cation could take place. As Figure 5.2 illustrates, the substrate is introduced via inlet 1 (anode) and the nucleophile solution through inlet 2 (cathode), the carbocation generated at the anode then diffused rapidly through the bulk solution where they were able to react to afford the target compound. As the electrodes are on opposite facing channel walls, oxidation of the nucleophile does not occur. Again the anodic oxidation of methyl pyrrolidinecarboxylate **1** was selected as a model, with the cation **4** reacted with allyltrimethylsilane **5** to afford methyl-2-allylpyrrolidine-1-carboxylate. Employing 2,2,2-trifluoroethanol as the reaction solvent, selected due to its known cation stabilizing properties, the authors were able to obtain the target carboxylate in 59% yield; compared with 36% conversion (6% yield) obtained during a bulk, preparative-scale reaction.

Further increases in yield, up to 91%, were subsequently obtained by screening a series of ionic liquids as the reaction media, with N,N-dimethyl-N-methyl-N-(2-methoxyethyl)ammonium bis(trifluoromethanesulfonyl)imide ([deme][TFSI]) proving to be the best reaction media. In comparison with the cation flow method reported by Yoshida and coworkers, by performing the generation and reaction of the cation in a single reactor, the authors were able to conduct the whole process at ambient temperatures, without observing any detrimental effects in selectivity and/or yield.

Friedel–Crafts Alkylation: While the Friedel–Crafts alkylation is conventionally performed using carbocations generated using Lewis acids, the formation is reversible and often very substrate dependent. To address this, Yoshida and coworkers [10] further developed their "cation pool" methodology to involve the irreversible generation of carbocations via the anodic oxidation of carbamates possessing a silyl group as the electroauxiliary. Using the Friedel–Crafts alkylation as a test case, the authors investigated if micro reactors could be used to separate issues of perceived chemical selectivity from mixing inefficiencies.

Using an H-type divided cell, containing a carbon-felt anode and a solution of silyl carbamates **6** (in Bu$_4$NBF$_4$ **2**) and a platinum cathode in a solution of TfOH **3**

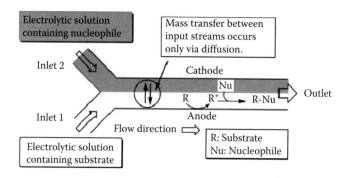

FIGURE 5.2 Schematic illustrating the flow reactor used to perform the anodic oxidation of carbamates and subsequent nucleophilic reaction of the *in situ* generated cationic species. (Reproduced with permission from Horii, D., Fuchigami, T., and Atobe, M. 2007, *J. Am. Chem. Soc.* 129: 11692–11693. Copyright (2007) American Chemical Society.)

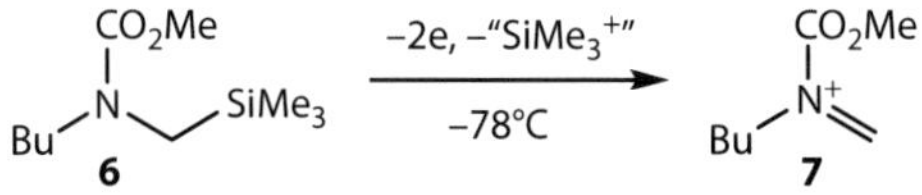

Electrochemical and Photochemical Applications

SCHEME 5.1 Illustration of the anodic oxidation of carbamates to afford highly reactive N-acyl iminium ions.

(in Bu_4NBF_4 **2**/DCM), under constant current electrolysis (30 mA), the authors formed the respective N-acyl iminium ion **7** (Scheme 5.1).

Employing the Friedel–Crafts alkylation of a series of aromatic substituents such as 1,3,5-trimethoxybenzene **8** (see Chapter 3, Liquid-Phase Micro Reactions for details), the authors rapidly identified that performing reactions under flow conditions did not directly improve chemical selectivity; however, by employing rapid mixing, using static mixers were achieved, polyalkylation was suppressed, and high yielding monoalkylations could be performed (Table 5.3).

TABLE 5.3
Comparison of the Product Distributions Obtained in Batch and under Flow Conditions for a Series of Friedel–Crafts Alkylation Reactions

Reactant	Nucleophilicity Parameter (N)	Reactor Type	Product Distribution	
			Mono-(%)	Di-(%)
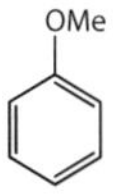 OMe (anisole)	−1.18	Batch	24	31
		Micromixer	26	0
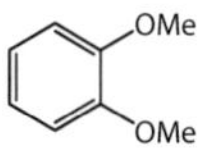 1,2-dimethoxybenzene	—	Batch	13	14
		Micromixer	34	0
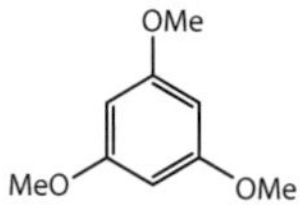 1,3,5-trimethoxybenzene **8**	3.40	Batch	37	32
		Micromixer	92	4
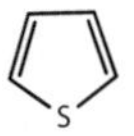 thiophene	−0.4	Batch	14	27
		Micromixer	84	0
 furan	1.45	Batch	11	5
		Micromixer	39	Trace
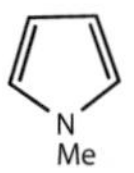 N-methylpyrrole	6.18	Batch	33	28
		Micromixer	60	6

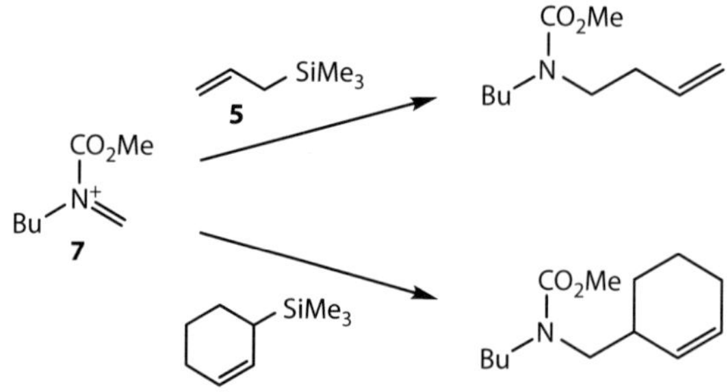

SCHEME 5.2 Schematic illustrating the Friedel–Crafts alkylation of allylsilanes.

In an extension to this, the substrate scope of the methodology was evaluated, with the authors investigating the reaction of N-acyl iminium ions with allylsilanes (Scheme 5.2), whereby the respective products were obtained in >70% isolated yield.

Unlike conventional reaction methodology, whereby fast reactions are often avoided due to processing difficulties associated with a lack of thermal control and poor mixing, the enhanced efficiency obtained within such microstructured devices affords the synthetic chemist the ability to reveal underlying chemical selectivity; resulting in a broadening of the synthetic techniques available for the preparation of monoalkylated compounds.

[4 + 2] Cycloaddition of N-Acyl Iminium Ions: Suga et al. [11,12] further developed the technique of irreversible "cation pool" formation, generated via the electrochemical oxidation of α-silyl carbamates, utilizing the N-acyl derivatives in a series of [4 + 2] cycloadditions, as illustrated in Scheme 5.3.

When performed using conventional batch techniques, the authors observed that the order of reactant addition played a large role in the proportion of cycloadduct formed versus polymeric by-products. As summarized in Table 5.4, in the case of substituted styrenes, moderate yields were obtained when the alkene was added to the N-acyl iminium ion **9**; however, inverting reactant addition dramatically reduced cycloadduct yield. While simultaneous addition of both reactants afforded increased cycloadduct formation, significant proportions of styrene polymerization were still observed. In comparison, by employing a micromixer (IMM, Germany), the authors were able to significantly increase the cycloadduct yield while simultaneously reducing the proportion of polymerized styrene derivative formed. As Table 5.4 illustrates, the enhancement was found to be a general one, with the effect observed for a range of dienophiles.

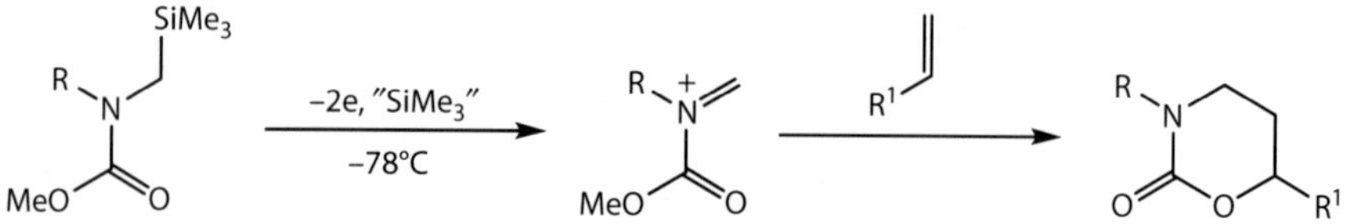

SCHEME 5.3 General reaction scheme illustrating the irreversible formation of N-acyl iminium ions and their subsequent reaction with dienophiles to afford [4+2] cycloadducts.

TABLE 5.4
Comparison of Various Reaction Methodologies for the [4+2] Cycloaddition of N-Acyl Iminium Ions to a Series of Dienophiles

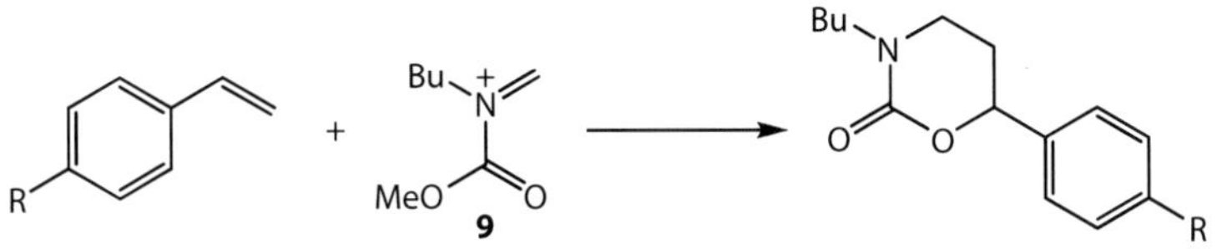

R	Method A[a]	Method B[b]	Method C[c]	Micromixing
H	57	20	55	79
Cl	43	12	54	70
Me	45	16	58	76

[a] Dienophile added to iminium ion **9**.
[b] Iminium ion **9** added to dienophile.
[c] Simultaneous addition.

Anodic Methoxylation: Utilizing a ceramic micro reactor, containing 40 Pt interdigitated electrodes orientated at 90° with respect to the seven microchannels (dimensions = 100 µm (deep)), Girault and coworkers [13] investigated the two-electron anodic methoxylation of methyl-2-furoate **10** (1×10^{-4} M) to methyl-2,5-dihydro-2,5-dimethoxy-2-furancarboxylate **11** in acidified (0.1 M) MeOH. Using online analysis by mass spectrometry (MS), the authors investigated the effect of flow rate (50–250 µl min^{-1}) on the reaction, at a fixed voltage (4 V), utilizing both single pass and recirculation techniques to optimize the formation of the target compound **11** and suppress side reactions (Scheme 5.4).

Electrochemical Iodination: With aromatic iodides used as precursors in a range of synthetically important reactions (see Chapter 3, Liquid-Phase Micro Reactions) utilized in the preparation of biologically active compounds, it is acknowledged that efficient methods are required for their synthesis. With this in mind, Midorikawa et al. [14] investigated the development of an electrochemical method for the generation of I$^+$ and demonstrated its use in the selective monoiodination of aromatic compounds under continuous flow conditions.

Following Miller's protocol, the authors initially generated I$^+$ from molecular iodine via anodic oxidation in Bu$_4$NBF$_4$ **2**/MeCN at 0°C; using a platinum plate

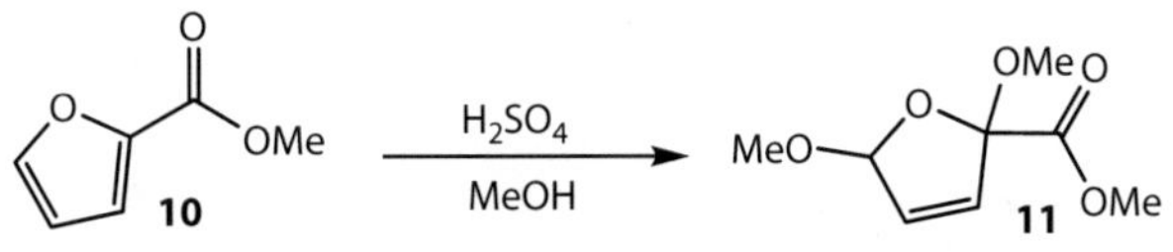

SCHEME 5.4 Illustration of the anodic methoxylation reaction performed in a ceramic electrochemical micro reactor, with online monitoring by mass spectrometry.

electrode. Once generated, the I⁺ was evaluated under batch conditions using 1,2-dimethoxybenzene **12**, 1,4-dimethoxybenzene **13**, and 1,3,5-trimethoxybenzene **8** as substrates. As Table 5.5 illustrates, in all cases batch experiments resulted in the formation of significant proportions of the diiodinated product. Concluding that the monoiodinated products should be less reactive than the starting materials, the authors believed that the low selectivity of the reaction was in fact attributed to inefficient mixing; as previously encountered for the Friedel–Crafts alkylation.

With this in mind, the reactions were repeated at 0°C within an IMM micromixer (Germany), as summarized in Table 5.5; using this approach, the authors were able to suppress the polyiodination of the aromatic substrates, obtaining the monoderivatives in higher yield and selectivity.

5.1.2 Electrolyte-Free Electroorganic Synthesis

In addition to the many physical advantages associated with the miniaturization of electrochemical flow reactors, namely high electrode surface-to-volume ratio and short interelectrode distance, several authors have reported the ability to efficiently conduct electroorganic syntheses in the absence of intentionally added electrolytes [15], resulting in an ease of production isolation and reduced operating costs. With these factors in mind, details of such transformations are provided in the following section.

Electrolyte-Free Anodic Methoxylation: Employing a thin-layered flow cell (interelectrode distance = 160 μm), containing a glassy carbon plate anode and a Pt plate cathode, Atobe and coworkers [16] investigated the methoxylation of furan **14** via the electrochemical reduction of MeOH (Scheme 5.5). Using this arrangement,

TABLE 5.5
Comparison of the Reaction Products Obtained for the Iodination of 1,2-Dimethoxybenzene 12, 1,4-Dimethoxybenzene 13, and 1,3,5-Trimethoxybenzene 14 in Batch and under Flow Conditions

Substrate	Mono-	Di-	Reactor Type	Ratio[a]
12			Batch	90:10
			Micro reactor	96.5:3.5
13			Batch	78:22
			Micro reactor	88:12
14			Batch	79:21
			Micro reactor	94:6

[a] Monoiodinated to diiodinated product distribution.

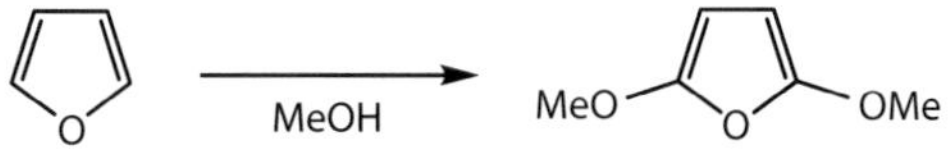

SCHEME 5.5 Illustration of the electrolyte-free methoxylation of furan **14** performed under continuous flow conditions.

the authors investigated the effect of current density on the consumption of furan **14** (0.01 M in MeOH), at a constant flow rate of 10 µL min^{-1}, observing 40% conversion at 0.1 mA cm^{-2} rising to 98% at current densities between 0.5 and 3 mA cm^{-2}, with product degradation observed >3 mA cm^{-2}. In order to increase the productivity of the reactor, the authors subsequently investigated the effect of flow rate on the reaction, with increases up to 100 µL min^{-1} maintaining quantitative methoxylation; further increases however led to a reduction in reaction efficiency (500 µL min^{-1} afforded 25% conversion). As this was a self-supported reaction, employing no added electrolyte, the authors were able to isolate reaction products by simply removing the reaction solvent by concentration *in vacuo*.

In a second example, using a microflow system comprising two carbon-felt electrodes—separated by a hydrophobic PTFE membrane spacer (dimensions = 75 µm)—Yoshida and coworkers [17] investigated the anodic methoxylation of 4-methoxytoluene **15** (Scheme 5.6). Pumping a solution of 4-methoxytoluene **15** (0.05 M) in MeOH into the anodic chamber, through the spacer membrane, into the cathode and out of the cathode chamber, the authors investigated the reaction under constant current (11 mA, 4 F mol^{-1}, cell voltage = 21–25 V) at a fixed flow rate of 2 mL min^{-1}. Under the aforementioned conditions, the authors obtained the target acetal **16** in 30% conversion, based on consumed 4-methoxytoluene **15**. Efforts to increase the proportion of **16** by increasing current up to 25 mA were met with degradation of the membrane at 25 mA; however, reducing the current to 22 mA enabled the authors to obtain the target acetal **16** in 69% conversion.

To extend the scope of the developed methodology, the authors also investigated the methoxylation of *N*-methoxycarbonyl pyrrolidine **1** and acenaphthalene **17** obtaining the target compounds in 40–65% conversion (Scheme 5.7).

Compared to earlier examples by Marken and coworkers [15], the reactor described employed an orientation which enabled parallel fluid and current flow; in addition, the whole of the electrochemical cell was filled with the carbon electrodes, affording

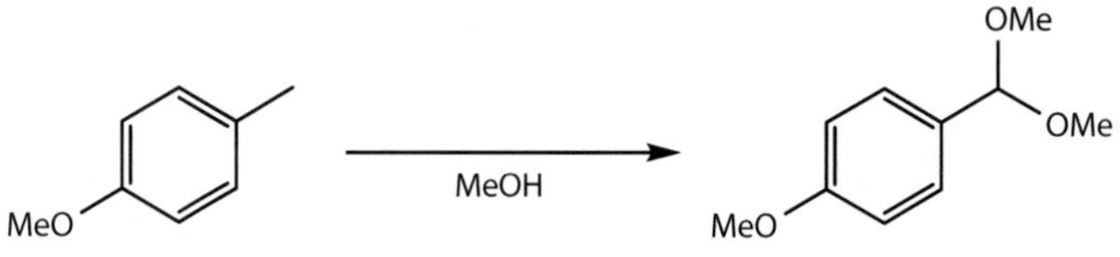

SCHEME 5.6 Illustration of the model reaction used to demonstrate electrolyte-free anodic methoxylation.

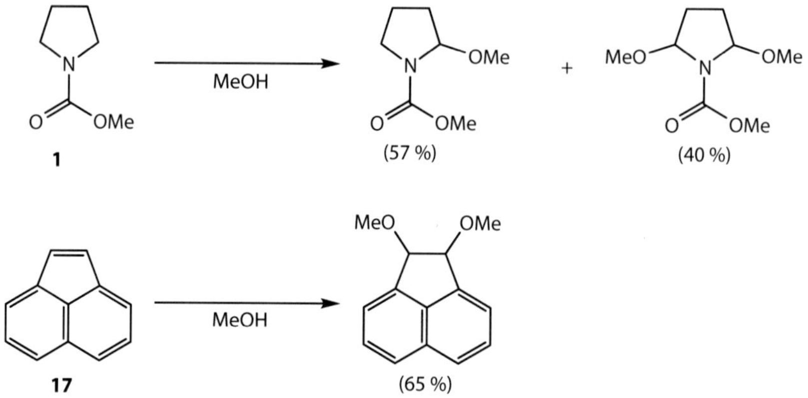

SCHEME 5.7 Illustration of additional anodic methoxylation reactions performed using an electrolyte-free microflow reactor.

a higher surface area and in turn a higher flow rate and current. The authors propose that this could be an advantage when considering scaling such reactors.

5.1.3 ELECTROCHEMICAL REDUCTIONS

Unlike oxidation reactions which have been extensively evaluated under microflow conditions, examples of reductions (see Chapter 3, Section 3.6), and in particular electrochemical reductions, have not been so widely investigated.

In an early example of electrochemical reductions under flow conditions, Marken and coworkers [15] evaluated the two electron/two proton reduction of tetraethylethylenetetracarboxylate (TEenTC) **18** to tetraethylethanecarboxylate (TEanTC) **19** utilizing a supporting electrolyte as the reaction is well known and proceeds with no side reactions (Scheme 5.8). Employing EtOH as the reaction solvent, the authors investigated the effect of flow rate on the reaction, observing increasing conversion to the alkane **19** with decreasing flow rate; with conversions approaching 90% under optimal conditions.

Electrosynthesis of Phenyl-2-propanone: Utilizing a one-step electrochemical acylation, He et al. [18] reported the direct electroreductive coupling of benzyl bromides and acetic anhydride **20** in a microgap flow reactor (interelectrode gap = 160 μm, reactor volume = 7.2 μL) containing Pt electrodes. Employing dimethylformamide (DMF) as the reaction solvent, the authors were able to perform the

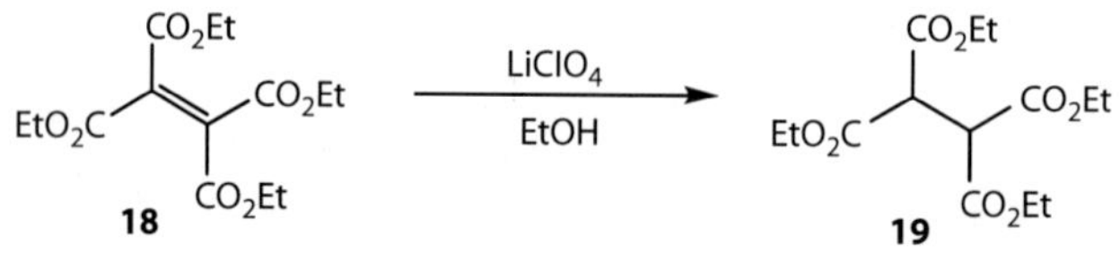

SCHEME 5.8 An early example of paired electrosynthesis performed using a microflow cell.

reaction in the absence of any intentionally added electrolyte, finding that the application of a constant current of 0.8 mA to a solution of benzyl bromide **21** (5×10^{-3} M) and Ac$_2$O **20** (10–40 eq.) enabled the conversion of ~90% of the benzyl bromide **21** to the target phenyl-2-propanone and the undesirable debromination by-product methyl benzene.

With this information in hand, the authors investigated the effect of current (0.8–1.1 mA) and the proportion of Ac$_2$O **20** (10–40 eq.) on the conversion of benzyl bromide **21** and the product distribution obtained. As Table 5.6 illustrates, with increasing current and Ac$_2$O **20** the authors were able to increase the conversion of benzyl bromide **21** to the desired product, phenyl-2-propanone, suppressing the formation of debrominated and dimer by-products.

In an extension, the authors investigated a series of substituted aromatic bromides, again obtaining the target ketones in high conversion and moderate selectivity.

5.1.4 ELECTROLYTE-FREE REDUCTIONS UNDER FLOW

He et al. [19] subsequently reported the use of a microgap flow cell (interelectrode gap = 160 or 320 μm) for the electrochemical reduction of 4-nitrobenzyl bromide **22** (1.0×10^{-2} M) in DMF:THF (3:1), in the absence of a supporting electrolyte, affording the homocoupling product 1,2-bis(4-nitrophenyl)ethane **23**. Using the microgap flow cell, the authors investigated the effect of the interelectrode gap,

---

TABLE 5.6
Summary of the Results Obtained for the Electrochemical Reductive Coupling of Aromatic Halides to Anhydrides Performed with a Reaction Time of 43 s

					Product Distribution	
R	R^1	Current (mA)	Anhydride:Halide (mol/mol)	Conversion (%)	Ketone (%)	Alkane (%)
H	H **21**	0.8	10	87	61	26
H	H	0.8	20	85	62	23
H	H	1.1	20	92	66	26
H	H	1.1	40	90	81	9
Me	H	1.1	40	93	87	6
H	Me	1.1	40	98	96	2
OMe	H	1.3	40	99	83	16
Br	H	1.1	40	73	51	22

TABLE 5.7
Summary of the Optimization Process used for the Electrochemical Reduction of 4-Nitrobenzyl Bromide 22 in the Absence of a Supporting Electrolyte

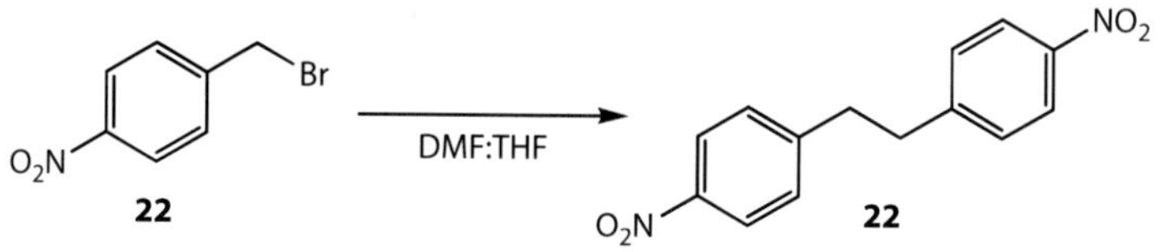

Interelectrode Gap (μm)	Current (mA)	Flow Rate (μL min^{-1})[a]	Conversion (%)[b]	Product Distribution (%)	
				R-R	R-H
160	0.8	20 (22)	99	68	32
160	1.3	40 (11)	95	69	31
320	0.6	20 (44)	70	93	7
320	1.2	20 (44)	91	91	9
320	0.6	40 (22)	58	94	6
320	2.5	40 (22)	92	91	9
320	2.5	40 (22)	100	76	24

[a] The number in parentheses is the calculated residence time (s).
[b] Determined by GC–MS analysis.

reactant flow rate, and current on the coupling reaction. As Table 5.7 illustrates, the authors readily optimized the process to afford 1,2-bis(4-nitrophenyl)ethane **23** in 94% selectivity.

The authors subsequently investigated the reductive coupling of benzyl bromides with a series of olefins developing an efficient, self-supported cathodic coupling technique [20] (Scheme 5.9). To perform reactions, the authors pumped a pre-mixed DMF solution containing the olefin (dimethyl malonate **23**, dimethyl-fumarate **24**, fumaronitrile **25** and maleic anhydride **26**) (5 × 10^{-3} M) and benzyl bromide derivative (5 × 10^{-3} M) through the microgap reactor (interelectrode gap = 320 μm) and applied a voltage of 4–4.4 V. Reaction products were collected in a sample tube, prior to offline analysis by gas chromatography-flame ionization detection (GC-FID) and as Table 5.8 illustrates, under the aforementioned conditions,

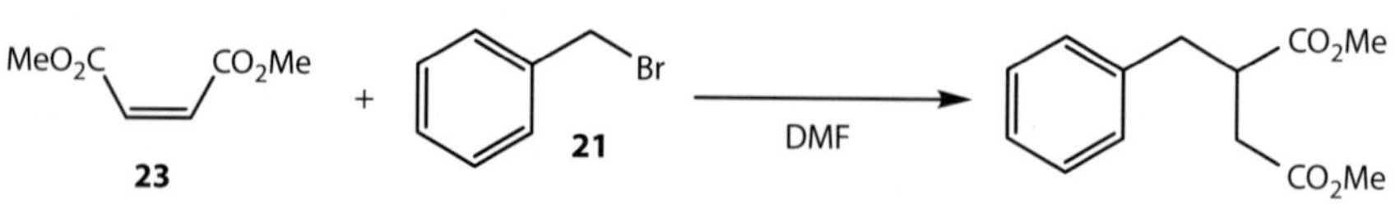

SCHEME 5.9 Schematic illustrating the C–C coupling reaction between dimethyl malonate **23** and benzyl bromide **21**.

TABLE 5.8
Summary of the Electrochemical C–C Bond-Forming Reactions Investigated in a Microgap Electrode Reactor

Olefin	Bromide	Current (mA)	Flow Rate (µL min⁻¹)	Conversion (%)[a]	Product Distribution (%) Coupling Product	By-Product[b]
MeO$_2$C–CH=CH–CO$_2$Me (23)	Benzyl 21	0.6	20	100	94	6
	4-Methoxybenzyl	0.6	10	100	94	6
	4-Methylbenzyl	0.6	10	100	94	6
	4-Bromo	0.6	10	100	99	1
	4-Iodo	0.6	10	100	99	1
	1-Phenylethyl	0.6	10	100	98	2
MeO–C(O)–CH=CH–C(O)–OMe (24)	Benzyl 21	0.6	10	100	98	2
N≡C–CH=CH–C≡N (25)	Benzyl 21	0.5	10	100	96	4
	4-Methylbenzyl	0.5	10	100	93	7
	4-Bromobenzyl	0.5	10	100	95	5
maleic anhydride (26)	Dibromide	0.3	10	82	84	16[c]

[a] Conversion was determined using offline GC-FID using internal standardization.
[b] By-products include dimerization of the olefin, debromination of the benzyl bromide.
[c] 1,2-Dimethylbenzene and 2-methylbenzyl bromide.

the target coupling products were obtained in high conversion (82–100%) and selectivity (84–99%).

5.1.5 Summary

Owing to the fact that small interelectrode gaps are possible within miniaturized electrochemical reactors, Joule heating is minimized and energy losses reduced [21]; as such, this methodology provides an extremely efficient method for the activation of molecules. Compared to conventional chemical methods of activation, electrochemical techniques are advantageous as they do not employ toxic reagents. In addition, mild reaction conditions are employed and reaction rate can be probed by adjusting the applied voltage or current. Furthermore, owing to the fact that intermediates are formed on an electrode surface, the reactions performed have the potential to be highly regio- and stereo-selective making the technique of great interest to the modern synthetic chemist who is always in search of novel and effective routes to target compounds.

5.2 PHOTOCHEMICAL SYNTHESIS UNDER CONTINUOUS FLOW

Although photochemistry is becoming more widely recognized as a useful method for the preparation of synthetically interesting and complex molecular architecture, the intricacies of the equipment required and problems associated with scaling the technique, however, often preclude its use within synthetic research laboratories and certainly within pilot plants. The high surface-to-volume ratio obtained within microflow reactors affords efficient light penetration and spatial homogeneity, affording reduced reaction times and increased product selectivities which are not currently accessible in batch reactors. Furthermore, the use of compact irradiation systems, as described in the following section, enables the construction of a low-energy apparatus which can be used for the construction of structurally diverse materials. With this in mind, the following section describes the advances made by researchers in the field of continuous flow technology, demonstrating an array of in-house fabricated devices for continuous flow photochemistry.

Benzopinacol Formation: An early example of miniaturized photochemistry was reported by Lu et al. [22] focused on the design and evaluation of a microfabricated reactor suitable for the efficient introduction of light. With this is mind, the authors constructed a silicon device (channel dimensions = 500 μm (wide) and 500 μm (deep)) with a quartz cover plate, housed within a stainless steel holder; containing high-pressure fluidic fittings and an optical cable for *in situ* UV detection. Light ($\lambda = 366$ nm) was introduced into the reactor via a recess in the holder and reactants pumped through the reactor using a syringe pump. To characterize the reactor, the authors selected the radical synthesis of benzopinacol **27** (Scheme 5.10) as a model reaction.

Prior to use, the reactant stock solution, benzophenone **28** in ^{i}PrOH (0.50 M) was degassed with N_2 to remove dissolved oxygen and the solution pumped through the reactor for 2 h to stabilize prior to analysis of the reaction products. Using this approach, the authors investigated the effect of flow rate (3–10 μL min^{-1}), and hence residence time, on the conversion of benzophenone **28** observing increasing benzopinacol **27** formation with decreasing flow rate (45–60%). Compared to conventional photochemical reactors, the authors found the use of micro reaction channels to be advantageous as the shallow nature of the reaction mixture enabled efficient irradiation of the whole reaction mixture; whereas in batch, only a few hundred micron depth would be irradiated.

Multipass Reactors: In addition to single-pass flow reactors, several authors have demonstrated the use of multipass reactors as a means of increasing photochemical

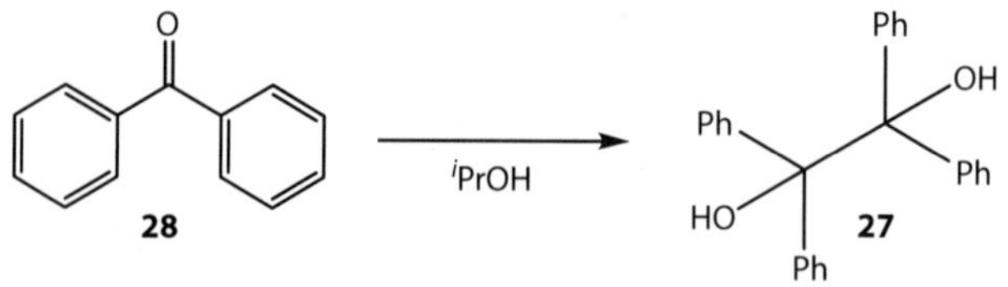

SCHEME 5.10 Schematic illustrating the photochemical formation of benzopinacol **27** under continuous flow conditions.

efficiency without the need for long reaction channels. Freitag et al. [23] demonstrated an automated laboratory plant based on this approach, where it was found to be advantageous compared to batch as the technique enabled solutions within the microchannel to efficiently absorb light without the need for low flow rates. In addition, through the incorporation of an inline IR sensor, the authors were able to monitor the status of the reaction and automatically open a release valve when the desired process conversion was detected.

5.2.1 Photocycloadditions under Continuous Flow

Investigating the intramolecular [2+2] and [2+3] photocycloaddition of 2-(2-alkenyloxymethyl)-naphthalene-1-carbonitriles, Maeda and coworkers [24] compared the efficiency of reactions performed under standard and flow conditions, building on their initial experience of photochemical transformations performed within PDMS devices [25]. Employing an in-house fabricated glass reactor, comprising two glass slides and an ionomer resin film (channel dimensions = 2.5 mm (wide) × 60 mm (long)), the authors investigated a series of intramolecular photochemical transformations under batch and continuous flow conditions, irradiating the reaction mixtures with a Xenon lamp (500 W, λ = 280 nm) (Table 5.9) and analyzing the reaction products by ^{1}H NMR spectroscopy.

In the case of adduct **29**, literature precedent stated that although the 1,2-naphthyl derivative **30** formed upon initial photoirradiation, due to [2+2] photocycloaddition, after prolonged irradiation the [3+2] photocycloadduct **31** dominated, due to photocycloreversion of **31** to **30**; consequently, under batch conditions only low conversions and poor selectivities were attainable. In comparison, performing the reaction under flow conditions, the authors were able to harness the advantages associated with uniform sample irradiation enabling a dramatic reduction in irradiation time from 240 to 1 min, thus minimizing the proportion of photocycloreversion observed. Using this approach, the authors were able to selectively perform the desired [2+2] photocycloaddition, affording the 1,2-adduct in excellent selectivity (96%), demonstrating that when fast reversible reactions and slow irreversible reactions coexist, micro reactors offer an efficient method for the synthesis of materials via the first reaction pathway. To demonstrate the generality of the method developed, the authors subsequently investigated the photocycloaddition of a series of other substituted carbonitriles and as can be seen in Table 5.9, excellent selectivities were obtained when benchmarked against standard batch reactions.

Paterno–Büchi Reaction: In addition to increasing the efficiency of standard light sources and devising methods for their use on a production scale, Ryu and coworkers [26] have recently demonstrated the use of black light and ultraviolet light emitting diode (UV-LED) light sources as a means of developing compact microflow systems suitable for photochemical applications. Employing the Paterno–Büchi reaction, the authors subsequently evaluated the reaction of benzophenone **28** with prenyl alcohol **32** to afford oxetane **33**, comparing the reaction efficiency of two light sources, a 300 W mercury lamp and a 15 W black light (Scheme 5.11). Employing benzene as the reaction solvent and a photo-micro reactor developed by Dainippon Screen Manufacturing Co. Ltd. (Japan) (channel dimensions = 1000 µm (wide) × 107 µm

TABLE 5.9
Comparison of Reactor Type on the Intramolecular [2+2] and [2+3] Photocycloaddition Reactions of 2-(2-Alkenyloxymethyl)-Naphthalene-1-Carbonitriles

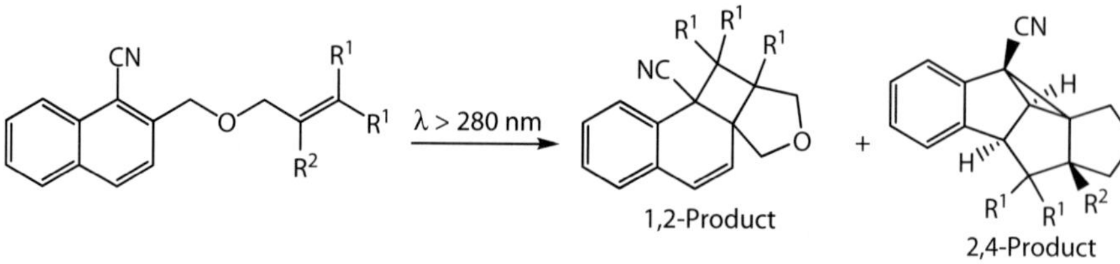

Reactor	R¹	R²	Solvent	Irradiation Time (min)[a]	Product Distribution (%) 1,2-	Product Distribution (%) 2,3-	Conversion (%)[b]
Batch	Me	Me 29	Benzene	240	55 30	45 31	65
Flow	Me	Me	Benzene	1	96	4	69
Batch	Me	H	Benzene	180	73	27	74
			MeCN	50	72	28	77
Flow	Me	H	Benzene	1	93	7	75
			MeCN	2.9	90	10	72
Batch	H	H	MeCN	90	3	97	33
Flow	H	H	MeCN	2.9	10	90	40

[a] In the case of flow reactions, irradiation time (min) was calculated based on the volume (mL) of the reactor per flow rate (mL min⁻¹).
[b] Determined by ¹H NMR spectroscopy.

(deep) × 2.2 m (long)), the authors obtained oxetane **33** in 91% yield upon irradiation with a 300 W mercury lamp for 1.2 h. In comparison, a yield of 84% **33** was obtained with a reaction time of 4 h, when using a black light (15 W, λ = 354 nm). As Table 5.10 illustrates, when comparing the energy efficiency (Wh) of the systems, the black light is six times more efficient and therefore superior to the mercury lamp for this mode of operation.

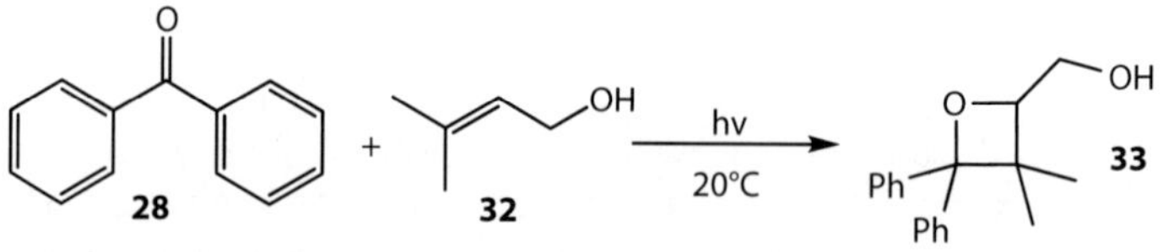

SCHEME 5.11 Illustration of the Paterno–Büchi reaction performed using a continuous flow photochemical micro reactor.

TABLE 5.10
Comparison of the Efficiency of Light Sources Employed for the Paterno–Büchi Reaction under Continuous Flow

Light Source	Residence Time (h)	Yield (%)	Wh	Yield/Wh
300 W (Hg)	1.2	91	360	0.25
15 W (BL)	4.0	81	60	1.40

Encouraged by this finding, the authors extended their investigation of light sources to encompass the use of UV-LED's ($\lambda = 365$ nm) coupled with a photo-micro reactor supplied by YMC Co. Ltd. (Japan), containing a stainless-steel channel dimensions of 1000 μm (wide) × 200 μm (deep) × 56.0 cm (long)) [27]. Using six UV-LEDs as the light source, the authors performed the [2+2] cycloaddition of cyclohexen-2-one **34** with vinyl acetate **35** to afford the cycloadduct **36** (Scheme 5.12), a synthetically powerful method for the construction of four-membered rings.

As Table 5.11 illustrates, using analogous conditions to those previously reported, a 200-fold increase in energy efficiency was obtained through the use of UV-LEDs as a light source compared to a conventional mercury lamp and a 10-fold increase with respect to black lights.

The methodology developed was not specific to vinyl acetate **35**, with other vinylic substrates—such as butylvinyl acetate **36** and isoprenylacetate—successfully employed, along with a series of substituted cyclohexanone derivatives (Table 5.12).

Large-Scale Photocycloaddition: Concomitantly, Booker-Milburn and coworkers [28] used a mercury immersion well light source (600 W) combined with a fluoropolymer tubular reactor (dimensions = 2.7 mm (i.d.) × 3.1 mm (o.d.), Volume =210 mL) to perform the [2+2] photocycloaddition of maleimide **37** and 1-hexyne **38** to afford the cyclobutane product, 6-butyl-3-azabicyclo[3.2.0]hept-6-ene-2,4-dione **39**, illustrated in Scheme 5.13.

Employing maleimide **37** in MeCN, the authors investigated the effect of reactant concentration (0.1–0.4 M) on the formation of cyclobutane derivative **39**, monitoring conversion by offline ^{1}H NMR spectroscopy in d-DMSO. Using a fixed residence time of 26 min (8 mL min^{-1}), the authors identified 0.4 M as the optimal reactant concentration, affording the target compound **39** in 83% conversion. Continuous operation of the flow reactor under the aforementioned conditions therefore enabled

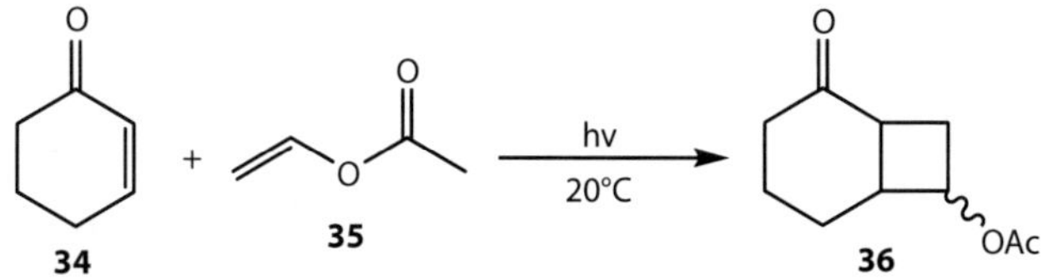

SCHEME 5.12 [2+2]-Cycloaddition of cyclohexen-2-one **34** to vinyl acetate **35b** performed in a photochemical micro reactor using UV-LEDs as a light source.

TABLE 5.11
Comparison of the Efficiency of Light Sources Employed for the [2+2]-Cycloaddition of Cyclohexen-2-one 5Ca to Vinyl Acetate 5Cb Performed under Continuous Flow

Light Source	Residence Time (h)	Yield (%)	Wh	Yield/Wh
300 W (Hg)	2.0	71	600	0.12
15 W (BL)	2.0	82	30	2.7
250 mW(UV-LED)	2.0	71	3.0	24.3

the authors to generate 6-butyl-3-azabicyclo[3.2.0]hept-6-ene-2,4-dione **39** at an impressive throughput of 685 g day^{-1}.

In a second example, the authors reported the [5+2] photocycloaddition of 3,4-dimethyl-1-pent-4-enylpyrrole-2,5-dione **40** to afford azepine **41** (Scheme 5.14). This time the authors employed a 0.1 M solution of dione **40** in MeCN and a flow rate of 8 ml min^{-1}, affording the target azepine,7,8-dimethyl-1,2,3,9a-tetrahydro-pyrrolo[1,2-*a*]azepine-6,9-dione **41**, in 80% yield. Compared to batch where typical yields of 66% were obtained, the continuous flow methodology afforded a facile route to the compound **41** at throughputs of 178 g day^{-1}.

TABLE 5.12
Summary of the Results Obtained for a Series of [2+2] Photocycloaddition Reactions Performed in a Continuous Flow Reactor

Enone	Vinylic Substrate	Cycloadduct	Irradiation Time (h)	Yield (%)
	OAc 35	OAc	2	70
	OAc 35	OAc	3.2	62
34	OAc	OAc	3.2	64
34	OBu	OBu	3.2	67

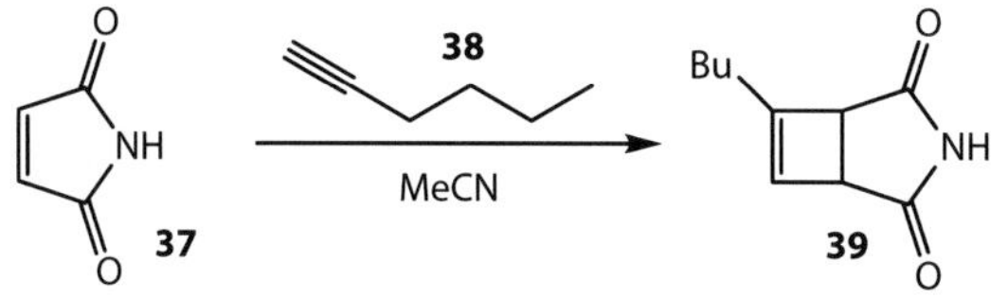

SCHEME 5.13 Illustration of the [2+2] photocycloaddition reaction performed on a 28.5 g h^{-1} scale utilizing an FEP tube reactor.

5.2.2 Photodecarboxylative Addition

Using a dwell device (channel dimensions = 2000 μm (wide) × 500 μm (deep) × 1.15 m (long)), supplied by Mikroglas (Germany), Oelgemöller and Coyle [29] and Oelgemöller and coworkers [30] reported the synthesis of bioactive molecules **42** via the photodecarboxylative addition of carboxylates **43** to phthalimides **44** in aqueous acetone, as depicted in Scheme 5.15. Irradiation of the reaction mixture within the micro reactor, at λ = 300 nm, enabled the authors to identify 21 min (0.8 mL min^{-1}) as the optimal reaction time for this transformation, affording 3-benzyl-3-hydroxy-2-methylisoindolin-1-one **42** in excellent conversion (97%).

5.2.3 Photocyanation

In 2002, Kitamura and coworkers [31] reported the photocyanation of pyrene **45** across an oil–water interface using sodium cyanide **46** as the cyanide source and 1,4-dicyanobenzene **47** as an electron acceptor. Employing a polystyrol micro reactor (Tamiya Inc., Japan), with a double Y-shaped reaction channel (channel dimensions = 100 μm (wide) × 20 μm (deep) × 35 cm (long)), the authors performed reactions by introducing pyrene **45** and 1,4-dicyanobenzene **47** (0.02 and 0.04 M respectively) in propylene carbonate and aq. NaCN **41** (1.0 M) into the reactor from separate inlets and irradiating the reaction mixture with a 300 W Hg lamp passing through a CuSO$_4$ solution filter (λ ~ 330 nm). The reaction products from the organic outlet were collected in a sample vial and analyzed offline using GC-FID in order to quantify the proportion of pyrene cyanated **48**.

Exploiting the longitudinal phase boundary formed within the microchannel, the authors proposed that the reaction would proceed as follows: photoinduced electron transfer would occur between pyrene **45** and **47** in the oil phase, with the resulting

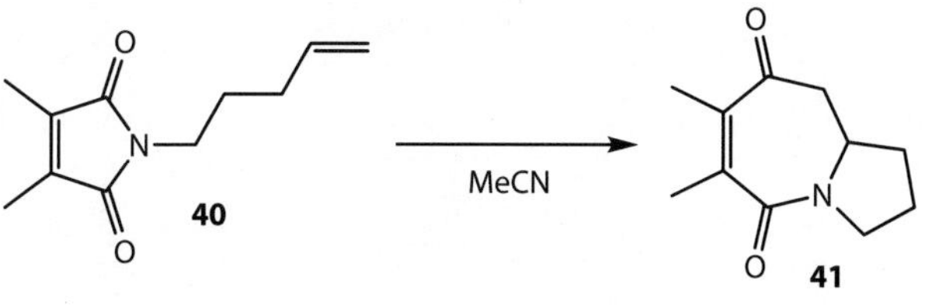

SCHEME 5.14 General schematic illustrating the [5+2] photocycloaddition used to synthesize azepines.

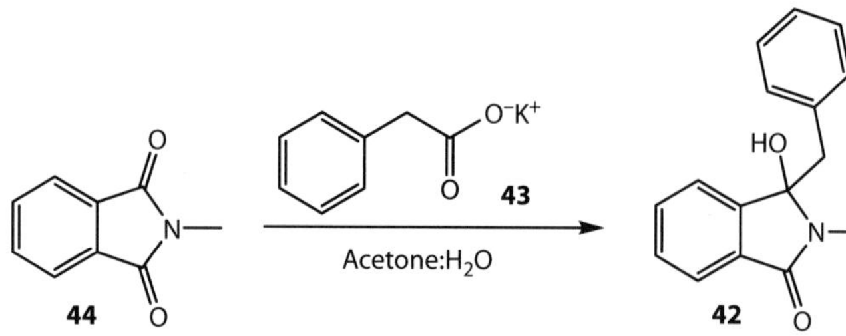

SCHEME 5.15 Schematic illustrating the photodecarboxylative addition of carboxylates to phthlimides investigated under continuous flow conditions.

radical undergoing nucleophilic attack by the cyanide anion at the oil–water phase boundary and the reaction product **48** remaining in the oil phase; affording facile separation via the Y-shaped outlet channel.

Using this approach, the authors were able to generate 1-cyanopyrene **48** in 28% yield with a reaction time of 210 s. This was readily increased to 73% (in 210 s) by performing the reaction in a three-phase reaction channel whereby the oil phase was flanked by aq. NaCN **46** (Scheme 5.16).

5.2.4 Photochemical Halogenations

Using a glass reactor (Mikroglass, Germany), Matsubara et al. [32] evaluated the bromination of a series of cycloalkanes under photoirradiation with black light (15 W, λ = 352 nm). Using mixtures of Br_2 **49** and alkanes, the authors observed slug flow due to the evolution of HBr as the reaction progressed. To avoid this, the authors employed a biphasic system whereby the cycloalkane and Br_2 **49** were introduced into the reactor from one inlet and deionized water from a second, the reaction products were then collected in aq. Na_2SO_3 (10%) prior to analysis.

Employing a reaction time of 19 min, the authors were able to obtain excellent selectivity toward the monobromide (Scheme 5.17); with further studies employing Cl_2 or $SOCl_2$ affording the respective chlorinated alkane.

See Section 5.3.5 for an example of photochemical chlorinations performed in a falling-film micro reactor.

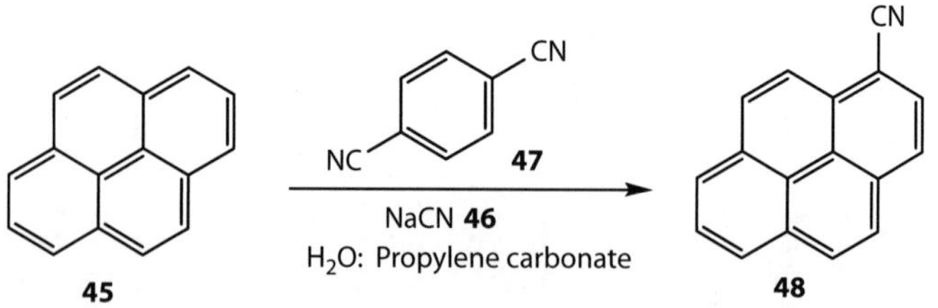

SCHEME 5.16 Schematic illustrating the biphasic photochemical cyanation performed in a polymeric micro reactor.

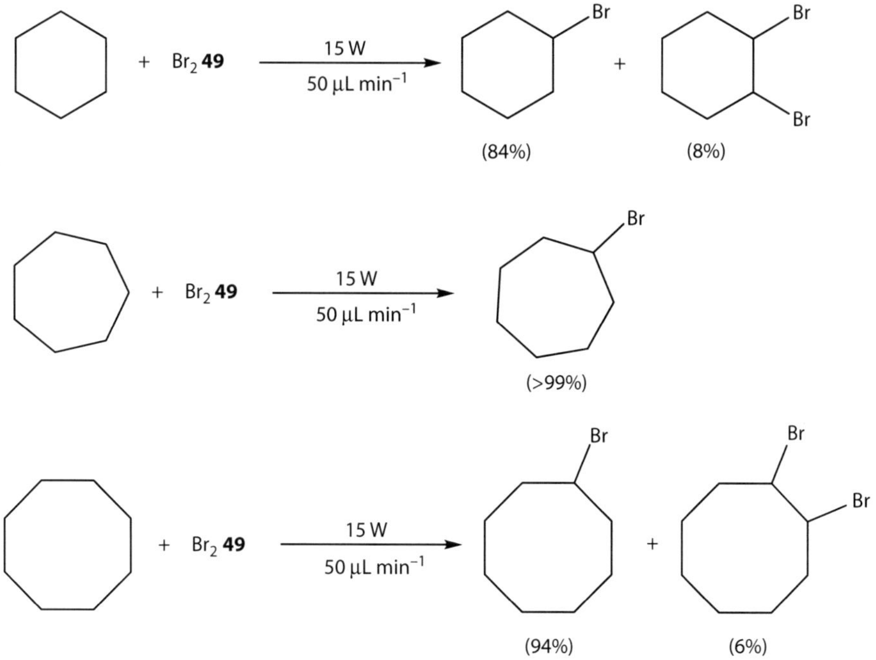

SCHEME 5.17 Illustration of the product selectivity obtained as a result of performing photochemical bromination reactions under continuous flow.

5.2.5 NITRITE PHOTOLYSIS UNDER FLOW CONDITIONS

As a means of accessing a structurally complex saturated alcohol **50**, Ryu and coworkers [33,34] investigated the Barton reaction (nitrite photolysis) of a steroidal substrate **51** within a glass-covered stainless steel micro reactor, owing to its use as a key intermediate in the production of myriceric acid A **52**; an Endothelin receptor antagonist (Scheme 5.18).

To minimize the quantities of material used to optimize the reaction, the authors initially performed the reaction in a device with channel dimensions of 1000 μm (wide) × 107 μm (deep) × 2.2 m (long) (volume = 200 μL). Maintaining a 7.5 cm gap between the 300 W high-pressure Hg lamp and the reactor, the authors pumped an acetone solution of the nitrite **51** (0.9×10^{-2} M) and pyridine (0.2 eq.) through the reactor at a flow rate of 33 μL min^{-1}, affording a residence time of 6 min. Under the aforementioned conditions, the authors obtained the rearranged product **50** in 59% yield; determined by HPLC analysis. After careful evaluation of the setup, the authors proposed that heating of the reaction mixture within the microchannels was preventing further increases in yield from being obtained. Consequently, the authors investigated a range of light sources, finding 15 W black lights to be suitable alternatives. This time a gap of 3.0 cm was introduced between the light source and the reactor, initially affording the target compound **50** in 21% yield. Increasing the residence time to 12 min, by reducing the reactant flow rate, the

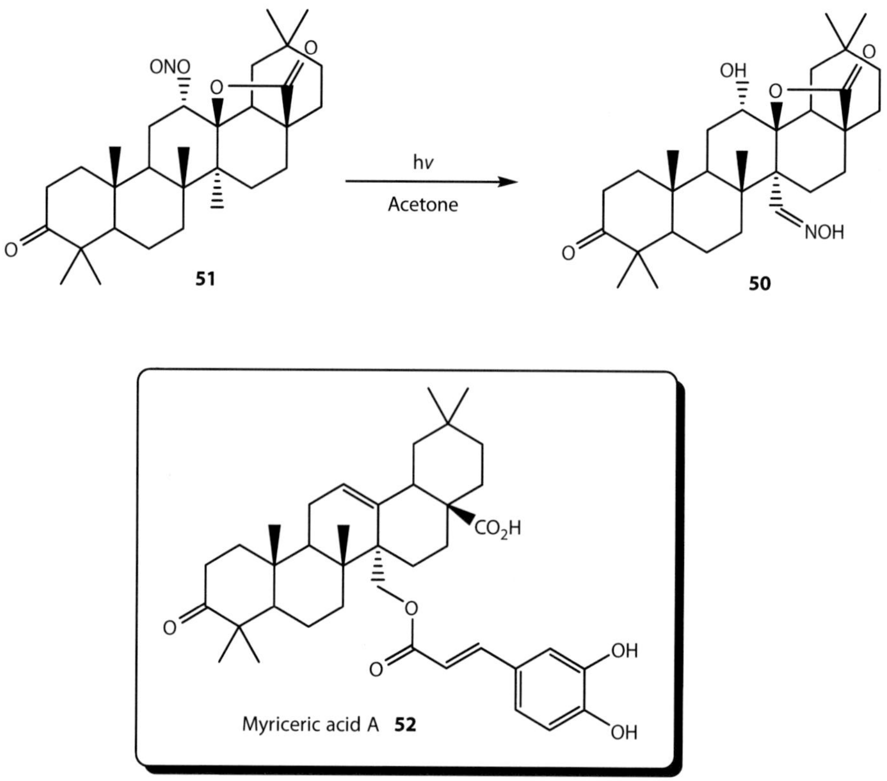

SCHEME 5.18 Barton nitrite photolysis, a key intermediate in the synthesis of myriceric acid A **52**.

authors obtained the product in 71% yield with almost a 10-fold increase in yield W^{-1} compared to the Hg lamp. To further increase the efficiency of the technique, the authors investigated the use of UV-LEDs (1.7 W) as the light source, which were advantageous as their size enabled them to be positioned much closer to the reaction channel (1.5 cm). Using this approach, the authors again obtained the product **50** in 70% yield; however, as Table 5.13 illustrates, the technique is almost 300 times more energy efficient than the initial investigation using an Hg lamp (300 W).

High-throughput Steroid Synthesis: Although initial investigations demonstrated the synthesis of **50** in acetone, the sparing solubility of the steroidal precursor **51** does not lend itself to high-throughput production. With this in mind, the authors screened a series of solvents as alternatives, finding the solubility of steroid **51** to be almost four times greater in DMF. Employing a more concentrated stock solution, 3.6×10^{-2} M, the authors repeated the flow synthesis, utilizing two serially connected micro reactors resulting in the synthesis of 3.1 g of oxime **50** in 24 h; after purification by silica gel chromatography.

TABLE 5.13
Comparison of the Energy Efficiencies of the Barton Reaction Performed under Continuous Flow Conditions

Light Source	Cover Plate	Flow Rate (mL h^{-1})^a	Yield (%)	Wh	Yield/Wh
300 W Hg lamp	Pyrex	2.0	6	21	0.7
300 W Hg lamp	Soda lime	2.0	6	56	1.89
15 W black light	Soda lime	2.0	6	15	10.0
15 W black light	Pyrex	2.0	6	29	19.3
15 W black light	Pyrex	1.0	12	71	23.7
1.7 W UV-LED	Pyrex	1.0	12	70	206.0

^{a} The residence time is given in parentheses (min).

5.2.6 PHOTOCHEMICAL DIMERIZATION

Although numerous examples of photochemistry performed in micro reactors have been reported within the literature, some photochemical reactions proved difficult to adapt to a single-pass approach due to the insolubility of products generated under photochemical irradiation; leading to clogging of the microchannel network.

To address this, Horie et al. [35] evaluated the combination of ultrasonic irradiation and a gas–liquid slug flow system. Utilizing the photochemical dimerization of maleic anhydride **26** in EtOAc (10% solution) to cyclobutane tetracarboxylic dianhydride **52** as a model reaction, the authors investigated the use of fluorinated ethylene propylene (FEP) or perfluoroalkoxy (PFA) tubular reactors coupled with a 400 W mercury immersion well lamp housed within an ultrasonic bath.

Controlling liquid and gas flow by the use of an HPLC plunger pump and mass flow controller, the authors were able to sweep the precipitated product **52** from the tube reactor at regular intervals. Using this approach, the authors were able to operate the flow reactor for more than 16 h without interruption. Evaluating the effect of channel size, the authors found that product conversion increased and the required reaction time decreased with smaller internal diameters, with 70% conversion to **52** obtained in 22 min using a 0.8 mm (i.d.) × 13.9 m (long) tube reactor. In comparison, <30% cyclobutane tetracarboxylic dianhydride **52** was obtained in a batch reactor operated under comparable conditions (Scheme 5.19).

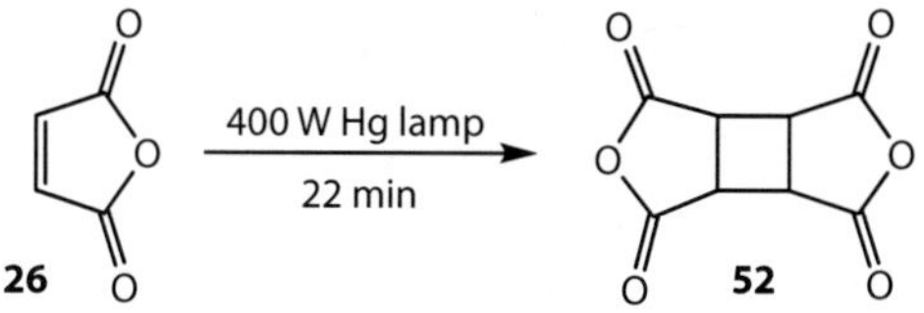

SCHEME 5.19 Illustration of the model reaction selected to demonstrate the combined advantages of ultrasonic irradiation and photochemistry within micro reaction systems.

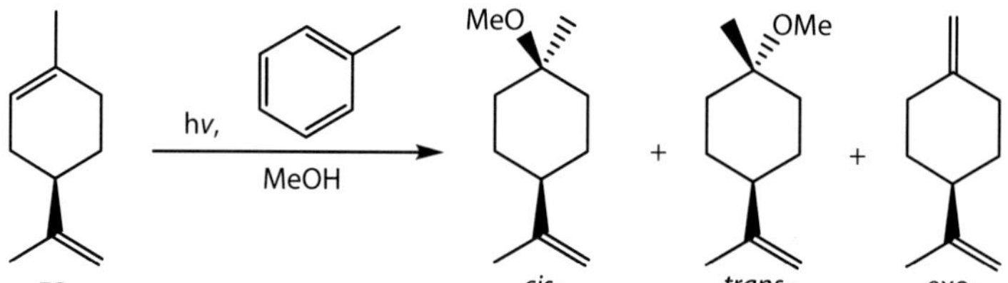

SCHEME 5.20 Model reaction used to demonstrate photosensitized diastereodifferentiation via MeOH addition.

5.2.7 Photosensitized Diastereo Differentiation

Using the photochemical addition of MeOH to (R)-(+)-(Z)-limonene 53 (Scheme 5.20), Ichimura and coworkers [36] evaluated the prospect of performing photochemical asymmetric synthesis under continuous flow conditions, comparing three quartz micro reactors with a standard laboratory cell (Table 5.14). Using a low-pressure Hg lamp ($\lambda = 254$ nm), the authors evaluated the effect of irradiation time on the conversion of (R)-(+)-(Z)-limonene 53, employing a reaction mixture of (R)-(+)-(Z)-limonene 53 (2.5×10^{-2} M) and toluene (1.0×10^{-2} M) in MeOH. Initial investigations were performed using a reactor with channel dimensions of 500 µm (wide) × 300 µm (deep) and illustrated a linear relationship between conversion and irradiation time, with decreasing *de* observed with prolonged exposure. In order to increase the photon efficiency of the system, the authors subsequently investigated the effect of channel depth on the process, with channels <40 µm deep affording increased illumination homogeneity; as illustrated in Table 5.14. Comparing the standard cell with the microchannel reactors, it can be seen that with a fixed irradiation time of 32 s, almost a fivefold increase in photon efficiency can be obtained within micro reactors, along with a slight increase in *de*, which the authors attribute to a suppression of side reactions within the micro reactors.

5.3 MULTIPHASE PHOTOCHEMICAL REACTIONS

As observed for more conventional reaction types, continuous flow photochemistry is not limited to homogenous reactions with examples of wall-coated, packed-bed,

TABLE 5.14
Comparison of the Photon Efficiencies Obtained within a Standard Batch Cell and a Series of Quartz Micro Reactors

Reactor	Dimensions	Photon Efficiency	de (%)
Batch	100 µm × 3 mm	0.06	28.7
Micro	500 µm × 300 µm	0.11	30.6
Micro	400 µm × 40 µm	0.27	29.4
Micro	200 µm × 20 µm	0.29	30.0

and nano-array reactors initially being utilized for the highly efficient photocatalytic degradation of small organic compounds such as phenol [37], 4-chlorophenol [38], and methylene blue [39–41].

More recently, researchers have begun to investigate these reactors for the performance of synthetic reactions, ranging from oxidations and reductions to cyclizations and halogenations; details of which are the focus of the following section.

5.3.1 Photocatalytic Reductions

Using a titanium dioxide wall-coated micro reactor, Matsushita and coworkers [42] investigated the photocatalytic reduction of benzaldehyde **54** and nitrotoluene **55**, as illustrated in Scheme 5.21, examining the effect of surface-to-volume ratio on the reactions. The reactor used consisted of a quartz microchannel (dimensions = 500 μm (wide) × 100 μm (deep) × 0.4 cm (long)) in which the bottom and sides were coated with TiO_2 (anatase), which afforded an illuminated area of 1.4×10^4 m² m⁻³; not taking into account the roughness of the TiO_2 surface. Irradiation of the reactor was achieved using a series of UV light-emitting diodes (λ = 365 nm, 2.2 m W/cm^{-2}) and reactants introduced into the reaction channel using pressure-driven flow.

In the first example, the authors investigated the reduction of benzaldehyde **54** (1.0×10^{-4} M) in the presence of an alcohol to provide the corresponding oxidation step observing rapid reduction to benzyl alcohol **56** with irradiation times of 60 s (10.7%). Based on the kinetic instability of the methoxy radical, EtOH was found to be the optimal solvent for the transformation, affording acetaldehyde as a volatile by-product. In addition, the authors found that by saturating the reaction mixture with N_2, thus excluding dissolved O_2 from the mixture, photocatalytic reduction was promoted as the electrons in the conduction band of the excited TiO_2 layer were not sequestered by any O_2 within the feedstock.

To demonstrate the tolerance of other functional groups, the authors subsequently evaluated the reduction of 4-nitrotoluene **55** (1.0×10^{-4} M), again employing EtOH

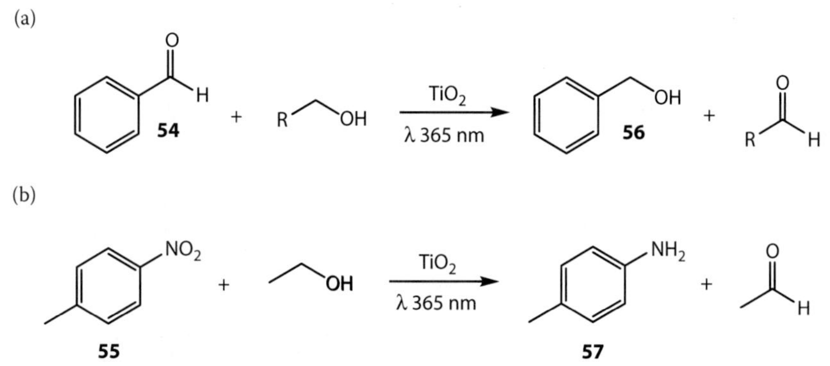

SCHEME 5.21 Illustration of the model reactions selected to demonstrate photocatalytic reduction in wall-coated micro reactors. (a) Benzaldehyde **54** and (b) 4-nitrotoluene **55** reduction.

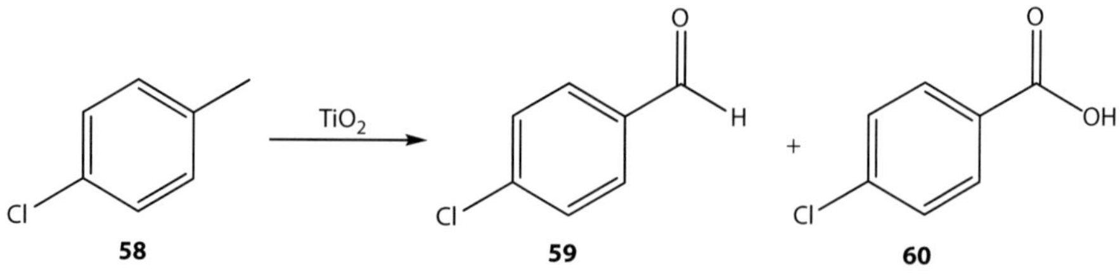

SCHEME 5.22 Model reaction used to demonstrate the selective photochemical oxidation of 4-chlorotoluene **58** to 4-chlorobenzaldehyde **59**.

as the reaction solvent. Employing variable flow rates, the authors investigated a range of irradiation times (0–60 s), observing no 4-aminotoluene **57** in the absence of light, 8.3% after 10 s and 45.7% when a reaction time of 60 s was employed. Based on these preliminary results, work is currently underway to optimize the excitation wavelength and micro reactor design in order to further increase the process efficiency.

5.3.2 Photocatalytic Oxidation Reactions

Employing wall-coated microchannels, Matsushita et al. [43,44] demonstrated the development of a selective process for the photochemical oxidation of 4-chlorotoluene **58**, as depicted in Scheme 5.22. Using micro reactors fabricated from Tempax or Quartz plates and a film of self-welding fluorinated polymer containing the microchannel network (channel dimensions = 100–500 μm (wide) × 20–500 μm (deep) × 5–20 cm (long)), the authors coated the channels with a layer of TiO_2 via a sol–gel method, followed by the photodeposition of a metal cocatalyst. Irradiation was achieved using UV-LED modules (λ = 310–365 nm) and reactant flow rates of 2–200 μL min^{-1} channel^{-1} were investigated, affording selective oxidation of 4-chlorotoluene **58** to 4-chlorobenzaldehyde **59**, with only 0.01% competing over-oxidation to **60** observed.

In a second example, the authors evaluated the conversion of nitro compounds to quinolines; selected due to their significance as biologically active materials. Using the synthesis of 2,5,7-trimethylquinoline **61** as a model (Scheme 5.23) the authors evaluated the photocatalytic conversion of 5-nitro-m-xylene **62**, observing 60% conversion to **61** with a reaction time of 150 s; demonstrating a dramatic increase in space time yield compared to batch processes [45].

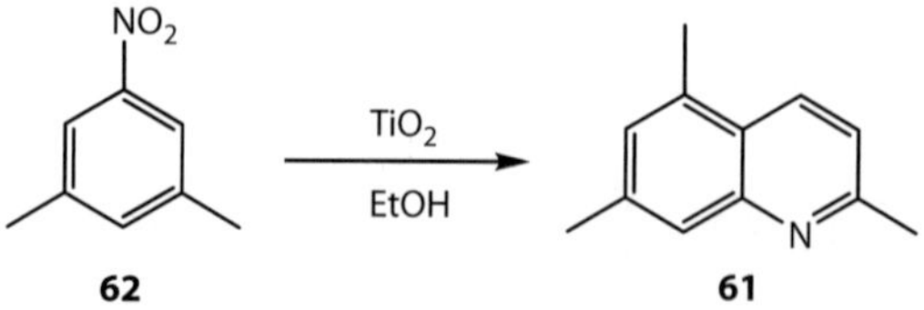

SCHEME 5.23 Illustration of the photocatalytic synthesis of quinolines performed under continuous flow conditions.

5.3.3 PHOTOCATALYTIC ALKYLATION REACTIONS

Using wall-coated microchannels (channel dimensions = 500 μm (wide) × 25 μm (deep) × 5.0 cm (long)), again fabricated from two Tempax plates and a thin film of self-welding fluorinated polymer, Matsushita et al. [46,47] evaluated the photocatalytic N-alkylation of a series of amines under continuous flow conditions. Literature precedent illustrated that the photocatalytic N-alkylation of aromatic amines had previously been performed with reaction times of 4 h affording 84% monoalkylated products; however, Pt-loaded TiO_2 was required with no reaction observed in the presence of TiO_2 only [48].

Using a process of spin coating $(Ti(O^iPr)_4)$ and calcination (770 K), the authors coated the surface of the microchannel with TiO_2, with evaluation by x-ray diffraction (XRD) analysis confirming the presence of the anatase form. The coating was then loaded with Pt by pumping an aqueous solution of chloroplatinic acid and MeOH through the channel network, controlling the catalyst loading with irradiation time.

As Scheme 5.24a illustrates, the reaction investigated within the micro reactor for the N-alkylation of amines involved the dehydrogenation of an alcohol, at the surface of the catalyst, to form H_2 and a carbonyl compound. The photoproduct was then converted into an imine by condensation with the amine and reduced by H_2 to afford the target N-alkyl derivative; in the case of secondary amines, dialkylation occurs readily (Scheme 5.24b).

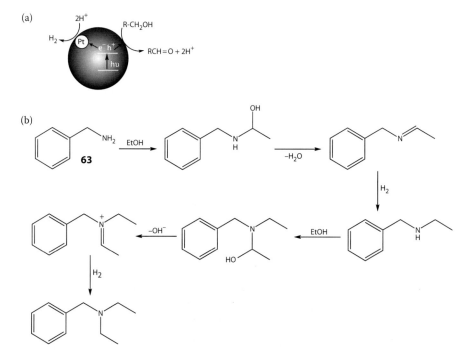

SCHEME 5.24 Schematic illustrating the reaction mechanism followed for the photochemical N-alkylation of amines: (a) reagent interaction with Pd-TiO_2 nanoparticles and (b) model reaction employed.

TABLE 5.15
Comparison of the Results Obtained for the Photochemical N-Alkylation of Amines under Flow and Batch Conditions

Amine	Method	Light Source	Reaction Time	Product Distribution	
				Mono-(%)	Di-(%)
Benzylamine 63	Batch flow	400 W (Hg)	5 h	84.4	2.4
		490 mW (7 × UV-LED)	6 s	68.0	3.8
Aniline 64	Batch flow	400 W (Hg)	20 h	8.1	0.0
		490 mW (7 × UV-LED)	90 s	57.0	1.5
Piperidine 65	Flow	490 mW (7 × UV-LED)	20 s	96.0	N/A

Investigating the alkylation of a series of amines including benzylamine 63 [49], aniline 64, and piperidine 65, the authors compared the reaction selectivity obtained under flow conditions with that of batch. As Table 5.15 illustrates, across the board, increases in product selectivity were observed as a result of utilizing a flow reaction; however, the most dramatic effect observed was the reduction in reaction times from 5–20 h down to 6–150 s. The authors attribute this reduction to an increased surface-to-volume ratio and a more uniform irradiation depth when compared to batch vessels.

5.3.4 Photocatalytic Cyclizations

In addition to the wide array of photocatalytic reductions and oxidations previously discussed, an area of great interest to the synthetic chemist is the ability to efficiently cyclize molecules as a means of accessing compounds of synthetic interest, particularly with respect to natural product synthesis.

With this in mind, Takei et al. [50] demonstrated the synthesis of L-pipecolinic acid 66 from an aqueous solution of L-lysine 67 (Scheme 5.25). To achieve this photocatalytic transformation, the authors fabricated a Pyrex® micro reactor in which the channel cover plate was coated with a 300 nm layer of TiO_2 anatase (100 nm particles), to afford a titania-coated micro reactor (TCM), the titania film was subsequently loaded with platinum (0.2 wt.%), by photodeposition, to enable the TCM to be used for redox-combined photosynthesis. A solution of L-lysine 67 (2.0 mM) was subsequently irradiated using a high-pressure mercury lamp (110 mW cm^{-1}) and the

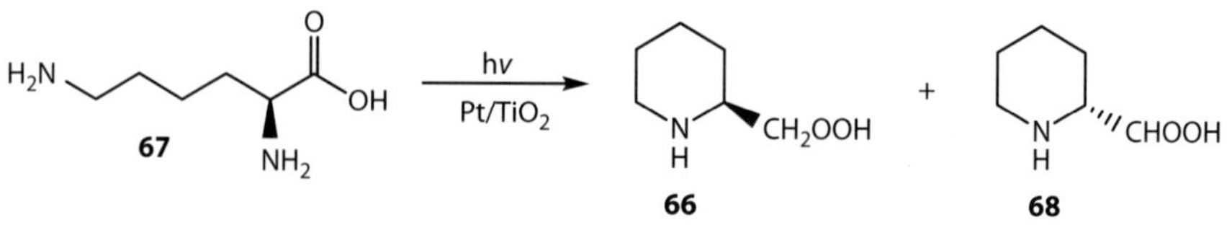

SCHEME 5.25 Illustration of the photocatalytic cyclisation of L-lysine 5Wb performed using a titania-coated micro reactor.

resulting reaction mixture analyzed by chiral HPLC in order to determine the proportion of L-lysine **67** converted to D- and L-pipecolinic acid **66/68**, respectively.

Employing a residence time of 0.86 min (1 µL min^{-1}), the authors obtained 87% conversion of L-lysine **67**, exhibiting 22% selectivity for L-pipecolinic acid **66**. For comparative purposes, the authors also performed the reaction in batch employing 2 wt.% Pt-loaded TiO_2 particles, which afforded the same surface-to-volume ratio of the catalyst as the TCM, where a 70 times longer reaction time (60 min) was required in order to obtain analogous results to the TCM. The authors concluded that the increased reaction efficiency, observed within the TCM, was attributed to the efficient irradiation of the reaction mixture; however, for a true comparison, they noted that a measurement of the quantum yield of each system would be required.

5.3.5 GAS–LIQUID TRANSFORMATIONS

In addition to biphasic reactions utilizing heterogeneous catalysts within microflow reactors, the authors have also demonstrated the performance of gas–liquid photochemical reactions, with safety advantages lending the technique to the *in situ* preparation of singlet oxygen and chlorinations.

Singlet Oxygen Generation: Singlet oxygen ($^1O_2$) is a clean and atom efficient reagent used for the introduction of oxygen into hydrocarbon skeletons; as such, it has found wide-ranging application in organic synthesis, with examples even reported in natural product synthesis. Singlet oxygen is commonly prepared by the photoexcitation of molecular oxygen in the presence of a photosensitizer or thermally upon treatment with H_2O_2 or sodium hypochlorite. In addition to the short-lived nature of the oxidant, the safety implications associated with the generation and handling of $^1O_2$ on a large scale has thus far precluded its common use. To address this, several authors have investigated not only the generation, but also the use of $^1O_2$ under continuous flow conditions.

In 2002, de Mello and coworkers [51] reported the fabrication of a glass micro reactor (channel dimensions = 150 µm (wide) × 50 µm (deep) × 5 cm (long)) suitable for the continuous photochemical generation of singlet oxygen, subsequently demonstrating its reaction in the synthesis of ascaridole **69** from α-terpinene **70** (Scheme 5.26).

Employing Rose Bengal **71** as the photosensitizer (5.0 × 10^{-3} g mL^{-1}) and MeOH as the reaction solvent, the authors introduced the organic reactants through inlet A at a rate of 1 µL min^{-1} and oxygen through inlet B at a flow rate of 15 µL min^{-1},

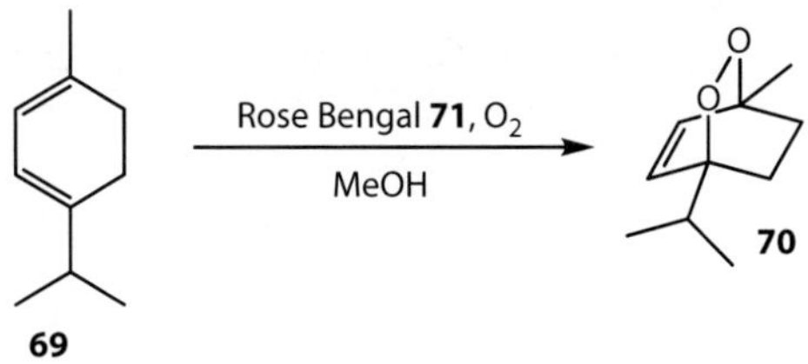

SCHEME 5.26 Schematic illustrating the model reaction used to demonstrate the generation and *in situ* reaction of singlet oxygen in a microfabricated reactor.

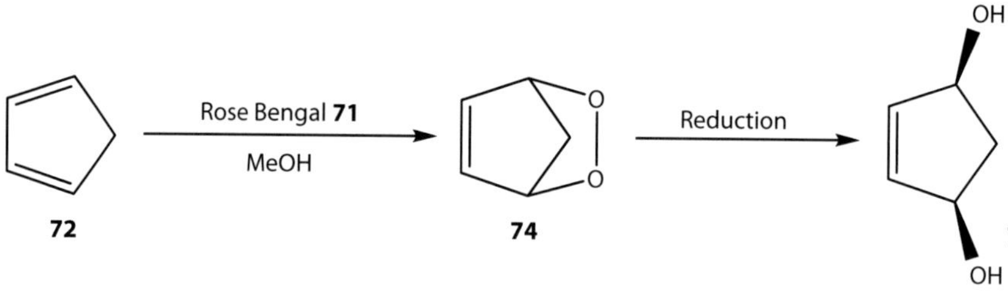

SCHEME 5.27 [4+2]-Photocycloaddition performed in a falling-film micro reactor.

irradiating the reaction mixture with a 20 W tungsten lamp. The reaction products were collected at the reaction outlet and prior to analysis; they were diluted in diethyl ether and purged with N_2 as a safety precaution. Conversion to ascaridole **69** was subsequently determined by offline GC analysis. Using this approach, the authors were able to confirm the formation of singlet oxygen via the formation of ascaridole **69** in 85% conversion.

From a safety perspective, the volume of sample being irradiated within the reactor at any one time was of the order of microliters and therefore represents a significant increase in process safety compared to conventional methodology.

In a subsequent example, Jähnisch and Dingerdissen [52] demonstrated the photochemical generation, and [4+2] cycloaddition, of singlet oxygen to cyclopentadiene **72** in a falling-film micro reactor (IMM, Germany) (Scheme 5.27). Again the authors employed MeOH as the reaction solvent and Rose Bengal **71** as the photosensitizer; however, unlike de Mello and coworkers, the authors developed a system suitable for the production of materials on a synthetically useful scale. With this in mind, the authors irradiated the reactor with a Xenon lamp and investigated the synthesis of 2-cyclopenten-1,4-diol **73** via the endoperoxide **74**, obtaining the pharmaceutically relevant intermediate with a substrate **72** throughput of 1 mL min^{-1}.

Industrially Relevant Oxidations: With (−)-rose oxide **75** selected as a synthetic target due to its use as a fragrance in the perfume industry, Meyer et al. [53] compared the efficiency of reactions performed in batch (40 mL) and under continuous flow conditions (Scheme 5.28). The micro reactor was fabricated from Borofloat® glass (Little Things Factory, Germany) with channel dimensions of 1 mm (wide) and a total reactor volume of 270 µL (Figure 5.3). The contents of the microchannel were irradiated using a diode array comprising 4 × 10 diodes (λ = 468 nm) and reactants pumped through the system.

To evaluate the micro reactor, the authors firstly purged a solution of (−)-β-citronellol **76** (0.1 M) and Ru(^{t}bpy)$_3$Cl$_2$ **77** (1 × 10^{-3} M in EtOH) with air at a flow rate of 0.4 l h^{-1}; after 20 min, the reaction mixture was cycled through the illuminated reactor using a peristaltic pump, with samples taken periodically and analyzed using offline HPLC analysis to determine the conversion of **76** to **78** and **79**. Comparing the reactor types, the authors were able to conclude that only 7.5 mL of the batch reactor (40 mL) was illuminated (15.2 cm²) at any one time, unlike the flow reactor where the whole 270 µL was illuminated (6.9 cm²). This consequently afforded an increased photonic efficiency which in turn led to a ninefold increase

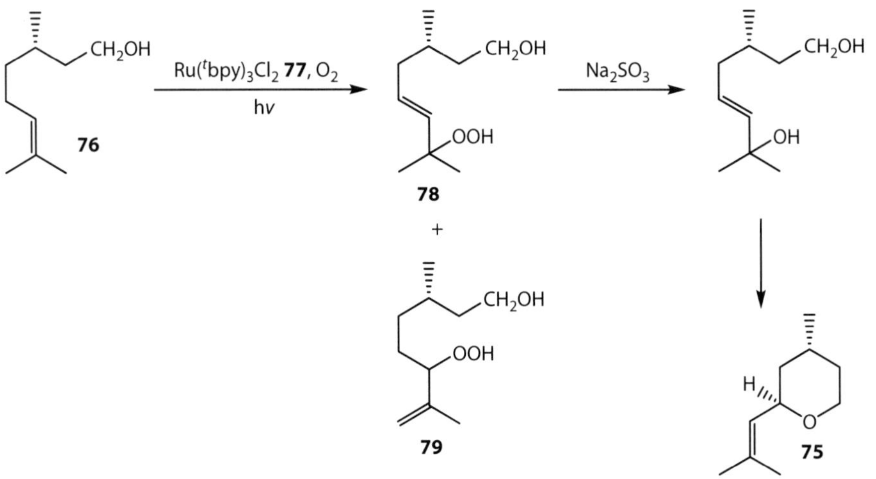

SCHEME 5.28 Illustration of the industrially relevant photooxidation of citronellol **76** to afford hydroperoxides **78** and **79** used in the synthesis of the fragrance (−)-rose oxide **75**.

in the space–time yield when compared with batch (0.9 mmol L^{-1} min^{-1} cf. 0.1 mmol L^{-1} min^{-1}).

Photooxidation Using Supercritical CO_2: While Meyer et al. [53] demonstrated a dramatic enhancement in space–time yield for the synthesis of (−)-rose oxide **75**, the technique developed was not suited for the large-scale synthesis of such aromatic compounds. With this in mind, Poliakoff and coworkers [54] more recently investigated the use of supercritical carbon dioxide (scCO_2) as the reaction solvent in the development of a clean and continuous photooxidation protocol. Unlike alcoholic

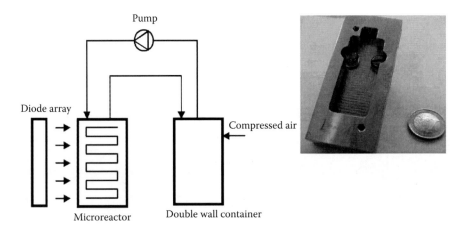

FIGURE 5.3 Illustration of the micro reactor utilized for the continuous flow synthesis of a (−)-rose oxide precursor. (Reproduced from *J. Photochem. Photobiol* A, Meyer, S. et al. Photosensitized oxidation of Citronellol in microreactors, 186: 248–253, Copyright (2007), with permission from Elsevier.)

solvents, the use of scCO$_2$ is advantageous as it has a low viscosity, high diffusivity, is nonflammable, and can solubilize high quantities of O$_2$, thus enabling the risks associated with the procedure to be further reduced.

Again focusing on the photooxidation of citronellol **76**, selected due to the synthetic utility of the product (−)-rose oxide **75**, the authors utilized a simple narrow sapphire tube reactor, housed within a stainless-steel holder, and investigated the irradiation of reaction mixtures with high-power LEDs at 90° to the direction of fluid flow. Using this approach, the authors were able to exploit a short pathlength, efficiently irradiating the reaction mixture and enabling the use of a single-pass method. To perform a reaction, the authors introduced liquid CO$_2$ into the reactor at 1 mL min^{-1} (−10°C, 48 bar) from one inlet, the organic reactant **76** and photosensitizer (5,10,15, 20-tetrakis(pentafluorophenyl)porphyrin) in dimethyl carbonate were introduced from a second inlet at 0.1 mL min^{-1} and O$_2$ (180 bar) introduced at a rate of 2 molar eq. with respect to the organic reactant; all components are soluble in the scCO$_2$. Under the aforementioned conditions, the authors obtained quantitative conversion of citronellol **76** to the target hydroperoxide in 52% selectivity, with the remaining 48% comprising the structural isomer. Owing to the unstable nature of the hydroperoxides generated, the reaction products were collected in aq. Na$_2$SO readily converting the intermediates into the respective diol. An acidic work-up followed, affording (−)-rose oxide **75** in 97.6% selectivity as determined by gas chromatography–mass spectrometry (GC–MS) analysis.

In order to benchmark their system further, the authors performed the oxidation of ascaridole **69**, finding that their system enabled the oxidation to be performed at a rate of 70 mmol L^{-1} min^{-1}, nearly two orders of magnitude greater than protocols previously reported under flow conditions.

Biphasic Chlorinations: Employing a falling-film micro reactor, Jähnisch and coworkers [55] demonstrated the ability to perform gas–liquid photochemical transformations other than the generation of singlet oxygen, reporting the selective photochlorination of toluene-2,4-diisocyanate **80**, obtaining 55% conversion to 1-chloromethyl-2,4-diisocyanobenzene **81** and toluene-5-chloro-2,4-diisocyanate **82** in 80% and 5% selectivity, respectively. Comparing the microprocess to a conventional batch protocol, whereby 1-chloromethyl-2,4-diisocyanobenzene **81** was obtained in 65% conversion and 45% selectivity, the use of a microstructured device afforded the authors with greater selectivities and a 308-fold increase in space–time yield (400 mol L^{-1} h^{-1}) (Scheme 5.29).

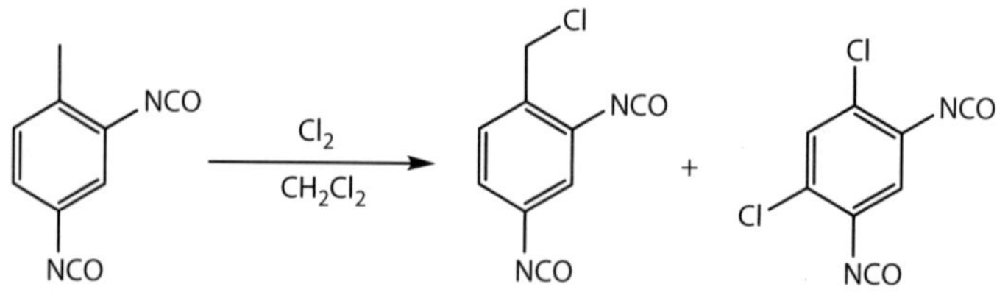

SCHEME 5.29 Illustration of the photochlorination reaction performed in a falling-film micro reactor.

5.3.6 GAS–LIQUID–SOLID REACTIONS

In efforts toward the reduction of carbon dioxide emissions, researchers have investigated the photochemical reduction of carbon dioxide to afford valuable materials such as carbon monoxide, formic acid, methane, and methanol. Using conventional reactor technology, low yields are however reported for this transformation precluding is widespread use. In order to address this, Fukazawa et al. [56] developed a micro reactor capable of performing the photocatalytic reduction of CO_2 to afford these synthetically valuable materials.

The micro reactor in question consisted of a Y-shaped mixer and an etched channel (200 μm (wide) × 40 μm (deep) × 5 cm (long)) coated with a thin layer of photocatalytic TiO_2 and a metal cocatalyst (Pt). To perform a reaction, a solution of distilled water and gaseous CO_2 were introduced into the reactor, affording pipe flow, and irradiated using a series of UV-LED's ($\lambda = 365$ nm); the reaction products were subsequently evaluated using GC.

Compared to batch reactors, employing powdered TiO_2, the pipe flow conditions obtained within the microchannel reactor afforded high interaction efficiencies and mass transfer between the three phases; resulting in a 700-fold increase in MeOH formation.

5.3.7 SUMMARY

From the examples discussed, it can be seen that the combination of micro reactors and photochemistry affords significant advantages compared to the conventional batch reaction methodology, including increased spatial illumination homogeneity, improved light penetration, and a high catalytic surface area. For the first time, these factors combined have the potential to enable the cost-effective implementation of photochemical transformations for the production of fine chemicals and pharmaceuticals. See Chapter 7 for relevant examples.

REFERENCES

1. Paddon, C. A., Atobe, M., Fuchigami, T., He, P., Watts, P., Haswell, S. J., Pritchard, G. J., Bull, S. D., and Marken, F. 2006, Towards paired and coupled electrode reactions for clean organic microreactor electrosyntheses, *J. Appl. Electrochem.* 36: 617–634.
2. Yoshida, J., Kataoka, K., Horcajada, R., and Nagaki, A. 2008, Modern strategies in electroorganic synthesis, *Chem. Rev.* 108: 2265–2299.
3. Yoshida, J. 2008, *Flash Chemistry Fast Organic Synthesis in Microsystems.* London: Wiley-VCH.
4. Löwe, H. and Ehrfeld, W. 1999, State-of-the-art in microreaction technology: Concepts, manufacturing and applications, *Electrochimica Acta* 44: 3679–3689.
5. Küpper, M., Hessel, V., Löwe, H., Stark, W., Kinkel, J., Michel, M., and Schmidt-Traub, H. 2003, Micro reactor for electroorganic synthesis in the simulated moving bed-reaction and separation environment, *Electrochimica Acta* 48: 2889–2896.
6. Ueno, K., Kim, H., and Kitamura, N. 2003, Characteristic electrochemical responses of polymer microchannel-microelectrode chips, *Anal. Chem.* 75: 2086–2091.
7. Tsujimoto, M., Tonomura, O., Kano, M., Hasebe, S., and Hinouchi, T. 2010, Design of micro electrolytic reactors towards acceleration of mass transfer, *11th International Conference on Microreaction Technology,* Kyoto, Japan, 320–321.

8. Suga, S., Okajima, M., Fujiwara, K., and Yoshida, J. 2001, "Cation-flow" method: A new approach to conventional and combinatorial electrochemical microflow systems, *J. Am. Chem. Soc.* 123: 7941–7942.
9. Horii, D., Fuchigami, T., and Atobe, M. 2007, A new approach to anodic substitution reaction using parallel laminar flow in a micro-flow reactor, *J. Am. Chem. Soc.* 129: 11692–11693.
10. Nagaki, A., Togai, M., Suga, S., Aoki, N., Mae, K., and Yoshida, Y. 2005, Control of extremely fast competitive consecutive reactions using micromixing. Selective Friedel–Crafts aminoalkyation, *J. Am. Chem. Soc.* 127: 11666–11675.
11. Suga, S., Tsutsui, Y., Nagaki, A., and Yoshida, J. 2003, N-acyliminium ion pool as a heterodiene in [4+2] cycloaddition reaction, *Org. Lett.* 5(6): 945–947.
12. Suga, S., Tsutsui, Y., Nagaki, A., and Yoshida, J. 2005, Cycloaddition of N-acyliminium ion pools with carbon–carbon multiple bonds, *Bull. Chem. Soc. Jpn.* 78: 1206–1217.
13. Mengeaud, V., Bagel, O., Ferrigno, R., Girault, H. H., and Haider, A. 2002, A ceramic electrochemical microreactor for the methoxylation of methyl-2-furoate with direct mass spectrometry coupling, *Lab Chip* 2: 39–44.
14. Midorikawa, K., Suga, S., and Yoshida, J. 2006, Selective monoiodination of aromatic compounds with electrochemically generated I$^+$ using micromixing, *Chem. Commun.* 3794–3796.
15. Paddon. C. A., Pritchard, G. J., Thiemann, T., and Marken, F. 2002, Paired electrosynthesis: micro-flow cell processes with and without added electrolyte, *Electrochem. Commun.* 4: 825–831.
16. Horii, D., Atobe, M., Fuchigami, T., and Marken, F. 2005, Self-supported paired electrosynthesis of 2,5-dimethoxy-2,5-dihydrofuran using a thin layer flow cell without intentionally added supporting electrolyte, *Electrochem. Commun.* 7: 35–39.
17. Horcajada, R., Okajima, M., Suga, S., and Yoshida, J. 2005, Microflow electroorganic synthesis without supporting electrolyte, *Chem. Commun.* 1303–1305.
18. He, P., Watts, P., Marken, F., and Haswell, S. J. 2007, Electrosynthesis of phenyl-2-propanone derivatives from benzyl bromides and acetic anhydride in an unsupported micro-flow cell electrolytic process, *Green Chem.* 9: 20–22.
19. He. P., Watts, P., Marken, F., and Haswell, S. J. 2005, Electrolyte-free electro-organic synthesis: The cathodic dimerisation of 4-nitrobenzylbromide in a micro-gap flow cell, *Electrochem. Commun.* 7: 918–924.
20. He, P., Watts, P., Marken, F., and Haswell, S. J. 2006, Self-supported and clean one-step cathodic coupling of activated olefines with benzyl bromide derivatives in a micro flow reactor, *Angew. Chem. Int. Ed.* 45: 4146–4149.
21. Belemont, C. and Girault, H. H. 1994, Coplanar interdigitated band electrodes for synthesis Part 1: Ohmic loss evaluation, *J. Appl. Electrochem.* 24(6): 475–480.
22. Lu, H., Schmidt, M. A., and Jensen, K. F. 2001, Photochemical reactions and online UV detection in microfabricated reactors, *Lab Chip* 1: 22–28.
23. Freitag, A., Dietrich, T. R., and Scholz, R. 2010, Energy efficient photochemistry in automated microreaction plants, *11th International Conference on Microreaction Technology*, Kyoto, Japan, 25–25.
24. Mukae, H., Maeda, H., Nashihara, S., and Mizuno, K. 2007, Intramolecular photocycloaddition of 2-(2-alkenyloxymethyl)-naphthalene-1-carbonitriles using glass-made microreactors, *Bull. Chem. Soc. Jpn.* 80(6): 1157–1161.
25. Maeda, H., Mukae, H., and Mizuno, K. 2005, Enhanced efficiency and regioselectivity of an intramolecular ($2\pi + 2\pi$) photocycloaddition of 1-cyanonaphthalene derivative using microreactors, *Chem. Lett.* 34: 66–67.
26. Fukuyama, T., Yonamine, Y., Kobayashi, M., Hino, Y., Kamata, N., Kajihara, Y., and Ryu, I. 2010, [2+2] Photocycloaddition reaction using an energy-saving photo-microflow system, *11th International Conference on Microreaction Technology*, Kyoto, Japan, 194–195.

27. Fukuyama, T., Hino, Y., Kamata, N., and Ryu, I. 2004, Quick execution of [2+2] type photochemical cycloaddition reaction by continuous flow system using a glass-made microreactor, *Chem. Lett.* 33: 1430–1431.
28. Hook, B. A., Dohle, W., Hirst, P. R., Pickworth, M., Berry, M. B., and Booker-Milburn, K. I. 2005, A practical flow reactor for continuous organic photochemistry, *J. Org. Chem.* 70: 7558–7564.
29. Coyle, E. and Oelgemöller, M. 2008, Micro-photochemistry: Photochemistry in microstructured reactors. The new photochemistry of the future?, *Photochem. Photobiol. Sci.* 7: 1313–1322.
30. Gallagher, S., Hatoum, F., Oelgemöller, M., and Griesbeck, A. G. 2008, The synthesis of bioactive compounds by photodecarboxylative addition of carboxylates to phthalimides, *Central European Conference on Photochemistry*, Austria, 69.
31. Ueno, K., Kitagawa, F., and Kitamura, N. 2002, Photocyanation of pyrene across an oil–water interface in a polymer microchannel chip, *Lab Chip* 2: 231–234.
32. Matsubara, H., Hino, Y., and Ryu, I. 2010, Highly selective radical halogenations of alkanes using a microflow reactor under photo-irradiation, *11th International Conference on Microreaction Technology*, Kyoto, Japan, 44–45.
33. Sugimoto, A., Sumino, Y., Takagi, M., Fukuyama, T., and Ryu, I. 2006, The Barton reaction using a microreactor and black light. Continuous-flow synthesis of a key steroid intermediate for an endothelin receptor antagonist, *Tetrahedron Lett.* 47: 6197–6200.
34. Sugimoto, A., Fukuyama, T., Sumino, Y., Takagi, M., and Ryu, I. 2009, Microflow photo-radical reaction using a compact light source: Application to the Barton reaction leading to a key intermediate for myriceric acid A, *Tetrahedron* 65: 1593–1598.
35. Horie, T., Sumino, M., Tanaka, T., Yoshida, J., Matsushita, Y., and Ichimura, T. 2010, Photo-dimerisation of maleic anhydride with micro flow system, *11th International Conference on Microreaction Technology*, Kyoto, Japan, 200–201.
36. Sakeda, K., Wakabayashi, K., Matsushita, Y., Ichimura, T., Suzuki, T., Wada, T., and Inuoe, Y. 2007, Asymmetric photosensitized addition of methanol to (R)-(+)-(Z)-limonene in a microreactor, *J. Photochem. Photobiol. A* 192: 166–171.
37. Kitamura, N., Yamada, K., Ueno, K., and Iwata, S. 2006, Photodecomposition of phenol by silica-supported porphyrin derivative in polymer microchannel chips, *J. Photochem. Photobiol. A* 184: 170–176.
38. Gorges, R., Meyer, S., and Kreisel, G. 2004, Photocatalysis in microreactors, *J. Photochem. Photobiol. A* 167: 95–99.
39. Teekateerawej, S., Nishino, J., and Nosaka, Y. 2006, Design and evaluation of photocatalytic micro-channel reactors using TiO_2-coated porous ceramics, *J. Photochem. Photobiol. A* 179: 263–268.
40. Li, X., Wang, H., Inoue, K., Uehara, M., Nakamura, H., Miyazaki, M., Abe, E., and Maeda, H. 2003, Modified micro-space using self-organized nanoparticles for reduction of methylene blue, *Chem. Commun.* 964–965.
41. He, Z., Li, Y., Zhang, Q., and Wang, H. 2010, Capillary microchannel based microreactors with highly durable ZnO/TiO_2 nanorod arrays for rapid high efficiency and continuous-flow photocatalysis, *Applied Catalysis B: Environmental* 93: 376–382.
42. Matsushita, Y., Kumada, S., Wakabayashi, K., Sakeda, K., and Ichimura, K. 2006, Photocatalytic reduction in microreactors, *Chem. Lett.* 35(4): 410–411.
43. Matsushita, Y., Sekine, Y., Sato, Y., Sakai, Y., Kimura, Y., Suzuki, T., and Ichimura, T. 2010, Highly selective and environmentally Benign photocatalytic reaction processes in microstructured devices, *11th International Conference on Microreaction Technology*, Kyoto, Japan, 106–107.
44. Matsushita, Y., Iwasawa, M., Suzuki, T., and Ichimura, T. 2009, Multiphase photocatalytic oxidation in a microreactor, *Chem. Lett.* 38(8): 846–847.

45. Hakki, A., Dillert, R., and Bahnemann, D. 2009, Photocatalytic conversion of nitroaromatic compounds in the presence of TiO_2, *Catal. Today* 144: 154–159.
46. Matsushita, Y., Ohba, N., Suzuki, T., and Ichimura, T. 2008, N-alkylation of amines by photocatalytic reaction in a microreaction system, *Catal. Today* 132: 153–158.
47. Matsushita, Y., Ohba, N., Kumada, S., Sakeda, K., Suzuki, T., and Ichimura, T. 2008, Photocatalytic reactions in microreactors, *Chem. Eng. J.* 135S: S303–S308.
48. Ohtain, B., Osaki, S., Nishimoto, S., and Kagiya, T. 1986, A novel photocatalytic process of amine N-alkylation by platinized semiconductor particles suspended in alcohols, *J. Am. Chem. Soc.* 180: 308–310.
49. Matsushita, Y., Ohba, N., Kumada, S., Suzuki, T., and Ichimura, T. 2007, Photocatalytic N-alkylation of benzylamine in microreactors, *Catal. Commun.* 8: 2194–2197.
50. Takei, G., Kitamori, T., and Kim, H. B. 2005, Photocatalytic redox-combined synthesis of L-pipecolinic acid with a titania-modified microchannel chip, *Catal. Commun.* 6: 357–360.
51. Wootton, R. C. R., Fortt, R., and de Mello, A. J. 2002, A microfabricated nanoreactor for safe, continuous generation and use of singlet oxygen, *Org. Proc. Res. Dev.* 6: 187–189.
52. Jähnisch, K. and Dingerdissen, U. 2005, Photochemical generation and [4+2]-cycloaddition of singlet oxygen in a falling-film microreactor, *Chem. Eng. Technol.* 28: 426–427.
53. Meyer, S., Tietze, D., Rau, S., Schäfer, B., and Kreisel, G. 2007, Photosensitized oxidation of Citronellol in microreactors, *J. Photochem. Photobiol. A* 186: 248–253.
54. Bourne, R. A., Han. X., Poliakoff, M., and George, M. W. 2009, Cleaner continuous photo-oxidation using singlet oxygen in supercritical carbon dioxide, *Angew. Chem. In. Ed.*, 48: 5322–5325.
55. Ehrich, H., Linke, D., Morgenschweis, K., Baerns, M., and Jähnisch, K. 2002, Application of microstructured reactor technology for the photochemical chlorination of alkylaromatics, *Chimia* 56(11): 647–653.
56. Fukazawa, Y., Matsushita, Y., Suzuki, T., and Ichimura, T. 2010, Photocatalytic reduction of carbon dioxide in a microreaction system, *11th International Conference on Microreaction Technology*, Kyoto, Japan, 236–237.

6 The Use of Microfluidic Devices for the Preparation and Manipulation of Droplets and Inorganic/Organic Particles

In addition to their employment in the development of synthetic continuous flow methodology, the predictable and controllable manipulation of fluids attainable within microflow devices has meant that they have begun to find application in the preparation of colloids, nanoparticles, and even liposome production. In addition to obtaining a high degree of fluidic control, the continuous nature of these devices means techniques can be developed which are suited to the production of highly reproducible droplets and particles; removing the current issues associated with batch-to-batch variation. In addition to the examples described herein, further reading on the subject can be found in the following reviews and articles; production of emulsions [1,2], polymeric particles [3], nanoparticles [4], crystals [5], lipid microstructures [6], and vesicles [7].

6.1 DROPLET FORMATION USING CONTINUOUS FLOW METHODOLOGY

As described in Chapter 1 and illustrated in Chapters 3 and 4, the formation of segmented flow within microchannel devices has the potential to enable the user to accelerate mixing within flow systems and to segregate reactant slugs as a means of performing multiple reactions in series. In an addition to this, the ability to manipulate and fuse multiple droplets of miscible fluids within an immiscible carrier has the potential to be applied to screening applications. Consequently, droplet-based transformations are opening up a new and exciting area of microfluidics for the evaluation of chemical and biochemical processes.

With this in mind, Lin and coworkers [8] developed a device whereby precise fluidic control was obtained through the use of microvalves. As Figure 6.1 illustrates,

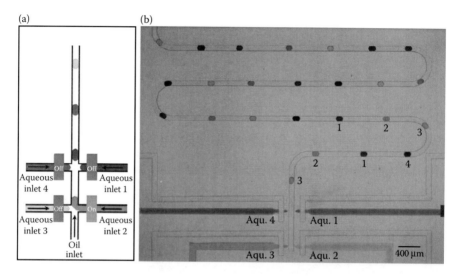

FIGURE 6.1 (a) Illustration of a microfluidic droplet generation technique and (b) its use for the formation of well-defined reaction vessels. (Zeng, S. et al. 2009. Microvalve-actuated precise control of individual droplets in microfluidics devices, *Lab Chip* 9: 1340–1343. Reproduced by permission of the Royal Society of Chemistry.)

the composition of each droplet could be readily altered, enabling the rapid screening of reaction conditions with ease.

High-throughput Droplet-based Reactions: From a more synthetic perspective, Garstecki and coworkers [9] demonstrated the use of a droplet generator for the serial performance of synthetic reactions in a high-throughput manner. Using what they termed as droplet-on-demand technique, the authors were able to scan combinations of three miscible solutions in 1.5 µL droplets, with fluidic dispensation controlled using off-device electromagnetic valves. Using this segmented approach, the authors reported the ability to scan 10,000 conditions h^{-1} employing only 10 mL of reactant solution. The throughput of the technique is however dependent on the analytical method of evaluating the process under investigation. Figure 6.2 illustrates the device in action using a series of colored aqueous dyes for illustrative purposes.

On-chip Incubation: Using the principle of droplet incubation, Griffiths and coworkers [10] demonstrated the development of a reliable microfluidic device for the incubation of droplets in delay lines, proposing application of the technique to microfluidic assays. Until now, it had been possible to hold droplets in single file within short channels; however, this was only suited to incubation times of less than 1 min. With this in mind, the authors developed a technique for stacking droplets within wide and deep microchannels as a means of enabling incubation times of minutes to hours to be accessed.

Double Emulsions: In addition to exploiting the regular droplet size and shape resulting from microfluidic droplet generation, researchers have also used the enhanced fluidic control as a means of preparing more complex double-emulsion structures. Nisisako et al. [11] investigated the challenges associated with the formation of monodisperse double emulsions, focusing on the preparation of water-in-oil-in-water

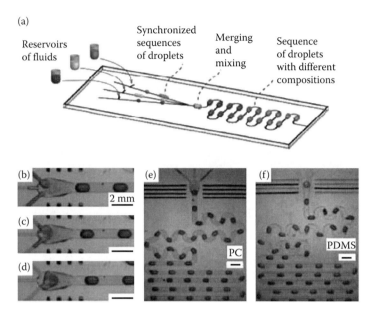

FIGURE 6.2 (a) Schematic illustrating the droplet on demand system used for the high-throughput screening of reactant conditions, (b–d) shows merging droplets (recorded at 125 fps), (e) in polycarbonate, and (f) in polydimethylsiloxane. (Churski, K., Korczyk, P., and Garstecki, P. 2010. High-throughput automated droplet microfluidic system for screening of reaction conditions, *Lab Chip* 10: 816–818. Reproduced by permission of the Royal Society of Chemistry.)

emulsions. Using a series of T-mixers, the authors initially prepared water-in-oil droplets by the introduction of water into a continuous oil phase. The solution containing the w/o droplets was then fed into a second continuous water phase, resulting in the formation of w/o/w droplets. Employing glass devices, selected due to the hydrophilicity of the surface, the authors were able to obtain double emulsions with an average size of 52 μm diameter, with a CV = 2.7%. Using the same technique, the authors were able to obtain a high degree of fluidic control which enabled them to prepare droplets containing a controlled number of droplets; ranging from 1 to 7. By introducing various chemical components into each of the phases, the authors propose that the technique could be used for screening or analyte encapsulation; if the droplets were polymerized *in situ*.

Subsequently, Zhao and Middleberg [12] reported the mass-transfer-controlled formation of complex multiple emulsions, also demonstrating the preparation of uniform w/o/w double emulsions.

Chemically Driven Assembly of Colloid Particles: In another example of the unique reaction conditions attainable within microfluidic devices, Kumacheva and coworkers [13] evaluated the assembly of colloids at the interfaces formed between a gas and a liquid. At a T-mixer, a particle dispersion, containing poly(styrene-*co*-acrylic acid) particles, met with CO_2 to form gas slugs within a liquid continuous phase. As the particles aligned at the interface, this afforded "armored" bubbles with a narrow size distribution. Imaging of the microtomed bubbles by SEM confirmed that the particles had formed a monolayer thick shell which stabilized the bubbles toward coalescence and disproportionation. The generality of the technique was illustrated for a series of anionic particles, including carboxylated silica and silica nanoparticles loaded with CdZe/ZnS core–shell quantum dots. The authors concluded that the technique could find application in the formation of high-quality stable bulk materials owing to the excellent control of bubble size.

6.1.1 Polymerization of Droplets under Flow

When conducted using conventional methodology, emulsification in the bulk solution is based on two competing properties, drop breakage and droplet coalescence, with the latter giving rise to polydispersity. In comparison, when such processes are performed under continuous flow whereby emulsions are formed by shearing one liquid into a second immiscible liquid phase, regular and reproducible droplet sizes, shapes and compositions are obtained. With this in mind, possibly one of the more commercial applications of droplet generation under continuous flow is their *in situ* polymerization to form highly regular polymeric beads for use in chromatographic and or solid-support roles [14].

Droplet Polymerization: An early example of the production of polymer beads using droplet polymerization was reported by Nisisako et al. [15] as an alternative to the conventional approach of suspension polymerization; where fractionation is required to obtain a narrow particle size distribution. Following extensive research into the area, the authors developed a technique for the preparation of monodisperse droplets using a microfluidic system, with polymerization performed offline in the collection vessel. Throughout these investigations, the authors employed quartz glass devices with droplets formed at either T- or Y-shaped junctions. Employing 1,6-hexanediol diacrylate, and a photoinitiator Darocur 1173, as the organic phase and aqueous PVA (2 wt.%) as the continuous phase, the effect of flow speed on droplet size was evaluated; enabling a linear correlation between aqueous flow speed and monomer droplet size to be identified. Compared to standard batch techniques where polydisperse particles are sorted postproduction, this technique is advantageous as monodisperse particles (CV = <2%) can be prepared with sizes ranging from 30 to 120 μm depending on the flow rates employed. As demonstrated with the projection photolithography techniques, bifunctional droplets can also be prepared using this methodology by simply employing two different monomer solutions; again with excellent particle size uniformity (Figure 6.3).

Multistep Polymer Bead Formation: Kumacheva and coworkers [16] subsequently reported the application of an "internal trigger" approach for multistep microfluidic polymerization reactions performed in droplets. With the hypothesis being that the heat generated from an exothermic free radical polymerization of an acrylate would be sufficient to trigger the polycondensation of a urethane oligomer, the authors investigated the continuous synthesis of polymer particles with an interpenetrating polymer structure. In addition to a recognized control over particle size distribution, the authors cited the motivation for the use of a microfluidic technique as being the speed with which monomer compositions could be evaluated.

With this in mind, the authors investigated the generation of droplets comprising tri(propylene glycol) diacrylate, poly(propylene glycol)toluene-2,4-diisocyanate,

The Use of Microfluidic Devices

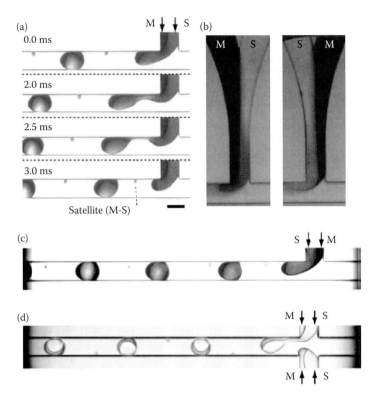

FIGURE 6.3 Illustration of (a) the formation of bifunctional droplets and the manipulation of droplet types by varying input position (b, c, and d) to afford bifunctional microparticles under continuous flow conditions.

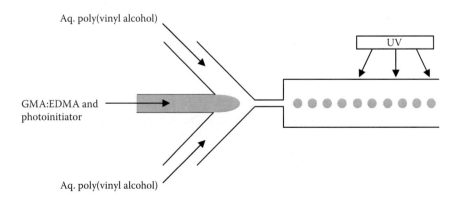

FIGURE 6.4 Schematic illustrating the microfluidic droplet generator (reactor dimensions = 10.2 cm × 7.6 cm, channel dimension = 150 μm (deep)) used to synthesize macroporous polymer beads.

diethanolamine, and a photoinitiator; 2-diethoxyacetophenone (Figure 6.4). Upon photoirradiation of the droplets the acrylate underwent polymerization and released sufficient heat (55–86 kJ mol^{-1}) to induce polyaddition and polycondensation reactions to take place within the droplet to yield the target polyTPGDA-polyurethane particles. Using poly(vinylalcohol) as the continuous phase, the authors were able to generate a series of spherical particles (90–130 nm) with a range of compositions and polydispersities of less that 3%. Using confocal microscopy, the authors examined the internal structure of the beads prepared which verified the uniform distribution of poly(propylene glycol)toluene-2,4-diisocyanate, and poly urethane.

Molecular Imprinting under Flow Conditions: Using acrylate-based formulations, Roeseling et al. [17] investigated the ability to prepare molecularly imprinted polymer beads with the aim of generation particles which could be used for the detection of explosives.

Basing their methodology on the formation of oil-in-water emulsions, generated via a process of segmented flow, whereby the "oil" was a functional monomer, cross-linking agent, porogen, and photoinitiator, the authors evaluated the *in situ* curing of the droplets using UV irradiation. In order to create a stable emulsion prior to photo-curing, the authors added a nonionic surfactant which prevented coalescence of the droplets. Using a flow focusing device, the authors evaluated the conditions required to generate droplets of the target size prior to incorporating a template into the oil phase, to generate the molecularly imprinted beads.

With the detection of explosives as their focus, the authors used trinitrotoluene (TNT) as the template to afford polymeric beads with TNT-selective sites. Using the aforementioned protocol, the authors were able to tune the bead size to afford monodisperse particles with sizes ranging from 15 to 140 µm at a throughput of 8 g h^{-1}. In order to develop an efficient sensing material, the authors found it necessary to increase the inner surface area of the beads, and this was achieved by varying the proportion of porogen used, resulting in the formation of beads with a surface area of 330 m^2 g^{-1} and an uptake of 0.2–0.3 ng of TNT per mg of polymer evaluated, representing an 8-fold increase in uptake compared to nonmolecularly imprinted beads.

6.2 PREPARATION OF INORGANIC NANOPARTICLES UNDER CONTINUOUS PROCESSING CONDITIONS

In 2004, Yang and coworkers [18] reported the development of a technique for the synthesis of silver nanoparticles via the thermal reduction of silver pentafluoropropionate to isoamyl ether and silver nanoparticles. Due to the uniform thermal regimes applied within continuous flow reactors, the authors were able to produce Ag nanoparticles with a narrow size distribution, with diameters ranging from 7.4 to 8.6 nm depending on the temperature used (100–140°C). In the same year, Wagner et al. [19] reported the use of a borohydride reduction reaction to generate gold, silver, and copper nanoparticles under flow conditions. Employing a modular micro reactor comprising micromixers and residence time units, the authors investigated the effect of flow rate on the particle size formation, observing that between 0.2 and 4 mL min^{-1}, the flow rate does not affect the size of the particles formed. They did however observe a link between solution concentration and size, identifying that larger particles were formed when dilute solutions were employed. In other examples, researchers have employed a citrate reduction [20] for the preparation of Au nanoparticles, an organometallic decomposition in scCO$_2$ for Pd nanoparticle preparation [21], while others have reported the *in situ* thiol-functionalization of Au nanoparticles [22].

Superparamagnetic Nanoparticles: Due to the diverse array of applications of superparamagnetic (Fe$_3$O$_4$) nanoparticles; such as magnetic resonance imaging, magnetic separation, and targeted drug delivery, Raston and coworkers [23] investigated the formation of nanoparticles smaller than 20 nm. Using a spinning disk reactor, as a means of generating rapid mixing under plug flow conditions, the authors were able to generate a process which afforded the target nanoparticles with low solvent wastage. The procedure used consisted of treating a thin film of Fe$^{2+/3+}$ with gaseous ammonia at room temperature. Using rotating disk speeds of 500–2500 rpm, the authors were able to obtain the target nanoparticles with a narrow size distribution over a tunable range of 5–10 nm; resulting in a material with high-saturation magnetizations (68–78 emu g^{-1}). In the same year, Hassan et al. [24] reported the fabrication of a device suited to the continuous flow synthesis of iron oxide nanoparticles. The authors focused on the development of coaxial flow in a millichannel which enabled them to react a solution of FeCl$_3$ and FeCl$_2 \cdot$ H$_2$O, in dilute and degassed HCl, to form Fe$_3$O$_4$ nanoparticles with an average size of 7 nm.

CdSe Nanoparticles: Using a Y-shaped micro reactor, de Mello and coworkers [25] developed a technique for the controlled synthesis of fluorescent CdSe nanoparticles, formed via a liquid-phase reaction between CdO and Se. Investigating the effect of reaction time, temperature (160–255°C) and Cd/Se ratio the authors were able to rapidly tune the processing conditions to alter the emission characteristics of the materials synthesized. In addition, Jensen and coworkers [26] demonstrated the use of gas–liquid segmented flow as a means of developing a continuous flow process for the formation of CdSe nanoparticles. By changing the ratio of Se/Cd, they too were able to demonstrate tuning of the band-gap and hence the fluorescent properties of the resulting materials. More recently, Moffitt and coworkers [27] described the controlled self-assembly of polymer-stabilized quantum dots using a two-phase segmented microfluidic reactor.

Droplet Generation of CdS Nanoparticles: Combining droplet generation and nanoparticle synthesis, Lee and coworkers [28] developed a technique based on controlled dynamic droplet fusion to enable the formation of CdS nanoparticles. Employing a PDMS device, the authors introduced solutions of Cd(NO$_3$) $\cdot$ 4H$_2$O (1 × 10^{-4} M) and Na$_2$S $\cdot$ 9H$_2$O (1 × 10^{-4} M) into a continuous phase of silicone oil; upon fusion of the two droplets, rapid mixing occurred and formed a super-saturated solution of CdS. Analysis of the resulting particles spectrophotometrically illustrated a blue shift compared to batch-prepared particles, indicating a smaller particle size.

CaCO$_3$ Nanoparticles: In addition to demonstrating the preparation of metallic nanoparticles under flow conditions, Luo and coworkers [29] recently demonstrated the preparation of calcium carbonate nanoparticles. Using the reaction of calcium hydroxide and CO$_2$ within a membrane and membrane-less reactor, the authors were able to control the particle size of the resulting calcium carbonate nanoparticles. The mixing performance obtained within the membrane reactor was found to be significantly enhanced when compared with a membrane-less reactor, enabling the formation of particles with a narrow size distribution and tunable particles sizes of 34–110 nm; representing a method suitable for industrialization, based on the demand for calcium carbonate.

BaSO$_4$ Precipitation: Employing a T-shaped silicon micromixer, Kockmann et al. [30] described an investigation into the development of a reactive particle precipitation technique for the controlled formation of barium sulfate nanoparticles. To evaluate the formation of BaSO$_4$, the authors employed a solution of sulfuric acid (0.33 M) and barium chloride (0.5 M) which upon mixing afforded the target BaSO$_4$. Investigating the effect of flow rate, the authors identified a clear link between particle diameter/size distribution with the Reynolds number; observing 100 nm crystals, with a narrow size distributions at $R_e > 500$. The particle nucleation was subsequently modeled with classic thermodynamic nucleation theory and the mixer used to investigate an azo-coupling under industrial conditions, which enabled them to design and develop a mixing device that was not prone to clogging or fouling.

Concomitantly, Chen and coworkers [31] described the development of a flow process for the synthesis of monodisperse BaSO$_4$ nanocrystals and extended their investigation to incorporate the neutralization of NaAlO$_2$ and Al$_2$(SO$_4$)$_3$, monitoring the effect of stoichiometry on the phase of alumina generated. Chen and coworkers [32] subsequently demonstrated the development of a process-intensified method for the production of BaSO$_4$ at a throughput of 9 L h^{-1}, affording an average particle size of 37 nm.

Silver Halide Production: Using a micro reactor based on annular microsegments, Nagasawa and Mae [33] demonstrated the development of a technique suited to the stable, continuous production of silver halide crystals with a high degree of control over particle size, distribution, and shape. In addition, a concept is given for achieving large-scale production with this methodology.

Zirconia Nanoparticles: Employing a continuous flow microwave reactor, Bondioli et al. [34] described the development of a protocol for the synthesis of monodisperse spherical nanoparticles of zirconia synthesized upon hydrolysis, and condensation, of tetra-*n*-propylzirconate. Investigating flow rates between 50 and 100 mL min^{-1}, the authors were able to identify the optimum conditions for the formation of particles with an average size of 100 nm.

In addition to the examples provided, numerous researchers have investigated the processing advantages associated with the production of nanoparticles under a continuous flow reactor, all the results of which point toward this becoming an industrially viable technique for the large-scale production of synthetically useful nanoparticles.

6.3 FORMATION OF ORGANIC PARTICLES WITHIN CONTINUOUS FLOW DEVICES

Owing to a growing array of synthetic applications that demand the use of monodisperse polymeric microspheres and particles, such as chromatography, particle image velocimetry, microparticle displays, and biopolymer preparation [35], many researchers have begun to investigate the use of microflow reactors for the synthesis of polymeric particles with a narrow size distribution. Serra and Chang [36] recently reviewed the area of microfluidic assisted polymer particle synthesis, differentiating the two techniques for the production of polymer particles as either (a) direct projection polymerization, or the more prevalent (b) the polymerization of immiscible monomer droplets.

Projection Photolithography: This is a technique which utilizes UV irradiation, through a microscope, to polymerize a polymer solution within a microchannel. Through the use of masks, placed within the microscope objective, the desired polymer shape can be printed within the fluid. The technique is however limited to polymers which polymerize rapidly, to prevent shape deformation, and the presence of a thin layer of oxygen at the fluid/wall boundary to prevent local polymerization from occurring. As such, this type of device is fabricated from PDMS, which is oxygen permeable, and enables the formation of such a barrier. An early example of this methodology was reported by Doyle and coworkers [37], whereby a series of mask patterns were used to polymerize an acrylate oligomer solution (PEG-DA) containing 5 v% Darocur 1173 (photoinitiator) within a rectangular PDMS microchannel (channel dimensions = 200,600, or 100 μm (wide) × 9, 6, 20, or 38 μm (deep)). Using this approach, the authors were able to obtain a spatial resolution of 3 μm, providing access to a range of exotic particle shapes, as depicted in Figure 6.5.

The authors subsequently extended their investigation to enable the formation of mixed polymer structures, by employing a Y-shaped mixer at which streams of a hydrophobic monomer (tri(methylol propane) triacrylate) and a hydrophilic monomer

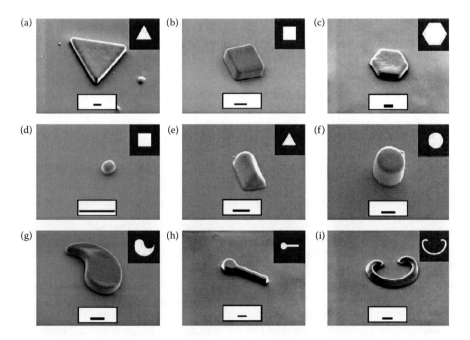

FIGURE 6.5 SEM images of a series of different shape (a–i) particles synthesized using projection photolithography. (Reproduced by permission from Macmillan Publishers Ltd. *Nat. Mater.* Dendukuri, D. et al. 2006. 5: 365–369. Copyright (2006).)

(65% PEG-DA in H_2O) met and coflowed [38]. Both solutions contained a photo initiator (Darocur 1173) and so when irradiated by UV, particles with a hydrophobic head and hydrophilic tail were synthesized. The amphiphilic nature of the particles enabled self-assembly; in water the particles afforded micelle-type structures and in emulsions, the particles migrated to the oil–water interface.

More recently, the authors developed a technique that employed hydrodynamic focusing lithography for the preparation of anisotropic multifunctional particles, for use as diagnostic tools for biomolecular screening [39].

Unlike flow lithography, the technique of hydrodynamic focusing lithography enables the fabrication of multifunctional particles with chemical anisotropy in the channel height domain. This is achieved by creating stacked monomer flows, in the z- and y-directions, and irradiating with UV perpendicular to the fluid interface, leading to the formation of more complex structures (2D arrays) than those previously obtained (1D arrays).

The device used is illustrated in Figure 6.6, and comprised a PDMS and glass device containing a narrow channel (channel dimensions = 40 μm (wide) × 40 μm (deep)) in which the monomer streams were stacked, and a wider channel (channel dimensions = 1 mm (wide) × 40 μm (deep)) where the particles were synthesized upon UV irradiation. Using this approach, the authors demonstrated the synthesis of bifunctional triangular PEG particles comprising an upper layer containing rhodamine acrylate and a bottom layer of 200 nm green fluorescent beads, finding that the relative thickness of each chemical region could be controlled by adjusting the ratio of inlet pressures or inlet flow rates. Under the aforementioned conditions and polymerization times of 50–300 ms, the authors were able to tune the height of a single layer between 32 and 4.5 μm, with high reproducibility. The technique was also illustrated for particles containing three layers, cross-shaped particles with dual-axis functionality and particles patterned with proteins based on surface functionality. This combination of flow stacking and microparticle synthesis therefore increases the complexity of particles that can be generated under continuous flow, opening up new possibilities in drug delivery, imaging, and 3D electrical circuitry.

In addition to enabling the fabrication of more complex particle structures, the technique is also amenable to the high-throughput production of such particles, unlike previous techniques such as projection photolithography.

Droplet-Based Nonspherical Particles: In addition to the synthesis of spherical polymer particles using continuous flow methodology discussed previously, Nisisako and coworkers [40,41] more recently demonstrated the production of biconcave polymer microparticles from triphasic emulsion droplets.

Using a shear-rupturing mechanism, droplets of uniform sizes were formed in a glass reactor, by employing 1,6-hexanediol diacrylate and silicone oil as the photocurable and noncurable organic phases and an aqueous solution of sodium dodecyl sulfate (SDS) (0.3 wt.%) as the continuous phase. Once again, droplet size was varied as a function of aqueous phase flow rate and at equilibrium the resulting droplets has a nonspherical shape comprising the curable monomer sandwiched between the noncurable organic; affording two curved internal interfaces. The droplets were collected offline where they underwent polymerization and washing to remove the noncurable organic phase.

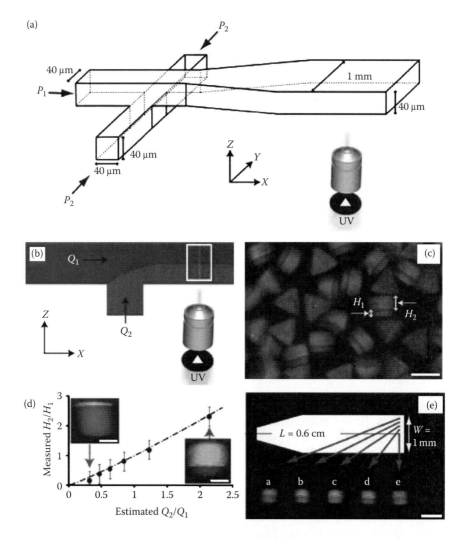

FIGURE 6.6 Schematic illustrating (a) the microfluidic setup used for the hydrodynamic focusing of polymer particles, (b) side view of flow focusing, (c) a fluorescent image of triangular particles, (d) a comparison of region thickness and flow ratio, and (e) uniformity of Janus particles synthesized at positions across the channel. (Bong, K. W. et al. Hydrodynamic focusing lithography. *Angew. Chem. Int. Ed.* 2009. 49(1): 87–90. Copyright Wiley-VCH Verlag GmbH & Co. KGaA. Reproduced with permission.)

By balancing the interfacial tension of the three phases, the authors were able to model and alter the resulting particle shape; however, at this stage, reproducible particles were not obtained; this is currently the focus of further studies based on *in-situ* photopolymerization.

Once the current challenges are overcome, the authors propose that particles of this type could potentially be applied as microoptic lenses, due to the tunable nature of the technique or components in paints, cosmetics, and clinical diagnostic tools.

6.4 THE USE OF MICRO REACTORS FOR THE POSTSYNTHETIC MANIPULATION OF ORGANIC COMPOUNDS

As observed with inorganic nanoparticles, the change in size and polydispersity can have a dramatic effect on the physical properties of the resulting material. With this in mind, several researchers have investigated the use of flow reactors as a means of modifying the physical properties of pigments, salts, and APIs.

In an attempt to reduce the polydispersity of their pigment, PR-254 **1**, researchers at Fujifilm Corporation (Japan) evaluated the use of an inhouse fabricated micro reactor system for the formation of nanopigment dispersion via a procedure of rapid heating and cooling.

Owing to the strong intermolecular hydrogen bonding of the solid PR-254 **1**, the material could not be dissolved even in high-polarity solvents. Consequently, in order to control particle size via precipitation, a latent pigment **2** must be used (Scheme 6.1) as a precursor to the target compound **1**. Using this approach, the latent pigment **2** is dissolved in a solvent, 1-methoxy-2-propanol acetate, which upon heating (150–180°C) decomposes to afford the target compound PR-254 **1**. In order to control nanoparticle size, the authors constructed a micro reactor capable of being heated to 325°C and rapidly cooled to −50°C. Using the aforementioned reactor, the latent pigment **2** was preheated to 100°C prior to mixing with the solvent at 325°C, to induce decomposition, followed by thermal quenching after 150 ms. Analysis of the resulting solution by laser scattering confirmed the formation of a clear solution containing 40 nm PR-254 **1** particles. Further investigations into this methodology enabled the researchers to tune the particle size based on the viscosity of the solvent employed, that is, 1000 cP afforded an average particle size of 17 nm.

Pharmaceuticals: The formation of nanocrystals of pharmaceutical agents is of current interest to researchers in a drive toward increasing the bioavailability of sparingly soluble drugs. Consequently, methods for the nanoprecipitation of pharmaceuticals are sought and the use of microchannel technology is one technique under investigation.

In 2007, Chen and coworkers [42] demonstrated the controlled precipitation of a hydrophobic pharmaceutical, Danazol **3** (Figure 6.7), within a micro reactor (dimensions = 300 μm (wide) × 300 μm (deep), reaction channel = 600 μm (wide) × 300 μm (deep)). Mixing a compound solution (S) with an antisolvent (AS), the authors investigated the effect of AS/S ratio on the resulting particle size and **3** solubility. In practice,

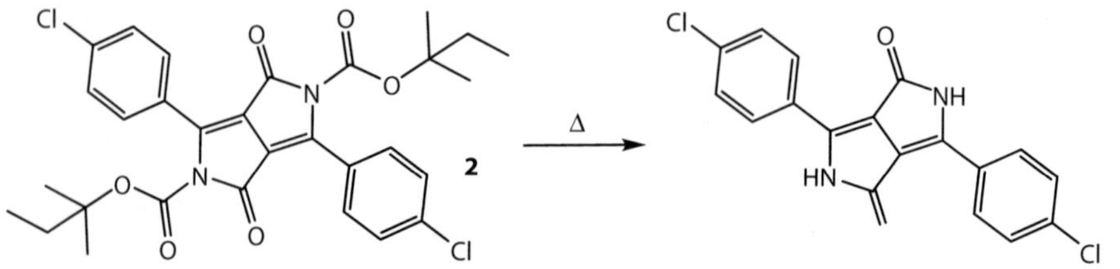

SCHEME 6.1 Illustration of the latent pigment 2 used as a precursor for PR-254 1.

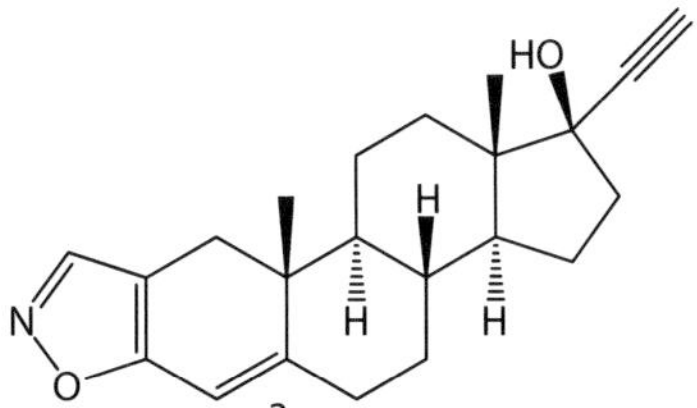

FIGURE 6.7 Danazol **3**, the hydrophobic pharmaceutical which underwent precipitation in a micro reactor.

this was achieved by dissolving the active ingredient **3**, in EtOH and employing various flow rates of deionized water (antisolvent) to afford a range of antisolvent/solution (AS/S) ratios. Upon contact of the two liquid streams, a precipitation of the active ingredient **3** was observed and the compound isolated upon filtration of the effluent (0.45 μm).

As Table 6.1 illustrates, decreasing S flow rate had a dramatic effect on Danazol **3** particle size affording an average reduction from 55 μm (20–120 μm) obtained in batch to 505 nm under continuous flow. Furthermore, reducing the reactor temperature to 4°C afforded a decrease in particle size to 364 nm and a dramatic reduction in particle size distribution. In all cases, the regularity and particle size distribution was greatly improved under continuous flow; however, the morphology of the Danazol **3** remained the same. Comparison of the dissolution profiles between the nanoparticles and the raw Danazol **3** illustrated 100% dissolution of the nanoparticles in <5 min compared with only 35% of the raw material **3** (complete dissolution = 40 min). Analysis of the active ingredient **3** by FT-IR and XRD confirmed that

TABLE 6.1
Illustration of the Effect of Antisolvent/Solution (AS/S) Ratio on the Particle Size and Concentration of Solubilized Danzol 3

AS/S Ratio (v/v)	Solubility of 3 (μg mL^{-1})	Average Particle Size (nm)
0	35,000.00	N/A[b]
1	1336.03	N/A[b]
2	79.30	N/A[b]
5	4.40	1250
10	2.83	900
20	2.16	510
40	1.63	505
20[a]	1.63	364

[a] Reactor temperature 4°C.
[b] No precipitation observed.

the physical characteristics of the compound were not affected by the antisolvent-based nanoization process. This enabled the authors to conclude that liquid antisolvent precipitation (LASP) within micro reactors is a facile method for controlling, and reducing, the particle size of hydrophobic drug molecules, thus presenting a facile method for improving the dissolution of sparingly soluble compounds and hence increasing bioavailability.

Subsequently, Cheng and coworkers [43] investigated the mixing properties within microdroplets as an environment for the controlled nanoparticles formation of pharmaceuticals; utilizing curcumin **4** as a model system (Figure 6.8). Using this approach, the authors found that curcumin **4** initially precipitated out as amorphous nanospheres (30–40 nm) which underwent aggregation (100–200 nm) before transforming into needle-shaped crystals. This work demonstrates a nonclassical crystallization pathway which has the potential to give rise to a new technique for the production of drug nanoparticles.

6.5 MIXED PARTICLE FORMATION

Employing PDMS microchannels, as depicted in Figure 6.9, Duraiswamy and Kahn [44] and Kahn and Duraiswamy [45] demonstrated the use of three-phase flow as a means of controlling the gold-shell thickness of metallodielectric core–shell nanoparticles. Using N_2 as the carrier gas and PDMS as the oil phase, functionalized silica, a reducing agent (hydroxylamine HCl) and the gold precursor (K_2CO_3 and tetrachloroaurate (III) trihydrate) were injected into the oil phase, where they mixed under chaotic advection to initiate shell growth. By manipulating the feedstock inlet speeds and concentrations, various shell thicknesses were obtained; as characterized by transmission electron microscopy (TEM). Further work is currently under investigation to develop this proof of concept work into a technique for the synthesis of nanoshells for applications in catalysis.

6.5.1 MICROENCAPSULATION OF ACTIVE PHARMACEUTICALS

An area of pharmaceutical science that has recently found interest in the use of microfluidics is that of nanomedicine, whereby the nanoprecipitation of biocompatible polymers and drugs has provided a novel vehicle for drug delivery. Current techniques which employ bulk mixing and subsequent precipitation of these materials however lack the required control; as it is thought that the size, shape, and homogeneity of such nanoparticles has an impact on the materials performance, improved methods have been sought for their mass production.

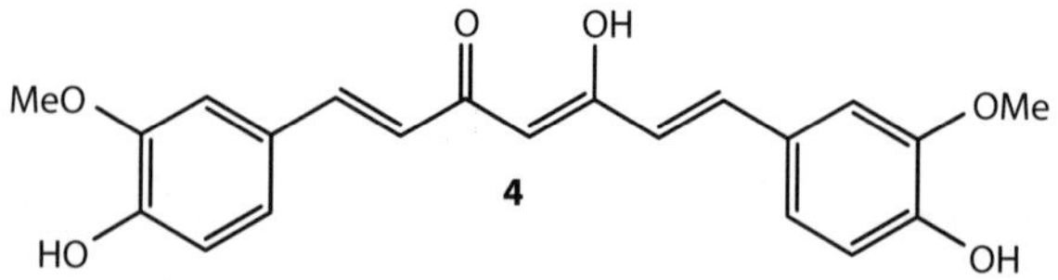

FIGURE 6.8 Illustration of curcumin **4** the principle bioactive curcuminoid found in the spice turmeric.

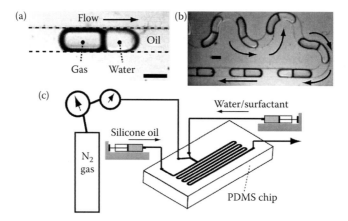

FIGURE 6.9 Schematic illustrating the technique used for the preparation of three-phase flow: (a) compound droplet comprising of a gas bubble and aqueous solution, (b) stable train of compound droplets passing through a microchannel, and (c) graphical representation of the micro reactor setup employed. (Kahn, S. A. and Duraiswamy, S. 2009. Microfluidic emulsions with dynamic compound drops, *Lab Chip* 9: 1840–1842. Reproduced by permission of the Royal Society of Chemistry.)

In 2008, Farokhzad and coworkers [46] evaluated the use of a microfluidic platform for the controlled nanoprecipitation of poly(lactic-co-glycolic acid)-b-poly(ethylene glycol) (PGLA-PEG) diblock copolymers as a model biomaterial for drug delivery; investigating the effect of parameters such as residence time, polymer composition, and polymer concentration on particle size. PLGA-PEG nanoparticles were selected as a model system for evaluation as they provide a biocompatible and biodegradable matrix which is suitable for the encapsulation and subsequent controlled release of pharmaceutical agents; typical sizes >150 nm.

Employing the PDMS micro reactor (microchannel dimensions = 20 μm (wide) × 60 μm (deep) × 10 mm (long)) illustrated in Figure 6.10, the authors utilized a technique known as hydrodynamic focusing as a means of reducing mixing time (0.04–0.4 ms) and increasing particle homogeneity. As Figure 6.10 illustrates, the use of coflowing water streams (10.0 μL min^{-1}) afford a narrow stream of PLGA$_{15K}$-PEG$_{3.4K}$ in MeCN (50 mg mL^{-1}, 0.5 μL min^{-1}), with offline analysis of the reaction mixture by TEM used to confirm the formation of PLGA-PEG nanoparticles (24 nm). To demonstrate the reproducibility of the technique, several experiments were conducted using the aforementioned conditions whereby an average particle size of 24 ± 1 nm was obtained; illustrating the robustness of the fluidic technique.

Having developed a technique capable of producing nanoparticles with a defined size, the authors investigated the effect of polymer composition on drug loading and release, evaluating Docetaxel **5** as a model therapeutic agent (Figure 6.11). Utilizing a flow rate of 100 μL min^{-1} for the aqueous stream and 5 μL min^{-1} for the polymer stream (25 mg mL^{-1} (containing 5 mg mL^{-1} Docetaxel **5**), the authors investigated the effect of polymer composition on the encapsulation and loading efficiency of **5**, whereby a solution containing 100% PLGA-PEG afforded 28% encapsulation and 4.1% drug **5**; this was however increased to 51% encapsulation and 6.8% loading when

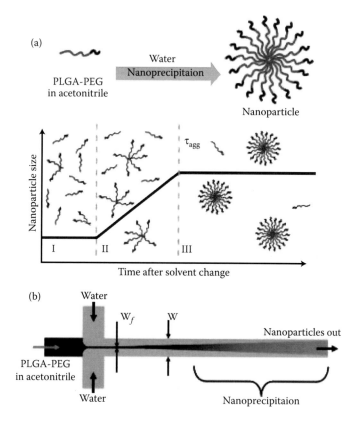

FIGURE 6.10 Schematic illustrating (a) the principle and (b) hydrodynamic focusing used to control the size of PLGA-PEG nanoparticles. (Reproduced with permission from Karnik, R. et al. 2008. *Nano Letters* 8(9): 2906–2912. Copyright (2008) American Chemical Society.)

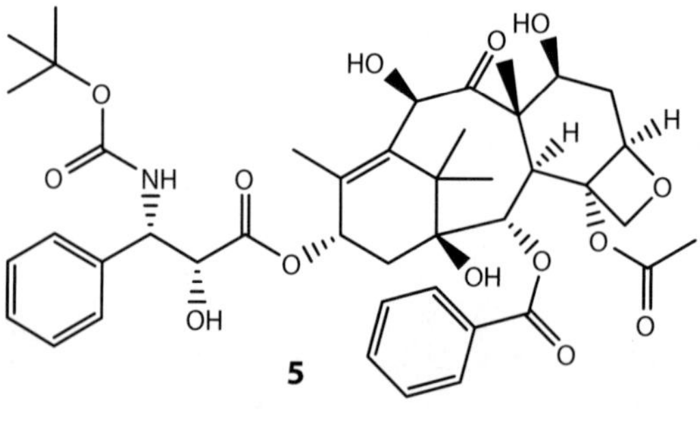

FIGURE 6.11 Docetaxel **5**, an antimitotic chemotherapy agent encapsulated in PLGA-PEG nanoparticles as a novel, controllable mode of drug delivery.

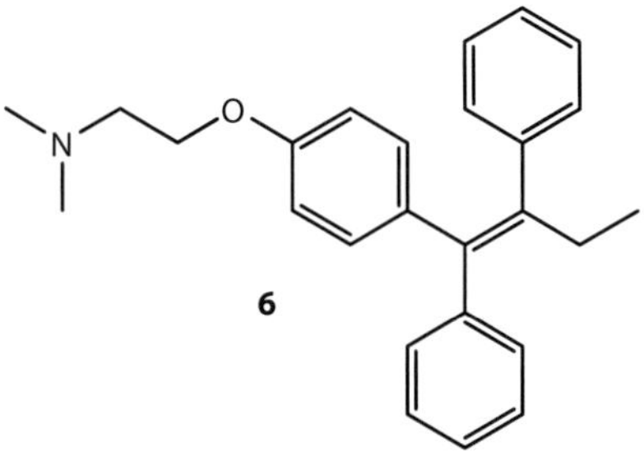

FIGURE 6.12 Tamoxifen **6**, the anticancer drug incorporated into a sophisticated drug delivery system prepared utilizing a continuous flow micro reactor.

20% free PLGA (hydrophobic) was introduced into the polymer solution. When comparing the release of **5** from the nanoparticles, the authors found that an increased drug release half-life of 19 h was obtained from the microfluidic nanoparticles *cf.* 11 h for batch-produced particles. Application of microfluidic technology therefore enabled the authors to develop a rapid and tunable method for the nanoprecipitation of homogeneous PLGA-PEG nanoparticles, which afforded increased control of size, surface characteristics, and drug loading *cf.* conventional synthetic protocols.

Tamoxifen Encapsulation: In an extension to the use of microfluidic technology for the development of novel drug carriers, Yang et al. [47] recently communicated a proof-of-concept investigation which focused on the encapsulation of CdTe quantum dots (QD), Fe_3O_4 nanoparticles, and Tamoxifen **6** (Figure 6.12) into polycaprolactone (PCL) microcapsules as a means of combining fluorescent imaging, magnetic targeting, and controlled drug release into a single system. Employing standard microfabrication techniques, the authors designed and fabricated a PDMS micro reactor containing a cross-junction, at which the microdroplets were formed (Figure 6.13a and b) and an observation channel (Figure 6.13c) through which the droplets were visually inspected. To generate the drug carriers described above, two different reactant solutions were introduced into the micro reactor, the first contained PCL, Tamoxifen **6**, CdTe QDs, and Fe_3O_4 nanoparticles. Using this approach, the authors investigated the relationship between flow velocities and the entrapped composition of the resulting capsules and size. Preparing capsules with sizes ranging from 50 to 200 μm, the authors subsequently investigated the release of Tamoxifen **6** from 60 μm capsules, observing 78% release in the first 24 h and 100% release after 2 days, with the release profile indicating an initial burst and then slow release, illustrating the techniques potential for the development of a smart drug delivery mechanism.

6.6 SUMMARY

From this section, it can be seen that the enhanced fluidic control obtained within microflow devices has the potential to revolutionize the way in which the materials of tomorrow are both researched and produced. The examples provided illustrate a diverse array of scientific fields that microfluidics has touched over the past decade, with an increasing number of applications being identified as interest in the technique grows.

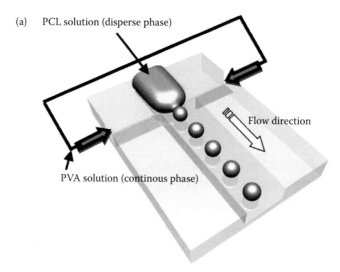

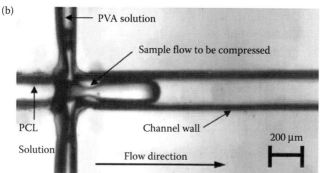

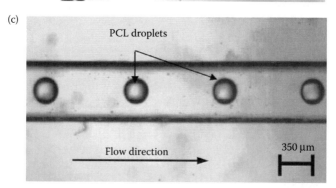

FIGURE 6.13 (a) Schematic illustrating the PDMS micro reactor used for the formation of PCL microcapsules, (b) photograph of the PCL droplet formation, and (c) illustration of PCL droplets containing Tamoxifen **6**, Fe_3O_4 nanoparticles and CdTe quantum dots. (Yang, C.-H., et al. 2009. Microfluidic assisted synthesis of multi-functional polycaprolactone microcapsules: Incorporation of CdTe quantum dots, Fe_3O_4 super-paramagnetic nanoparticles and tamoxifen anticancer drugs, *Lab Chip* 9: 961–965. Reproduced by permission of the Royal Society of Chemistry.)

REFERENCES

1. Nisisako, T. 2008. Microstructured devices for preparing controlled multiple emulsions, *Chem. Eng. J.* 31: 1091–1098.
2. Fukiu, K. B., Kobayashi, I., Uemura, K., and Nakajima, M. 2010. Temperature effect on microchannel emulsification: Generation characteristics of uniform oil droplets, *11th International Conference on Micro reaction Technology*, Kyoto, Japan, pp. 342–343.
3. Martin-Banderas, L., Flores-Mosquera, M., Riesco-Chueca, P., Rodriguez-Gil, A., Cebolla, A., Chavez, S., and Ganan-Calvo, A. M. 2005. Flow focusing: A versatile technology to produce size-controlled and specific-morphology microparticles, *Small* 1(7): 688–692.
4. Jahn, A., Reiner, J. E., Vreeland, W. N., DeVoe, D. L., Locascio, L. E., and Gaitan, M. 2008. Preparation of nanoparticles by continuous-flow microfluidics, *J. Nanopart. Res.* 10: 925–934.
5. Wang, J., Zhang, J., and Han, J. 2010. Synthesis of crystals and particles by crystallization and polymerization in droplet-based microfluidic devices, *Chem. Eng. China* 4: 26–36.
6. West, J., Manz, A., and Dittrich, P. S. 2008. Massively parallel production of lipid microstructures, *Lab Chip* 8: 1852–1855.
7. Ota, S., Yoshizawa, S., and Takeuchi, 2009. Microfluidic formation of monodisperse, cell-sized, and unilamellar vesicles, *Angew. Chem. Int. Ed.* 48: 6533–6537.
8. Zeng, S., Li, B., Su, X., Qin, J., and Lin, B. 2009. Microvalve-actuated precise control of individual droplets in microfluidics devices, *Lab Chip* 9: 1340–1343.
9. Churski, K., Korczyk, P., and Garstecki, P. 2010. High-throughput automated droplet microfluidic system for screening of reaction conditions, *Lab Chip* 10: 816–818.
10. Frenz, L., Blank, K., Brouzes, E., and Griffiths, A. D. 2009. Reliable microfluidic on-chip incubation of droplets in delay lines, *Lab Chip* 9: 1344–1348.
11. Nisisako, T., Okushima, S., and Torii, T. 2005. Controlled formation of monodisperse double emulsions in a multiple-phase microfluidic system, *Soft Matter* 1: 23–27.
12. Zhao, C.-X. and Middelberg, A. P. J. 2009. Microfluidic mass-transfer control for the simple formation of complex multiple emulsions, *Angew. Chem. Int. Ed.* 48: 7208–7211.
13. Park, J. I., Nie, Z., Kumachev, A., Abdelrahman, A. I., Binks, B. P., Stone, H. A., and Kumacheva, E. 2009. A microfluidic approach to chemically driven assembly of colloidal particles at gas–liquid interfaces, *Angew. Chem. Int. Ed.* 48: 5300–5304.
14. Serra, C. A. and Chang, Z. 2008. Microfluidic-assisted synthesis of polymer particles, *Chem. Eng. Technol.* 31: 1099–1115.
15. Nisisako, T., Torii, T., and Higuchi, T. 2004. Novel microreactors for functional polymer beads, *Chem. Eng. J.* 101: 23–29.
16. Li, W., Pham. H. H., Niew, Z., MacDonald, B., Guenther, A., and Kumacheva, 2008. Multi-step microfluidic polymerization reactions conducted in droplets: The internal trigger approach, *J. Am. Chem. Soc.* 130: 9935–9941.
17. Roeseling, D., Tuercke, T., Krause, H., and Loebbecke, S. 2009. Microreactor-based synthesis of molecularly imprinted polymer beads used for explosive detection, *Org. Proc. Res. Dev.* 13: 1007–1013.
18. Lin, X. Z., Terepka, A. D., and Yang. H. 2004. Synthesis of silver nanoparticles in a continuous flow tubular microreactor, *Nano Lett.* 4: 2227–2232.
19. Wagner, J., Tshikhudo, T. R., and Köhler, J. M. 2008. Microfluidic generation of metal nanoparticles by borohydride reduction, *Chem. Eng. J.* 135S: S104–S109.
20. Wagner, J., Kirner, T., Mayer, G., Albert, J., and Kohler, J. M. 2004. Generation of metal nanoparticles in a microchannel reactor, *Chem. Eng. J.* 101: 251–260.
21. Desportes, S., Tidona, B., and von Rohr, P. R. 2010. Palladium nanoparticles synthesis in a reactive two phase flow and under supercritical conditions in a microreactor, *11th International Conference on Micro reaction Technology*, Kyoto, Japan, 142–143.

22. Shalom, D., Wootton, R. C. R., Winkle, R. F., Cottam, B. F., Vilar, R., de Mello, A. J., and Wilde, C. P. 2007. Synthesis of thiol functionalized gold nanoparticles using a continuous flow microfluidic reactor, *Mater. Lett.* 61: 1146–1150.
23. Chin, S. F., Iyer, K. S., Raston, C. L., and Saunders, M. 2008. Size selective synthesis of superparamagnetic nanoparticles in thin fluids under continuous flow conditions, *Adv. Funct. Mater.* 18: 922–927.
24. Hassan, A. A., Sandre, O., Cabuil, V., and Tableing, P. 2008. Synthesis of iron oxide nanoparticles in a microfluidic device: Preliminary results in a coaxial flow millichannel, *Chem. Commun.* 1783–1785.
25. Krishnadasan, S., Brown, R. J. C., de Mello, A. J., and de Mello, J. C. 2007. Intelligent routes to the controlled synthesis of nanoparticles, *Lab Chip* 7: 1434–1441.
26. Yen, B. K. H., Günther, A., Schmidt, M. A., Jensen, K. F., and Bawendi, M. G. 2005. A microfabricated gas–liquid segmented flow reactor for high-temperature synthesis: The case of CdSe quantum dots, *Angew. Chem. Int. Ed.* 44: 5447–5451.
27. Schabas, G., Wang, C.-W., Oskooei, A., Yusuf, H., Moffitt, M. G., and Sinton, D. 2008. Formation and shear-induced processing of quantum dot colloidal assemblies in a multiphase microfluidic chip, *Langmuir* 24: 10596–10603.
28. Hung, L.-H., Choi, K. M., Tseng, W.-Y., Tan, Y.-C., Shea, K. J., and Lee, A. P. 2006. Alternatively droplet generation and controlled dynamic droplet fusion in microfluidic device for CdS nanoparticle synthesis, *Lab Chip* 6: 174–178.
29. Wang, K., Wang, Y. J., Chen, G. G., Luo, G. S., and Wang, J. D. 2007. Enhancement of mixing and mass transfer performance with a microstructure minireactor for controllable preparation of $CaCO_3$ nanoparticles, *Ind. Eng. Chem. Res.* 46: 6092–6098.
30. Kockmann, N., Kastner, J., and Woias, P. 2008. Reactive particle precipitation in liquid microchannel flow, *Chem. Eng. J.* 135S: S110–S116.
31. Ying, Y., Chen, G., Zhao, Y., Li, S., and Yuan, Q. 2008. A high-throughput methodology for continuous preparation of monodispersed nanocrystals in microfluidic reactors, *Chem. Eng. J.* 135: 209–215.
32. Wang, Q.-A., Wang, J.-X., Li, M., Shao, L., Chen., J.-F., Gu, L., and An, Y.-T. 2009. Large-scale preparation of barium sulfate nanoparticles in a high-throughput tube-in-tube microchannel reactor, *Chem. Eng. J.* 149: 473–478.
33. Nagasawa, H. and Mae, K. 2006. Development of a new microreactor based on annular microsegments for fine particle production, *Ind. Chem. Eng. Res.* 45: 2179–2186.
34. Bondioli, F., Corradi, A. B., Ferrari, A. M., and Leonelli, C. 2008. Synthesis of zirconia nanoparticles in a continuous-flow microwave reactor, *J. Am. Ceram. Soc.* 91: 3746–3748.
35. Amici, E., Tetradis-Meris, G., Pulido de Torres, C., and Jousse, F. 2008. Alginate gelation in microfluidic channels, *Food Hydrocolloids* 22: 97–104.
36. Serra, C. A. and Chang, Z. 2008. Microfluidic-assisted synthesis of polymer particles, *Chem. Eng. Technol.* 31(8): 1099–1115.
37. Dendukuri, D., Pregibon, D. C., Collins, J., Hatton, T. A., and Doyle, P. S. 2006. Continuous-flow lithography for high-throughput microparticle synthesis, *Nat. Mater.* 5: 365–369.
38. Dendukuri, D., Hatton, T. A., and Doyle, P. S. 2006. Synthesis and self-assembly of amphiphilic polymeric microparticles, *Langmuir* 23: 4669–4674.
39. Bong, K. W., Bong, K. T., Pregibon, D. C., and Doyle, P. S. 2009. Hydrodynamic focusing lithography, *Angew. Chem. Int. Ed.* 49(1): 87–90.
40. Nisisako, T., Ando, T., and Hatsuzawa, T. 2010. Microfluidic fabrication of biconcave polymer microparticles from triphasic emulsion droplets, *11th International Conference on Microreaction Technology*, Kyoto, Japan, 80–81.
41. Nisisako, T. and Hatsuzawa, T. 2010. A microfluidic cross-flowing emulsion generator for producing biphasic droplets and anisotropically shaped polymer particles, *Microfluid Nanofluid* 9: 427–437.

42. Zhao, H., Wang, J. X., Wang, Q. A., Chen, J. F., and Yun, J. 2007. Controlled liquid antisolvent precipitation of hydrophobic pharmaceutical nanoparticles in a microchannel reactor, *Ind. Eng. Chem. Res.* 46: 8229–8235.
43. Liu, Z., He, Y., Wang, W., Huang, Y., and Cheng, Y. 2010. Visualization of liquid–liquid mixing in micro-channels and micro-droplets for controlled nanoparticle formation of drugs, *11th International Conference on Microreaction Technology,* Kyoto, Japan, 264–265.
44. Duraiswamy, S. and Kahn, S. A. 2010. Continuous-flow synthesis of metallodielectric core–shell nanoparticles using three-phase microfluidics, *11th International Conference on Microreaction Technology*, Kyoto, Japan, 78–79.
45. Kahn, S. A. and Duraiswamy, S. 2009. Microfluidic emulsions with dynamic compound drops, *Lab Chip* 9: 1840–1842.
46. Karnik, R., Gu, F., Basto, P., Cannizzaro, C., Dean, L., Kyei-Manu, W., Langer, R., and Farokhzad, O. C. 2008. Microfluidic platform for controlled synthesis of polymeric nanoparticles, *Nano Letters* 8(9): 2906–2912.
47. Yang, C.-H., Huang, K.-S., Lin, Y.-S., Lu, K., Tzeng, C.-C., Wang, E.-C., Lin, C.-H., Hsu, W.-Y., and Chang, J.-Y. 2009. Microfluidic assisted synthesis of multi-functional polycaprolactone microcapsules: Incorporation of CdTe quantum dots, Fe_3O_4 superparamagnetic nanoparticles and tamoxifen anticancer drugs, *Lab Chip* 9: 961–965.

7 Industrial Interest in Micro Reaction Technology

With the scientific literature dominated in the years since 2005 with pharmaceutical and fine chemical applications of micro reaction technology [1], researchers have more recently begun to expand the types of materials produced using this technology to incorporate consumer goods, with examples in cream formulation, microencapsulation, and pigments being reported [2,3].

7.1 MRT IN PRODUCTION ENVIRONMENTS

From an industrial perspective, any new technology is only of interest to process chemists if it provides an economic or safety advantage. With this in mind, micro reaction technology has been embraced by industrial researchers as it offers both improvements in process safety while having the potential to reduce operating costs via isothermal operation, increased productivities, and reduced waste generation. In addition to these factors, the ability to readily scale synthetic protocols identified within the research laboratory all the way through to multitonne-scale production without the need for reoptimization of the underlying process is a particularly attractive prospect to an industry plagued by the expensive problem of process scalability [4].

Micro Reactor Parallelization: Utilizing the principle of numbering-up, Togashi et al. [5] demonstrated the parallelization of a quartz micro reactor (channel dimensions = 250 μm (wide) × 150 μm (deep)) housed within a corrosion-resistant Hastelloy-C276 holder, which provided connections to the reactant inlets and product outlet, fitted to PTFE tubes which acted as a residence time unit. Prior to demonstrating parallel operation, the authors evaluated the applicability of the reaction setup toward a series of chemical reactions, as summarized in Scheme 7.1, controlling reactor temperature via immersion in a thermostatted bath.

To perform a bromination reaction, two reactant solutions, 3,5-dimethylphenol **1** (0.82 M) and bromine **2** (0.82 M) in DCM were brought together in the micro reactor where they mixed and reacted (Scheme 7.1a). Employing residence times ranging from 3.6 to 17.8 s and reactor temperatures of 0–20°C, the authors optimized the formation of 4-bromo-3,5-dimethylphenol **3** to afford the target compound in 77.2% conversion compared to 56.1% in batch. In addition, the use of a flow reactor resulted in a decrease in the formation of the dibrominated by-product **4** from 31.9% in batch to 17.3% (at 20°C) (Table 7.1). In a second example, the authors investigated the

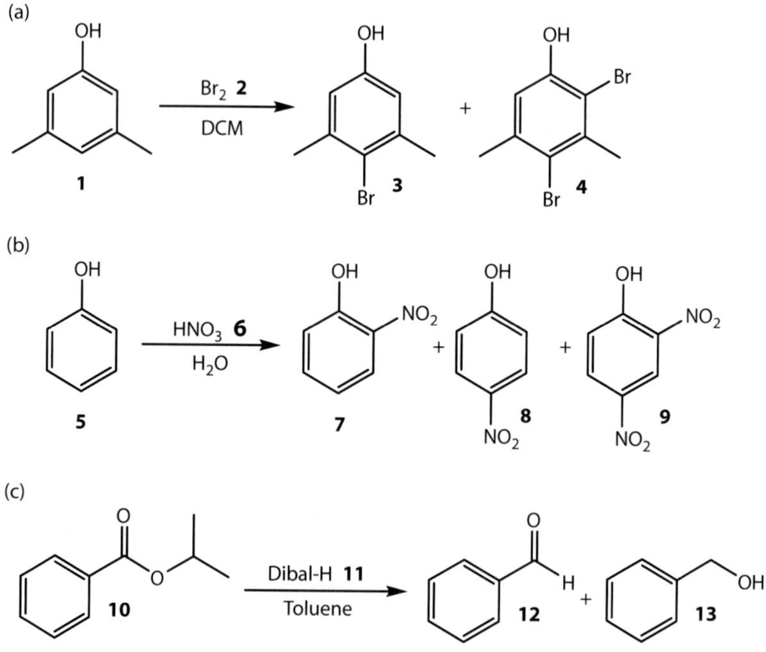

SCHEME 7.1 Schematic illustrating the model reactions used to evaluate the quartz micro reactors suitability for pilot-scale operation: (a) selective bromination, (b) aromatic nitration, and (c) ester reduction.

nitration of phenol **5** (0.9 M) and nitric acid **6** (15.8 M) in DI H_2O, evaluating reaction times of 5–256 s and reactor temperatures of 0–20°C. Using this approach, the authors identified 160 s, at 20°C, as the optimal condition affording o-nitrophenol **7** and p-nitrophenol **8** in 86.3% yield with only 2.3% 2,4-dinitrophenol **9** formed (Scheme 7.1).

In their final example of single reactor evaluation, the authors investigated the reduction of benzoic acid isopropyl **10** (0.1 M) using Dibal-H **11** (0.1 M) in

TABLE 7.1
Comparison of the Model Reactions Conducted in Batch with Those Performed in a Single Micro Reactor

	Bromination		Nitration		Hydride Reduction	
	Product 3 (%)	By-Product 4 (%)	Product 7/8 (%)	By-Product 9 (%)	Product 12 (%)	By-Product 13 (%)
Batch	56.1	31.9	77.0	7.7	25.2	27.0
Micro	77.2	17.3	86.3	2.3	38.1	29.2

Note: Product distribution determined by offline HPLC analysis.

TABLE 7.2
Comparison of the Yields Obtained for the Nitration of Phenol 5 Performed in a Batch Reactor, Single Micro Reactor, and Parallel Micro Reactor

Reactor Type	Mononitrated Products (7/8) (%)	2,4-Dinitrophenol 9 (%)
Batch	77.0	7.7
Single micro reactor	86.3	2.3
Pilot microplant	88.1	1.7

toluene, finding a residence time of 600 s afforded consumption of Dibal-H **11** and formation of benzaldehyde **12** and benzyl alcohol **13** in 38.1% and 29.2% yield, respectively.

Having confirmed the synthetic utility of their microfabricated reactor, the authors investigated the construction of a pilot plant (1.5 m (wide) × 0.9 m (long) × 1.5 m (high)) consisting of 20 single micro reactors; operated in four banks of five. Flow to the reactors was provided by means of two nonpulsatile pumps and each reactor had its own thermocouple to enable reactor temperature to be monitored. Once satisfied that efficient flow distribution was obtained (accuracy of ±3%), the system was subsequently evaluated chemically using the nitration of phenol **5**. Under the aforementioned conditions, the authors were able to increase the productivity of the system by a factor of 20; obtaining the target compounds **7** and **8** in comparable yield and purity (Table 7.2). Using this approach, the authors calculated that if operated continuously, the pilot plant would be capable of producing the mononitrated products in 72 tonne annum^{-1}.

In one of the few discussions within the open literature, Tekautz and Kirchneck [6] describe the incorporation of a micro reactor into an existing production plant (Microinnova, Austria), installed with the aim of doubling the current production capacity of two-step batch processes. Using a StartLam 3000 micro reactor (IMM, Germany), the authors discussed the successful implementation of the technology into the first reaction step, reporting the development of a process with a production capacity of 3.6 ton h^{-1}; which at the time of the chapter going to press had been in operation for a year with no signs of corrosion observed upon inspection after 10 months of continuous operation.

7.2 SYNTHESIS OF FINE CHEMICALS USING MICRO REACTORS

As the target end point of any chemical process development is ideally implementation into a production process, the ability to cost effectively and efficiently scale a process is of great interest to the fine chemicals industry. With this in mind, several industrial and academic researchers have evaluated the use of flow reactors for the production of g to kg-scale quantities of material; consequently, the following section describes a series of industrially relevant targets, selected to demonstrate the advantages of flow reactor technology to the fine chemicals industry.

7.2.1 SYNTHESIS OF CARBAMATES UNDER CONTINUOUS FLOW CONDITIONS

Bannwarth and coworkers [7] investigated the synthesis of (1R,2S,4S)-(7-oxabicyclo[2.2.1]hept-2-yl)-carbamic acid ethyl ester **14**, an intermediate to a receptor antagonist compound, from (1R,2S,4S)-7-oxabicyclo[2.2.1]heptane-2-carboxylic acid ethyl ester **15** via an exothermic hydrazide **16** formation (ΔH: -102 kJ mol^{-1}) and an acylazide intermediate **17** (ΔH: -599 kJ mol^{-1}), as depicted in Scheme 7.2. In addition to the exothermic nature of several reaction steps, the authors also found the instability of the acylazide **17** to be problematic and aimed to address these factors by performing the reactions in a series of single-step micro reactions; utilizing a CYTOS® Lab System (CPC, Germany), comprising a stainless-steel micromixer and residence time unit.

Initially focusing on the formation of the hydrazide **16**, the authors employed a biphasic reaction mixture comprising of ester **15** and hydrazine monohydrate **18** (1.1 eq.) and investigated the effect of reaction time (2–106 min) and temperature (88–93°C) on the formation of the desired hydrazide **16**. Using offline analysis by HPLC, the authors identified that a residence time of 106 min coupled with a reactor temperature of 93°C afforded the target hydrazide **16** in >96% conversion. Upon cooling to 0°C, the hydrazide **16** crystallized and was obtained in excellent yield and purity upon filtration.

When performed in batch, the subsequent acylazide **17** was generated using hydrochloric acid; however, due to the use of a stainless-steel reactor, an alternative was sought. Acidifying an aqueous solution of hydrazide **16** within the micro reactor, using 2 eq. of 85% phosphoric acid **19** at 2°C, for 2 min, followed by the addition of aqueous sodium nitrite **20** (10%) and reaction for an additional 2 min, the authors were gratified to obtain the azide **17** in yields ranging from 83% to 94%. After extraction into DCE, the azide **17** underwent a Curtius rearrangement, achieved via the recirculation of the reaction mixture through the micro reactor for a period of 20 h. Using this approach, the authors obtained the target urethane **14** in an isolated yield of 84%, upon evaporation of the reaction solvent.

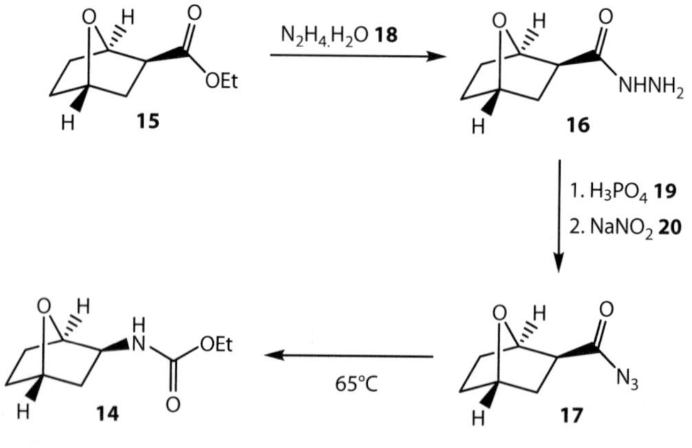

SCHEME 7.2 Illustration of the reaction sequence employed for the synthesis of (1R,2S,4S)-(7-oxabicyclo[2.2.1]hept-2-yl)-carbamic acid ethyl ester **14** under continuous flow conditions.

When compared with the previous batch protocol, the development of flow methodology for the three reaction steps enabled the authors to accelerate the hydrazide **16** formation dramatically, reducing the reaction time from 24 h to 106 min. In addition, the authors comment that this example illustrates the potential of micro reactor technology to perform highly exothermic transformations, affording safe handling and reaction of potentially hazardous compounds.

7.2.2 Production-Scale Synthesis of Ionic Liquids

Löwe and coworkers [8] demonstrated an extension to their previous investigations (see Chapter 3) by developing a system suitable for the on demand preparation of ionic liquids, focusing on those derived from 1-methylimidazole **21** (Table 7.3). The flow reactor used for this investigation comprised a stainless-steel interdigital micromixer (SIMM-V2) coupled to a stainless-steel capillary tube reactor (dimensions = 500 μm i.d., reactor volume = 10 or 30 mL) and the whole system was tempered within an oil bath (140°C). To synthesize an ionic liquid, the alkyl bromide under investigation and 1-methylimidazole **21** were delivered to the heated micromixer from separate inlets, using HPLC pumps, where they mixed and reacted prior to collection and analysis by ^{1}H NMR spectroscopy. The reaction time was varied as a function of the reactor volume and flow rate to afford residence times ranging from 5 to 10 min.

Using this approach, the authors were able to readily derive ionic liquids from commonly used alkyl halides such as 1-bromohexane **22**, 2-bromoethyl benzene **23**, and 1-bromo-3-phenylpropane **24**, along with reporting for the first time the imidazolium derivatives of 1-(3-bromopropyl)-4-nitrobenzene **25** and 1-(3-bromophenyl)-2-nitrobenzene **26**. As Table 7.3 illustrates, a higher system throughput is obtained for aliphatic alkylating agents (**22**) compared to aromatic derivatives; however, in all cases, the ionic liquids produced under continuous flow were of high purity and could be utilized without the need for further purification steps which is not the case for batch procedures.

Löwe et al. [9] subsequently reported the use of a micro reactor, comprising a microstructured stainless-steel plate (dimensions = 500 μm (wide) × 300 μm (deep) × 40 cm (long), channel volume = 600 μL), PEEK cover plate, and graphite gaskets (Figure 7.1), for the synthesis of the ionic liquid 1,3-dimethylimidazolium-triflate.

Although the reaction pathway appears simple, the *N*-alkylation of imidazole-type substrates are often fast and highly exothermic reactions; resulting in overheating of the ionic liquid which can subsequently lead to discoloration of the reaction product; consequently, specialized equipment is required in order to safely prepare such compounds in the desired high yield and purity. With this in mind, a large amount of process development has been conducted into the development of safe and efficient protocols for the preparation of ionic liquids.

As described earlier, while the efficient dissipation of the heat of reaction has been widely reported within microfabricated reaction systems, in this case the reaction was still found to proceed uncontrollably within a micro reactor leading to volatilization of the reactants with a reaction time of only 30 s. As a means of increasing not only the reaction control, but also the throughput of such exothermic reactions, the authors employed the microstructured reactor depicted in Figure 7.1, coupled

TABLE 7.3
Illustration of the Range of Ionic Liquids Synthesized under Continuous Flow, Providing Access to Ionic Liquids On-Demand

R	Imidazole 21 (mmol min^{-1})	Alkylating Agent (mmol min^{-1})	Product	T_g (°C)
C_6H_{13} **22**	12.5	12.5	hexyl-methylimidazolium bromide	−68.6
$(CH_2)_2Ph$ **23**	12.5	12.5	phenethyl-methylimidazolium bromide	−47.3
$(CH_2)_3Ph$ **24**	10.0	10.0	phenylpropyl-methylimidazolium bromide	−39.1
4-O_2N-C$_6$H$_4$-(CH$_2$)$_2$ **25**	7.5	7.5	4-nitrophenethyl-methylimidazolium bromide	−8.9
2-O_2N-C$_6$H$_4$-(CH$_2$)$_2$ **26**	7.5	7.5	2-nitrophenethyl-methylimidazolium bromide	−6.3

Note: All Reactions Conducted at 140°C.

with a heat pipe (Figure 7.2) and fan which increased the transfer of thermal energy (120 W) away from the reactor.

Using pressure-driven flow (HPLC pumps), the raw materials, 1-methylimidazole **21** and methyl triflate, were introduced into the reactor from separate inlets at flow rates selected to afford a 1:1 molar ratio, that is, 0.356 mL (**21**):0.250 mL (methyl triflate). The reactants were then mixed within the microstructured device using a caterpillar static micromixer [10], reacted and collected at the outlet prior to analysis by ^{1}H NMR spectroscopy. Due to the volatility of methyl triflate, product conversions were determined with respect to residual 1-methylimidazole **21** as solutions in $CDCl_3$ or d-DMSO. Using this approach, the authors were able to maintain the reactor temperature at 20°C while conducting the quantitative *N*-alkylation of 1-methylimidazole **21**, compared to a reactor temperature of 140°C in the absence of the heat pipe. Unlike the micro reactor alone, the heat pipe system enabled thermal control with liquid velocities of 20 mL min^{-1} (reaction time = 1.8 s). Therefore, by efficiently removing the heat of formation from the reactor, the system throughput could be increased along with the product quality. As such, optimization of the model reaction is currently underway, the results of which will be used to develop and automated pilot-scale system for the production of ionic liquids.

7.2.3 Scalable Technique for the Synthesis of Diarylethenes

Utilizing a stainless-steel micro reactor, comprising two T-mixers and tubular residence time units, Ushiolgi et al. [11] reported a facile method for the synthesis of

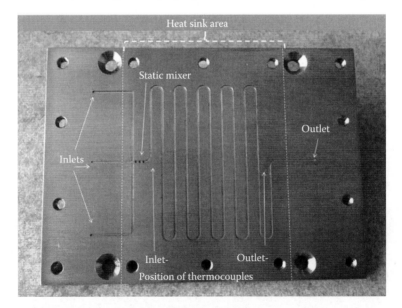

FIGURE 7.1 Illustration of the microstructured reactor plate (outer dimensions = 58 mm × 80 mm) and the positions of the thermocouples and heat sink. (Reproduced from *Chem. Eng. J.*, 155, Löwe, H. et al., Heat pipe controlled syntheses of ionic liquids in microstructured reactors, 548–550, Copyright (2009), with permission from Elsevier.)

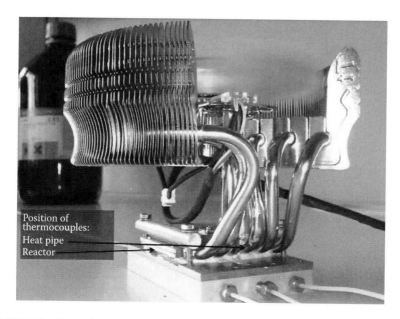

FIGURE 7.2 Illustration of the heat pipe system mounted on a microstructured reactor. (Reproduced from *Chem. Eng. J.*, 155, Löwe, H. et al., Heat pipe controlled syntheses of ionic liquids in microstructured reactors, 548–550, Copyright (2009), with permission from Elsevier.)

unsymmetrical diarylethenes, a process which is notoriously difficult to control under standard batch conditions even when performed at reduced reaction temperatures (−78°C). Focusing their initial investigations on identifying any effect of reactor temperature on reaction yield, the authors investigated the halogen–lithium exchange reaction between an aryl bromide (0.3 M, 7.5 mL min^{-1}) and *n*-butyllithium **27** (1.5 M, 1.5 mL min^{-1}) (Scheme 7.3a). Conventionally performed at −78°C to prevent decomposition of the aryllithium intermediate, the authors found that by performing the reaction under continuous flow, they were able to isolate the diarylethenes in yields ranging from 47% to 87% depending on the aryl bromide used.

Employing a convergent approach, the authors were subsequently able to adapt the flow process to enable the synthesis of unsymmetrical diarylethenes in moderate isolated yields by initially preparing the monoarylated product **28** in reactors 1 and 2, followed by the aryllithium intermediate **29** in a third reactor. By combining the output streams of reactors 2 and 3, the authors were able to synthesize the target unsymmetrical diarylethene **30** in an isolated yield of 53% (Table 7.4).

Owing to the different photochromic properties of such compounds, the synthetic methodology presented affords a facile route to the fine tuning of compounds physical properties, while providing a method capable of producing such materials on a large scale.

7.2.4 Continuous Flow Synthesis of Light-Emitting Materials

A subsequent example demonstrating the synthetic utility of lithium–halogen exchange reactions and their potential use for the production of industrially relevant

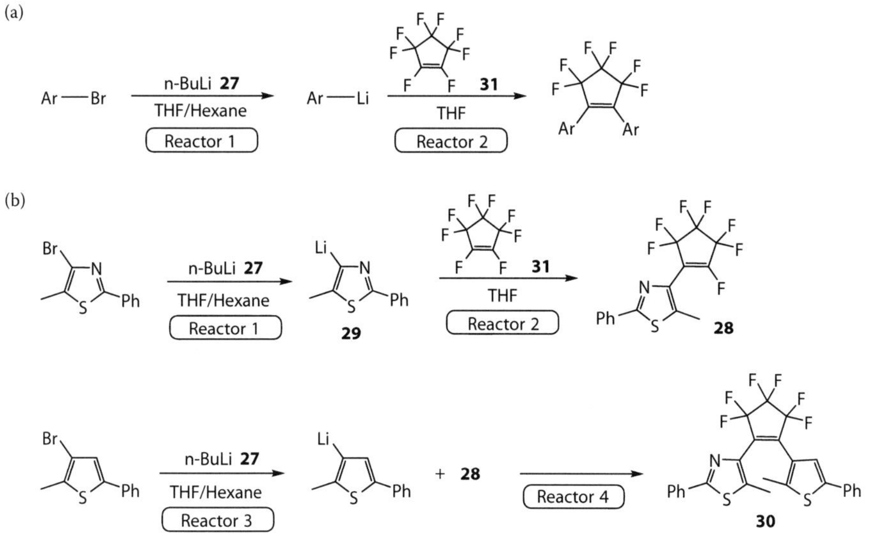

SCHEME 7.3 Illustration of the stepwise approach used for the synthesis of diarylethenes: (a) bis-alkylation and (b) selective, stepwise monoalkylation.

materials was recently reported by Song and coworkers [12]. Using a two-stage continuous micro reactor, consisting of two caterpillar split-recombine micromixers (CPSRMM R-840, IMM Germany) connected in series, the authors investigated the effect of reactant flow rate on the lithiation of bromonaphthalene **32** (0.2 M), to afford naphthyl lithium **33**, which subsequently underwent reaction

TABLE 7.4
Comparison of the Product Distributions Obtained as a Result of Performing the Arylation of Octafluorocyclopentene 31 in Batch and under Flow Conditions

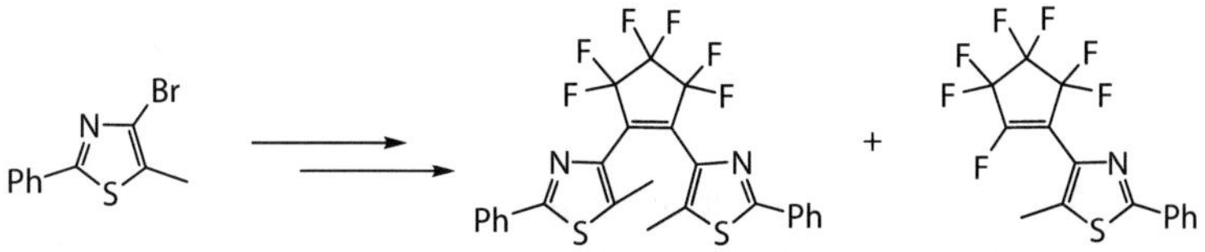

Reactor	Reaction Temperature (°C)	Isolated Yield (%)	
		Diarylethene	Monoarylethene
Batch	−78 (3 h)[a]	10	52
Micro	0 (9 s)[a]	51	11

[a] The values within parentheses represent the reaction times employed.

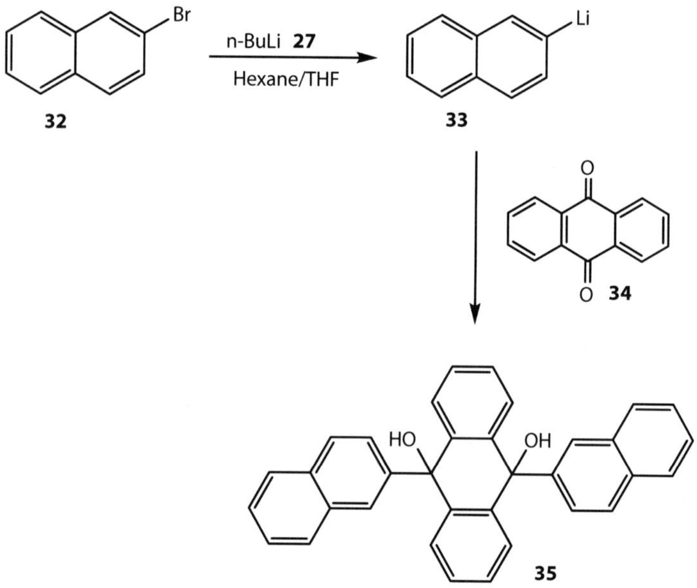

SCHEME 7.4 Synthetic protocol employed for the preparation of light-emitting diode materials.

with 9,10-anthraquinone **34** to afford the target light-emitting diode material **35**, as summarized in Scheme 7.4.

Due to the exothermic nature of the reaction, the reaction is conventionally performed at −78°C with dropwise addition of naphthyl lithium **33** to 9,10-anthraquinone **34** in order to reduce any temperature increase and avoid hot-spot formation, critical for ensuring reaction selectivity. Conducting the reaction under continuous flow, the authors were able to thermostat the reactor at −20°C and obtain the target compound **35** in 85% yield and 97% purity; as determined by HPLC analysis. Due to the excellent heat transfer and mixing properties of the micro reactor, the authors were able to perform the reaction at a throughput of 6 L h^{-1} compared to batch where it is reported to take 40 min to transfer 100 mL of the naphthyl lithium **33** intermediate into the second reactor, increasing to 4 h for a 5 L reaction volume.

7.2.5 2-(2,5-Dimethyl-1*H*-Pyrrol-1-yl)Ethanol Synthesis

As discussed throughout, in addition to the challenges associated with the manipulation of large volumes of hazardous chemicals, the performance of exothermic reactions can also be problematic when considering large-scale production, due to issues with efficient removal of heat from stirred reactor vessels. As a solution to this problem, Schwalbe et al. [13] demonstrated an early example of large-scale production using a continuous flow reactor, selecting the Paal–Knorr synthesis as a model reaction (Scheme 7.5). Using the CYTOS™ college system, the authors were able to react 2-aminoethanol **36** with a 1,4-diketone, acetonylacetone, **37** to afford

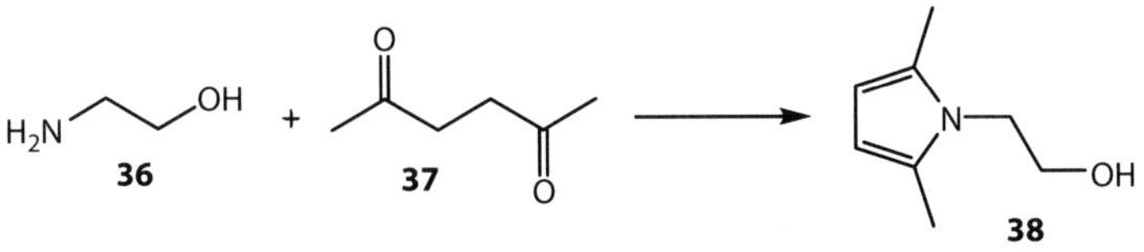

SCHEME 7.5 Illustration of the Paal–Knorr synthesis performed under continuous flow conditions.

2-(2,5-dimethyl-1H-pyrrol-l-yl)ethanol **38** in 91% yield. Employing a residence time of only 5.3 min, the authors were able to generate the target compound **38** with a throughput of 210 g h^{-1}, from a single stainless-steel tube reactor.

Using the X-cube Flash™ (ThalesNano Inc., Hungary), a stainless-steel micro reactor, capable of heating and pressurizing reactions up to 350°C and 200 bar, Darvas et al. [14] subsequently reported the Paal–Knorr reaction, obtaining 2-(2,5-dimethyl-1H-pyrrol-l-yl)ethanol **38** in 99% conversion and selectivity, when performing the reaction in acetic acid at 150°C (200 bar), with a residence time of 8 min. The reaction has more recently been demonstrated by Rutjes et al. [15] in glass-based micro reactors (Future Chemistry BV, The Netherlands) where the authors demonstrated the rapid scale-up from a single micro reactor to four meso reactors, accessing production volumes of 55 g h^{-1}.

7.2.6 Synthesis of Pigments under Flow Conditions

In 2004, Wille et al. [16] published an article disclosing developments made into the synthesis of pigments using a three-stage micro reactor pilot plant. Employing azo-chemistry (see Chapter 3), the authors reported the synthesis of a diazo-pigment within a laboratory-scale micro reactor. Using this approach, the authors were able to compare the quality of the pigment produced with that obtained using conventional batch methodology. In addition, the authors reported the successful transfer of their laboratory data from a single reactor (CPC, Germany) to an MRT-pilot plant, based on a numbering-up concept, obtaining a pigment of comparable quality, as illustrated in Table 7.5, in a process that took less than 1 week. Compared to batch

TABLE 7.5
Comparison of Coloristic Properties Obtained in the Across Lab and Pilot-Scale Scenarios Employing Standard Batch and Micro Reactors

Coloristic Property	Pigment Production Method				
	MRT-Lab	MRT-Pilot Plant	Lab-Scale	Pilot Plant	Production Plant
Color strength (%)	140	145	150	135	135
Color shade (dH)	5.5	5.8	6.0	5.5	4.6

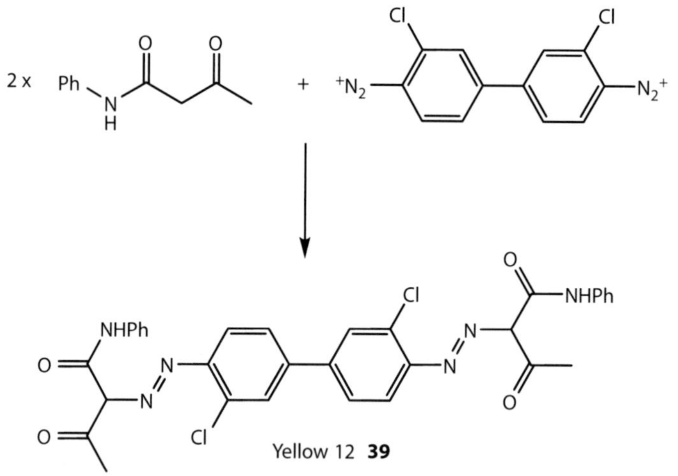

SCHEME 7.6 Illustration of the azo-coupling reaction used to synthesize the pigment, Yellow 12 **39**.

protocols, the authors were able to scale the flow process without observing a decrease in color strength and shade, unlike batch scale-up where reductions in coloristic properties are obtained with each increase in batch size.

In a separate investigation, Pennemann et al. [17] investigated the improvement of the dye properties of the azo pigment Yellow 12 **39** as a result of employing micromixer technology. In addition to reducing the pigment particle size and size distribution, conducting the azo-coupling (Scheme 7.6) within an IMM interdigital micromixer (Germany), the authors were able to improve pigment glossiness (73%) and transparency (66%) without altering the tinctorial power.

Owing to the current limitations associated with scale-up protocols employed for the production of pigments, it can be seen from both of these studies that MRT has vast potential for the production of high-quality pigments.

7.2.7 Production of Thermally Labile Compounds under Flow Conditions

When considering the up-scaling of a discovery route for the synthesis of NBI-75043 **40** (Scheme 7.7), a histamine receptor (H1) antagonist, Gross et al. [18] identified a scale limitation of a key reaction step—a lithium–halogen exchange—which required high dilution (0.04 M), and led to low yields 22–28% when performed on 2.1–2.6 mol scales; due to instability of a reactive intermediate. While small-scale reactions were shown to be rapid, their exothermic nature had the potential to be problematic when conducted at scale; with this in mind, the authors investigated the reaction under continuous flow conditions.

Employing a stainless-steel tubular reactor (dimension = 0.25 in. (i.d.)), submerged in a dry ice/acetone bath, the authors initially investigated the precooling of

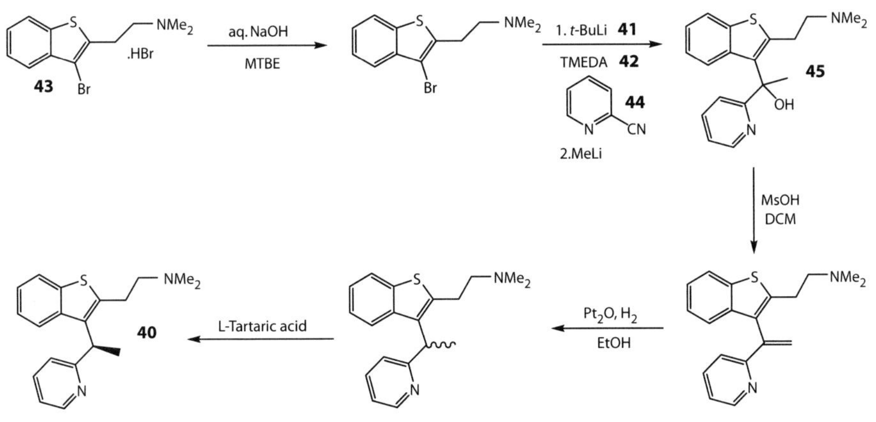

Industrial Interest in Micro Reaction Technology

SCHEME 7.7 Illustration of the synthesis of a H1 antagonist **40** using a flow reaction for the key halogen–lithium exchange step.

a solution of *t*-BuLi **41** and TMEDA **42** in THF, prior to mixing with a precooled solution of bromobenzothiophenone hydrobromide **43** and reaction to afford the lithiated intermediate. A third reactant stream comprising 2-chloropyridine **44** and Ti(O^{i}Pr)$_4$ was then added, where it reacted to afford 1-(2-(2-(dimethylamino)ethyl) benzo[*b*]thiophen-3-yl)-1-(pyridine-2-yl)ethanol **45**. The reaction products were subsequently quenched, upon exiting the reactor by stirring in a solution of MeOH, prior to analysis by HPLC. Employing too short a reaction time, the authors detected the formation of a by-product arising from the reaction of *t*-BuLi **41** and 2-chloropyridine **44**; therefore, the authors varied the flow rates of all reactant solutions until the temperatures within the reactor remained stable and within range (target −65°C), enabling complete intermediate formation and suppression of the competing side reactions.

Improving the reactor by reducing the internal diameter of the cooling coils, the authors were able to generate the alcohol **45** in similar conversions (66.2%) to those obtained in small batch-scale experiments. Employing 2 eq. of the starting bromide **43**, which could readily be removed from the reaction product, the authors were able to increase the conversion to 92.6% product **45**, demonstrating the successful development of a continuous flow method for this previously scale-retarded synthetic step.

Harnessing the heat exchange properties of microstructured reactors, Loebbecke and coworkers [19] constructed a micro reaction plant, complete with remotely controlled operation, demonstrating the synthesis and downstream processing of liquid nitrate esters at a technical scale (see Chapter 3).

These energetic materials are deemed hazardous at a production scale due to their huge exothermicity and thermolability. By constructing an automated system, the authors were able to safely investigate the synthesis of an undisclosed material via systematic variation of reaction parameters, providing access to previously uninvestigated reactant combinations such as high temperatures and concentrations.

Once identified, the optimal conditions were transferred to their automated multipurpose plant in which the continuous synthesis of liquid explosive materials was investigated. At this stage, downstream processing (washing and extraction) was incorporated into the system and compounds such as ethylene glycol dinitrate **46**, butanetriol trinitrate **47**, methyl nitrate **48**, and trinitroglycerin **49** were produced in quantities up to 150 g min^{-1} at pharma-grade purity (Figure 7.3).

7.2.8 Peracetic Acid Production Using an On-Site Microprocess

In a recent example, Ebrahimi et al. [20,21] demonstrated one of the most underutilized principles of micro reaction technology, that of increased process safety, by means of investigating on-site raw material production—to minimize handling and transportation risks. Using the synthesis of unstable percarboxylic acids, selected due to their widespread use as oxidizers such as disinfectants and bleaches, the authors investigated the development of a microprocess capable of producing such materials.

In addition to evaluating the chemical reaction, the authors also considered the feasibility of constructing a microstructured reactor for the preparation of materials at the site of use, along with the safety implications associated with its long-term application.

Employing the synthesis of peracetic acid **50** as a test case, the authors used four evaluation methods in order to compare the microprocess with conventional batch methodology and aid in the design/validation of the new methodology; these were (a) reaction matrix, (b) Dow's fire and explosive index, (c) inherent safety index, and (d) worst-case and consequence analysis.

As Scheme 7.8 illustrates, peracetic acid **50** is synthesized via an acid catalyzed, H_2SO_4 **51**, reaction between acetic acid **52** and hydrogen peroxide **53**; with great care taken with respect to temperature and concentration due to the inherent detonation risk.

To transfer the reaction to a flow process, the authors divided the reaction into a series of unit operations, the first involved the mixing of acetic acid **52** with sulfuric acid **51**, varied between 3 and 9 wt.%, followed by the addition of H_2O_2 **53** and reaction. The reactor employed consisted of a multichannel unit with submillimeter channels (channel dimensions = 250 μm) and a total volume of < 10 dm^3. Using this approach, the authors were able to develop a process with a production capacity of 10 kg h^{-1}; based on a reaction time of 300 s.

As Table 7.6a illustrates, when considering the use of a continuous flow process for the synthesis of peracetic acid **50**, the fire and explosion index rates the damage that may result from an incident in the process plant as intermediate (112) compared

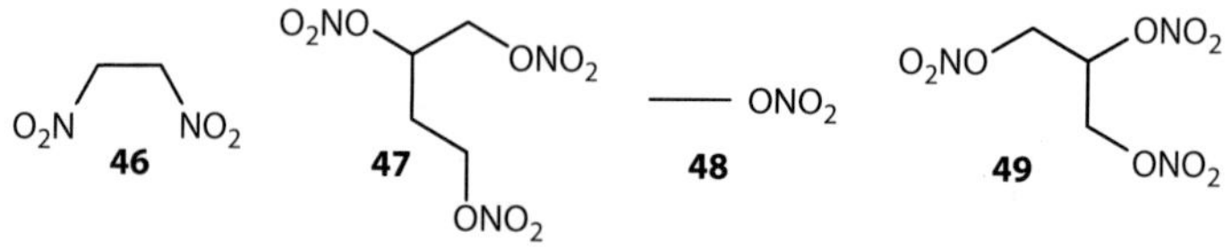

FIGURE 7.3 Illustration of some of the potentially explosive materials synthesized in an automated, remotely controlled micro reactor plant.

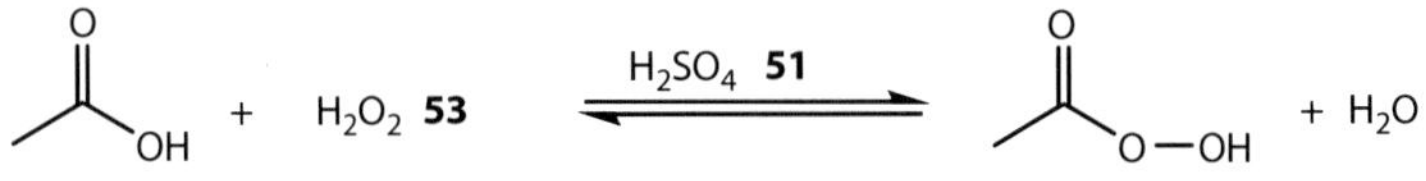

Industrial Interest in Micro Reaction Technology

SCHEME 7.8 Illustration of the equilibrium reaction of acetic acid **52** and H_2O_2 **53** to afford the oxidizing agent, peracetic acid **50**.

to a severe rating (226) for the conventional batch process. Furthermore, should catastrophic failure of the microprocess occur, due to the reduced damage incurred, processing downtime is dramatically reduced. The worst-case and consequence analysis also concluded that the use of an on-site microprocess facility is advantageous compared to a centralized batch facility (Table 7.6b).

Using the aforementioned evaluation criteria, the authors were able to conclude that the use of onsite microprocessing offers the user substantial reductions in risk

TABLE 7.6
Summary of the Safety Aspects Associated with the Batch and Microscale Synthesis of Peracetic Acid 50

	Batch	Microprocess
(a) Dow's Fire and Explosion Index (F&EI)		
Process Unit Risk Analysis		
Material factor for **50**	40	40
F&EI	226	112
Radius of exposure (m)	58	29
Area of exposure (m²)	1.05×10^4	0.26×10^{-4}
Value of area of exposure ($)	5×10^6	5×10^5
Damage factor (max. =1)	0.95	0.80
Maximum probable property damage ($)	5×10^6	4×10^5
Maximum days outage (day)	75	35
(b) Worst Case and Consequences		
Categories[a]		
Material release	4	2
Operating conditions	2	2
Process equipment	4	3
Transportations	5	1
Unpredictable factors	4	2
Total	19	10

[a] 5 = Very serious hazard, 4 = serious hazard, 3 = medium hazard, 2 = small hazard and 1 = no hazard.

compared to batch facilities and should therefore be considered as an alternative to the centralized production and transportation of hazardous materials.

7.2.9 THE *IN SITU* SYNTHESIS AND USE OF DIAZOMETHANE

As can be seen from the plethora of examples described in Chapters 2 through 4, micro reaction technology offers a safe and efficient technique for the use of hazardous reactants and intermediates. In addition to this, various authors have demonstrated the development and use of synthetic protocols for the performance of synthetic transformations that would otherwise be prohibited. One such example was recently communicated by Stark and coworkers [22] and described a continuous flow reactor (Little Things Factor, Germany) for the generation and subsequent reaction of diazomethane **54**.

Diazomethane **54** is commonly generated from the commercially available N-methyl-N-nitroso-p-toluenesulfonamide (Diazald®) **55** upon treatment with a strong base; however, the toxic and hazardous nature of the resulting material **54** means that its use beyond the laboratory is largely avoided. With its applicability for the formation of a diverse array of compounds, restricted due to the toxic and explosive nature of the precursor **54**, Stark and coworkers embarked upon the development of a continuous flow approach which would have the potential to enable the scale-independent application of diazomethane **54**.

Identifying the need to develop a protocol based on a homogeneous reaction mixture, to prevent clogging of the reaction channels, the authors initially evaluated a series of solvent combinations in order to identify one that would be suitable for long-term operation. Using this approach, the authors concluded that di(ethylene glycol) ethyl ether (carbitol) afforded minimal particulate formation and importantly enabled the dissolution of starting materials in high concentration, making the process commercially interesting.

In order to demonstrate the synthetic utility of this technique, the authors selected the methylation of benzoic acid **56**, to afford methyl benzoate **57**, as a model reaction (Scheme 7.9). Employing a precursor **55**:base **58**:substrate **56** of 1.0:1.5:4.0, the authors obtained a complete conversion of Diazald **55** to diazomethane **54**, with 60% methyl benzoate **57** formed when 2-propanol was used as the base **58** solvent and residence times of >5 s were employed in conjunction with reactor temperatures of 0–50°C; temperatures of 50–85°C afforded decreased product **57** formation due to decomposition.

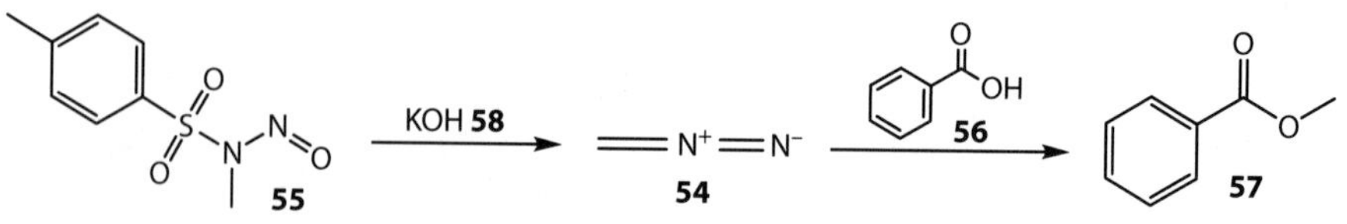

SCHEME 7.9 Illustration of the *in situ* generation and reaction of diazomethane **54** in a continuous flow micro reactor.

The technique did however demonstrate the feasibility of continuously generating diazomethane **54** at the site of reaction, making the technique applicable to the production of fine and specialty chemicals, along with pharmaceutical agents.

7.3 SYNTHESIS OF PHARMACEUTICALS AND NATURAL PRODUCTS USING CONTINUOUS FLOW METHODOLOGY

With lead times averaging 10–12 years between compound identification and development into a saleable medicinal product, the pharmaceutical industry are particularly interested in emerging technologies that have the potential to reduce the lead time taken to identify prospective lead compounds, that is, microwave, combinatorial, and flow chemistry. In addition to this step, the financial costs associated with taking a prospective lead from the R&D laboratory into process scale-up and beyond are vast and due to the associated risks of failure to scale anything that can speed this process up is desirable. With this in mind, the attraction to flow chemistry is twofold as it enables rapid lead compound identification (see Chapters 2 through 4) and facile process scaling (Chapter 1). Consequently, in addition to those researchers active within academia, the past five years have seen rapid growth in the number of publications originating from pharmaceutical companies, describing the implementation of microflow steps for the synthesis of pharmaceutically interesting compounds at both a research and development and process stage.

7.3.1 Ciprofloxacin and Its Analogs

As one of the pioneers of commercial micro reaction technology, Schwalbe et al. demonstrated numerous synthetic transformations within tubular reactors (CYTOS or SEQUOS®, CPC (Germany)), none as impressive as the synthesis of Ciprofloxacin **59**, a fluoro-quinoline antibiotic (a potent gyraze inhibitor), and a series of 28 analogs [23]. As Scheme 7.10 illustrates, the synthetic route employed involved the acylation of a β-dimethylaminoacrylate **60** with trifluorobenzoyl chloride **61** to afford the first building block **62**; from this point, forward diversification was implemented. Using a Michael addition step, followed by ring closure, the target fluoro-quinoline derivatives were obtained. Performing these steps, for the synthesis of Ciprofloxacin **59** under continuous flow, the authors were able to obtain all intermediates in high to excellent yield, as summarized in Table 7.7.

Gratified with the results obtained so far, the authors subsequently investigated diversifying the molecule and as illustrated in Table 7.8, all compounds were obtained on a several hundred mg-scale with yields ranging from 5% to 99% and purities of 23–96%. While the technique demonstrated was limited to the serial production of compounds, the screening process could be rapidly intensified by the operation of multiple, parallel reaction units continuously operated.

7.3.2 Synthesis of Pristane

Extending their earlier investigation into the efficient dehydration of β-hydroxyketones (Chapter 3, Section 3.4.1), Fukase and coworkers [24] evaluated the continuous flow

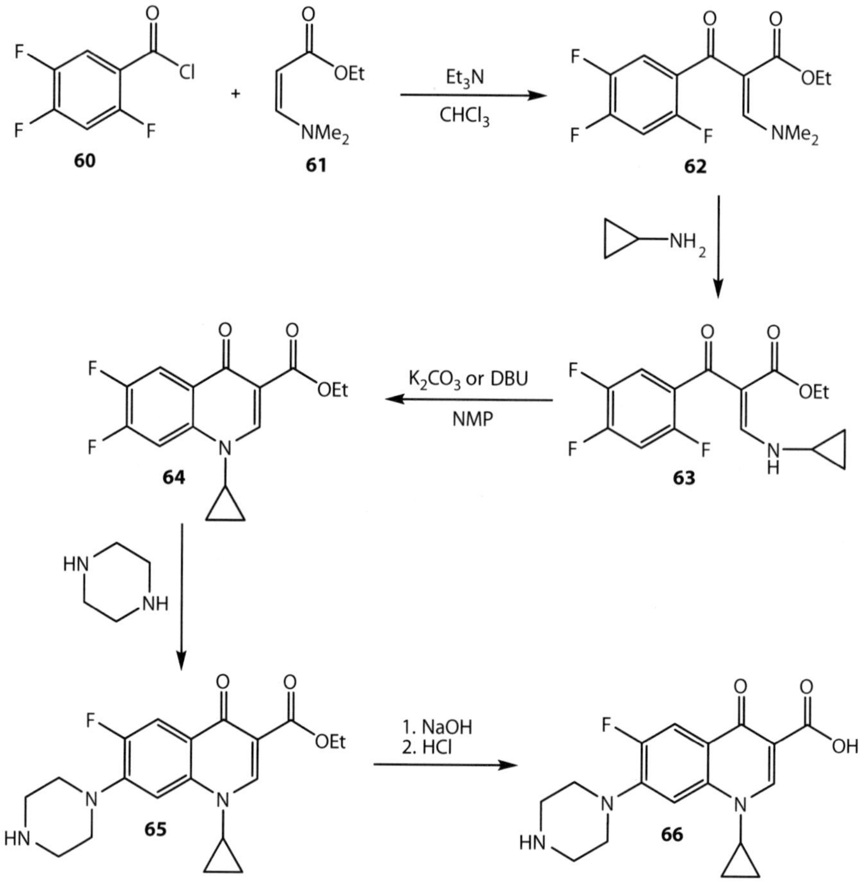

SCHEME 7.10 Schematic illustrating the linear synthetic approach utilized for the continuous flow synthesis of Ciprofloxacin and its analogs.

TABLE 7.7
Summary of the Yields and Throughputs Obtained for the Synthesis of a Series of Reactive Intermediates Performed under Continuous Flow Conditions

Compound	Yield (%)	Purity (%)	Throughput (g h^{-1})
62	99	>99	0.38
63	84	98	0.4
64	75	99	1.3
65	>99	>99	1.7
66	92	99	1.8

TABLE 7.8
Illustration of the Structural Diversity Readily Attainable through the Use of Sequential Continuous Flow Reaction Methodology

R¹	R²	Yield (mg)	Yield (%)	Purity (%)
Cyclopropyl	Piperazine	138	75	92
	Morpholine	146	79	61
	Pyrrolidine	138	75	71
	Diethanolamine	—	<10	—
	1,2,4-Triazole	—	<5	—
	4-Phenyl-piperazin-1-yl	151	68	23
	Thiomorpholine	123	71	26
n-Propyl	Piperazine	139	75	91
	Morpholine	117	63	85
	Piperidine	127	69	95
	Pyrrolidine	113	64	89
Isoamyl	Piperazine	145	73	89
	Morpholine	122	61	90
	Piperidine	147	74	96
	Pyrrolidine	135	68	90
Benzyl	Piperazine	155	74	94
	Morpholine	132	63	94
	Piperidine	140	67	96
	Pyrrolidine	149	73	—
Cyclohexyl	Piperazine	143	70	91
	Morpholine	121	59	81
	Piperidine	202	99	69
	Pyrrolidine	199	95	81

synthesis of the immunoactivating natural product, pristane (2,6,10,14-tetramethylpentadecane) **66**, which was only commercially available in limited quantities. In order to meet the current market demands of 5 kg week^{-1}, the authors proposed the development of a continuous flow process as a facile means of obtaining a method suitable for the production of pristane **66**. With this in mind, the first step of the reaction involved treating farnesol **67** with MnO_2 to afford the respective aldehyde, which subsequently underwent reaction with isobutylmagnesium chloride to afford the allylic alcohol **68** illustrated in Scheme 7.11. The alcohol **68** (1.0 M in THF) was subsequently dehydrated under continuous flow upon mixing with p-TsOH **69** (1.0 M)

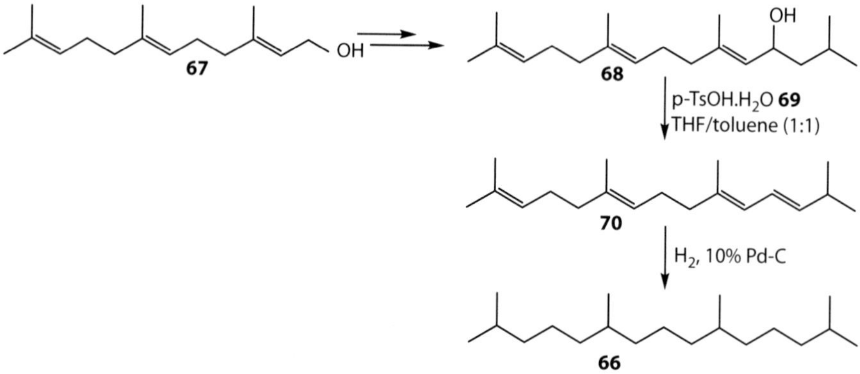

SCHEME 7.11 Schematic illustrating the synthetic strategy employed for the continuous flow synthesis of the Immunoactivating natural product, pristane **66**.

in tetrahydrofuran (THF)/toluene in a micromixer (Comet X-01) at a total flow rate of 600 µL min^{-1} and a reaction temperature of 90°C. The unsaturated product **70** subsequently underwent catalytic hydrogenation to afford the target compound **66** in 80% yield from farnesol **67**. Compared to the conventional batch route where multiple distillations were required to isolate **66** in sufficient purity, conducting the dehydration under flow was found to be advantageous as the material could be purified in a single step, due to reduced by-product formation.

7.3.3 Synthesis of Imatinib under Flow Conditions

Hopkin et al. [25] recently demonstrated an academic investigation into the development of a polymer-assisted route to the synthesis of the active pharmaceutical ingredient Imatinib **71** (4-[(4-methylpiperazin-1-yl)methyl]-N-4-methyl-3-[(4-pyridin-3-ylpyrimidin-2-yl)amino]phenylbenzamide) marketed as Gleevec, by Novartis AG, and used in the treatment of chronic myelogenous leukemia and gastrointestinal stromal tumors among other cancers (Figure 7.4).

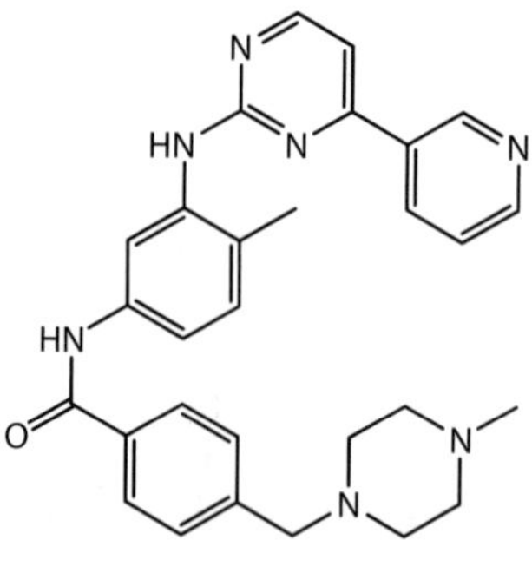

FIGURE 7.4 Imatinib **71** the active pharmaceutical ingredient of Gleevec.

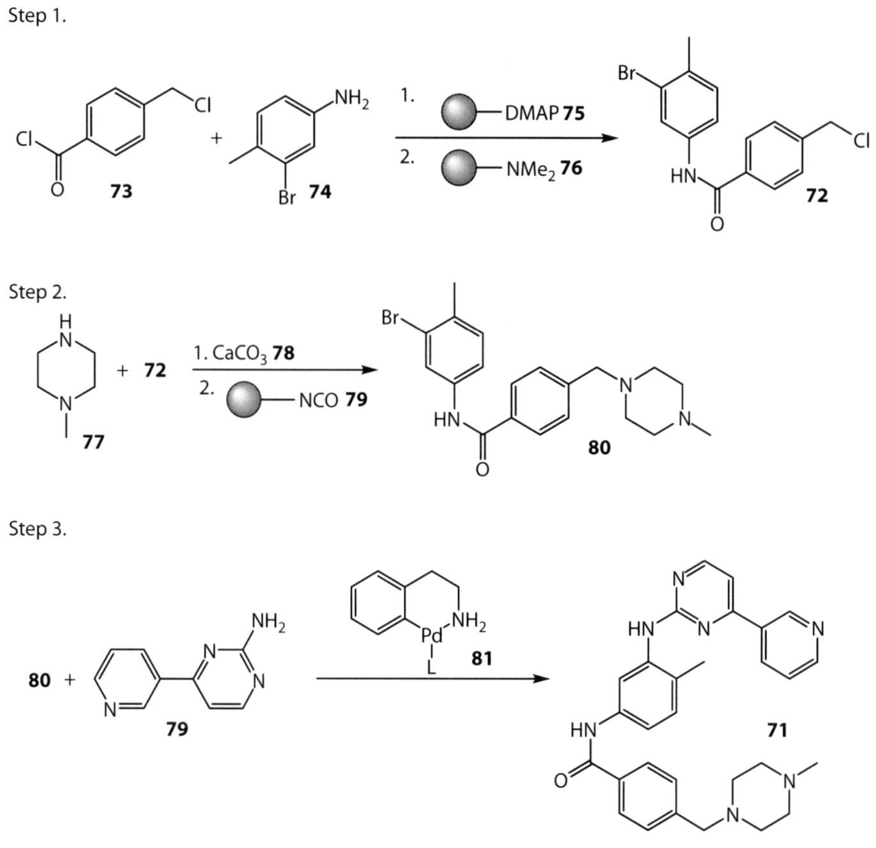

Industrial Interest in Micro Reaction Technology 367

Step 1.

Step 2.

Step 3.

SCHEME 7.12 Flow strategy employed for the synthesis of Imatinib **71**.

As Scheme 7.12 illustrates, the flow strategy employed comprised three steps, with the first step involving the synthesis of amide **72**, derived from the reaction of acid chloride **73** (01.3 M in DCM) and aniline derivative **74** (0.2 M in DCM); in the presence of polymer-supported DMAP **75**. To achieve this, the acid chloride **73** was preloaded onto PS-DMAP **75** at a flow rate of 100 µL min^{-1}, washing with DCM then removed any unactivated **73** from the packed bed. The aniline derivative **74** was then pumped through the reactor at a flow rate of 400 µL min^{-1}, where it reacted and released amide **72**, prior to passing through a column containing QuadraPure dimethylamine **76**; which scavenged any hydrolysis products. Using this approach, the compound **72** was isolated in >95% purity and 78% yield (53 mg) upon evaporation of the reaction solvent. In a separate second step, the amide **72** (7.5 × 10^{-3} M) underwent S$_N$2 displacement of the chloride with N-methylpiperazine **77** (1.5 × 10^{-2} M) in DCM/DMF in the presence of CaCO$_3$ **78** at 80°C. The reaction products were subsequently passed through a packed-bed containing polymer-supported isocyanate **79** to scavenge any unreacted N-methyl piperazine **77** from the reaction product **80** in 70% conversion (38 mg). In step three, a Buchwald–Hartwig coupling between amide **80** and 4(pyridine-3-yl)pyrimidin-2-amine **79** in the presence of BrettPhos Pd precatalyst **81** (10 mol%) performed at 150°C with a residence time of 30 min, afforded Imatinib **71**.

The target compound was subsequently purified by automated silica gel chromatography to remove unreacted materials and a proto-halogenated compound; affording Imatinib **71** in 69% yield (32 mg). Using this approach, the authors were able to synthesize the API in an overall yield of 32% (95% purity). This technique therefore has the potential to enable a facile method for the synthesis of structural variations of the API.

Although this approach represents a continuous flow route to the synthetic target **71**, the technique would require improvement, such as increasing precursor concentration, in order to facilitate the cost-effective preparation of Imatinib **71** at scale.

7.3.4 Synthesis of Aspirin and Vanisal Sodium

In addition to harnessing the advantages associated with increased processing control, users of flow reactor technology also require flexible operation. As such, the equipment needs to be easy to clean in order to exchange manufacturing processes, where relevant. With this in mind, Ricardo and Xiongwei [26] reported the adaptability of a continuous oscillatory baffled reactor (COBR), firstly for the synthesis of vanisal sodium **82** (Scheme 7.13) and aspirin **83** (Scheme 7.14), and secondly monitored the percentage loss of product during cleaning.

A COBR is a continuous flow reactor comprising a tube reactor containing spaced orifice baffles (perpendicular to the flow) which maintain oscillatory motion, caused by motors at either end of the device, with eddies formed as fluid strikes the baffles.

As plug flow dominates within this system, excellent residence time distribution is obtained, which affords a highly reproducible process for the production of fine chemical and pharmaceutical agents.

In their first example, the authors demonstrated the synthesis of vanisal sodium **82** (a second-order reaction), a vanillin **84** derivative used in soaps and cosmetics. As Scheme 7.13 illustrates, the material **82** is produced via the treatment of vanillin **84** with a sulfating agent **85** (40% w/w) at 60°C. Using the COBR system, samples were taken from the reactor every 6 h and analyzed by HPLC and NMR spectroscopy to assess the product purity. Over the course of 7 days, 54 samples were taken and an average purity of 99.94% was obtained.

After the 7-day campaign, the authors investigated the time required to clean the reactor, prior to performing a second reaction, using a single path cleaning in place

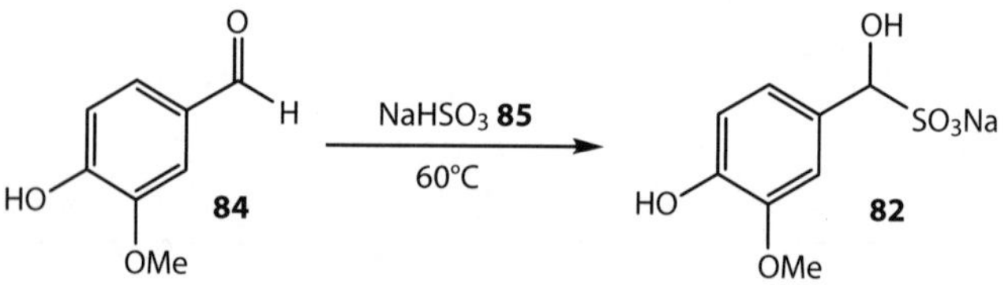

SCHEME 7.13 Illustration of the reaction protocol employed for the synthesis of the fragrance vanisal sodium **82** using a COBR.

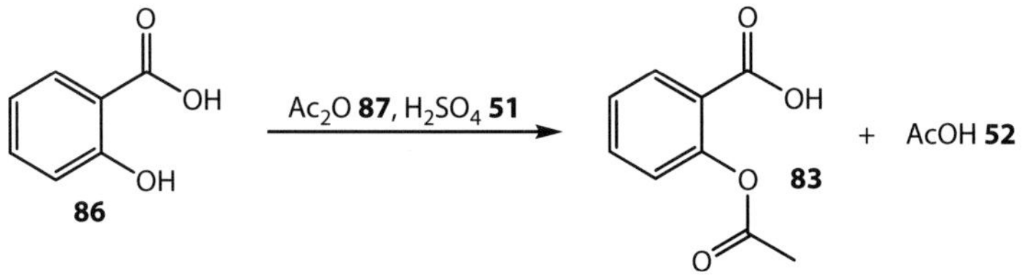

SCHEME 7.14 Schematic illustrating the synthetic route employed for the synthesis of aspirin **83** under continuous flow.

(CIP) procedure; common practice within the industry. Employing tap water (pH = 7.3) at 60°C initially, then Liquinox (1% solution), and finally United States Pharmacopeia (USP) water, the authors found 0.001% residual vanisal sodium **82**; significantly lower than the industry standard of 0.1–0.2%.

Satisfied with the cleaning, the authors investigated the synthesis of aspirin **83** via the treatment of salicylic acid **86** with acetic anhydride **87** in the presence of sulfuric acid **51**, affording acetylsalicylic acid (aspirin) **83** and acetic acid **52** as the by-product (Scheme 7.14). To perform a reaction, reactant solutions were preheated to 90°C and the first four sections of the reactor maintained at 60°C; upon cooling, the reaction products crystallized and the quality of the product **83** assessed by HPLC, NMR, XRD, and SEM. Over a period of 7 days, the authors obtained aspirin **83** in 99.5% purity and confirmed the formation of **83** as form 1, with no evidence of polymorphism.

Using an analogous cleaning procedure as employed for vanisal sodium **82**, the authors quantified the percentage API loss as 0.005% of the total material produced and in both cases less than 11 L of waste was generated from a system with a total reactor volume of 2 L.

7.3.5 Synthesis of Suberoylanilide Hydroxamic Acid

The formation of C–N bonds via the reaction of esters and hydroxylamine derivatives was discussed earlier, in Section 3.1.11, as an efficient method for the synthesis of hydroxamic acid derivatives. In an extension to this, Martinelli and coworkers [27] investigated the synthesis of suberoylanilide hydroxamic acid **88**, an early histone deacetylase (HDAC) inhibitor utilized in anticancer therapy, again in a PTFE flow reactor (reactor 1 = 10 mL, reactor 2 = 10 mL, and scavenger cartridge = 6.6 mm (i.d.)).

As illustrated in Scheme 7.15, the first step of the reaction involves the alkylation of aniline **89** (1.0 M) with suberoyl chloride **90** (1.0 M) in the presence of $NaHCO_3$ **91** (1.0 M) to afford the amide **92**. Treatment of the crude amide **92** with 50% aq. hydroxylamine **93** and sodium methoxide **94** (0.5 M) at 90°C afforded the target, suberoylanilide hydroxamic acid **88**. The crude reaction products were subsequently passed through a scavenger cartridge, containing a silica-supported quaternary amine compound (ISOLUTE PE-AX, 1.5 g), in order to sequester any carboxylic acid by-products formed. The resulting reaction mixture then underwent evaporation to remove any

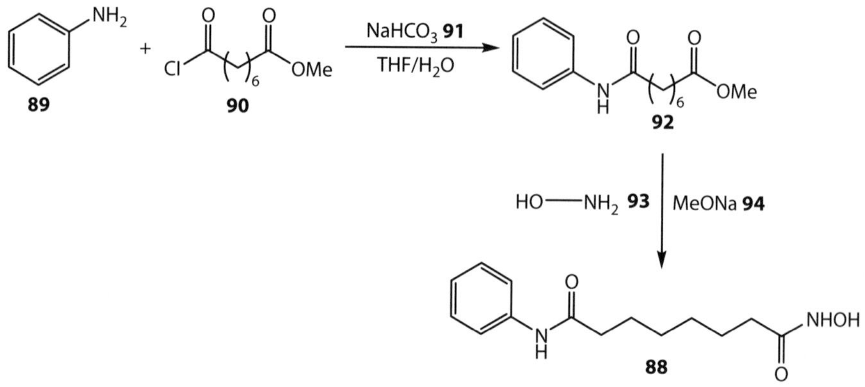

SCHEME 7.15 Illustration of the reaction protocol employed for the continuous flow synthesis of suberoylanilide hydroxamic acid **88**, in a continuous flow process.

organic solvents, prior to dissolution in water and neutralization (pH 7) using acetic acid **52**, which resulted in the precipitation of the target compound, suberoylanilide hydroxamic acid **88** in 80% yield over the two steps and 99% purity.

7.3.6 Synthesis of Rimonabant and Efaproxiral Using AlMe₃

Using a continuous flow micro reactor (Syrris Ltd., UK), volume = 16 mL, Gustafsson et al. [28] developed an efficient, functional group tolerant method for the trimethylaluminum **95**-mediated amide bond formation. Initially demonstrating the reaction of simple aromatic and aliphatic esters, the authors were able to identify those reaction conditions suitable for the efficient synthesis of amides. In an extension to this, the authors applied their technique to the synthesis of Rimonabant **96** (Scheme 7.16), an antiobesity drug, and Efaproxiral **97** (Scheme 7.17) a radiation therapy enhancer.

In the case of Rimonabant **96**, the synthetic sequence employed involved the treatment of 4-chloropropiophenone **98** with LiHMDS **99** for 1 min, prior to the addition of ethyl oxalate **100** at 50°C. After reacting for 5 min, the reaction products were subjected to an offline work-up and purification, which afforded the β-ketoester **101** in 70% yield. Treatment of the β-ketoester **101** with 4-chlorophenylhydrazine hydrochloride **102** in acetic acid at 125°C afforded the pyrazole derivative **103** in 80% yield with a residence time of 16 min. In the final step, the amide bond formation was performed by treatment of the pyrazole **103** with AlMe₃ **95** in the presence of 1-aminopiperidine **104** at 125°C. Employing a reaction time of 2 min, the authors were able to obtain the target compound **96** in 80% yield, affording an overall yield of 49%.

In the synthesis of their second pharmaceutical agent, Efaproxiral **97**, the authors combined a series of batch and flow reactions, due to the heterogeneous nature of the reaction mixture utilized for the alkylation of the phenol derivative **105**. As Scheme 7.17 illustrates, the first step of the reaction involved the alkylation of phenol **105** with the *tert*-butyl ester of 2-bromo-2-methylpropionic acid **106** to afford a methyl ester **107** (75% yield). Using the previously developed AlMe₃ **95**-mediated amide

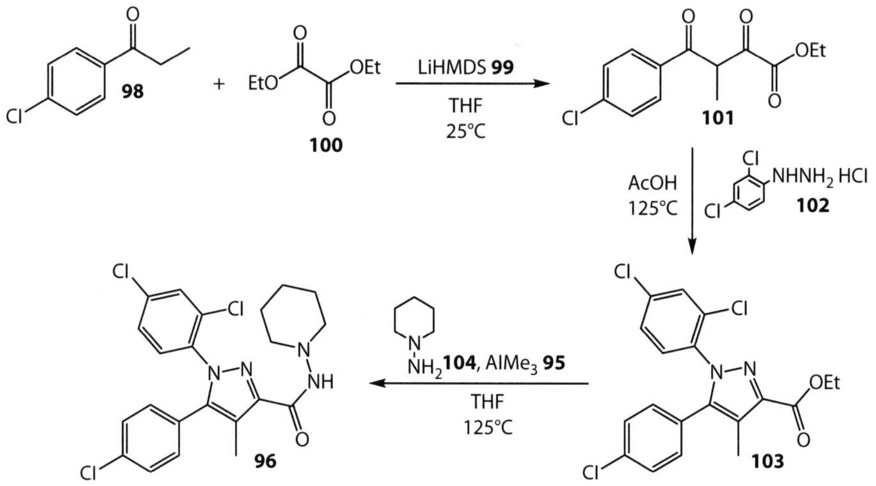

Industrial Interest in Micro Reaction Technology

SCHEME 7.16 Schematic illustrating the reaction sequence employed for the flow synthesis of rimonabant **96**.

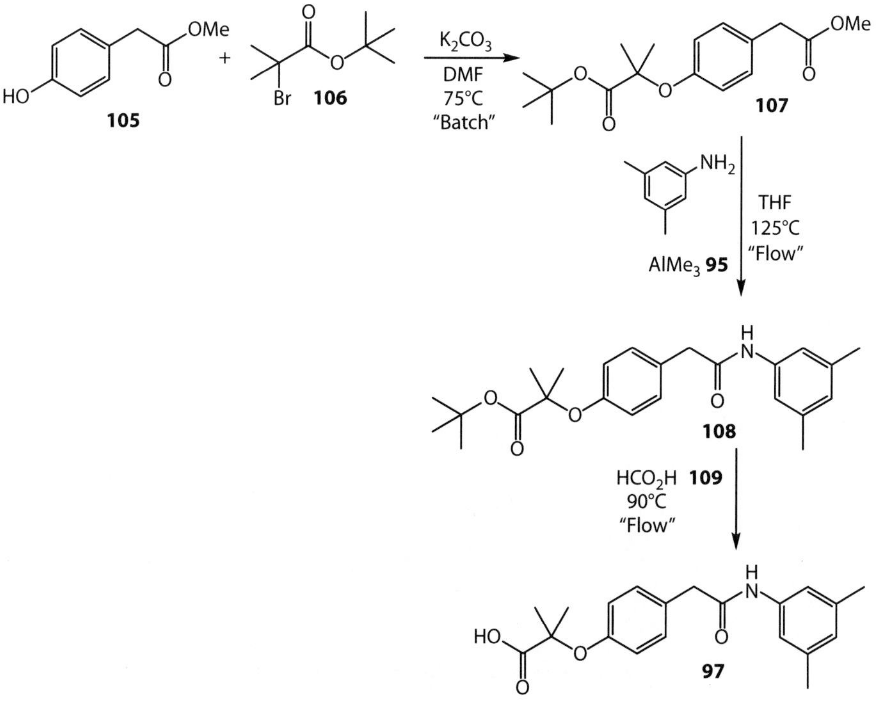

SCHEME 7.17 Combination of batch and flow protocols utilized for the synthesis of efaproxiral **97**.

bond formation, the authors synthesized *tert*-butyl-2-(4-(2-(3,5-dimethylphenylamino)-2-oxoethyl)phenoxy)-2-methylpropanoate **108** under continuous flow using a residence time of 2 min and a reactor temperature of 125°C. In the final reaction step, the authors performed an ester hydrolysis using formic acid **109** to afford Efaproxiral **97** at a throughput of 24 mmol h^{-1}.

7.3.7 Continuous Flow Synthesis of Sildenafil

With electrophilic nitration reactions featuring widely within the industrial process, and a large proportion of them being regarded as temperature sensitive, Taghavi-Moghadam and coworkers [29] embarked upon performing a temperature-sensitive nitration reaction within a stainless-steel micro reactor (CPC, Germany). Selecting the lifestyle drug Sildenafil® **110** as a synthetic target (Scheme 7.18)—marketed as Viagra—for preparation under continuous flow conditions, the authors investigated the ability to control the formation of 1-methyl-2-propyl-1*H*-pyrazole-5-carboxylic acid **111** and suppress the decomposition pathway. Employing a residence time of 35 min and a reactor temperature of 90°C, the authors were able to synthesize the nitro-derivative **111** in 73% yield, at a throughput of 5.5 g h^{-1}. As the reactor temperature was accurately controlled, no thermal decomposition products were observed and the authors were able to perform the nitration in a safe and efficient manner, unlike batch reactions whereby CO_2 evolution led to foaming and the potential for thermal runaway.

7.3.8 Synthesis of 6-Hydroxybuspirone

LaPorte et al. [30] at Bristol-Myers Squibb, employed three consecutive continuous flow processes for the synthesis of 6-hydroxybuspirone **112** (Scheme 7.19), as a

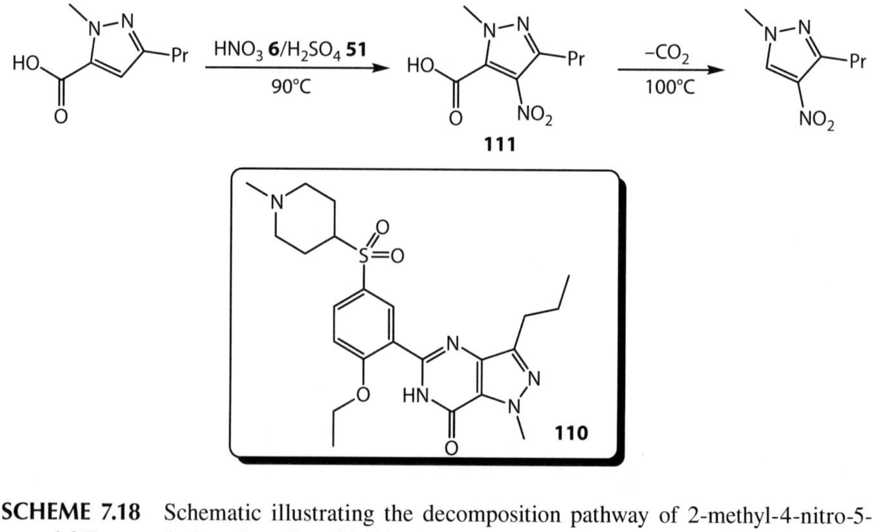

SCHEME 7.18 Schematic illustrating the decomposition pathway of 2-methyl-4-nitro-5-propyl-2*H*-pyrazole-3-carboxylic acid **111**, observed during the batch synthesis of a key intermediate of sildenafil **110**.

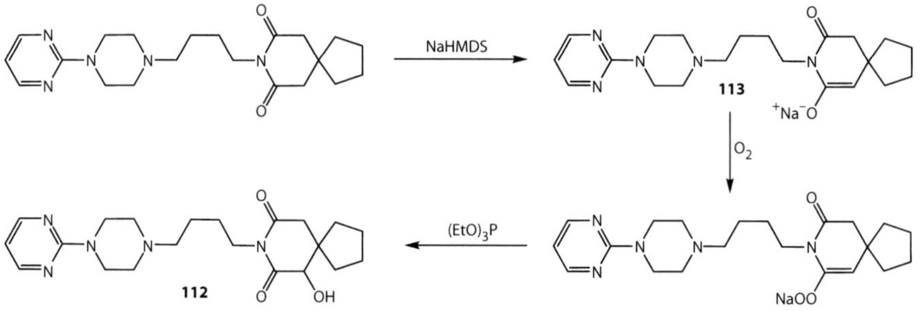

SCHEME 7.19 Illustration of the reaction protocol employed for the continuous flow synthesis of 6-hydroxy buspirone **112**.

means of overcoming potential manufacturing issues associated with the protracted timescales required to perform a key oxidation step at the pilot plant stage of scale-up. Owing to the need to perform the oxidation over a period of 16–24 h at −70°C in batch, the authors investigated a flow reaction between the preformed enolate **113** (3 mL min⁻¹) and oxygen (0.3 L min⁻¹). Cooling the micro reactor to only −10°C, afforded the target compound **112** in 65–70% conversion. On addition of a second micro reactor, whereby the reaction product was mixed with a second stream of oxygen, the authors were able to obtain the target **112** on 85–92% conversion; employing a reaction time of 5–6 min afforded a processing volume of 300 g day⁻¹. The reaction was finally scaled, to a multikilo scale, and implemented in a pilot plant in order to provide the active pharmaceutical ingredient on a sufficient scale to service clinical trials, formulation development, and drug safety evaluations.

7.3.9 A Key Step in the Synthesis of (rac)-Tramadol

As an extension to research into the reaction of carbonyl containing compounds with Grignard reagents (Chapter 3) to prepare alcohols under continuous flow, Rencurosi and coworkers [31] demonstrated the synthesis of (*rac*)-Tramadol, a centrally active analgesic, within a tubular flow reactor (Vapourtec, UK) (Scheme 7.20).

Employing an analogous synthetic route to that used commercially, the authors reacted (*rac*)-2-((dimethylamino)methyl)-cyclohexanone **114** (0.248 M) with 3-methoxyphenylmagnesium bromide **115** (1.2 eq.) in anhydrous THF at room temperature. Using a PTFE tube reactor, connected to a column containing polymer-supported benzaldehyde, the authors employed a residence time of 33 min, affording the product in 96% yield, as a diastereoisomeric mixture of **116** and **117** (8:2). Treatment of the reaction products with HCl afforded Tramadol HCl as a white crystalline solid; a full characterization of the compound was performed and the results were found to be in agreement with literature reports.

7.3.10 Claisen Rearrangement to Afford 2,2-Dimethyl-2H-1-Benzopyrans

During the development of a potassium channel activator drug candidate (BMS-180448) (Figure 7.5) Polomski and coworkers [32] identified a large release of heat,

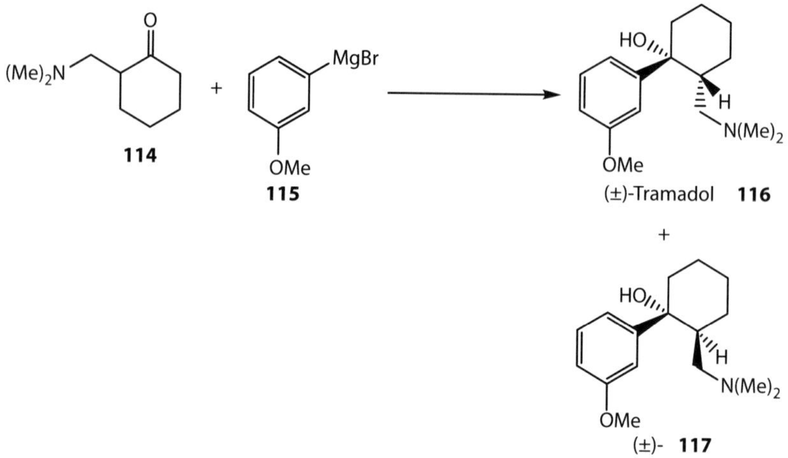

SCHEME 7.20 Illustration of the synthetic protocol employed for the flow synthesis of (±)-tramadol **116**.

which posed a significant problem from a production perspective, with an increase in reactor temperature from 180 to 445°C reported on one occasion over a period of 33 s. In addition to the safety risks, this represents when reactions are performed at scale, the large heat release can also give rise to a quality issue, resulting from thermal degradation of the product (Figure 7.5). Using the thermally induced Claisen rearrangement of 4-cyanophenyl-1,1-dimethylpropargyl ether **118**, the authors investigated the use of continuous flow reactors for the scalable synthesis of 6-cyano-2,2-dimethylchromene **119**, an intermediate in the synthesis of **120** (Scheme 7.21). Utilizing a plug flow reactor, constructed from stainless-steel tubing (dimensions = 0.00625 in (i.d.) × 10 m (long)) immersed in a heated oil bath (220°C), the authors pumped the neat 4-cyanophenyl-1,1-dimethylpropargyl ether **118** through the reactor over a range of flow rates so as to afford residence times in the range of 3.5–17.7 min to be evaluated.

At all flow rates, the authors obtained ≥96% conversion of the propargyl ether **118** to the chromene **119**, as determined by HPLC analysis.

In order to synthesize quantities of material suitable for subsequent process development, the reactor was scaled to one with dimensions of 0.125 in (i.d.) × 25 ft (long).

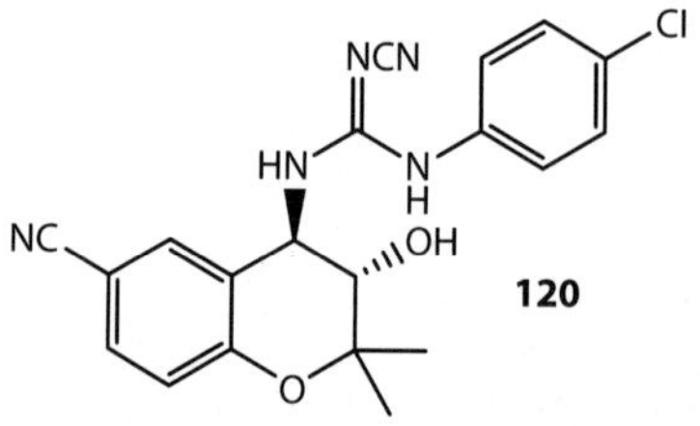

FIGURE 7.5 A benzopyran-based potassium channel activator **120** developed by Bristol-Myers Squibb.

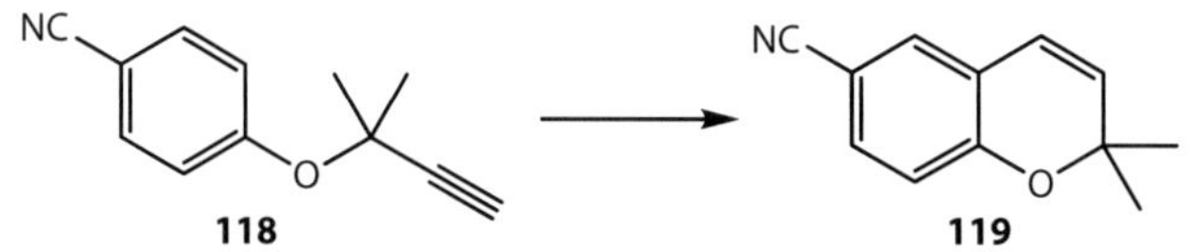

SCHEME 7.21 Schematic illustrating the Claisen rearrangement of 4-cyanophenyl-1,1-dimethylpropargyl ether **118** to 6-cyano-2,2-dimethylchromene **119**.

Employing a residence time of 4 min (20 mL min^{-1}), the authors produced the target compound **119** in 91% yield (95.92% purity) at a bath temperature of 195–200°C. Over subsequent batches ($n = 4$), the authors processed 39.09 g of ether **118**, affording 37.76 g of chromene **119** with yields ranging from 94.7% to 97.9% and purities, determined by HPLC, in the range of 96.7–97.9%. The small losses observed were attributed to residual material within the flow reactor and manual handling steps. In a final upscaling step, the authors investigated the use of a jacketed flow reactor (0.375 in. (i.d.) × 10 ft (long)), in which 6-cyano-2,2-dimethylchromene **119** was able to be synthesized at a throughput of 7 kg h^{-1}. Based on the success of this trial, the authors subsequently investigated the reaction of a series of propargyl ethers, obtaining the respective benzopyrans in yields ranging from 74% to 98%, after purification by column chromatography (Table 7.9).

7.3.11 SYNTHESIS OF A 5HT$_{1B}$ ANTAGONIST

Using a series of "catch and release" steps, Qian et al. [33] demonstrated the flow synthesis of a potent 5HT$_{1B}$ antagonist, 6-methoxy-8-(4-methyl-1,4-diazepan-1yl)-*N*-(4-morpholinophenyl)-4-oxo-1,4-dihydroquinoline-2-carboxamide **121** (Figure 7.6),

TABLE 7.9
Summary of the Results Obtained for the Synthesis of a Series of Benzopyran Derivatives Utilizing the Claisen Rearrangement under Continuous Flow Conditions

R	Temperature (°C)	Residence Time (min)	Yield (%)[a]
4-CN **119**	220	3.5	98
4-OCH$_3$	240	11.7	78
4-COCH$_3$	240	5.5	80
4-NO$_2$	240	5.5	96
4-I	220	5.5	78
3-CF$_3$	260	17.7	86
3-NO$_2$	230	5.5	82
2-CHO	240	5.5	74

[a] Yields refer to purified product.

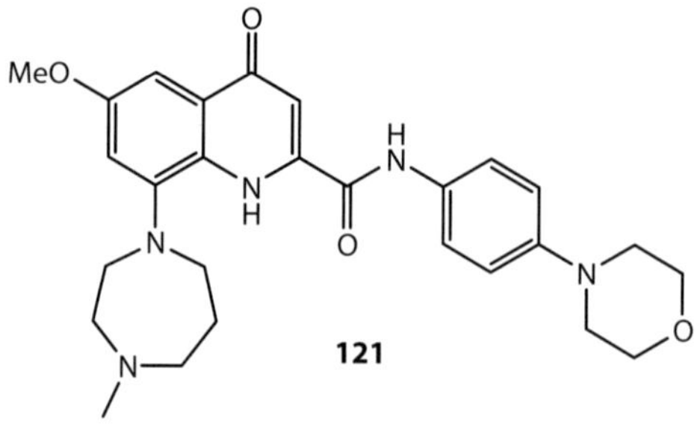

FIGURE 7.6 Illustration of the potent 5HT$_{1B}$ antagonist **121** developed by AstraZeneca, synthesized using a series of flow processes.

in an overall yield of 18% and >98% purity; compared to a previous batch synthesis whereby the target compound **121** was obtained in 7% [34].

To perform the multistage synthesis of the quinoline derivative **121**, the authors initially synthesized 4-methoxy-2-(4-methyl-1,4-diazepan-1-yl)aniline **122** via the alkylation of 1-methylhomopiperazine (0.5 M in EtOH) with 3-fluoro-4-nitroanisole **123** (0.5 M in EtOH) at 135°C for 10 min (Scheme 7.22). The resulting reaction products were then pumped through a glass column, containing QuadraPure-benzylamine

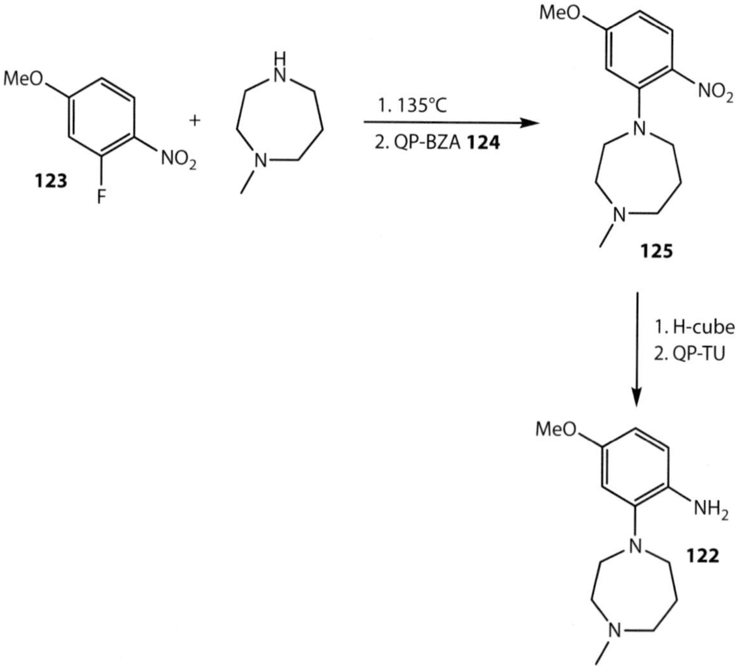

SCHEME 7.22 Schematic illustrating the first step employed in the synthesis of the 5HT$_{1B}$ antagonist **121**.

124 (5 eq.) to scavenge any hydrofluoric acid generated during the nucleophilic aromatic substitution. The purified reaction product **122** was hydrogenated using the ThalesNano Inc. H-cube™ and directed through a second scavenger cartridge, containing QuadraPure-thiourea **124**, to remove any leached Pd; affording the target amine **122** in quantitative yield upon evaporation of the reaction solvent.

In the second flow process, toluene was employed as the reaction solvent and the authors demonstrated the reaction of amine **122** (0.2 M) with dimethyl acetylenedicarboxylate **125** (0.24 M, 1.2 eq.), at 130°C, with a residence time of 12.5 min. As Scheme 7.23 illustrates, the reaction products were subsequently purified by passing through a QP-BZA **124** (5 eq.), to remove any residual dicarboxylate **125**, followed by a column containing potassium carbonate (2.5 g) to remove any traces of water prior to performing a second hydrogenation. The anhydrous reaction mixture was then subjected to a high-temperature cyclo-condensation reaction, 250°C for 13 min, and the reaction products cooled to ambient temperature upon the addition of aqueous THF (3%). The reaction products were subsequently pumped through a column containing Ambersep 900 (hydroxide form) **126**, which acted twofold; first to hydrolyze the ester, and second to sequester the carboxylic acid product **127**.

In the final reaction step, a solution of *O*-(benzotriazol-1yl)-*N,N,N',N'*-tetramethyluronium tetrafluoroborate (TBTU) **128** (2.5 eq.) and 1-hydroxybenzotriazole (HOBt) **129** (2.5 eq.) in DMF was pumped through the packed-bed containing the sequestered carboxylic acid, both activating and releasing the carboxylic acid which was coupled with morpholine **130** with a residence time of 50 min. In a final "catch

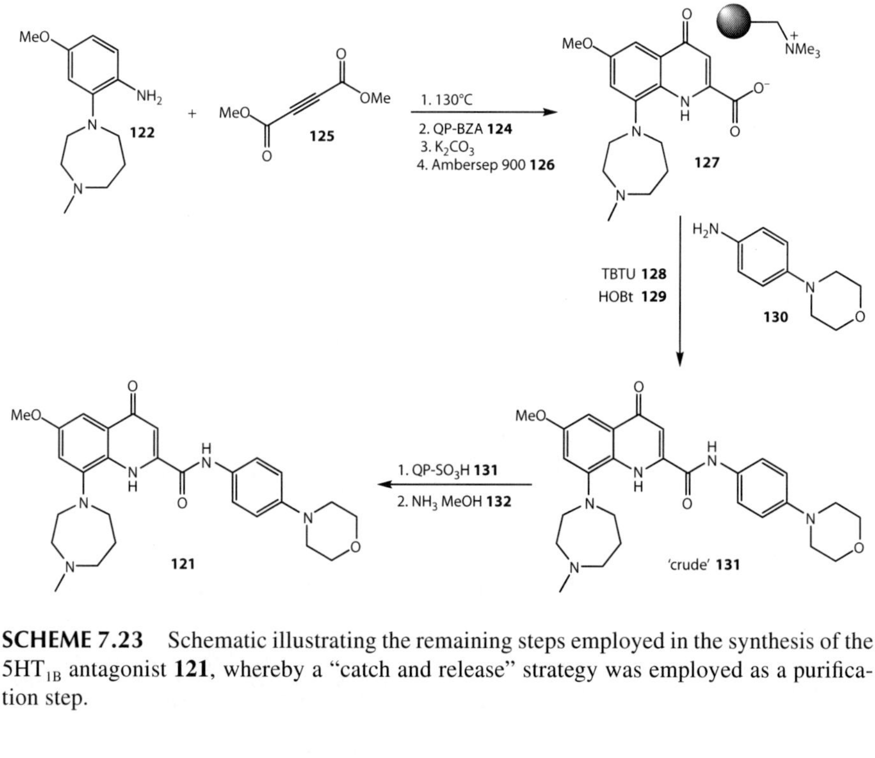

SCHEME 7.23 Schematic illustrating the remaining steps employed in the synthesis of the 5HT$_{1B}$ antagonist **121**, whereby a "catch and release" strategy was employed as a purification step.

and release" step, the reaction product **121** was sequestered using QuadraPure-sulfonic acid (QP-SO$_3$H) **131**, washed and released using a solution of methanolic ammonia **132** (2.0 M, 5 eq.). The reaction products were concentrated *in vacuo* and the crude product recrystallized from MeOH, to afford 6-methoxy-8-(4-methyl-1,4-diazepan-1yl)-*N*-(4-morpholinophenyl)-4-oxo-1,4-dihydroquinoline-2-carboxamide **121** in an 18% yield; representing a 2.6-fold increase in yield compared to previous synthetic strategies.

7.3.12 Serial Approach to a Novel Anticancer Agent Using Flow Reactors

As part of an investigation into the synthesis of novel anticancer agents, Tietze and Liu [35] evaluated the use of micro reactors for the serial preparation of intermediates used in the synthesis of an aminonaphthalene derivative **133** (Scheme 7.24)—a key intermediate in the synthesis of prodrug **134** (Figure 7.7). Using a CYTOS (CPC, Germany) reactor, the authors performed nine separate micro reactions, ranging from the synthesis of esters to the Wittig–Horner olefination and a Friedel–Crafts acylation. Throughout the process the authors also performed the reactions in batch enabling a true comparison of any advantages or disadvantages associated with the use of a flow reactor. This mode of operation allowed the authors to calculate the empirical accelerating factor (F) for each step, enabling them to conclude that the use of a micro reactor resulted in a safer and faster process with acceleration of $F = 3$ to 10 depending on the reaction employed.

7.3.13 Synthesis of Grossamide under Flow Conditions

In 2005, Ley and coworkers [36] employed a serial flow reaction approach to the enantioselective synthesis of the Neolignan natural product, Grossmide **135**. As Scheme 7.25 illustrates, the first step of the reaction involved the polymer-supported HOBt **136** coupling of ferrulic acid **137** with tyramine **138** to afford amide **139**; analysis of the reaction products by LC–MS confirmed the formation of the amide **139** in 90% conversion. As it was imperative to remove the residual tyramine **138** before the next reaction step, the reaction products were passed through a second column, containing a polymer-supported sulfonic acid **131** resin, affording the amide **139** in excellent purity.

In the second step, the amide **139** was premixed with the H$_2$O$_2$–urea complex **140** in aqueous buffer (pH 4.5) and pumped through a packed column containing silica-supported peroxidase **141**, affording the target compound **135** in excellent yield and purity.

7.3.14 Synthesis of the Natural Product (±)-Oxomaritidine

Baxendale et al. [37] subsequently reported the serial use of solid-supported reagents, catalysts, and scavengers, for the synthesis of (±)-oxomaritidine **142** via a previously published synthetic sequence. As Scheme 7.26 illustrates, the initial stages of the

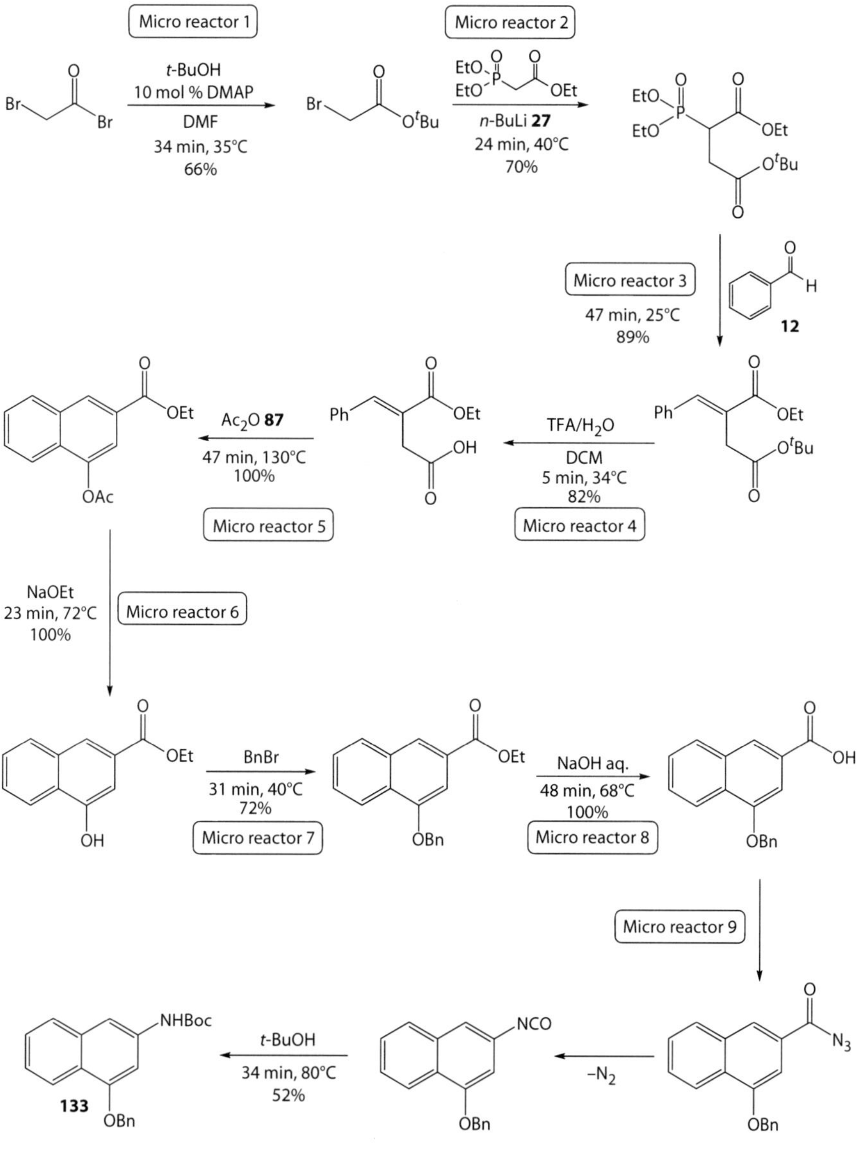

SCHEME 7.24 Illustration of the nine micro reactions performed in the continuous flow synthesis of aminonaphthalene **133**.

synthetic process were based on the convergent synthesis of an imine **143**, derived from an azide **144** and aldehyde **145**; independently synthesized using an excess of a solid-supported azide **146** (20 eq., 70°C) and oxidizing agent **147** (10 eq., 25°C), respectively (AFRICA®, Syrris, UK). Reduction of the imine **143** followed using the commercially available H-cube® (ThalesNano Inc., Hungary), affording the 2° amine as a solution in THF.

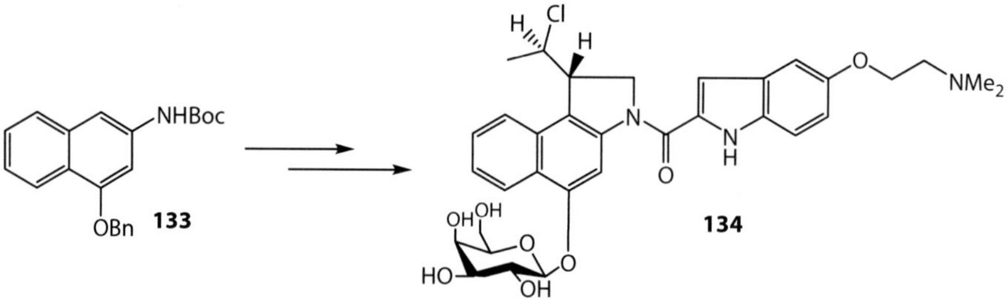

FIGURE 7.7 Illustration of the aminonaphthalene target **133** a key intermediate in the synthesis of prodrug **134**.

Prior to performing the next reaction step, the authors removed the reaction solvent using a solvent evaporator (V-10, Biotage) and dissolved the residue in DCM (10 min). The phenolic compound **148** was then trifluoroacetylated (5 eq. **149**) within a micro reactor at 80°C to afford amide **150** which in the presence of polymer-supported (ditrifluoroacetoxyiodo)benzene **151** underwent oxidative phenolic coupling to afford the seven-membered tricyclic derivative **152**. In the final step, a solution of the reaction product was pumped through a column reactor containing a polymer-supported base **153** which cleaved the amide bond and enabled spontaneous

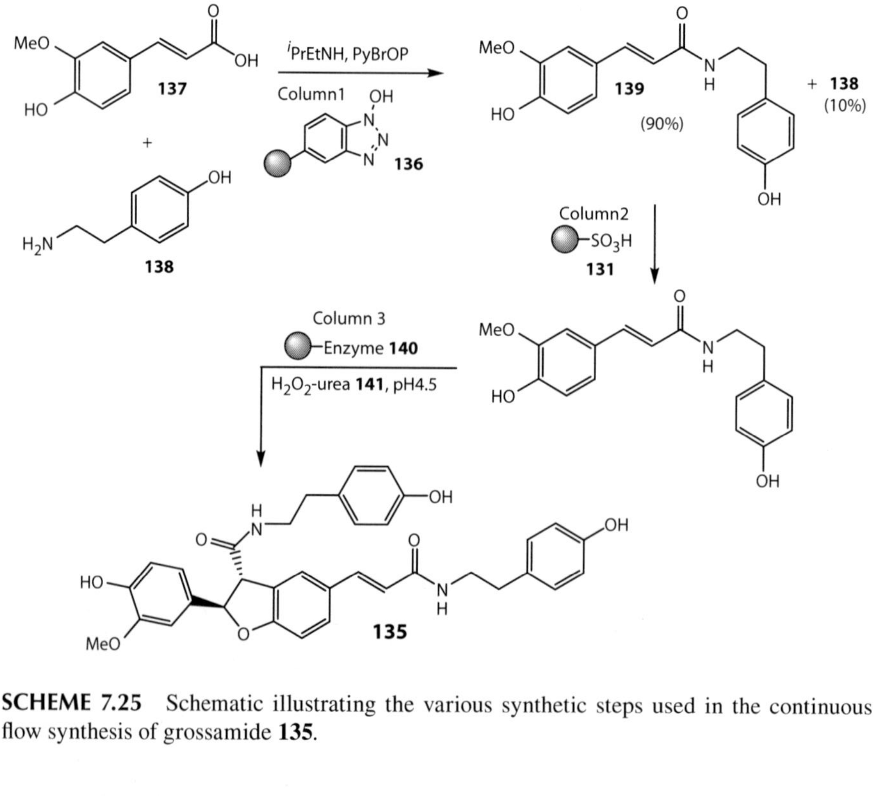

SCHEME 7.25 Schematic illustrating the various synthetic steps used in the continuous flow synthesis of grossamide **135**.

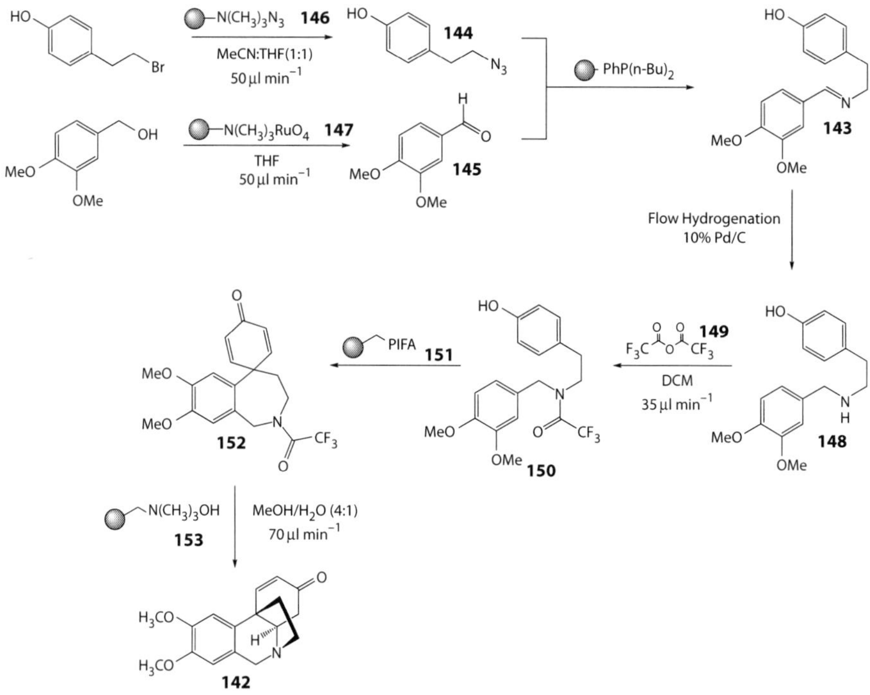

Industrial Interest in Micro Reaction Technology

SCHEME 7.26 Illustration of the serial flow reaction methodology used for the continuous flow synthesis of (±)-oxomaritidine **142**.

1,4-conjugate addition to occur, affording the target compound (±)-oxomaritidine **142** in 90% purity.

7.3.15 SYNTHESIS OF FUROFURAN LIGANS

Building on their experience gained in the *in situ* generation and subsequent reaction of radical intermediates within an array of continuous flow reactors (see Chapters 3 and 5 for details), Ryu and coworkers [38] extended their investigations to encompass the chemical generation of radicals under flow conditions. Using a MiChS-α micromixer (channel dimensions = 200 μm) coupled to a tubular residence time unit (dimensions = 1 mm (i.d.) × 1 m (long)), the authors investigated the gram-scale synthesis of a tetrahydrofuran derivative **154**, used as a key intermediate in the synthesis of furofuran ligans such as paulownin **155** and samin **156**.

As Scheme 7.27 illustrates, the first step of the reaction involved mixing tris(trimethylsilyl)silane (TTMSS) **157** and an initiator in toluene, followed by the downstream introduction of the α-bromo-unsaturated ester **158** (0.2 M), also in toluene. Employing a reaction temperature of 95°C and a residence time of 1 min, the authors obtained 7.6 g of **154** over a period of 3 h; after purification by silica-gel chromatography (74% yield).

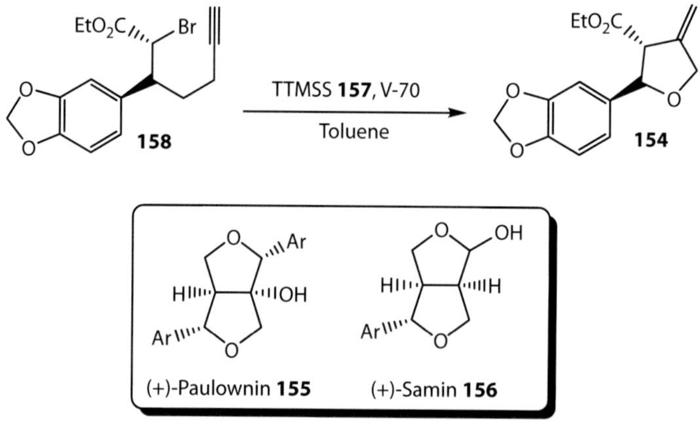

SCHEME 7.27 Schematic illustrating the methodology used for the continuous flow synthesis of a furofuran ligan **154**.

7.4 SYNTHESIS OF SMALL DOSES OF RADIOPHARMACEUTICALS

The emerging area of miniaturized PET radiosynthesis demonstrates a niche application of micro reaction technology [39,40], which has the potential to revolutionize not only the way that PET tracers are manufactured but also how they are administered. Owing to the fact that positron emitters, by their very nature, are short lived ($t_{1/2}$ for Carbon-11 = 20.4 min and Fluorine-18 = 109.7 min), efficient synthetic methods are required for their safe and efficient production. As micro reactions are inherently safer and more accurately controlled than their batch counterparts, it is widely envisaged that reactions could be initially evaluated using cold reagents and then exchanged for hot reactants when synthetic methodology is optimized. Using this approach, it is hoped that process optimization will be more thorough and reproducible enabling the generation of rapid and efficient synthetic routes to PET tracer molecules. With this in mind, researchers believe that the resulting materials could be administered in smaller doses due to the increased specific radioactivity, with the potential to even process the compounds at the site of use, further reducing the costs associated with the therapy [41].

Using the most common positron emission tomography probe as a model, Lee et al. [42] investigated the advantages associated with performing the synthesis of ^{18}F-fluorodeoxyglucose (FDG) **159** under continuous flow conditions.

Employing a PDMS micro reactor, consisting of a complex array of reaction channels (dimensions = 200 μm (wide) × 45 μm (deep)), valves and packed beds, the authors investigated a sequence of five reaction steps to transform mannose triflate **160** to ^{18}FDG **159** (Scheme 7.28). The reaction protocol devised consisted of an initial [^{18}F]fluoride concentration (500 μCi) and a solvent exchange from H$_2$O to MeCN. This was followed by the [^{18}F]fluoride substitution of mannose triflate **160** (324 ng), performed at 100°C for 30 s and 120°C for 50 s, to afford the labeled intermediate **161**. The reaction mixture then underwent a second solvent exchange from MeCN to H$_2$O and acid hydrolysis performed at 60°C, to afford ^{18}FDG **159**. Using this

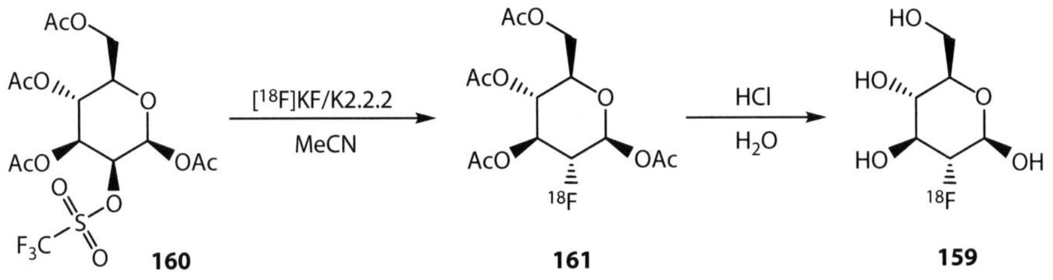

SCHEME 7.28 Illustration of the reaction protocol employed for the flow synthesis of FDG **159**.

approach, the authors were able to synthesize the probe **159** in 38% radiochemical yield, with a purity of 97.6%; as determined by radio-TLC. Having successfully performed the continuous synthesis of ^{18}FDG **159**, the authors compared the process developed with conventional techniques, identifying the dramatic reduction in processing time, from 50 to 14 min, as the key advantage to the miniaturization of the synthetic procedure.

In a more recent example of the use of continuous flow synthesizers toward this niche areas of synthesis was reported by Lu et al. [43], for the preparation of small doses of the radiopharmaceutical [^{18}F]-Fallypride **162** (Scheme 7.29). Utilizing the commercially available Nanotek (Advion, US) tube reactor (dimensions = 100 μm (i.d.) × 2 m (long)), the authors investigated the flow synthesis of [^{18}F]-Fallypride **162** as a means of developing a method suitable for the preparation of small doses (0.5–1.5 mCi) for micro-PET studies of brain dopamine subtype-2 (D_2) receptors in rodents.

Employing equal flow rates for the tosylate **163** (0.99 μmol in 255 μL) and [^{18}F]-fluoride (10–200 mCi), the authors evaluated the effect of reactor temperature (95–175°C) on the formation of [^{18}F]Fallypride **162**; quantification performed by radio-TLC or radio-HPLC.

Using this approach, the decay-corrected radiochemical yield (RCY) increased from zero to 65% with results found to compare when analogous reactions were performed at different research centers. As the methodology required only 4 min to perform the radiosynthesis (residence time = 96 s), this method afforded the user rapid access to the preparation of [^{18}F]-Fallypride **162** in small does for rodent studies. By increasing the volume of reagents pumped through the reactor, from 10 to

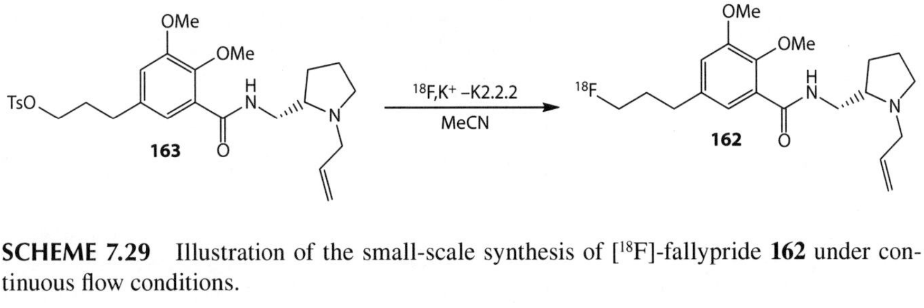

SCHEME 7.29 Illustration of the small-scale synthesis of [^{18}F]-fallypride **162** under continuous flow conditions.

200 µL, the authors were able to rapidly increase the production volumes to those sufficient for human injection (5–20 mCi).

7.5 SUMMARY

In addition to the array of synthetic examples discussed in detail, many more industrial research groups have communicated their results and findings of syntheses performed using a wide array of continuous flow equipment ranging from homemade [44] to commercially available reactors [45], with some technology companies aligning themselves with academic institutions [46] or chemical producers [47] to develop new reactor types. From these references alone, it can be seen that the field of MRT has much to offer not only the modern research chemist, but also those researchers working in pilot-scale and production environments [48,49].

REFERENCES

1. Weilier, A., Wille, G., Kaiser, P., and Wahl, F. 2009. Micro reactors offer improved solutions for liquid multipurpose bulk and fine chemical production, *Chem. Today* 27(3): 38–39.
2. Kock, M. V., VandenBusshe, K. M., and Chrisman, R. W. (Eds) 2007. *Micro Instrumentation for High Throughput Experimentation and Process Intensification—A Tool for PAT*. Wiley-VCH, New York and Kyoto, Japan.
3. Hessel, V., Renken, A., Schouten, J. C., and Yoshida, J. (Eds) 2008. *Micro Process Engineering: A Comprehensive Handbook Volume 2: Devices Reactions and Applications*. Wiley-VCH, New York and Kyoto, Japan.
4. Knockmann, N., Gottsponer, M., and Roberge, D. M. 2010. Scale-up concept of single-channel microreactors from process development to industrial production, *11th International Conference on Microreaction Technology*, Kyoto, Japan, 94–95.
5. Togashi, S., Miyamoto, T., Asano, Y., and Endo, Y. 2009. Yield improvement of chemical reactions by using a microreactor and development of a pilot plant using the numbering-up of microreactors, *J. Chem. Eng. Jpn*. 42: 512–519.
6. Kirchneck, D. and Tekautz, G. 2007. Integration of a microreactor in an existing production plant, *Chem. Eng. Technol*. 30: 305–308.
7. Rumi, L., Pfleger, C., Spurr, P., Klinkhammer, U., and Bannwarth, W. 2009. Adaptation of an exothermic and acylazide-involving synthesis sequence to microreactor technology, *Org. Proc. Res. Dev*. 13: 747–750.
8. Wilms, D., Klos, J., Kilbinger, A. F. M., Löwe, H., and Frey, H. 2009. Ionic liquids on demand in continuous flow, *Org. Proc. Res. Dev*. 13: 961–964.
9. Löwe, H., Axinte, R. D., Breuch, D., and Hofmann, C. 2009. Heat pipe controlled syntheses of ionic liquids in microstructured reactors, *Chem. Eng. J*. 155: 548–550.
10. Hessel, V., Hardt, S., and Löwe, H. 2004. *Chemical Microprocess Engineering-Fundamentals, Modelling and Reactions*. Weinheim: Wiley-VCH.
11. Ushiolgi, Y., Hase, T., Iinuma, Y., Takata, A., and Yoshida, J. 2007. Synthesis of photochromatic diarylethenes using a microflow system, *Chem. Commun*. 2947–2949.
12. Choe, J., Seo, J. H., Kwon, Y., and Song, K. H. 2008. Lithium-halogen exchange reaction using microreaction technology, *Chem. Eng. J*. 135S: S17–S20.
13. Schwalbe, T., Autze, V., and Wille, G. 2002. Chemical synthesis in microreactors, *Chimia* 56: 636–646.
14. Darvas, F., Dormán, G., Lengyel, L, Kovács, I., Jones, R., and Ürge, L. 2009. High pressure, high temperature reactions in continuous flow: Merging discovery and process chemistry, *Chimica Oggi* 27(3): 40–43.

15. Rutjes, F. P. J. T., Segers, R., Nieuwland, P., Koch, K., Lelivelt, H. G., van den Berg, J. F. D., and van Hest, J. C. M. 2010. Fast scale up using microreactors: From micro to production scale, *11th International Conference on Microreaction Technology*, Kyoto, Japan, 104–105.
16. Wille, Ch., Gabski, H. –P., Haller, Th., Kim, H., Unverdorben, L., and Winter, R. 2004. Synthesis of pigments in a three-stage microreactor pilot plant—An experimental technical report, *Chem. Eng. J.* 101: 179–185.
17. Pennemann, H., Forster, S., Kinkel, J., Hessel, V., Löwe, H., and Wu, L. 2005. Improvement of dye properties of the azo pigment Yellow 12 using a micromixer-based process, *Org. Proc. Res. Dev.* 9: 188–192.
18. Gross, T. D., Chou, S., Bonneville, D., Gross, R. S., Wang, P., Campopiano, O., Ouellette, M. A. et al. 2008. Chemical development of NBI-75043. Use of a flow reactor to circumvent a batch-limited metal–halogen exchange reaction, *Org. Proc. Res. Dev.* 12: 929–939.
19. Loebbecke, S., Boskovic, D., Tuercke, T., Mendl, A., and Antes, J. 2010. Employing microstructured reactors for the synthesis and downstream processing of liquid nitrate esters at a technical scale, *11th International Conference on Microreaction Technology*, Kyoto, Japan, 82–83.
20. Ebrahimi, F., Kolehmainen, E., and Turunen, I. 2009. Safety advantages of on-site microprocesses, *Org. Proc. Res. Dev.* 13: 965–969.
21. Ebrahimi, F., Kolehmainen, E., Oinas, P., Hietapelto, V., and Turunen, I. 2010. Production of unstable percarboxylic acids in a microstructured reactor, *11th International Conference on Microreaction Technology*, Kyoto, Japan, 404–405.
22. Struempel, M., Ondruschka, B., Daute, R., and Stark, A. 2008. Making diazomethane accessible for R&D and industry: Generation and direct conversion in a continuous micro-reactor set-up, *Green Chem.* 10: 41–43.
23. Schwalbe, T., Kadzimirsz, D., and Jas, G. 2005. Synthesis of a library of ciprofloxacin analogues by means of sequential organic synthesis in micro reactors, *QSAR Comb. Sci.* 24: 758–768.
24. Tanaka, K., Motomatsu, S., Koyana, K., Tanaka, S., and Fukase, K. 2007. Large-scale synthesis of immunoactivating natural product, Pristane, by continuous flow microfluidic dehydration as the key step, *Org. Lett.* 9: 299–302.
25. Hopkin, M. D., Baxendale, I. R., and Ley, S. V. 2010. A flow-based synthesis of imatinib: The API of Gleevac, *Chem. Commun.* 2450–2452.
26. Ricardo, C. and Xiongwei, N. 2009. Evaluation and establishment of a cleaning protocol for the production of vanisal sodium and aspirin using a continuous oscillatory baffled reactor, *Org. Proc. Res. Dev.* 13: 1080–1087.
27. Riva, E., Gagliardi, S., Mazzoni, C., Passarella, D., Rencurosi, A., Vigo, D., and Martinelli, M. 2009. Efficient continuous flow synthesis of hydroxamic acids and suberoylanilide hydroxamic acid preparation, *J. Org. Chem.* 74: 3540–3543.
28. Gustafsson, T., Pontén, F., and Seeberger, P. H. 2008. Trimethylaluminium mediated amide bond formation in a continuous flow microreactors as key to the synthesis of Rimonabant and Efaproxiral, *Chem. Commun.* 1100–1102.
29. Panke, G., Schwalbe, T., Stirner, W., Taghavi-Moghadam, S., and Wille, G. 2003. A practical approach of continuous processing to high energetic nitration reactions in microreactors, *Synthesis* 2827–2830.
30. LaPorte, T. L., Hamedi, M., DePue, J. S., Shen, L., Watson, D., and Hsieh, D. 2008. Development and scale-up of three consecutive continuous reactions for production of 6-hydroxybuspirone, *Org. Proc. Res. Dev.* 12: 956–966.
31. Riva, E., Gagliardi, S., Martinelli, M., Passeralla, D., Vigo, D., and Rencurosi, A. 2010. Reaction of Grignard reagents with carbonyl compounds under continuous flow conditions, *Tetrahedron* 66: 3242–3247.

32. Bogaert-Alvarez, R. J., Demena, P., Kodersha, G., Polomski, R. E., Soundararajan, N., and Wang, S. S. Y. 2001. Continuous processing to control a potentially hazardous process: Conversion of aryl 1,1-dimethylpropargyl ethers to 2,2-dimethylchromenes (2,2-dimethyl-2*H*)-1-benzopyrans, *Org. Proc. Res. Dev.* 5: 636–645.
33. Qian, Z., Baxendale, I. R., and Ley, S. V. 2010. A flow process using microreactors for the preparation of a quinoline derivative as a potent $5HT_{1B}$ antagonist, *Synlett* 4: 0505–0508.
34. Horchler, C. L., McCauley, J. P., Hall, J. E., Snyder, D. H., Moore, W. C., Hudzik, T. J., and Chapdelaine, M. J. 2007. Synthesis of novel quinoline and quinoline-2-carboxylic acid (4-morpholin-4-yl-phenyl)amines: A late stage diversification approach to potent $5HT_{1B}$ antagonists, *Bioorg. Med. Chem.* 15: 939–950.
35. Tietze, L. F. and Liu, D. 2008. Continuous-flow microreactor multi-step synthesis of an aminonaphthalene derivative as starting material for the preparation of novel anticancer agents, *Arkivoc* viii: 193–210.
36. Baxendale, I. R., Griffiths-Jones, C. M., Ley, S. V., and Tranmer, G. K. 2006. Preparation of the neolignan natural product grossamide by a continuous flow process, *Synlett* 3: 427–430.
37. Baxendale, I. R., Deeley, J., Griffiths-Jones, C. M., Ley, S. V., Saaby, S., and Tranmer, G. K. 2006. A flow process for the multi-step synthesis of the alkaloid natural product oxomaritidine: A new paradigm for molecular assembly, *Chem. Commun.* 2566–2568.
38. Fukuyama, T., Kobayashi, M., Rahman, M. T., Kamata, N., and Ryu, I. 2008. Spurring radical reactions of organic halides with tin hydride and TTMSS using microreactors, *Org. Lett.* 10(4): 533–536.
39. Gilles, J. M., Prenant, C., Chimon, G. N., Smethurst, G. J., Dekker, B. A., and Zweit, J. 2006. Microfluidic technology for PET radiochemistry, *Appl. Rad. Isotopes.* 64: 333–336.
40. Steel, C. J., O'Brien, A. T., Luthra, S. K., and Brady, F. 2007. Automated PET radio-syntheses using microfluidic devices, *J. Label Compd. Radiopharm.* 50: 308–311.
41. Lu, S. Y. and Pike, V. W. 2007. *PET Chemistry: 10 Micro-Reactors for PET Tracer Labelling*, Berlin: Springer, 271–287.
42. Lee, C. C., Sui, G., Elizarov, A., Shu, C. J., Shin, Y. –S., Doley, A. N., Huang, J. et al. 2005. Multistep synthesis of a radiolabeled imaging probe using integrated microfluidics, *Science* 310: 1793–1796.
43. Lu, S., Giamis, A. M., and Pike, V. W. 2009. Synthesis of [^{18}F]-fallypride in a microreactor: Rapid optimization and multiple production in small doses for micro-PET studies, *Curr. Radiopharm.* 2: 49–55.
44. Roberge, D. M., Gottsponer, M., Eyholzer, M., and Kockmann, N. 2009. Industrial design, scale-up and use of microreactors, *Chem. Today* 27: 8–11.
45. Moseley, J. D. and Woodman, E. K. 2008. Scaling-out pharmaceutical reactions in an automated stop-flow microwave reactor, *Org. Proc. Res. Dev.* 12: 967–981.
46. Hessel, V., Hofmann, C., Löb, P., Löwe, H., and Parals, M. 2007. Microreactor processing for the aqueous Kolbe–Schmitt synthesis of hydroquinone and phloroglucinol, *Chem. Eng. Technol.* 30: 355–362.
47. Braune, S., Pöchlauer, P., Reintjens, R., Steinhofer, S., Winter, M., Lobet, O., Guidat, R., Woehl, P., and Guermeur, C. 2008. Selective nitration in a microreactor for pharmaceutical production under cGMP conditions. *Chem. Today* 26(5): 1–4.
48. Schwalbe, T., Kursawe, A., and Sommer, J. 2005. Application report on operating cellular process chemistry plants in fine chemical and contract manufacturing industries, *Chem. Eng. Technol.* 28: 408–419.
49. Asano, Y., Togashi, S., Tsudome, H., and Murakami, S. 2010. Microreactor technology: Innovations in production processes, *Pharm. Eng.* 32–42.

8 Microscale Continuous Separations and Purifications

8.1 INTRODUCTION

It can be seen throughout that the use of micro reaction technology offers many advantages to the synthetic chemist for the performance of organic reactions; however, the time saved at this stage can often be eroded by the need to collect sufficient material to enable conventional batch-type postreaction work-ups and purifications to be performed prior to analysis and characterization. With this in mind, researchers have recently begun to investigate methods for the performance of these unit operations under continuous flow and as such, a brief overview into current techniques available to the researcher form the basis of this chapter. For detailed discussions on this subject see reviews by Aota et al. [1], Tia and Herr [2], and Hartman and Jensen [3].

8.2 LIQUID–LIQUID EXTRACTIONS

One of the most common purification techniques used in synthetic chemistry is that of liquid–liquid extraction (LLE), a technique that has been shown to be suited to continuous microprocessing due to the high interfacial areas obtained between phases within microfluidic systems [4]. Using both experimental and numeric evaluation, Kuban et al. [5] investigated the effect of wetting properties on the resulting flow regime, observing that at low flow rates segmented flow dominated within microchannels however, at larger flow rates (velocities >10 mm s^{-1}) side-by-side, or stratified, flow resulted; affording a large interfacial area. In the case of solvents with moderate surface tension (3.5×10^{-4} Nm^{-1}) and low viscosity (<10^3 Pa s), such as CHCl$_3$ and hexane, only segmented flow is observed within open channel networks.

From previous discussions (Chapters 1 and 3) it can be seen that both stratified and segmented flow have their advantages with respect to the performance of synthetic reactions, and with it widely acknowledged that the side-by-side contacting of immiscible fluids enables ease of phase separation but provides low interfacial surface areas and system throughputs, compared with segmented flow [6]. In addition to the use of conventional solvent systems for continuous LLE, the authors have also demonstrated the efficient use and recycle of less well-known solvents such as

$CF_3C_6F_{11}$ [7] and ionic liquids [8]. With this in mind, the following section describes examples of both stratified and segmented techniques for the purification of reactant streams under continuous conditions.

8.2.1 Side-by-Side (Stratified) Contacting

As previously touched upon, the formation of a stable liquid–liquid interface within microfluidic devices is dependent on many physical parameters such as channel dimensions, reactant stream composition, and flow rate. Consequently, researchers have investigated a series of stabilization methods, ranging from surface derivatization [9,10] to the fabrication of supports and guide structures [11], as a means of affording the spatial control required for the performance of stratified LLEs [12,13].

Standard Microchannels: An early example of this approach was reported by Kitamori and coworkers [14], who utilized a quartz glass device for the solvent extraction of a Co-2-nitroso-5-dimethylaminophenol complex from DI H_2O. Employing microchannels comprising of a Y-shaped inlet and a Y-shaped outlet (channel dimensions = 250 μm (wide) × 250 μm (deep) × 2 mm (long)), the authors monitored the extraction efficiencies obtained within their microfluidic system using thermal lens microscopy (TLM) of the organic phase. Employing equal flow rates for the aqueous Co-2-nitroso-5-dimethylaminophenol complex (1×10^{-3} M) and toluene, the authors investigated the effect of residence time on the extraction efficiency; identifying an extraction time of 60 s within the microfluidic system. For comparative purposes the authors performed an analogous extraction in a separating funnel (using mechanical agitation) whereby an extraction time of 10 min was required to reach equilibration; demonstrating a 10-fold reduction in extraction time within the micro reactor, a result that was attributed to the large specific interfacial area obtained. The authors simultaneously demonstrated that extraction times were at least one order of magnitude greater than those obtained using conventional methodologies, for the ion-pair extraction of Fe(II) with 4,7-diphenyl-1,10-phenanthroline-disulfonic acid and tri-*n*-octylmethylammonium chloride achieved within 45 s [15–17] illustrating the generality of the technique.

From a more synthetic perspective, Kitamori and coworkers [18,19] subsequently demonstrated the exploitation of the large interface formed between an aqueous solution containing 4-nitrobenzene diazonium tetrafluoroborate **1** (1×10^{-4} M) and 5-methylresorcinol **2** (1×10^{-3} M) in ethyl acetate to simultaneously perform a diazotization reaction (Scheme 8.1) and product **3** isolation. Using a glass micro reactor, with a Y-shaped inlet (channel dimensions = 250 μm (wide) × 100 μm (deep) × 3.0 cm (long)), the authors employed a total volumetric flow rate of 20 μL min^{-1} affording a residence time of 2.3 s. Under the aforementioned conditions, the diazo product **3** was obtained in quantitative conversion with respect to 4-nitrobenzene diazonium tetrafluoroborate **1**. In addition to the rapid nature of the reaction, this approach proved advantageous as the *bis*-azo adduct was not formed, as observed in analogous batch reactions. This was attributed to efficient removal of the reaction product **3** from the aqueous layer to the organic phase, thus preventing additional reaction from occurring which would give rise to by-product formation.

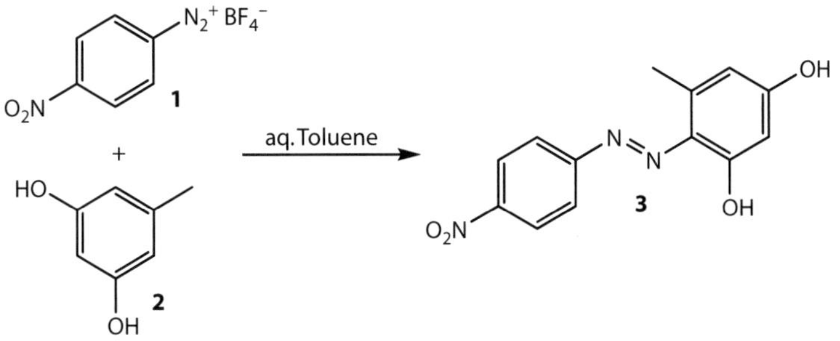

SCHEME 8.1 Schematic illustrating the diazotization reaction used to demonstrate the ability to perform both syntheses and inline extractions within a microfluidic reactor.

Surface Functionalization: Employing a PDMS/glass microdevice with a rectangular microchannel (dimensions = 200 μm (wide) × 40 μm (deep)), Xiao et al. [20] investigated the extraction of Ephedrine from an aqueous solution into an organic solvent as a means of sample preparation, prior to analysis by gas chromatography. Due to the formation of segmented flow within the reaction channels, under the conditions evaluated, the authors found it necessary to modify half the channel wall, using octadecyltrichlorosilane, in order to obtain a stable stratified organic–aqueous interface; which was straight and parallel to the side walls.

Employing various ephedrine concentrations (50–250 μg mL^{-1}), the authors evaluated the extraction efficiency of their two-phase laminar flow reactor, observing that a Y-shaped outlet efficiently separated the two phases prior to analysis by GC. Under the aforementioned conditions, the authors were gratified to obtain excellent repeatability (CV = 4.8%) between reactors and excellent linearity with respect to concentration; finding complete extraction was achieved in only 12 min. Additional studies into the effect of extraction solvent (EtOAc, BuOAc, and cyclohexane) enabled the authors to identify EtOAc as the most efficient solvent for this application affording 92.1% ephedrine isolation with a contact time of 2.1 s compared with 11.7% with cyclohexane.

During the development of a tool for the integrated 11α-hydroxylation of progesterone **4** to 11α-hydroxyprogesterone **5** (Figure 8.1), Žnidaršič-Plazl and Plazl [21] focused on the extraction of steroids in a Y-shaped microfabricated reactor (channel dimensions = 220 μm (wide) × 50 μm (deep) × 33.2 cm (long)) as a means of

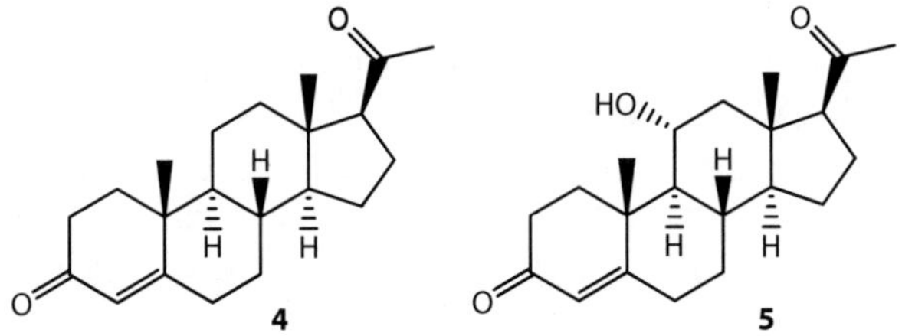

FIGURE 8.1 Illustration of the two steroids efficiently separated using a three-phase separation device.

replacing a batch separation process used in the industrial production of corticosteroidal compounds.

Employing water as the aqueous phase and EtOAc as the organic phase, again the authors found it necessary to treat the microchannel walls with octadecylsilane in order to stabilize the liquid–liquid interface using a previously reported technique [22], with the extraction efficiency monitored at a range of flow rates (aqueous: 10–100 µL min^{-1}, organic: 21–210 µL min^{-1}).

At aqueous flow rates of 20 µL min^{-1} and organic flows of 42 µL min^{-1} respectively, the authors were able to demonstrate the efficient extraction of progesterone **5** and 11α-hydroxyprogesterone **4** from the aqueous layer into the organic fraction; affording analogous results to those obtained in a mathematical model. Although incomplete extraction was observed, as predicted, the mathematical model enabled the authors to determine the channel length required in order to develop a 100% efficient LLE device suitable for increasing the capacity of the system. Further work is therefore underway into the development of complimentary downstream processes such as solvent removal and filtration; enabling the development of a completely automated process.

Guide Structures: As previously discussed, depending on the characteristics of the binary phase employed, it can be difficult to maintain a stable longitudinal interface between the two immiscible phases. With this in mind, Tagawa et al. [23] developed a micro reactor containing a guideline structure which served to stabilize the fluidic interface and prevent the formation of slug or segmented flow without the need for surface functionalization. Employing a microchannel device with overall dimensions of 95.6 µm (wide) × 17.9 µm (deep) × 3.0 cm (long), containing a guideline pillared structure down the center of the channel (dimensions = 100 µm (long)) at an interval spacing of 100 µm, the authors evaluated the technique using the hydrolysis of benzoyl chloride **6** as a model reaction.

Using a two-phase flow system, whereby the benzoyl chloride **6** was dissolved in toluene (containing an internal standard) and DI H$_2$O was used as the immiscible aqueous phase, the authors investigated the effect of flow rate and temperature on both the reaction and interface obtained. Employing the pillared structures, the authors were able to stabilize the interface over a range of flow rates, observing no temperature effect on the interface formed. In addition to the guideline structures maintaining a stable interface, the authors were also gratified to observe efficient separation of the two phases via a Y-shaped channel at the reactor outlet. Comparing the space–time yields obtained in flow with a conventional batch reaction, the authors obtained a 13 times enhancement in conversion; however, further work is needed to optimize the system chemically.

In addition to their significant contribution to the development of miniaturized continuous flow LLE's, Kitamori and coworkers [24] demonstrated an elegant example of Co(II) analysis which combined LLE with a series of additional microunit operations. Employing a Pyrex reactor manifold, containing guide structures to stabilize the liquid–liquid interface (Figure 8.2), the authors investigated the ability to perform the steps of Co(II) chelation, extraction, decomposition, and removal in a single integrated device.

To perform the complete wet analysis of Co(II) under continuous flow, the sample (0.2 µL min^{-1}) was introduced into the micro reactor alongside a solution of

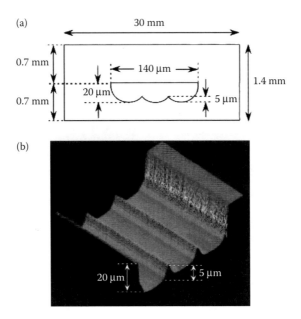

FIGURE 8.2 (a) Schematic illustrating the cross-sectional view of the guide structures fabricated by wet-etching and (b) 3D image of the guide structures. (Reprinted with permission from Tokeshi M. et al. 2002. *Anal. Chem.* 74: 1565–1571. Copyright (2002) American Chemical Society.)

2-nitroso-1-naphthol (NN) in NaOH (1.0 M, 0.2 µL min^{-1}) and *m*-xylene (0.4 µL min^{-1}) enabling the Co(II) to be chelated and extracted into the *m*-xylene; the extraction equilibrium was determined to occur at a microchannel length of 130 mm. Concentrated HCl (12.0 M, 0.2 µL min^{-1}) was subsequently introduced into the reactor, to the left of the *m*-xylene layer and NaOH (1.0 M, 0.2 µL min^{-1}) to the right, affording the decomposition and aqueous extraction of any coexisting metal chelates. TLM was then used to quantify the proportion of Co(II) (488 nm) within the *m*-xylene layer, with measurements taken 3 mm from the phase confluence point of HCl, *m*-xylene and NaOH (Figure 8.3).

Using the aforementioned methodology, the authors obtained stable laminar flow in all sections of the micro reactor enabling the rapid determination of Co(II) in an admixture of Cu(II) with a detection limit of 1.8×10^{-8} M with a detection volume of only 7.2 fl.

Reduced Solvent Consumption: Although initial examples of continuous flow LLE have demonstrated increased extraction efficiencies, compared to batch techniques, the authors have found the use of a continuously flowing organic stream to be disadvantageous and sought to reduce the volume of solvent required to perform these LLEs [25]. In this case, the glass device developed comprised of a shallow microchannel (dimensions = 12 µm (deep), volume = 1.2 nL) which contacted a deeper microchannel (dimensions = 59 µm (deep)); with selective surface functionalization (hydrophobic) of the shallow channel enabling the organic phase to be held within the channel, while the aqueous phase flowed continuously within the deeper channel

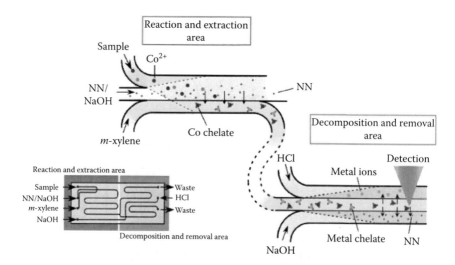

FIGURE 8.3 Schematic illustrating the continuous flow wet analysis of Co(II). (Reprinted with permission from Tokeshi M. et al. 2002. *Anal. Chem.* 74: 1565–1571. Copyright (2002) American Chemical Society.)

(Figure 8.4). In order to demonstrate the extraction capacity of the device, the authors evaluated the extraction of methyl red (3×10^{-6} M, 2 µL min^{-1}) from an aqueous phase into toluene. Using this approach, a maximum concentration of methyl red (3.5×10^{-5} M) in toluene was reached after 20 min; demonstrating a 12-fold increase in analyte concentration and a dramatic reduction in the volume of solvent required in order to execute the extraction.

In 2006, Aota et al. [26] evaluated the use of countercurrent laminar flow streams as a means of increasing the separation efficiency obtained in systems utilizing stratified liquid contacting. Employing a reaction channel whereby the upper half of the reaction channel was treated with ODS, rendering it hydrophobic and the lower channel hydrophilic (after treatment with base), the authors once again investigated the extraction of Co-2-nitroso-5-dimethylaminophenol complex (1×10^{-3} M) from one phase to the other. Employing equal organic and aqueous flow rates, ranging from 0.15 to 1.0 µL min^{-1}, the authors monitored the extraction efficiency using TLM, confirming 98.6% extraction under optimized conditions. Compared with coflowing techniques previously reported, the use of counterflows was found to be 4.6 times more efficient, leading to a reduction in the solvent volumes required for postreaction processing.

See Chapter 3 for a discussion of other biphasic reactions whereby inline separations have been performed including a series of aldol reactions by Mikami et al. [27] and phase transfer alkylations by Kobayashi and coworkers [28].

8.2.2 Three-Phase Microextractions

When considering the purification of more complex reaction mixtures, it can be necessary to perform three-phase extractions (back extractions), where the analyte

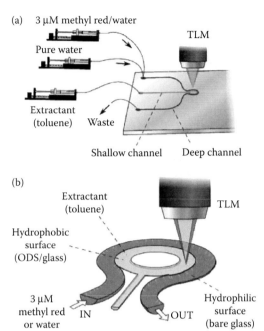

FIGURE 8.4 Schematic illustrating (a) a low volume micro extraction device and (b) a close-up of the detection mechanism used. (With kind permission from Springer Science+Business Media: *Microchim. Acta*, Circulation microchannel for liquid–liquid microextraction, 164, 2009, 241–247, Kikutani Y. et al.)

of interest is extracted from the aqueous phase (feed phase) into an organic phase (transport phase) and then back into a second aqueous phase (acceptor phase), a microfluidic example of this was originally reported by Hibara et al. [29]. As discussed previously, Tokeshi et al. [24] subsequently demonstrated the extraction of 2-nitroso-1-naphthol metal complexes and Goto and coworkers [30] reported the forward and back extraction of yttrium and zinc ions using this approach.

In an extension to these analytical examples, the technique has subsequently been applied to the purification of reaction mixtures and natural product extracts, demonstrating the applicability of the technique toward more conventional synthetic product purification.

A more synthetic example reported by van Beek and coworkers [31] described the development of a micro reaction unit in which they could perform the simultaneous extraction and back extraction of the weakly basic alkaloid Strychnine **7** (Figure 8.5). To achieve this, the authors employed a glass reactor with a contact length of 3.56 cm (channel dimensions = 100 μm (wide) × 40 μm (deep)), containing a series of semi-toroidal pillar structures between the three channels (Figure 8.6), a nonpolar coating was applied to the central channel. Using the aforementioned device, the authors investigated the purification of a crude alkaloid **7** solution (0.2 mM), known to contain some polar and nonpolar impurities, and monitored the cleanup efficiency by offline HPLC analysis. Employing a basic aqueous feed phase

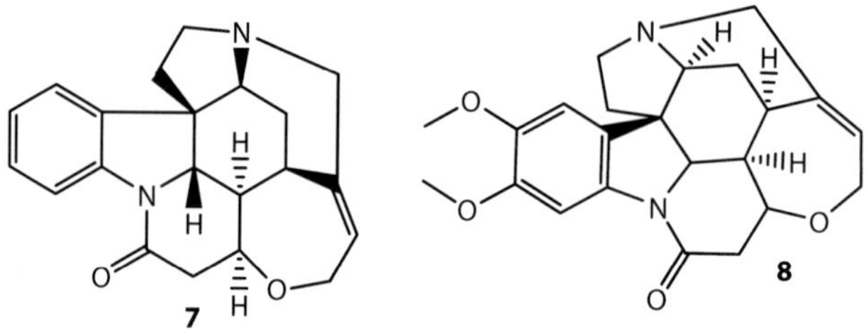

FIGURE 8.5 Illustration of Strychnine **7** and Brucine **8**, the alkaloids extracted from an extract of Strychnos seeds under continuous flow.

(0.5 µL min^{-1}), a transport phase of CHCl$_3$ (1.0 µL min^{-1}) and an acceptor phase of aqueous formic acid (0.5 µL min^{-1}), the crude alkaloid **7** was extracted from the feed phase (91%) into the transport phase and then back-extracted into the acceptor phase (93%) with a residence time of 25 s.

Having demonstrated the successful isolation of Strychnine **7** from a simulated mixture, the authors subsequently investigated the cleanup of an alkaloid mixture extracted from Strychnos seeds; known to contain the alkaloids Strychnine **7** and Brucine **8**. Using flow rates of 1:2:1 µL min^{-1}, the authors were able to separate the two alkaloids **7** and **8** into an acidic acceptor phase, leaving any polar impurities in the feed phase and any nonpolar impurities within the transport phase. Analysis of the initial feed phase and resulting acceptor phase by offline HPLC analysis enabled the authors to confirm successful purification of Strychnine **7** and **8** from the plant extract. Based on these findings, the authors are focusing their attention on interfacing the micro flow separator with nano ESI-MS to enable purification and online detection of alkaloids deriving from plant extracts.

From these examples it can be seen that in addition to the advantages associated with performing reactions under continuous flow, further increases in system efficiency can be harnessed through the coupling of an inline continuous separation, which is no longer limited to a single liquid–liquid phase extraction.

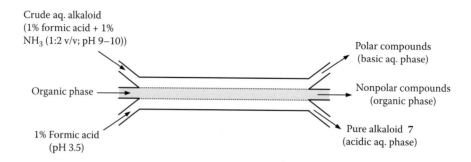

FIGURE 8.6 Schematic illustrating the micro reactor used for the rapid sample clean up of Strychnine **7** from plant extracts (where "——" represent pillars within the microchannel and "·····" represents the hydrophobic microchannel).

8.2.3 Segmented Flow

While numerous examples have reported LLEs conducted under continuous flow, when it comes to the use of segmented or slug flow, separation of the two phases is normally conducted offline within a collection vessel [32]; as at the microscale, gravitational forces are small compared to surface forces. As such it can be difficult to perform phase separations using conventional differences in fluid density (Chapter 1) [33] although examples have been reported [34,35]. With the powerful nature of this separation technique forming a fundamental processing step within synthetic organic chemistry, Jensen and coworkers [36] investigated the exploitation of surface forces as a means of efficiently separating biphasic systems at the microscale. Utilizing a thin, porous fluoropolymer membrane (pore size = 0.1–1.0 µm) that selectively wetted nonaqueous solvents, the authors were able to separate aqueous slugs from a continuous organic phase, over a total flow rate range of 10–2000 µL min^{-1}, utilizing

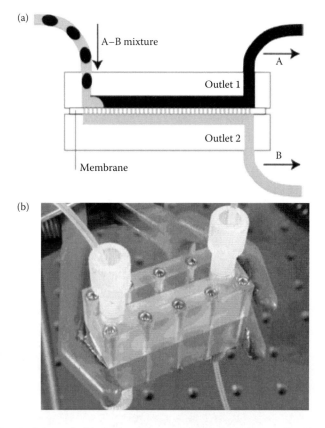

FIGURE 8.7 (a) Schematic illustrating the cross-sectional view of the liquid–liquid extractor containing a porous membrane (A = aqueous solution, B = organic phase) and (b) a photograph of the separator housed within a polycarbonate holder (dimensions = 10 mm (wide) × 50 mm (long) × 20 mm (deep)). (Kralj, J. G., Sahoo, H. R., and Jensen, K. F. 2007. Integrated continuous microfluidic liquid–liquid extraction, *Lab Chip* 7: 256–263. Reproduced by permission of the Royal Society of Chemistry.)

a polycarbonate reactor. As depicted in Figure 8.7, the immiscible liquids are pumped into a channel above the membrane, where selective wetting and capillary pressure is used to both induce and maintain separation of the two phases. To ensure optimum separation efficiency within this system, it is imperative that the immiscible phases possess sufficiently different surface wetting properties in order to drive the phase separation. As water does not wet the membrane, and hence cannot pass through it, the system was found to be suitable for the rapid separation of aqueous phases (Figure 8.7, outlet 1) from organic/fluorous materials (Figure 8.7, outlet 2); as visually confirmed through the use of red and blue dyed aqueous buffers (pH 4 and 10).

To enable integration of this technology into microchemical systems without a loss of performance, due to the presence of large dead volumes at interconnects, the authors subsequently fabricated a silicon micro reactor which contained both mixing, and reaction and separating zones (Figure 8.8, Table 8.1).

Employing typical flow rates of 50 µL min^{-1} for the organic and aqueous phases respectively, the authors demonstrated the extraction of DMF from DCM into an aqueous phase. To monitor the efficiency of the technique, in particular the effect of miscible components on the processing ability, the proportion of DMF in the nonpolar organic phase was varied between 1 and 20 mol%. Measuring the concentration of DMF in the aqueous phase, enabled the authors to determine the extraction efficiency (60%) of their system to be equivalent to that obtained in a conventional single "shake-flask" extraction; concluding that 94% extraction of the undesired

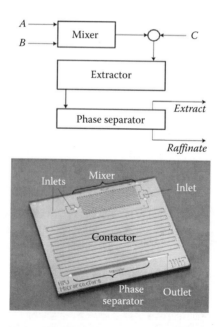

FIGURE 8.8 Photograph illustrating a silicon device containing a membrane phase separator (dimensions = 35 mm (wide) × 30 mm (long) × 1.5 mm (deep)). (Kralj, J. G., Sahoo, H. R., and Jensen, K. F. 2007. Integrated continuous microfluidic liquid–liquid extraction, *Lab Chip* 7: 256–263. Reproduced by permission of the Royal Society of Chemistry.)

TABLE 8.1
Dimensions of Each Component within the Silicon Microfluidic Liquid–Liquid Extraction Device

Section	Depth (μm)	Width (μm)[a]	Length (mm)
Flow ports	500	750	N/A
Mixer	71	100	350
Contactor	156	300	400
Separator	648	1000	21

[a] Width at the top surface of the silicon wafer.

species, in this case DMF, could be removed using a three-stage separation. Using the aforementioned technique, the authors have developed a continuous processing tool capable of separating organic-aqueous and fluorous-aqueous liquid–liquid systems containing high fractions of miscible components, to attain extraction efficiencies that are in agreement with the literature.

Curtius Rearrangement: To further demonstrate the synthetic utility of this continuous processing technique, the authors integrated the LLE step into the multistep synthesis of carbamates [37].

As Scheme 8.2 illustrates, the model reaction selected was the synthetically useful azidation of benzoyl chloride **6** to afford benzoyl azide **9**. Employing a residence time of 200 min, within an integrated silicon micro reactor (Figure 8.8), the authors obtained 99% conversion of benzoyl chloride **6** (0.36 M in toluene) to benzoyl azide

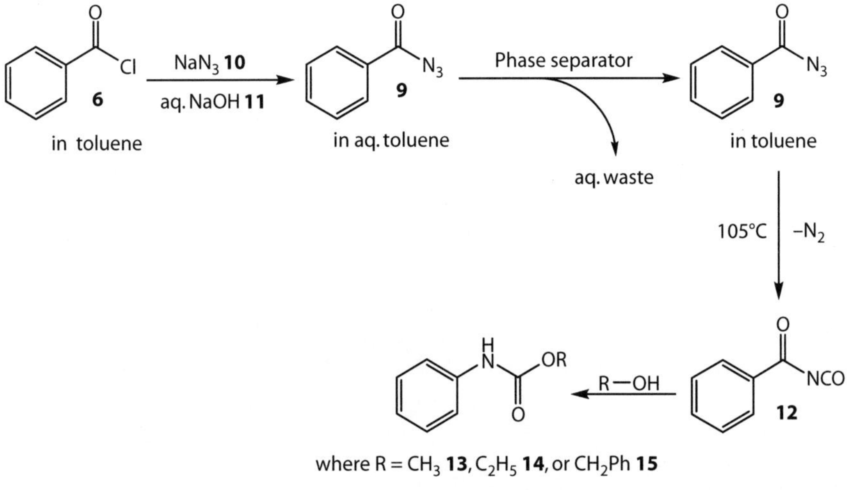

SCHEME 8.2 General reaction scheme illustrating the multistep synthesis of carbamates, used to demonstrate the application of an online liquid–liquid separator for the removal of water.

9, using a slight excess of aq. sodium azide **10** (0.4 M in NaOH **11**). This reaction step was then followed by a LLE, utilizing the phase separation membrane, to afford a reactant stream containing anhydrous benzoyl azide **9** in toluene. This step was then followed by the decomposition of the organic azide **9** to the respective isocyanate **12** (99% conversion), within a second silicon micro reactor heated to 105°C. After complete decomposition of the azide **9**, N_2 was removed in a gas–liquid separator [38], whereby the liquid passed through a membrane which retarded gas penetration, and the final reaction step, a biphasic reaction of the isocyanate **12** with an alcohol, was performed. To demonstrate the versatility of the reactor, the authors investigated the reaction of three alcohols with phenyl isocyanate **12** to afford the methyl- **13**, ethyl- **14** and benzyl-carbamate **15** with typical throughputs of 3–5 mg h^{-1}. A detailed discussion of the incorporation of the phase separation principle into a synthetic process can be found in Chapter 3.

8.2.4 FLUOROUS PHASE EXTRACTIONS

In efforts toward advances in glycobiology, Goto and Mizuno [39] and Kawakami et al. [40] evaluated the coupling of micro reactors and fluorous tag methodology for the synthesis and extraction of carbohydrates. Using the synthesis of monosaccharide **16**, a glycosyl acceptor, the authors evaluated the ability to reduce the time taken to perform the synthetic sequence as a result of employing continuous flow reactor technology.

Utilizing a Y-shaped micromixer (0.5 mm i.d.) and a Teflon tube reactor (dimensions = 0.5 mm (i.d.) × 5 or 10 cm (long)), the authors investigated each reaction step, depicted in Scheme 8.3, separately; however, the postreaction processing was the same for all reaction steps. Using a second micromixer, a solution of a fluorous solvent ($MeOC_4F_9$-FC72) and organic solvent (aq. MeCN or aq. MeOH) were mixed prior to combining with the reaction products to terminate the reaction. The biphasic reaction mixture then passed through a third micromixer prior to collection offline within a separating funnel. The fluorous layer was then evaporated and reduced under pressure to afford the precursor for the next step of the reaction, which was used without further purification.

Employing this serial approach, the authors were able to perform the flow synthesis and purification of monosaccharide **16** which required only purification via a single silica-gel chromatography column as the final step. Based on 0.15 mmol of fluorous tag **17** and using a reaction and separation time of 6 h, the authors obtained the target **16** in 55% yield (11% based on per-*O*-acetyl-β-d-galactopyranose **18**) with a total synthesis time of 9 h including the final purification step.

Suzuki–Miyaura Reaction: In a novel extension to the use of inline phase separations, Theberge et al. [41] demonstrated the performance of Suzuki–Miyaura coupling reactions in aqueous microdroplets within a fluorous continuous phase which served to form a catalytically active fluorous interface; capable of being recycled using a specially modified tube reactor (Figure 8.9). Using this approach, the authors were able to essentially anchor the catalyst within the carrier phase so it could catalyze reactions within the aqueous microdroplets; coupling the advantages of fluorous phase chemistry with those of microfluidics.

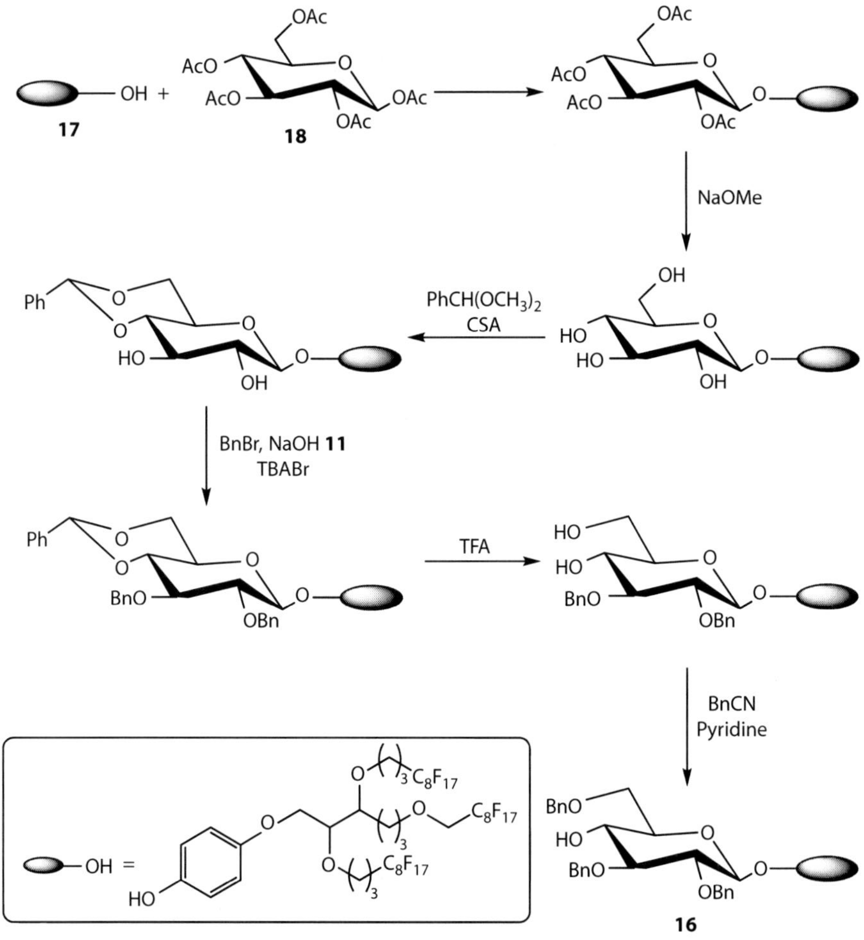

SCHEME 8.3 Illustration of the reaction sequence used to demonstrate the advantages associated with coupling of fluorous tag methodology and continuous flow reactors.

Employing a PTFE tubular reactor (dimensions = 0.75 mm (i.d.)) and pressure driven flow, the authors employed a T-junction at which they generated monodisperse droplets of aqueous reagents within the fluorous catalyst (Figure 8.9a). The coupling reaction was then performed in a length of PTFE tubing prior to collection of the reaction products in aqueous HCl; followed by analysis of the aqueous phase by off-line HPLC (with UV detection). Under the aforementioned conditions the authors were able to conduct, for the first time, the Suzuki–Miyaura reactions at ambient temperatures, obtaining moderate to excellent yields of the target biaryl compounds; with reaction times ranging from 0.75 to 8 h; depending on the reactor length employed (Table 8.2).

In an attempt to increase the synthetic utility of the reaction system, the authors exploited the continuous nature of the technique and the phase separation obtained

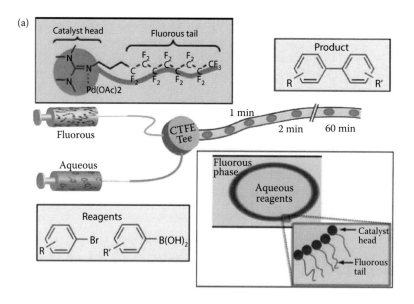

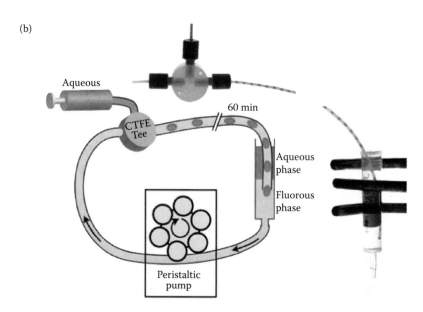

FIGURE 8.9 Schematic illustrating the microdroplets formed with (a) catalytically active interfaces and (b) derived from fluorous tagged palladium catalysts and aqueous reagents. (Theberge, A. B. et al., 2009. Suzuki–Miyaura coupling reactions in aqueous microdroplets with catalytically active fluorous interfaces, *Chem. Commun.* 6225–6227. Reproduced by permission of the Royal Society of Chemistry.)

TABLE 8.2
Summary of the Suzuki–Miyaura Coupling Reactions Performed within the Catalytic Microdroplets[a]

$$R\text{—}Br + R^1\text{—}B(OH)_2 \xrightarrow[Pd(OAc)_2\ \mathbf{19},\ FC\text{-}77]{K_2CO_3} R\text{—}R^1$$

Aryl Bromide	Boronic Acid	Residence Time[b] (h)	Yield[c] (%)
4-HO-C₆H₄-Br	C₆H₅-B(OH)₂	3	90
4-HO-C₆H₄-Br	4-HO₂C-C₆H₄-B(OH)₂	3	77
4-HO₂C-C₆H₄-Br	C₆H₅-B(OH)₂	0.75	99
3-HO₂C-C₆H₄-Br	4-CH₃-C₆H₄-B(OH)₂	1	91
5-Br-furan-2-CO₂H	C₆H₅-B(OH)₂	8[c]	63

[a] The aqueous phase contained K_2CO_3 (0.14 M), boronic acid (5.5×10^{-2} M), aryl halide (4.5×10^{-2} M) and the fluorous phase contained $Pd(OAc)_2$ **19** (1.36×10^{-4} M) and ligand (2.72×10^{-4} M).
[b] Unless otherwise stated a flow rate ratio of 11.5:6.9 µL min⁻¹ was employed.
[c] Analyzed by HPLC-UV.

between the aqueous and fluorous phases to afford continuous recycling of the catalyst. As Figure 8.9b illustrates, by employing a wider tube at the exit of the reaction channel, phase separation of the reaction products and catalyst were observed, with the denser fluorous phase then pumped with a peristaltic pump back to the T-junction where it was used to generate more catalytic droplets. Using such a closed-loop synthesizer, the authors were able to fully automate the coupling reactions and efficiently recycle the catalytic material, $Pd(OAc)_2$ **19**, with no observed reduction in catalytic activity.

8.2.5 Comparison of Liquid–Liquid Extraction Efficiencies

Fries et al. [42] compared side-by-side (stratified) contacting and segmented flow for the industrially relevant LLE of Vanillin **20** (Figure 8.10) in rectangular microchannels of varying dimensions; summarized in Table 8.3. The microchannels employed herein were fabricated from PDMS and sealed using a glass cover plate resulting in hydrophilic microchannels.

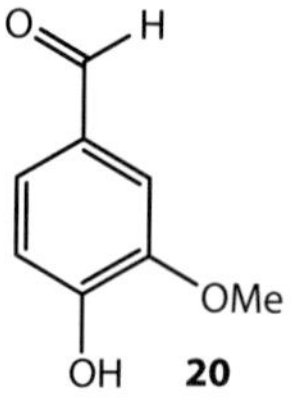

FIGURE 8.10 Vanillin **20**, the analyte used to demonstrate the enhanced extraction efficiencies obtained using segmented flow compared with side-by-side liquid contacting; used as a flavoring in foodstuffs and pharmaceutical agents.

In order to compare the two flow regimes, solutions of deionized H_2O and toluene (containing 1 g L^{-1} Vanillin **20**) were delivered to the reactors under investigation, using pressure driven flow; where flow rates were varied between 10 and 150 µL min^{-1}. Upon exiting the micro reactor, the organic phase was analyzed by GC-MS, using 4-ethylresorcinol as an internal standard, and the aqueous phase evaluated by UV–VIS spectroscopy (λ = 280 nm). Using this approach, the authors investigated the effect of residence time (0.5–7 s) on the LLE, identifying increasing extraction efficiency (E) as a function of increasing residence time and decreasing channel width whereby E = 0.8 at 400 µm and E = 1.00 at 200 µm. The use of narrower channels and longer residence times however only serves to reduce the volumetric throughput of the extraction module, reducing industrial applicability of such a technique. With this in mind, the authors investigated the use of segmented flow where due to the hydrophobic nature of the reaction channels, the organic phase formed the continuous phase and the dispersed phase was the aqueous solution. Under this flow regime, the authors obtained extraction efficiencies of 1 with residence times ≥0.5 s and since the interfacial area is increased 1.8-fold, overall mass transfer (k_La) increased by 2.5 times; resulting in increased extraction efficiency.

TABLE 8.3
Summary of the Microchannel Configurations Used to Compare Side-by-Side and Segmented Flow Regimes

Channel Configuration[a]						
Width (µm)	Depth (µm)	Inlet Type	Outlet Type	Flow Regime	Surface-to-Volume Ratio (L m^{-1})	k_La (L/s)
200	140	Y	Y	Side-by-side	10,000	10.8
300	140	Y	Y	Side-by-side	6,667	1.9
400	140	Y	Y	Side-by-side	5,000	1.1
300	180	T	CFS[b]	Segmented	20,000	5.3

[a] All channels had a length of 4.4 cm.
[b] CFS = capillary force separator.

More recently, the authors have demonstrated the ability to increase extraction efficiencies further by the addition of an inert gas phase, attributing enhancements to the resulting increase in surface area between the liquid phases [43] and via the development of large capacity microchannel reactors capable of processing 10,000 metric ton year^{-1} of material [44].

8.3 GAS–LIQUID SEPARATION

As discussed previously, when considering separations on the microscale, gravitational forces are small when compared to surface forces, as such conventional methods of separation, based on densities for example do not prove useful; surface tension effects do however dominate gravitational and viscous forces, affording many novel separation possibilities.

8.3.1 MEMBRANE SEPARATORS

Singh et al. [45] recently demonstrated the fabrication of a gas–liquid separator unit based on a partition wall, containing 5 µm openings, positioned between gas and liquid phases. Using this reactor configuration, the authors investigated the separation of $CO_2(g)$ from a biphasic reaction system containing CO_2/He and a mixture of distilled water and amine. During this investigation, the authors did however find it necessary to ensure pressure balance between the inlets and outlets in order to prevent liquid from entering the gas channel.

The use of segmented gas–liquid flow has been shown to be a facile method of increasing the mixing efficiency of miscible liquids within micro reaction channels (Chapters 1 and 2) however safe, rapid, and efficient separation of the resulting phases has proved problematic. With this in mind, Jensen and coworkers [46] developed a planar capillary separator based on interfacial forces, suitable for incorporation into microfabricated reactors. Comprising of a molded PDMS layer and a glass cover plate, the reactor contained two fluidic inputs, a single gas inlet, separate liquid/gas outlets and a 16-channel capillary separator (Figure 8.11). Using the aforementioned device, gas–liquid separation is achieved at the end of the microchannel by applying a differential pressure, smaller than the capillary pressure, across the capillary separator, typically 0.072 atm for water and 0.022 atm for EtOH, which results in liquid flow through the separator; leaving gas within the microchannel and liquid in the separator.

8.3.2 MICROFLUIDIC DISTILLATIONS

Based on the exploitation of differences in volatility, distillations are a powerful synthetic route to the isolation and purification of many industrially relevant compounds, in particular low boiling feedstocks and recently biofuels [47]. Distillation is a continuous method of purifying liquid components and involves the generation of vapor above a heated liquid followed by cooling of the vapor phase to afford a condensate. Within a microspace, gravitational forces do not dominate resulting in uncontrollable boiling on this length scale; as such methods for making and

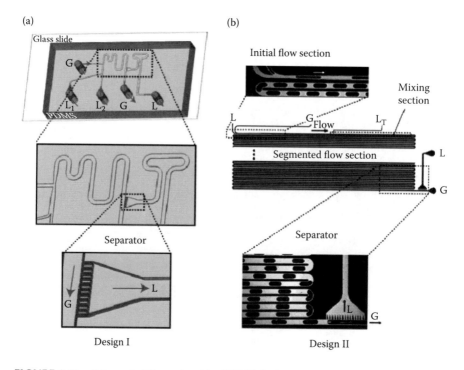

FIGURE 8.11 Schematic illustrating (a) a PDMS device used for the separation of liquids from (b) gas–liquid segmented flow. (Reprinted with permission from Günther A. et al., 2005. *Langmuir* 21: 1547–1555. Copyright (2005) American Chemical Society.)

transporting vapor are required that do not involve boiling. With this in mind, several authors have demonstrated the development of miniaturized distillation equipment suitable for integration within microchemical systems.

An early report of such equipment was the subject of a communication by Wootton and de Mello [48] whereby a carrier gas (He) was employed to induce evaporative transport [49], a well-established technique for low-pressure distillations, the authors designed the glass reactor manifold illustrated in Figure 8.12 (microchannels = 100–500 μm (wide) × 50 μm (deep)). As Figure 8.12 illustrates, the device consisted of three functional compartments ((i) to (iii)), whereby the gas (25 psi) and liquid phases (75 μL min^{-1}) were brought together and heated (60°C) in section (i) and recirculated to afford a vapor-saturated carrier gas. Within this section, the chamber floor contained "chevron"-shaped ridges which served to preserve laminar integrity by inducing viscous drag on any droplets formed, resulting in recombination with the liquid lamina (exploded section, Figure 8.12). As the gas moved through the condensation channel (dimensions = 150 μm (wide) × 50 μm (deep)), toward section (ii), heat transfer resulted in cooling of the channel and promoted condensation; see the magnified section in Figure 8.12. The final section (iii) acted as a separation module, which enabled the condensate to be removed from the carrier gas and collected.

To demonstrate the capabilities of the device, the authors investigated the separation of a 50:50 v/v mixture of MeCN:DMF, whereby heating section (i) to 60°C

Microscale Continuous Separations and Purifications 405

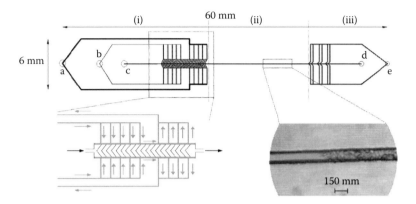

FIGURE 8.12 Schematic illustrating the laminar evaporation device developed for low pressure, continuous distillations. (Wootton, R. C. R. and de Mello, A. J. 2004, Continuous laminar evaporation: Micron-scale distillation, *Chem. Commun.* 266–267. Reproduced by permission of the Royal Society of Chemistry.)

afforded a ninefold enrichment in MeCN concentration in the distillate after a single pass (0.72 theoretical plate); however, the distillate can readily be recirculated through the device in order to increase the purity further. Interestingly, evaporation of MeCN was induced at a temperature 22°C below the solvents boiling point, demonstrating the devices potential application for the purification of thermally unstable compounds.

In an extension to their previous liquid-phase microunit operations, Hibara et al. [50] reported the fabrication of a microfluidic system containing a gas–liquid separator suitable for performing microdistillations. The reactor consisted of two fused silica wafers (dimensions = 30 mm (wide) × 70 mm (long) × 0.7 mm (deep)) into which microchannels were wet-etched (top plate = 250 µm (wide) × 30 µm (deep), bottom plate = 140 µm (wide) × 10 µm (deep)) to afford the gas–liquid separator. In the second wafer, square nanopillars (dimensions = 900 nm (wide) × 900 nm (long) × 250 nm (deep)), with a 300 nm inter pillar spacing, were fabricated using electron-beam lithography and plasma etching [51], and formed the condensation zone.

Once thermally bonded, the surface of the shallow channel was subsequently modified with a fluoride resin to afford a contact angle of 90°C at temperatures of ≤100°C. As proof of concept, the authors demonstrated the evaporation and subsequent condensation of water within the system. Heating the device externally using a temperature control stage (100°C) they observed condensation of water on the nanopillars 1 min after its introduction. The system was further evaluated using a 9 wt.% aq. EtOH solution whereby employing a reactor temperature of 78°C and a flow rate of 2.0 µL min^{-1} afforded 19 wt.% EtOH in the condensed phase within the nanopillars; additional optimization of the system is however required to attain efficient distillation.

Jensen and coworkers [52] subsequently reported the development of a vapor–liquid separator (Figure 8.13) and its subsequent integration into a silicon micro reactor (Figure 8.14) to afford a device capable of performing microfluidic distillations. In both systems, vapor–liquid equilibrium are obtained through the use of segmented flow, as depicted in Figure 8.13, which is subsequently separated using a PTFE

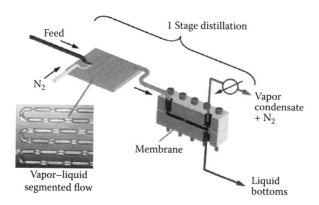

FIGURE 8.13 Schematic illustrating the proof of concept apparatus used to perform single-stage distillations under continuous flow. (Hartman, R. L. et al., 2009. Distillation in microchemical systems using capillary forces and segmented flow, *Lab Chip* 9: 1843–1849. Reproduced by permission of the Royal Society of Chemistry.)

membrane separator (pore size = 0.5 μm). Using this approach, the liquid phase wets the membrane, passing through to outlet 2, while the vapor-inert gas (N_2) does not pass through the membrane and is collected at outlet 1. The integrity of the membrane was evaluated for the flashing of MeOH, whereby induction of segmented flow, in the presence of N_2, followed by heating of the reactor to 73°C (boiling point of MeOH = 65°C) resulted in the formation of MeOH vapor and collection of MeOH vapor condensate.

Having demonstrated the proof of concept, the authors fabricated a silicon micro reactor (Figure 8.14), which contained three reactant inlets, a condenser (400 μm (wide) × 400 μm (deep) × 87.5 cm (long)), a vapor–liquid separator and a vapor outlet, housed within a temperature-controlled holder; coating the microchannels with 1*H*,1*H*,2*H*,2*H*-perfluorodecyltrichlorosilane, to reduce the solvent wettability of the channels surface and prevented drying out of the membrane pores.

Within this reactor, gas and liquids were combined within the condenser to attain segmented flow, prior to entering the heated part of the reactor where the liquid slugs were flashed prior to entering the vapor–liquid separator. The authors investigated the separation of MeOH/toluene (50:50) and DCM/toluene (50:50), at 70°C, whereby mole fractions of 0.22 ± 0.03 (liquid):0.79 ± 0.06 (vapor) and 0.16 ± 0.07 (liquid): 0.63 ± 0.05 (vapor) were obtained, respectively. In both cases, data obtained were found to be consistent with phase equilibrium predictions, confirming the techniques potential as a tool for the continuous flow separation of liquid mixtures.

More recently, Kato and coworkers [53] described the development of a vacuum membrane distillation device with a temperature gradient, capable of performing the rectification of water and MeOH mixtures. The polymer device consisted of a cooling channel, used to generate the temperature gradient along the distillation channel, which is separated into two compartments, one for liquid flow and the second for the vapor phase (Figure 8.15). The temperature gradient within the device was controlled by the cooling water flow rate and the temperature applied via a hotplate. Under optimum conditions, a theoretical plate number of 1.8 was achieved; whereby a value

Microscale Continuous Separations and Purifications

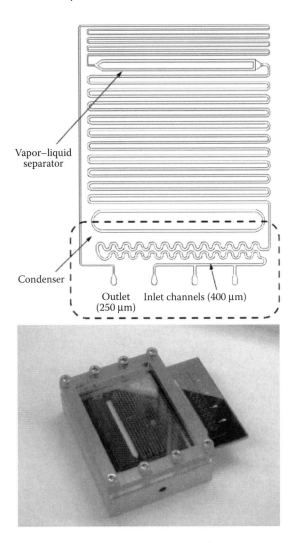

FIGURE 8.14 Schematic of the reactor manifold used to perform continuous flow distillations (top) and a photograph of the device once housed within a holder assembly (bottom). (Hartman, R. L. et al., 2009. Distillation in micro-chemical systems using capillary forces and segmented flow, *Lab Chip* 9: 1843–1849. Reproduced by permission of the Royal Society of Chemistry.)

of 2 represents a completely separated system. In order to further optimize the device the authors acknowledge the need to improve the membrane pore size, with investigations currently underway.

8.4 SOLVENT EXCHANGE AND SOLVENT REMOVAL

While many of the examples investigated using micro reaction technology have demonstrated significant advantages compared to standard batch protocols, there are

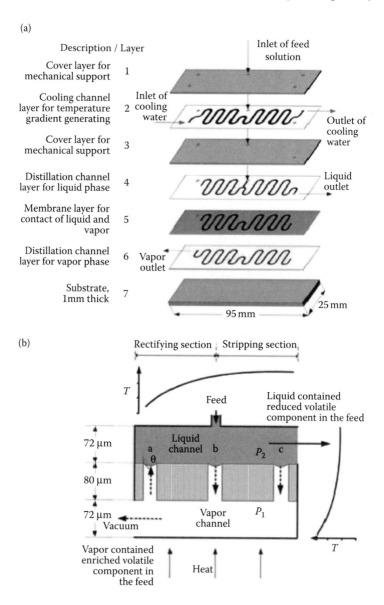

FIGURE 8.15 Illustration of a polymeric microdistillation device: (a) exploded view and (b) description of the process. (Zhang, Y., Kato, S., and Anazawa, T. 2010. Vacuum membrane distillation by microchip with temperature Gradient, *Lab Chip* 10: 899–908. Reproduced by permission of the Royal Society of Chemistry.)

limitations to the technique when it comes to performing multistep synthesis in a continuous, automated manner. As many synthetic reactions can be profoundly affected by the solvent that they are performed in, it is desirable to be able to change reaction solvents between synthetic steps. In batch, this is usually performed when purifying a material before continuing the synthetic route and in flow researchers have simply linked up reaction steps by reacting the intermediate without isolation; as such up to now it has been imperative that all reaction steps be performed in the same reaction solvent. Obviously this can be disadvantageous as it is often imperative that the solvent be exchanged prior to performing incompatible reaction steps.

With this in mind, a small number of researchers have demonstrated the development of techniques for the removal and replacement of reaction solvents during continuous flow processes. An early example of this was reported by Timmer et al. [54], who developed a microevaporation concentrator, used to increase the concentration of electrolytes; however, showing application for solvent removal within synthetic microdevices.

Tseng and coworkers [55] demonstrated the need to remove water from their reaction mixture, replacing it with MeCN and subsequently back again as part of a continuous, multistep process for the synthesis of ^{18}F-FDG (see Chapter 7 for details). In the same year, Ley and coworkers [56] demonstrated the ability to exchange reaction solvents during the synthesis of (±)-oxomaritidine, employing a commercially available solvent evaporator by Vapourtec (UK) to remove THF and replace it with DCM prior to performing the next reaction step (see Chapter 7).

More recently, Buchwald and coworkers [57] have demonstrated the development of a device capable of performing a reaction, followed by a LLE, solvent removal (DCM) and solvent exchange (DMF), enabling the performance of a second reaction in a different reaction solvent (Scheme 8.4). Using the aforementioned protocol, the authors were able to investigate the continuous flow synthesis of a triflate derivative **21** in DCM, remove the residual base **22** via LLE and remove the reaction solvent by microdistillation. Exchange of the reaction solvent for DMF, prior to performing a Pd-catalyzed **19** Heck reaction with *n*-butyl vinyl ether **23**, enabled the authors to obtain the target alkene **24** as a solution in DMF at a throughput of 32 mg h^{-1}.

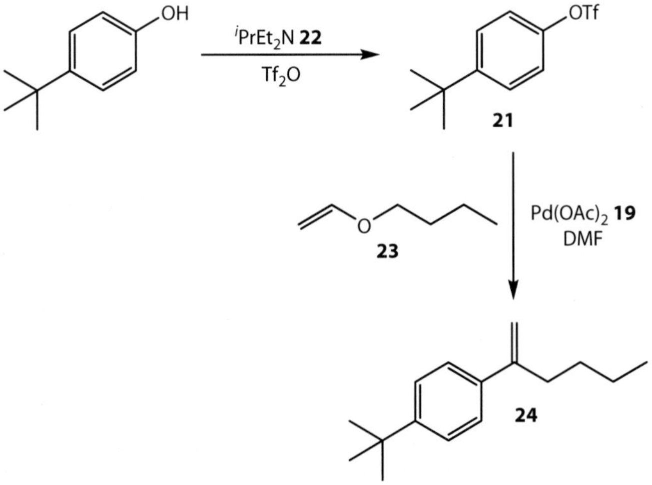

SCHEME 8.4 Illustration of the synthetic route used to demonstrate the efficient solvent removal and exchange under continuous flow conditions.

Although these techniques are in their infancy, the ability to perform inline solvent exchanges will be imperative if the pharmaceutical industries are to fully exploit the advantages that micro reaction technology has to offer.

8.5 THE USE OF SCAVENGER RESINS FOR PRODUCT PURIFICATION UNDER FLOW

When developing a production protocol, particularly in the case of API's, it is imperative to consider the residual quantities of by-products tolerated within the final material. Owing to the stringent regulations in place for the production of pharmaceuticals, continuous flow reactors are finding novel application not only in the synthesis of API's, but also in their purification.

8.5.1 Trace Metal Removal

Using a series of model reactions to demonstrate the synthetic utility of flow reactors for the postproduction purification of reaction products, Pitts and coworkers [58] illustrated the ability to remove trace metals such as Pd, Cu, and Rh from reaction mixtures derived from Suzuki, Rosemund von-Braun, Michael, and hydrogenation reactions. As Table 8.4 illustrates, employing a residence time of typically 8 min, excellent sequestration efficiencies were obtained, reducing the trace metal impurities to levels acceptable.

Ley and coworkers [59] subsequently demonstrated the inline sequestration of Cu while performing 14 click reactions under continuous flow (Table 8.5). Using a modular flow reactor, the authors were able to couple an immobilized copper (I) iodide reagent **25** with two immobilized scavengers, QuadraPure-TU (thiourea) **26** and a phosphane resin **27**, enabling the [3+2] cycloaddition of a series of azides (0.15–0.20 M) and terminal alkynes (0.1 M) to afford 1,4-disubstituted-1,2,3-trizaoles in high to excellent yield. Through the use of inline scavengers, the authors were able to produce the target compounds in excellent purity, removing the need for additional purification steps.

TABLE 8.4
Summary of the Scavenging Efficiencies Obtained for a Series of QuadraPure Resins Employed under Continuous Flow Conditions

			Metal Concentration (ppm)		
Reaction Type	Contaminant	Resin Type[a]	Initial	Final	Sequestered (%)
Suzuki	Pd	QuadraPure TU	60	4	93.3
Rosemund von-Braun	Cu	QuadraPure IDA	345	<1	99.7
Michael	Fe	QuadraPure AMPA	61	<0.2	99.7
Hydrogenation	Rh	QuadraPure AMPA	309	4	98.7

[a] TU = thiourea; IDA = iminodiacetate; AMPA = aminomethylphosphonic acid.

TABLE 8.5
A Selection of the 1,4-Disubstituted-1,2,3-Triazoles Synthesized, and Purified under Continuous Flow

$$R-N_3 + R^1-\equiv-H \xrightarrow{\substack{1.\ \bullet-NMe_2 \cdot CuI\ \mathbf{25} \\ 2.\ \bullet-NH-C(S)-NH_2\ \mathbf{26} \\ 3.\ \bullet-PPh_3\ \mathbf{27}}} \underset{R^1}{\text{triazole}}\!-\!R$$

Triazole	Yield (%)
HO–CH₂CH₂–(triazole)–C₆H₄–NO₂	93
HO–CH₂–(triazole)–CH₂–Ph	85
Ph–(triazole)–CH₂–Ph	85
(4-MeC₆H₄SO₂)–(triazole)–CH₂–(4-CF₃C₆H₄)	91
(sugar/acetonide with OH)–(triazole)–Ph; benzyl-CF₃ group	88

8.5.2 Removal of Unreacted Starting Materials

While some researchers prefer to optimize micro reactions using stoichiometric quantities of reagents, others have demonstrated the use of activating agents in particular in excess; in other examples, reagents can generate residues upon reaction. As such the residual reagents/by-products need to be removed from the reaction product to enable isolation of the target compound in high purity. With this in mind, several research groups have demonstrated the incorporation of inline scavenger columns as a means of obtaining pure reaction products under continuous flow conditions.

Synthesis of Yne-Ones: By employing a series of inline scavenger cartridges, Ley and coworkers [60] were able to devise a continuous flow strategy for the synthesis of pyrazoles, derived from *in situ* generated yne-ones in moderate to high yield and excellent purity using the commercially available R2 + /R4 unit from Vapourtec (Table 8.6).

Initially, the authors performed a screen of reaction conditions including reaction time, temperature, and reactant concentration which enabled them to identify, Pd(OAc)$_2$ **19** (1.0 M), Hünig's base **22** (1.2 eq.), acid chloride (1.2 M) and terminal acetylene (1.0 eq.) as the reactant concentrations, 100°C as the reaction temperature, and a reaction time of 30 min. Using this approach, the authors were able to synthesize 17 yne-ones with yields ranging from 41% to 95%; demonstrating a tolerance to a wide range of functional groups including aromatic and aliphatic acid chlorides.

TABLE 8.6
Illustration of the Two-Step Protocol Employed for the Synthesis and Reaction of Yne-Ones under Flow Conditions

R¹	R²	R³	Yield (%)
Ph	Ph	H	86
cyclopropyl	Ph	H	69
Ph	Ph	Me	74
ᵗBu	Ph	Me	74
cyclopropyl	Ph	Ph	78
Ph	Ph	Ph	65
benzothiophen-2-yl	3-MeO-C₆H₄	Me	53
3-Br-C₆H₄	Ph	Me	57
3-Br-C₆H₄	C₄H₉	Me	57

In all cases, no additional product purification was required after removal of the reaction solvent, in this case DCM.

As an extension to this investigation, the authors added a third reaction stream, after the IRA-743 **28** scavenger column, enabling the addition of a hydrazine derivative (5.0 eq.) and the reaction of *in situ* generated yne-ones to afford the respective pyrazole (20–30 min at 25–100°C). As summarized in Table 8.6, using this approach the authors were able to synthesize nine pyrazoles in moderate to high yield; however due to the large excess of hydrazine derivative employed, a series of scavenger cartridges were required, including calcium carbonate **29**, QuadraPure sulfonic acid **30** and QuadraPure thiourea **26**.

Fluorinations Using DAST: Having demonstrated the use of DAST for the fluorination of aldehydes, Baumann et al. [61] extended their investigation to incorporate the electrophilic α-fluorination of activated carbonyl compounds, to the respective fluoride, using the reagent Selectfluor® **31**. Using this approach, a cyclic amine is generated as a by-product and as such the reaction products must be purified to obtain the target compounds in high purity.

In order to remove the need to perform an aqueous extraction followed by conventional purification via column chromatography, the authors devised a continuous flow approach capable of performing the electrophilic fluorination and purification to remove unreacted starting materials and by-products; affording the target fluoride upon evaporation of the reaction solvent.

Employing a Vapourtec flow reactor (dimensions = 1000 μm (i.d.), volume = 9 mL), the authors mixed a solution of activated carbonyl compound (0.1 M) and Selectfluor **31** (0.1 M) at a T-mixer prior to passing into the flow reactor which was heated to 100–120°C. Using a total flow rate of 300 μL min^{-1}, equating to a reactant residence time of 27 min, the authors were able to efficiently monofluorinate a series of activated carbonyl compounds. Upon exiting the reactor, the reaction mixture was pumped through a packed-bed reactor (1 cm (i.d.) × 15 cm (long)) containing QuadraPure-SA **30** and QuadraPure DME **32**. These scavengers removed any unreacted carbonyl compounds, Selectfluor **31** and by-products to afford the target fluoride in high yield, with typical purities of 95%; determined by ^{1}H NMR spectroscopy and LC-MS (Table 8.7). In the case of highly reactive substrates such as trifluoroacetyl-derivative **33**, difluorination was observed; therefore in order to drive the reaction to completion, two equivalents of Selectfluor **31** were employed affording 2,2,4,4,4-pentafluoro-1-(naphthalene-2-yl)butane-1,3-dione **34** in 83% yield.

For more examples of the use of solid-supported reagents as scavengers and purification aids, see Chapter 4, where sequestering along with "catch and release" methodologies are described.

8.6 CONTINUOUS FLOW RESOLUTIONS

Chiral compounds find widespread use in synthetic organic chemistry due to their application in pharmaceuticals, fragrances, and flavors, with synthetic complexity

TABLE 8.7
Summary of the Results Obtained for the Continuous Flow Fluorination of Activated Carbonyl Compounds Using Selectfluor 31

Starting Material	Product	Yield (%)
Phenyl-CH(CN)-C(O)-	Phenyl-C(CN)(F)-C(O)-	90
Aryl amide with Cl	α-fluoro aryl amide with Cl	82
MeO-aryl-C(O)-CH₂-C(O)OEt	MeO-aryl-C(O)-CHF-C(O)OEt	93
Phenyl-C(O)-CH₂-C(O)OEt	Phenyl-C(O)-CHF-C(O)OEt	89
Cyclopentanone acetyl	α-fluoro cyclopentanone acetyl	82
Naphthyl-C(O)-CH₂-C(O)CF₃ (33)	Naphthyl-C(O)-CF₂-C(O)CF₃ (34)	83[a]

[a] Employing 2 equivalents of Selectfluor **31**.

---

increasing with the number of stereo-centers incorporated into a molecule. As such chiral building blocks are an important precursor to the preparation of many industrially relevant compounds and efficient methods for their preparation and purification are often sought.

In addition to classical resolution techniques, where enantiomerically pure products are synthesized by carefully tuning of a chemical catalyst or selection of a biocatalyst, racemization provides the opportunity to synthesize racemic products via inexpensive synthetic routes and subsequently increase the proportion of target compound obtained from 50% to quantitative yield.

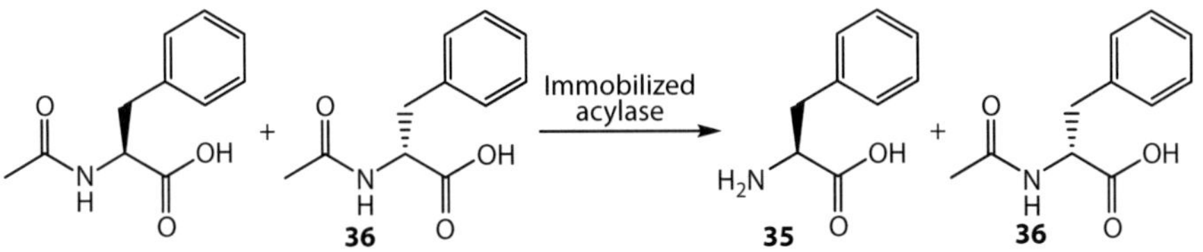

SCHEME 8.5 Illustration of the model reaction used to demonstrate the continuous optical resolution of a racemic amino acid.

8.6.1 BIOCATALYTIC RESOLUTIONS

Combining biocatalytic transformations with an inline LLE, Maeda and coworkers [62] reported the use of a micro reactor device capable of achieving the optical resolution of racemic amino acids and efficient turnover of the biocatalyst. Using the model reaction illustrated in Scheme 8.5, the authors firstly performed the enzyme-catalyzed enantioselective hydrolysis of acetyl-D,L-phenylalanine using a tubular reactor (dimensions = 500 µm (i.d.)), wall-coated with aminoacylase from *Aspergillus*, whereby L-phenylalanine **35** was obtained in 99.2–99.9% ee. Residual acetyl-D-phenylalanine **36** was subsequently separated from the reaction mixture via acidification of the reaction products and inline extraction into EtOAc. Using this approach, the authors reported an extraction efficiency of 84–92% under optimized conditions of 0.5 µL min^{-1}, for the enzymatic reaction and 2.0 µL min^{-1} for the LLE enabled resolution of 240 nmol h^{-1} of the racemate.

In a subsequent example, Ürge and coworkers [63] reported the use of a continuous flow packed-bed reactor for the large-scale enantiomer selective acylation of (*rac*)-alcohols (Scheme 8.6). Using *Candida antarctica* Lipase B (CaLB) **37** as the biocatalyst, retained within a tubular reactor (X-cube™, Hungary), and a reactant solution of (*rac*)-phenyl-1-ethanol **38** (7.8 × 10^{-2} M) in a mixed solvent system comprising of hexane:THF:vinyl acetate **39** (2:1:1), the authors investigated the effect of reactant flow rate. Employing a reactor temperature of 25°C and a flow rate of 100 µL min^{-1}, the authors obtained (*R*)-acetate **40** in 50% conversion and 99.2% ee and the residual (*S*)-alcohol **41** in 98.9% ee; as quantified by chiral GC analysis. In comparison to batch which took 24 h, the flow reactor afforded optimal acetylation in only 8.2 min. Pleased with the rapid nature of their approach, the authors investigated the kinetic resolution of a series of secondary (*rac*)-alcohols, pumping 20 mL of reactant solution through the biocatalyst **37** over a period of 3.3 h. Using this approach, the authors

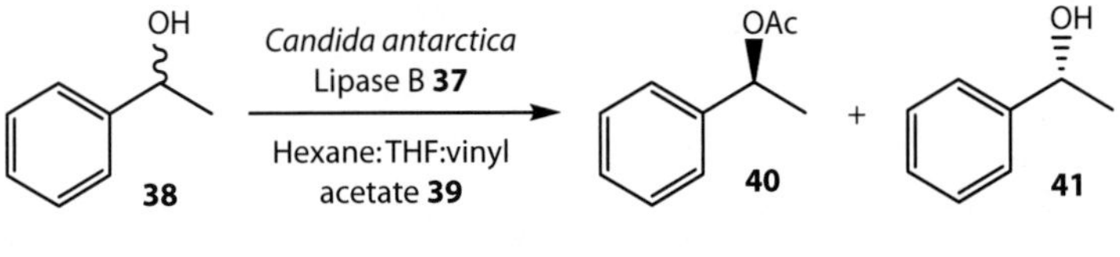

SCHEME 8.6 Schematic illustrating the enantiomer selective acylation of racemic alcohols utilizing a packed-bed reactor.

obtained the results illustrated in Table 8.8, which demonstrate the successful application of their technique to the performance of kinetic preparative scale resolutions. Importantly, throughout their investigation the authors employed the same catalyst cartridge, thus illustrating the stability of the enzyme and the synthetic utility of the approach for large-scale isolation of chiral compounds.

8.6.2 Chemical Racemization

Hulshof and coworkers [64] investigated the racemization of N-acetylindoline-2-carboxylic acid **42**, using acetic anhydride **43**, as a means of isolating the

TABLE 8.8
Summary of the Results Obtained for a Series of Preparative Scale Kinetic Resolutions Performed under Flow Conditions

Compound	Yield (%)[a]	ee[b]	$[\alpha]_D^{25}$ [c]	E[d]
(R)-**40** (OAc)	40	98.5	−62.8	
(S)-**41** (OH)	48	99.1	+125.3	>>>200
(R)- (OAc, cyclohexyl)	26	77.4	+2.0	
(S)- (OH, cyclohexyl)	41	99.0	+7.1	>200
(R)- (OAc, benzyl)	34	56.4	+4.9	
(S)- (OH, benzyl)	41	85.1	−23.3	22

[a] Products isolated from the reactor output stream.
[b] Determined by enantioselective GC.
[c] Specific rotations (c 1.0, $CHCl_3$).
[d] Due to sensitivity to experimental errors, enantiomer selectivity values calculated in the range 25–500 are reported as >200 and those above 500 as >>> 200.

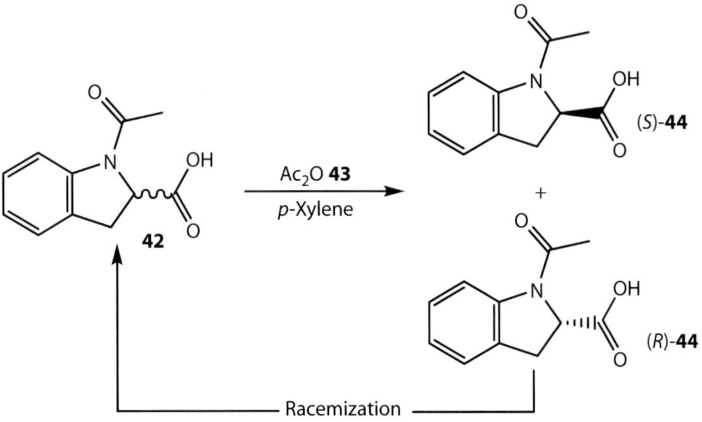

Microscale Continuous Separations and Purifications

SCHEME 8.7 Schematic illustrating the classical approach for the racemization of N-acetylindoline-2-carboxylic acid **42**.

(S)-enantiomer **44** (Scheme 8.7) which is a key intermediate in the synthesis of the ACE inhibitor Perindopril **45** (Figure 8.16).

Initial investigations were conducted using conventional batch reactor methodology where the effect of cosolvent (p-xylene), acetic acid stoichiometry and temperature were evaluated. The initial findings of this study led the authors to investigate the effect of microwave heating as a means of enhancing the racemization rate. Using this approach, the authors observed that changing the degree of heterogeneity was found to directly influence the magnitude of the microwave effect, as a result of the reaction occurring at the phase boundary and most likely in the vicinity of solid particles. Although promising results were obtained, the authors acknowledged the use of microwaves to be limited to small-scale reactors due to minimal penetration depth of the reactor. In order to address this production shortfall the authors

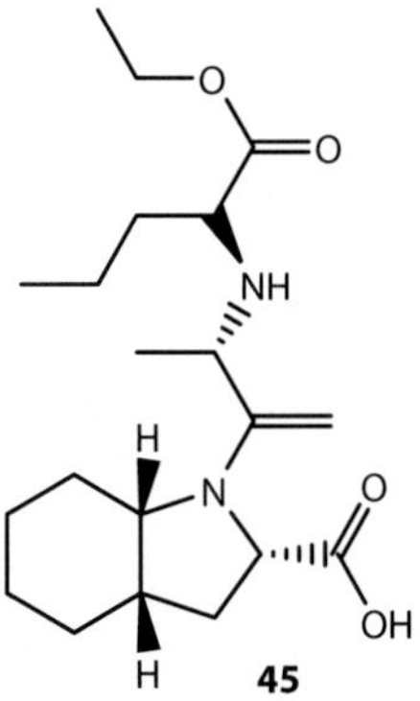

FIGURE 8.16 Perindopril **45**, an ACE inhibitor synthesized using the key intermediate (S)-N-acetylindoline-2-carboxylic acid **44**.

employed a continuous flow loop reactor, as a means of increasing the quantity of material generated using this process; employing an average power usage of 280 W analogous results were obtained compared to the small-scale batch reaction with a dramatic increase in reaction rate from 2.05 ee min^{-1} (batch) to 2.43 ee min^{-1} (flow); with product **44** isolation achieved via precipitation.

8.7 PRODUCT ISOLATION

One of the most widely employed separation techniques used for the isolation of products is solution crystallization and while on the surface it may be perceived to be a technique that is unsuitable for performance within continuous flow tubular reactors, several authors have demonstrated the successful generation and manipulation of particles within such systems. In fact, this mode of operation has been shown to afford many operational advantages such as improved particle size distribution (see Chapter 6, Continuous Particle Formation), along with a positive impact on the polymorphic form isolated and crystal morphology); not to mention the fact that the techniques not scale limited unlike batch procedures. Owing to the fact that product isolation is merely a single unit operation in a vast process, controlling how the product is formed can have a knock on effect on the remaining steps of a process, namely filtration, milling, and formulation.

With continuous crystallizations identified as a targeted area, key in improving the manufacture of fine chemicals and pharmaceuticals, this area has started to attract interest from process chemists [65].

8.7.1 Antisolvent Precipitation

In 2007, Chen and coworkers [66] evaluated the performance of liquid antisolvent precipitation (LASP) within a microchannel reactor as a means of obtaining increased process control over the chemical composition and particle size of pharmaceutical nanoparticles obtained. Using the controlled precipitation of Danazol **46** as a model, Figure 8.17, the authors evaluated the use of a microchannel reactor (inlet channels = 300 μm (wide) × 300 μm (deep), reaction channel = 600 μm (wide) × 300 μm (deep)) to perform continuous LASP.

As mentioned previously, while the precipitation of materials within microchannels is perceived to be problematic, the authors herein report the use of large reaction

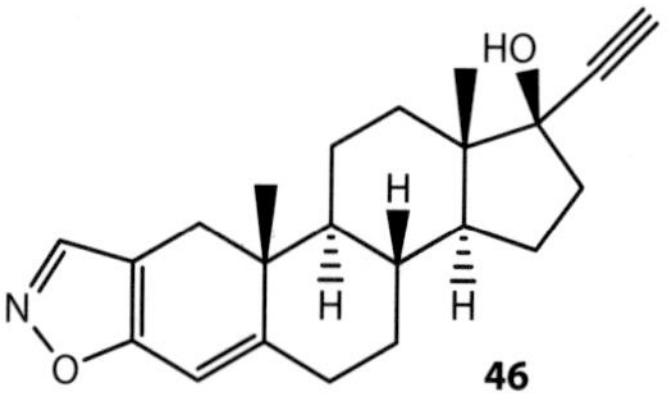

FIGURE 8.17 Illustration of Danazol **46**, the pharmaceutical agent used to demonstrate controlled precipitation under continuous flow.

channels relative to the particle size, that is, 0.15% of the channel depth; as a result, no problems with channel blocking or clogging are observed. Owing to the fact that the driving force for precipitation via this mode is the supersaturation of a solution by mixing the drug molecule (S) with an antisolvent (AS), the authors investigated the effect of AS/S ratio on the resulting particle size **46** and subsequent solubility.

To achieve this, the active ingredient **46** was dissolved in EtOH (S) and pumped through the reactor, varying the proportion of DI H_2O (AS) based on flow rate. At the interface between the coflowing liquid streams, the authors observed precipitation of **46** which was subsequently collected at the reactor outlet and isolated upon filtration of the effluent through a 0.45 μm filter.

As Table 8.9 illustrates, with increasing AS/S ratio (at 30°C), a marked effect on Danazol **46** particle size is observed, affording a 109-fold reduction in particle size from 55 μm to 505 nm by performing the procedure under continuous flow. The authors also observed a further reduction in particle size upon reducing the reactor temperature from 30°C to 4°C affording particles of 364 nm and improved particle size distribution. It is important to note however that in all cases, the morphology of the API **46** remained the same with the authors reporting an improvement in the particle size and regularity as a result of employing a continuous flow process.

The impact of this was subsequently demonstrated with respect to API **46** solubility, with 100% dissolution of the nanoparticles in 5 min compared to only 35% of the raw material; with 1 h required to obtain comparable solubility. Analysis of the resulting nanoparticles **46** by x-ray diffraction confirmed that the physical characteristics of the material were not affected by the precipitation technique enabling the authors to conclude that the LASP technique under continuous flow affords facile access to uniform particle sizes compared to standard techniques. In the case of hydrophobic drug molecules, this is particularly advantageous as it facilitates the dissolution of sparingly soluble compounds in aqueous media, increasing bioavailability.

TABLE 8.9
Summary of the Results Obtained for the LASP of Danazol 46 Performed under Continuous Flow

AS/S Ratio (v/v)	Solubility of 46 (μg mL^{-1})	Average Particle Size (nm)
0	35,000	NA[a]
1	1336.03	NA[a]
2	79.30	NA[a]
5	4.40	1250
10	2.83	900
20	2.16	510
40	1.63	505
20[b]	1.63	364

[a] No precipitation observed.
[b] Reactor temperature 4°C.

8.7.2 Lysozyme Crystallization

As can be seen throughout Chapter 6, the formation of droplets within microfabricated domains has a wide range of applications including particle production, polymer formation, and product crystallization. To date, these techniques have been limited to continuous droplet formation, with the aim of minimizing coalescence in order to obtain high levels of control over particle size (see Chapter 1 also). In order to increase the applications of droplets within such systems, Maeki et al. [67] developed a method of performing controlled droplet fusion; using a reactor containing an enlarged channel, which facilitated droplet fusion. The reactor in question comprised of a double Y-shaped mixer and a microchannel (dimensions = 200 μm (wide) × 200 μm (deep)) followed by the fusion section (dimensions = 600 μm (wide) × 200 μm (deep) × 2 mm (long)) and prior to use the microchannel walls were treated with trichloro-(1H,1H,2H,2H-perfluorooctyl)silane. Using solutions of fluorinert (FC40) as a continuous phase and aqueous glycerol (68 and 24 wt.%), the effect of flow rate and phase ratio was evaluated. Under the aforementioned conditions, the authors were able to identify a trend of decreasing droplet size as a function of increasing viscosity.

Encouraged by these results, the authors evaluated the fusion of droplets containing a protein solution with those comprising of a precipitant solution and demonstrated the controlled crystallization of lysozymes. Further work is currently underway to screen protein crystallization conditions using the platform described.

8.7.3 Solution Crystallization

More recently, Ni and coworkers [68] reported a method for the continuous crystallization of a model API using a continuous oscillatory baffled crystallizer to perform solution crystallization.

By conducting the two stages of solution crystallization under continuous flow, that is, nucleation and crystal growth, the authors were able to harness the excellent heat transfer rates obtained in such systems to enable rapid, scalable cooling of solutions; compared with batch reactors where the specific area reduces with increasing reactor size and thus heating efficiency. Theorizing that operating in a plug flow regime would ensure consistent fluid conditions and heat transfer rates, the authors had two choices of reactor, a series of continuous stirred tank reactors (CSTR) or a tubular reactor operated under turbulent flow. Based on this requirement, the authors developed a COBC reactor, which contained periodically spaced orifice baffles which, superimposed oscillatory motion on the net flow, reducing the length of reactor required compared to conventional tubular reactors. With repeating cycles of vortices, radial motion is created which leads to uniform mixing and plug flow along the length of a column which would normally result in a laminar flow regime. The continuous oscillatory baffled crystallizer (COBC) reactor was fabricated from DN25 jacketed glass tubes, containing baffles of polyvinylidenefluoride (PVDF) and coated in insulation material to minimize undesirable heat loss to the environment; the use of a temperature control unit enable cooling rates of 0.25–15°C min^{-1} to be accessed.

Using an unnamed API, provided by AstraZeneca, the author's demonstrated proof of concept basing their investigation on the conditions employed for manufacture of the API, that is, crude API (6% w/w) dissolved in a solvent and held at reflux followed by particulate removal and cooling to 20°C (10°C h^{-1}) with agitation; agitation was then reduced and the solution cooled to 10°C. Under the aforementioned conditions, crystallization was found to take a total of 9 h 40 min after which the crystals are filtered, washed and dried prior to milling to afford the correct crystal size.

Within the COBC reactor, API solutions were manipulated using two peristaltic pumps and found to afford the API with a narrow size distribution and of the correct morphology, as determined by XRPD analysis. Using this approach, the authors were able to isolate the model API in 12 min compared to 9 h 40 min in a batch process; demonstrating significant savings with respect to processing time and operational costs; largely attributed to the ability to obtain the desired particle size without the need for milling ~50% reduction in capital costs.

8.8 SUMMARY

Until recently, micro reaction technology was viewed as a means of increasing the efficiency of reactions performed on a small scale, useful mainly for the screening of reaction conditions rather than for the production of useful volumes of synthetic intermediates and or products. With growth in the area of continuous flow purifications, it has been seen that the technology has the opportunity to revolutionize the way the synthetic chemistry is performed both within the laboratory and an industrial setting. Where it had been previous thought that reactions performed in series were limited to those that could be coerced into occurring within the same reaction solvent, more than likely suboptimal for all steps, it can now be seen with the advent of scavenger modules, microdistillation equipment and membrane reactors, that multistep process can be performed utilizing different reaction conditions and solvent systems. These approaches therefore negate the need for offline batchwise purifications; clearly illustrating the flexibility associated with current continuous flow synthesis and going some way toward increasing the reaction capabilities of continuous flow reaction technology.

REFERENCES

1. Aota, A., Mawatari, K., Takahasi, S., Matsumoto, T., Kanda, K., Anraku, R., Hibara, A., Tokeshi, M., and Kitamori, T. 2009. Phase separation of gas–liquid and liquid–liquid microflows in microchips, *Microchim. Acta* 164: 249–255.
2. Tia, S. and Herr, A. E. 2009. On-chip technologies for multi-dimensional separations, *Lab Chip* 9: 2524–2536.
3. Hartman, R. L. and Jensen, K. F. 2009. Microchemical systems for continuous-flow synthesis, *Lab Chip* 9: 2495–2507.
4. Aota, A., Mawatari, K., and Kitamori, T. 2009. Parallel multi-phase microflows: Fundamental physics, stabilization methods and applications, *Lab Chip* 9: 2470–2476.
5. Kuban, P., Berg, J., and Dasgupta, P. K. 2003. Vertically stratified flows in microchannels. Computational simulations and applications to solvent extraction and ion exchange, *Anal. Chem.* 75: 3549–3556.
6. Shui, L., Eijkel, J. C. T., and van den Berg, A., 2007. Multiphase flow in microfluidic systems—Control and applications of droplets and interfaces, *Adv. Colloid Interface Sci.* 133: 35–49.

7. Yoshida, A., Hao, X., and Nishijkido, J. 2003. Development of the continuous-flow reaction system based on the Lewis acid-catalyzed reactions in a fluorous biphasic system, *Green Chem.* 5: 554–557.
8. Liu, S., Fukuyama, T., Sato, M., and Ryu, I. 2004. Continuous microflow synthesis of butyl cinnamate by a Mizoroki–Heck reaction using a low-viscosity ionic liquid as the recycling reaction medium, *Org. Proc. Res. Dev.* 8: 477–481.
9. Zhao, B., Moore, J. S., and Beebe, D. J., 2001. Surface-directed liquid flow inside microchannels, *Science* 291: 1023–1026.
10. Hibara, A., Nonaka, M., Hisamoto, H., Uchiyama, K., Kikutani, Y., Tokeshi, M., and Kitamori, T. 2002. Stabilization of liquid interface and control of two-phase confluence and separation in glass microchips by utilizing octadecylsilane modification of microchannels, *Anal. Chem.* 74: 1724–1728.
11. Maruyama, T., Kaji, T., Ohkawa, T., Sotowa, K., Matsushita, H., Kubota, F., Kamiya, N., Kusakabe, K., and Goto, M. 2004. Intermittent partition walls promote solvent extraction of metal ions in a microfluidic device, *Analyst* 129: 1008–1013.
12. Miyaguchi, H., Tokeshi, M., Kikutani, Y., Hibara, A., Inoue, H., and Kitamori, T. 2006. Microchip-based liquid–liquid extraction for gas-chromatography analysis of amphetamine-type stimulants in urine, *J. Chromatogr. A* 1129: 105–110.
13. Smirnova, A., Mawatari, K., Hibara, A., Proskurnin, M. A., and Kitamori, T. 2006. Micro-multiphase laminar flows for the extraction and detection of carbaryl derivative, *Anal. Chim. Acta* 558: 69–74.
14. Tokeshi, M., Minagawa, T., and Kitamori, T. 2000. Integration of a microextraction system. Solvent extraction of a Co-2-nitroso-5-dimethylaminophenol complex on a microchip, *J. Chromatogr. A* 894: 19–23.
15. Tokeshi, M., Minagawa, T., and Kitamori, T. 2000. Integration of a microextraction system on a glass chip: Ion-pair solvent extraction of Fe(II) with 4,7-diphenyl-1,10-phenanthrolinedisulfonic acid and tri-n-octylmethylammonium chloride, *Anal. Chem.* 72: 1711–1714.
16. Hisamoto, H., Horiuchi, T., Uchiyama, M., and Kitamori, T., 2001. On-chip integration of sequential ion-sensing system based on intermittent reagent pumping and formation of two-layer flow, *Anal. Chem.* 73: 5551–5556.
17. Hisamoto, H., Horiuchi, T., Tokeshi, M., Hibara, A., and Kitamori, T. 2001. On-chip integration of neutral ionophore-based ion pair extraction reaction, *Anal. Chem.* 73: 1382–1386.
18. Hisamoto, H., Saito, T., Tokeshi, M., Hibara, A., and Kitamori, T. 2001. Fast and high conversion phase-transfer synthesis exploiting the liquid–liquid interface formed in a microchannel chip, *Chem. Commun.* 2662–2663.
19. Smirnova, A., Shimura, K., Hibara, A., Proskurnin, M. A., and Kitamori, T. 2007. Application of a micro multiphase laminar flow on a microchip for extraction and determination of derivatised carbamate pesticides, *Anal. Sci.* 23: 103–107.
20. Xiao, H., Liang, D., Liu, G., Guo, M., Xing, W., and Cheng, J. 2006. Initial study of two-phase laminar flow extraction chip for sample preparation for gas chromatography, *Lab Chip* 6: 1067–1072.
21. Žnidaršič-Plazl, P. and Plazl, I. 2007. Steroid Extraction in a microchannel system: Mathematical modeling and experiments, *Lab Chip* 7: 883–889.
22. Maruyama, T., Uchida, J., Ohkawa, T., Futami, T., Katayama, K., Nishizawa, K., Sotowa, K., Kubota, F., Kamiya, N., and Goto, M. 2003. Enzymatic degradation of *p*-chlorophenol in a two-phase flow microchannel system, *Lab Chip* 3: 308–312.
23. Tagawa, T., Aljbour, S., Matouq, M., and Yamada, H. 2007. Micro-channel reactor with guideline structure for organic-aqueous binary system, *Chem. Eng. Sci.* 62: 5123–5126.

24. Tokeshi, M., Minagawa, T., Uchiyama, K., Hibara, A., Sato, K., Hisamoto, H., and Kitamori, T. 2002. Continuous-flow chemical processing on a microchip by combining microunit operations and a multiphase flow network. *Anal. Chem.* 74: 1565–1571.
25. Kikutani, Y., Mawatari, K., Hibara, A., and Kitamori, T. 2009. Circulation microchannel for liquid–liquid microextraction, *Microchim. Acta* 164: 241–247.
26. Aota, A., Nonaka, M., Hibara, A., and Kitamori, T. 2007. Countercurrent laminar microflow for highly efficient solvent extraction, *Angew. Chem. Int. Ed.* 46: 878–880.
27. Mikami, K., Yamanaka, M., Islam, M. N., Kudo, K., Seino, N., and Shinoda, M. 2003. Fluorous nanoflow system for the Mukaiyama Aldol reaction catalyzed by the lowest concentration of lanthanide complex with bis(perfluorooctanesulfonyl)amide ponytail, *Tetrahedron* 59: 10593–10597.
28. Ueno, M., Hisamoto, H., Kitamori, T., and Kobayashi, S. 2003. Phase-transfer alkylation reactions using microreactors, *Chem. Commun.* 936–937.
29. Hibara, A., Tokeshi, M., Uchiyama, K., Hisamoto, H., and Kitamori, T. 2001. Integrated multilayer flow system on a microchip, *Anal. Sci.* 17: 89–93.
30. Maruyama, T., Matsushita, H., Uchida, J., Kubota, F., Kamiya, N., and Goto, M. 2004. Liquid membrane operations in a microfluidic device for selective separation of metal ions, *Anal. Chem.* 76: 4495–4500.
31. Tetala, K. K. R., Swarts, J. W., Chen, B., Janssen, A. E. M. and van Beek, T. A. 2009. A three-phase microfluidic chip for rapid sample clean-up of alkaloids from plant extracts, *Lab Chip* 9: 2085–2092.
32. Okubo, Y., Maki, T., Aoki, N., Khoo, T. H., Ohmukai, Y., and Mae, K. 2008. Liquid–liquid extraction for efficient synthesis and separation by utilizing micro spaces, *Chem. Eng. Sci.* 63: 4070–4077.
33. Günther, A. and Jensen, K. F. 2006. Multiphase microfluidics: From flow characteristics to chemical and materials synthesis, *Lab Chip* 6: 1487–1503.
34. Dietrich, T. R., Freitag, A., Link, S., and Scholz, R. 2010. Continuous micro separation modules and processes, *11th International Conference on Microreaction Technology*, Kyoto, Japan, 112–113.
35. Kolehmainen, E. and Turunen, I. 2007. Micro-scale liquid–liquid separation in a plate-type coalescer, *Chem. Eng. Proc.* 46: 834–839.
36. Kralj, J. G., Sahoo, H. R., and Jensen, K. F. 2007. Integrated continuous microfluidic liquid–liquid extraction, *Lab Chip* 7: 256–263.
37. Sahoo H. R., Kralj, J. G., and Jensen, K. F. 2007. Multi-step continuous flow microchemical synthesis involving multiple reactions and separations, *Angew. Chem. Int. Ed.* 46: 5704–5708.
38. Günther, A., Jhunjhunwala, M., Thalmann, M., Schmidt, M. A., and Jensen, K. F. 2005. Micromixing of miscible liquids in segmented gas–liquid flow, *Langmuir* 21: 1547–1555.
39. Goto, K. and Mizuno, M. 2007. Synthesis of monosaccharide units using fluorous method, *Tetrahedron Lett.* 48: 5605–5608.
40. Kawakami, H., Goto, K., and Mizuno, M. 2010. Multi-step synthesis of a protected monosaccharide unit in microreactors by fluorous liquid-phase extraction, *11th International Conference on Microreaction Technology*, Kyoto, Japan, 172–173.
41. Theberge, A. B., Whyte, G., Frenzzel, M., Fidalgo, L. M., Wootton, R. C. R., and Huck, W. T. S. 2009. Suzuki–Miyaura coupling reactions in aqueous microdroplets with catalytically active fluorous interfaces, *Chem. Commun.* 6225–6227.
42. Fries, D. M., Voitl, T., and Rudolf von Rohr, P. R., 2008. Liquid extraction of vanillin in rectangular microreactors, *Chem. Eng. Technol.*, 31: 1182–1187.
43. Assmann, N., Desportes, S., and Rudolf von Rohr, P. 2010. Extraction in microreactors: Enhanced mass transfer by addition of an inert gas phase, *11th International Conference on Microreaction Technology*, Kyoto, Japan, 336–337.

44. Sugiyama, Y., Tokuda, Y., Nishimura, M., Yoshida, T., and Noishiki, K. 2010. The development of the large capacity microchannel reactor to the extraction process, *11th International Conference on Microreaction Technology*, Kyoto, Japan, 124–125.
45. Singh, C., Nijhuis, T. A., Rebrov, E. V., and Schouten, J. C. 2010. Gas–liquid absorption in a partition wall microchannel, *11th International Conference on Microreaction Technology*, Kyoto, Japan, 262–263.
46. Günther, A., Jhunjhunwala, M., Thalmann, M., Schmidt, M. A., and Jensen, K. F. 2005. Micro-mixing of miscible liquids in segmented gas–liquid flow, *Langmuir* 21: 1547–1555.
47. Romàn-Leshkov, Y., Barrett, C. J. Liu, Z. Y., and Dumesic, J. A. 2007. Production of dimethylfuran for liquid fuels from biomass-derived carbohydrates, *Nature* 477: 982–985.
48. Wootton, R. C. R. and de Mello, A. J. 2004. Continuous laminar evaporation: Micronscale distillation, *Chem. Commun.* 266–267.
49. Timmer, B. H., van Delf, K. M., Olthius, W., Bergveld, P., and van den Berg, A. 2003. Micro-evaporation electrolyte concentrator, *Sens. Actuators B* 91: 342–346.
50. Hibara, A., Toshin, K., Tsukahara, T., Mawatari, K., and Kitamori, T. 2008. Microfluidic distillation utilizing micro-nano combined structure, *Chem Lett.* 37(10): 1064–1065.
51. Tamaki, E., Hibara, A., Kim, H. B., Tokeshi, M., Ooi, T., Nakao, M. and Kitamori, T. 2006. Liquid filling method for nanofluidic channels utilizing the high solubility of CO2, *Anal. Sci.* 22: 529–532.
52. Hartman, R. L., Sahoo, H. R., Yen, B. C., and Jensen, K. F. 2009. Distillation in microchemical systems using capillary forces and segmented flow, *Lab Chip* 9: 1843–1849.
53. Zhang, Y., Kato, S., and Anazawa, T. 2010. Vacuum membrane distillation by microchip with temperature Gradient, *Lab Chip* 10: 899–908.
54. Timmer, B. H., van Delft, K. M., Olthuis, W., Bergveld, P., and van den Berg, A. 2003. Micro-evaporation electrolyte concentrator, *Sens. Actuators B.* 91: 342–346.
55. Lee, C. C., Sui, G., Elizarov, A., Shu, C. J., Shin, Y. –S., Doley, A. N., Huang, J. et al. 2005. Multistep synthesis of a radiolabeled imaging probe using integrated microfluidics, *Science* 310: 1793–1796.
56. Baxendale, I. R., Deeley, J., Griffiths-Jones, C. M., Ley, S. V., Saaby, S., and Tranmer, G. K. 2006. A flow process for the multi-step synthesis of the alkaloid natural product oxomaritidine: A new paradigm for molecular assembly, *Chem. Commun.* 2566–2568.
57. Hartman, R. L., Naber, J. R., Buchwald, S. L., and Jensen, K. F. 2010. Multistep microchemical synthesis enabled by microfluidic distillation, *Angew. Chem. Int. Ed.* 122: 911–915.
58. Hinchcliffe, A., Hughes, C., Pears, D. A., and Pitts, M. R. 2007. QuadraPure cartridges for removal of trace metal from reaction mixtures in flow, *Org. Proc. Dev.* 11: 477–481.
59. Smith, C. D., Baxendale, I. R., Lanners, S., Hayward, J. J., Smith, S. C., and Ley, S. V. 2007. Cycloaddition of acetylenes with azides to give 1,4-disubstituted 1,2,3-triazoles in a modular flow reactor, *Org. Biomol. Chem.* 5: 1559–1561.
60. Baxendale, I. R., Schou, S. C., Sedlemeier, J., and Ley, S. V. 2010. Multi-step synthesis by using flow reactors: The preparation of yne-ones and their use in heterocycle synthesis, *Chem. Eur. J.* 16: 89–94.
61. Baumann, M., Baxendale, I. R., Martin, L. J., and Ley, S. V. 2009. Development of fluorination methods using continuous-flow microreactors, *Tetrahedron* 65: 6611–6625.
62. Honda, T., Miyazaki, M., Yamaguchi, Y., Nakamura, H., and Maeda, H. 2007. Integrated microreaction system for optical resolution of racemic amino acids, *Lab Chip* 7: 366–372.
63. Csajági, C., Szatzker, G., Töke, E. R., Ürge, L., Darvas, F., and Poppe, L. 2008. Enantiomer selective acylation of racemic alcohols by lipases in continuous-flow bioreactors, *Tetrahedron: Asymm.* 19: 237–246.

64. Dressen, M. H. C. L., van de Krujis, B. H. P., Meuldijk, J., Vekemans, J. A. J. M., and Hulshof, L. A. 2009. From batch to flow processing: Racemization of N-acetylamino acids under microwave heating, *Org. Proc. Res. Dev.* 13: 888–895.
65. Pellek, A. and Arnum, P. V. 2008. Continuous processing: Moving with or against the manufacturing flow, *Pharm. Technol.* 9: 52:58.
66. Zhao, H., Wang, J.-X., Wang, Q.-A., Chen, J.-F., and Yun, J. 2007. Controlled liquid antisolvent precipitation of hydrophobic pharmaceutical nanoparticles in a microchannel reactor, 46: 8229–8235.
67. Maeki, M., Yamaguchi, H., Miyazaki, M., Nakamura, H., and Maeda, H. 2010. Continuous droplet formation and creation of stream fusion behavior by using microfluidic chip, *11th International Conference on Microreaction Technology*, Kyoto, Japan, 234–235.
68. Lawton, S., Steel, G., Shering, P., Zhao, L., Laird, I., and Ni, X.-W. 2009. Continuous crystallization of pharmaceuticals using a continuous oscillatory baffled crystallizer, *Org. Proc. Res. Dev.* 13: 1357–1363.

Index

A

Addition reactions
   acylation, 77
   aldol, 12, 83, 136–139
   base promoted, 78
   Friedel-Crafts, 124–125
   Knoevenagel, 139–142, 227–229
   Michael, 229–230
   oxidation, 290, 295
   photochemical, 315
   photochemical cycloaddition, 303–306
   radical, 209–210
Aliphatic nucleophilic substitution. *See* nucleophilic substitution
Anodic methoxylation, 294, 296
Antisolvent precipitation, 418
Aromatic substitution. *See* nucleophilic substitution
Aspirin, 368
Azo coupling, 133–136

B

Beads. *See* multiphase
Biocatalytic
   epoxidations, 248
   esterifications, 221
   hydrolysis, 90, 267
   resolutions, 415
Bromination. *See* electrophilic substitution

C

Carbamates, 350
Carbonylations
   gas-liquid, 48, 52, 53, 54
   gas-liquid-solid, 62
Catch and release, 277
Cation pool. *See* electrochemistry
Ceramic devices, 4
Chlorinations
   gas-phase, 44
   photochemical, 44
Chromatography, 22, 24
Ciprofloxacin, 363
Claisen re-arrangement. *See* re-arrangements
Commercially available
   instrumentation, 24

Coupling reactions
   Buchwald–Hartwig, 186
   electrochemical, 290
   Heck, 51, 58, 175–183, 252–255
   Kumada-Corriu, 154
   Sonogashira, 183–185, 257
   Stille, 185
   Suzuki-Miyaura, 175, 250–252, 398
   Ullmann, 259
Crystallization
   lysozyme, 420
   solution, 420
Cycloadditions, 144–145, 153, 237–240, 294–303

D

Diastereoselective, 93
Dehydration
   oxidative, 42
Dehydrogenation
   oxidative, 42
Detection, 14–24
Diazo-couplings, 133, 357–358
   multi-faceted, 23
Diazomethane, 362
Diels–Alder, 144–146, 232
Droplets, 325
   coalescence, 327–328, 330
   coefficient of variation (CV), 327–328
   core-shell, 328, 338
   double droplet layer, 326
   interface, 327
   polymerisation, 328–330
   synthesis, 336

E

Efaproxiral, 370
Electrochemistry, 289–301
   electrolyte-free, 296, 299
   oxidations, 290–296
   reductions, 298
Electrolyte. *See* electrochemistry, 158, 241
Electroosmotic flow (EOF), 4, 77, 85
Electrophilic substitution
   bromination, 225, 348
   bromo-lithium exchange. *See* reactions lithium-halogen exchange
   diazotisations, 133–136
   Friedel–Crafts, 124–125, 377

Electrophilic substitution (*Continued*)
  nitrations, 126, 133, 348, 360–362
  phosgene, 226
  sulfonations, 136
Elimination reactions
  dehalogenation, 160–163, 242
  dehydration, 158, 241
Emulsion
  double, 326
  molecular imprinting, 330
  triple, 334
Enantioselective, 262
  biochemical, 266–271
  chemical, 83, 262–266
Epoxidations, 57, 168–172, 248–249
Esterification. *See* nucleophilic substitution
Etching
  chemical, 2
  deep reactive ion etching (DRIE), 3
  lamination, 3
  metals, 3
  photochemical, 2
  wet chemical, 2

F

Fischer–Tropsch, 43
Flow types, 6,7
  annular, 12
  cation, 290
  co-flowing laminar, 393
  counter-current, 394
  electroosmotic. *See* electroosmotic flow (EOF)
  internal circulation, 127, 179
  multiphase. *See* multiphase
  pressure-driven, 6, 7
  segmented. *See* multiphase
Fluorination
  gas-liquid, 45
  liquid-liquid, 109–113, 413
Fluorous extractions, 398
Fourier Transform Infrared Spectroscopy, 16–21, 24
Friedel–Crafts. *See* electrophilic substitution

G

Gas-liquid phase carbonylations, 48
  chlorinations, 44
  fluorinations, 45
  Misc., 57
  ozonolysis, 45, 47, 48
  photochemistry, 317–320
  separations, 403
  transfer hydrogenations, 55

Gas-liquid separation, 403
Gas-liquid-solid phase
  carbonylations, 62
  hydrogenations, 63
  Misc., 69, 321
  oxidations, 61
  slurries, 66
Gas phase oxidations, 38, 43
  carbonylations, 48
  dehydrations, 42
  dehydrogenations, 42
  Fischer–Tropsch synthesis, 43
  Hydrogenations, 41
Glass devices, 2
Grossamide, 378

H

Halogen–lithium exchange. *See* reactions lithium–halogen exchange
Heck reaction. *See* coupling reactions
Horner–Wadsworth–Emmons, 146
Hydrodynamic focussing, 339
Hydrogenation
  gas-liquid, 56
  gas-liquid-solid, 41, 42, 63, 64, 65
  Pd-catalyzed, 42
  slurries, 66–68
  transfer, 55, 57
Hydrophilic
  microchannel, 394, 403
Hydrophobic
  microchannel, 394, 404, 421

I

Ibuprofen, 201
Imatinib, 366
Indacterol, 104, 108
Indoles, 206–207
Infra-red spectroscopy, 16–21, 24
Inorganic nanoparticles, 330
Interface
  gas-liquid, 55, 57–58
  oil-water, 330, 334
  reactor to world, 2, 4
  water-oil-water, 326
Iododeamination, 206
Ionic liquids, 92, 351–353
Isothermal, 9, 22

K

Knoevenagel condensation. *See* nucleophilic addition
Kumada coupling, 255

Index

L

Laser
  ablation, 3
  selective laser melting, 3
Light-emitting, 354
Liquid chromatography LC
  mass spectrometry, 25
Liquid-liquid extractions
  efficiencies, 401
  fluorous, 398
  segmented, 395
  stratified, 388
  three-phase, 392

M

Membranes, 397, 400, 403
  chemically functionalised, 259
  separation, 405–412
Metal catalysed reactions
  Heck, 177, 252
  Misc., 185, 255
  Sonogashira, 183
  Suzuki-Miyaura, 175, 250
Metal-coated, 206
Metallic devices, 3
Metaprolol, 105
Micro reactors
  ceramic, 4
  fabrication, 2
  glass, 2
  metallic, 3
  polymeric, 3
  silicon, 3
Microdistillation, 403
Microencapsulation, 338–341
Microwave, 91, 131, 199, 206
Mixed particles, 338
Mixing, 5
Monoliths, 221–222, 224, 228, 234, 267
Multi-phase, 222–281, 312, 325–342
Multi-step reactions, 198–211, 272–281

N

Nanocrystals, 336
Nanoparticles
  inorganic, 330–332
  polymeric, 332–336
Nuclear magnetic resonance
    spectroscopy, 21
Nucleophilic addition
  acetalizations, 240
  Aldol condensation, 136
  Aldol reaction, 12, 136, 234
  alkylation, 151
  aza-Michael addition, 149
  azoles, 157–158, 412
  carbon-hetero, 152, 157
  Diels–Alder, 144–146, 232
  enantioselective, 148
  Henry reaction, 230
  Horner–Wadsworth–Emmons, 146
  Knoevenagel condensation,
    139–142, 227
  Michael addition, 142, 229
  thioacetalizations, 241
  triazoles, 152–157, 237, 303, 410
  trifluoromethylation, 234
Nucleophilic substitution
  acylation, 77, 93, 94, 95
  alkylation, 78, 90, 93, 98
  arylation, 98–99
  azidation, 99–102, 224–225
  carbon-fluorine, 109
  enantioselective, 83
  epoxide aminolysis, 103
  epoxide hydrolysis, 90
  esterification, 22, 85–89, 221–224
  etherification, 89
  hydroxamic acid, 102
  nitration, 126–130
Numbering-up, 13, 347, 357

O

Oligosaccharides, 205, 398
On-site processing, 11
Organic nanoparticles, 332
Organic particles
  pigments, 336
  synthesis, 420, 332–336
Oxidation, 163, 242–249
  anodic, 292–295
  Au-catalyzed, 61
  deprotection, 172
  electrochemical, 290
  epoxidations, 57, 168, 247–250
  gas-phase, 37, 38, 39, 43
  inorganic, 163
  Oxone, 168
  Pd-catalyzed, 40, 41
  peracetic acid production, 360
  photochemical, 314
  selective, 242
  Swern-Moffat, 164
  TEMPO, 167
Oxomaritidine, 378–379
Ozonolysis, 16–18
  gas-liquid, 45–50

P

Peracetic acid, 360
Phase transfer
　alkylation, 78–80
Photochemistry benzopinacol, 302
　alkylations, 315–316
　cyanation, 307–308
　cyclisations, 316–317
　cycloadditions, 303–306
　diastereodifferentiation, 312
　dimerisation, 311
　halogenation, 308–309
　nitrite photolysis, 309–311
　oxidations, 314–315
　reductions, 313–314
Pigments, 357
Polymer-assisted solution-based (PASS), 232, 252
Polymeric devices, 3
Positron Emission Tomography (PET), 86
Precipitation, 418
　antisolvent, 12
　biphasic, 420
　lysozyme, 420
Pristane, 363
Process intensification, 13

R

Radical additions, 209
Radiochemistry. *See* radiopharmaceuticals
Radiopharmaceuticals, 382–384
Raman spectroscopy, 14–16, 24
Reaction
　addition. *See* nucleophilic and electrophilic addition
　aldol reaction, 83, 136–139, 234–237
　aliphatic substitution. *See* nucleophilic and electrophilic substitution
　aromatic substitution. *See* nucleophilic and electrophilic substitution
　benzoin condensation, 232–234
　brominations. *See* electrophilic substitution
　chlorination. *See* electrophilic substitution
　coupling. *See* coupling reactions
　cyanations. *See* photochemistry
　Diazo-coupling, 133–135, 357–358
　Diels–Alder, 144–146, 232
　elimination, 158, 160–163, 241–242
　epoxide hydrolysis, 90
　esterification. *See* nucleophilic substitution
　etherification. *See* nucleophilic substitution
　Fischer–Tropsch, 43
　fluorination, 109–113
　Friedel–Crafts reactions. *See* electrohilic substitution
　Grignard, 118
　Heck aminocarbonylation, 49
　Horner–Wadsworth–Emmons, 146, 377
　iodinations. *See* aromatic substitution
　Knoevenagel condensation. *See* nucleophilic addition
　Knorr, 158, 356
　Kolbe-Schmitt, 384
　Kumada, 255
　lithium–halogen exchange, 114–122, 353–356
　Mannich, 84
　Michael addition, 142
　nitrations. *See* aromatic substitution
　peptide, 277–279
　phase transfer. *See* multi-phase
　pyrrole, 356–357
　Sandmeyer, 134
　Sulfonation, 136
　Suzuki–Miyaura. *See* coupling reactions
　transfer hydrogenation, 171
　trifluoromethylation, 234
　Wittig, 377
Re-arrangements
　Claisen, 187–191, 373–375
　Curtius, 196–197
　Dimroth, 197
　Fisher indolization, 195–196
　Hofmann, 193–195
　Misc., 261
　Newman–Kwart, 191–193
Reductions
　Dibal-H, 173, 348
　electrochemical, 299
　photochemical, 313
　Transition-metal free, 173
Regioselective, 78
Resolutions
　biocatalytic, 415
　chemical, 416
Rimonabant, 370–372

S

Safety, 7
Scale-up, 7, 13, 357, 363
Scavengers, 274, 410–411
Segmented flow, 395–410
Sensors, 23
Sildenafil, 372
Silicon, 2
Sonogashira. *See* coupling reactions
Sulfonations, 136
Supercooled, 8

Index

Supercritical
    carbon dioxide, 10
    methanol, 10

T

Tramadol, 373
Tri-phasic
    carbonylation, 62
    hydrogenation, 63–66
    oxidation, 61

U

Ultra-violet (UV) detection, 11

V

Vanisyl sodium, 368